Computational Plasma Physics

With Applications to Fusion and Astrophysics

Computational Plasma Physics

With Applications to Fusion and Astrophysics

Toshiki Tajima
Kansai Research Establishment
Japan Atomic Energy Research Institute, Kyoto

Advanced Book Program

CRC Press is an imprint of the
Taylor & Francis Group, an **informa** business

First published 2004 by Westview Press

Published 2018 by CRC Press
Taylor & Francis Group
6000 Broken Sound Parkway NW, Suite 300
Boca Raton, FL 33487-2742

A Cataloging-in-Publication data record for this book is available from the Library of Congress.

ISBN 13: 978-0-8133-4211-5 (pbk)

Visit the Taylor & Francis Web site at
http://www.taylorandfrancis.com

and the CRC Press Web site at
http://www.crcpress.com

Frontiers in Physics

David Pines, Editor

Volumes of the Series published from 1961 to 1973 are not officially numbered. The parenthetical numbers shown are designed to aid librarians and bibliographers to check the completeness of their holdings.

Titles published in this series prior to 1987 appear under either the W. A. Benjamin or the Benjamin/Cummings imprint; titles published since 1986 appear under the Westview Press imprint.

No.	Author(s)	Title
1.	N. Bloembergen	Nuclear Magnetic Relaxation: A Reprint Volume, 1961
2.	G. F. Chew	S-Matrix Theory of Strong Interactions: A Lecture Note and Reprint Volume, 1961
3.	R. P. Feynman	Quantum Electrodynamics: A Lecture Note and Reprint Volume, 1961
4.	R. P. Feynman	The Theory of Fundamental Processes: A Lecture Note Volume, 1961
5.	L. Van Hove N. M. Hugenholtz L. P. Howland	Problem in Quantum Theory of Many-Particle Systems: A Lecture Note and Reprint Volume, 1961
6.	D. Pines	The Many-Body Problem: A Lecture Note and Reprint Volume, 1961
7.	H. Frauenfelder	The Mössbauer Effect: A Review—with a Collection of Reprints, 1962
8.	L. P. Kadanoff G. Baym	Quantum Statistical Mechanics: Green's Function Methods in Equilibrium and Nonequilibrium Problems, 1962
9.	G. E. Pake	Paramagnetic Resonance: An Introductory Monograph, 1962 [cr. (42)—2nd edition]
10.	P. W. Anderson	Concepts in Solids: Lectures on the Theory of Solids, 1963
11.	S. C. Frautschi	Regge Poles and S-Matrix Theory, 1963
12.	R. Hofstadter	Electron Scattering and Nuclear and Nucleon Structure: A Collection of Reprints with an Introduction, 1963
13.	A. M. Lane	Nuclear Theory: Pairing Force Correlations to Collective Motion, 1964
14.	R. Omnès M. Froissart	Mandelstam Theory and Regge Poles: An Introduction for Experimentalists, 1963
15.	E. J. Squires	Complex Angular Momenta and Particle Physics: A Lecture Note and Reprint Volume, 1963
16.	H. L. Frisch J. L. Lebowitz	The Equilibrium Theory of Classical Fluids: A Lecture Note and Reprint Volume, 1964
17.	M. Gell-Mann Y. Ne'eman	The Eightfold Way (A Review—with a Collection of Reprints), 1964
18.	M. Jacob G. F. Chew	Strong-Interaction Physics: A Lecture Note Volume, 1964
19.	P. Nozières	Theory of Interacting Fermi Systems, 1964
20.	J. R. Schrieffer	Theory of Superconductivity, 1964 (revised 3rd printing, 1983)
21.	N. Bloembergen	Nonlinear Optics: A Lecture Note and Reprint Volume, 1965

22.	R. Brout	Phase Transitions, 1965
23.	I. M. Khalatnikov	An Introduction to the Theory of Superfluidity, 1965
24.	P. G. deGennes	Superconductivity of Metals and Alloys, 1966
25.	W. A. Harrison	Pseudopotentials in the Theory of Metals, 1966
26.	V. Barger D. Cline	Phenomenological Theories of High Energy Scattering: An Experimental Evaluation, 1967
27.	P. Choquàrd	The Anharmonic Crystal, 1967
28.	T. Loucks	Augmented Plane Wave Method: A Guide to Performing.Electronic Structure Calculations—A Lecture Note and Reprint Volume, 1967
29.	Y. Ne'eman	Algebraic Theory of Particle Physics: Hadron Dynamics in Terms of Unitary Spin Currents, 1967
30.	S. L. Adler R. F. Dashen	Current Algebras and Applications to Particle Physics, 1968
31.	A. B. Migdal	Nuclear Theory: The Quasiparticle Method, 1968
32.	J. J. J. Kokkede	The Quark Model, 1969
33.	A. B. Migdal V. Krainov	Approximation Methods in Quantum Mechanics, 1969
34.	R. Z. Sagdeev A. A. Galeev	Nonlinear Plasma Theory, 1969
35.	J. Schwinger	Quantum Kinematics and Dynamics, 1970
36.	R. P. Feynman	Statistical Mechanics: A Set of Lectures, 1972
37.	R. P. Feynman	Photon-Hadron Interactions, 1972
38.	E. R. Caianiello	Combinatorics and Renormalization in Quantum Field Theory, 1973
39.	G. B. Field H. Arp J. N. Bahcall	The Redshift Controversy, 1973
40.	D. Horn F. Zachariasen	Hadron Physics at Very High Energies, 1973
41.	S. Ichimaru	Basic Principles of Plasma Physics: A Statistical Approach, 1973 (2nd printing, with revisions, 1980)
42.	G. E. Pake T. L. Estle	The Physical Principles of Electron Paramagnetic Resonance, 2nd Edition, completely revised, enlarged, and reset, 1973 [cf. (9)—1st edition]

Volumes published from 1974 onward are being numbered as an integral part of the bibliography.

43.	R. C. Davidson	Theory of Nonneutral Plasmas, 1974
44.	S. Doniach E. H. Sondheimer	Green's Functions for Solid State Physicists, 1974
45.	P. H. Frampton	Dual Resonance Models, 1974
46.	S. K. Ma	Modern Theory of Critical Phenomena, 1976
47.	D. Forster	Hydrodynamic Fluctuations, Broken Symmetry, and Correlation Functions, 1975
48.	A. B. Migdal	Qualitative Methods in Quantum Theory, 1977
49.	S. W. Lovesey	Condensed Matter Physics: Dynamic Correlations, 1980
50.	L. D. Faddeev A. A. Slavnov	Gauge Fields: Introduction to Quantum Theory, 1980
51.	P. Ramond	Field Theory: A Modern Primer, 1981 [cf. 74—2nd ed.]

52. R. A. Broglia, A. Winther — Heavy Ion Reactions: Lecture Notes Vol. I: Elastic and Inelastic Reactions, 1981
53. R. A. Broglia, A. Winther — Heavy Ion Reactions: Lecture Notes Vol. II, 1990
54. H. Georgi — Lie Algebras in Particle Physics: From Isospin to Unified Theories, 1982
55. P. W. Anderson — Basic Notions of Condensed Matter Physics, 1983
56. C. Quigg — Gauge Theories of the Strong, Weak, and Electromagnetic Interactions, 1983
57. S. I. Pekar — Crystal Optics and Additional Light Waves, 1983
58. S. J. Gates, M. T. Grisaru, M. Rocek, W. Siegel — Superspace *or* One Thousand and One Lessons in Supersymmetry, 1983
59. R. N. Cahn — Semi-Simple Lie Algebras and Their Representations, 1984
60. G. G. Ross — Grand Unified Theories, 1984
61. S. W. Lovesey — Condensed Matter Physics: Dynamic Correlations, 2nd Edition, 1986
62. P. H. Frampton — Gauge Field Theories, 1986
63. J. I. Katz — High Energy Astrophysics, 1987
64. T. J. Ferbel — Experimental Techniques in High Energy Physics, 1987
65. T. Appelquist, A. Chodos, P. G. O. Freund — Modern Kaluza-Klein Theories, 1987
66. G. Parisi — Statistical Field Theory, 1988
67. R. C. Richardson, E. N. Smith — Techniques in Low-Temperature Condensed Matter Physics, 1988
68. J. W. Negele, H. Orland — Quantum Many-Particle Systems, 1987
69. E. W. Kolb, M. S. Turner — The Early Universe, 1990
70. E. W. Kolb, M. S. Turner — The Early Universe: Reprints, 1988
71. V. Barger, R. J. N. Phillips — Collider Physics, 1987
72. T. Tajima — Computational Plasma Physics, 1989 (updated 2004)
73. W. Kruer — The Physics of Laser Plasma Interactions, 1988 (updated 2003)
74. P. Ramond — Field Theory: A Modern Primer, 2nd edition, 1989 [cf. 51—1st edition]
75. B. F. Hatfield — Quantum Field Theory of Point Particles and Strings, 1989
76. P. Sokolsky — Introduction to Ultrahigh Energy Cosmic Ray Physics, 1989 (updated 2004)
77. R. Field — Applications of Perturbative QCD, 1989
80. J. F. Gunion, H. E. Haber, G. Kane, S. Dawson — The Higgs Hunter's Guide, 1990
81. R. C. Davidson — Physics of Nonneutral Plasmas, 1990
82. E. Fradkin — Field Theories of Condensed Matter Systems, 1991
83. L. D. Faddeev, A. A. Slavnov — Gauge Fields, 1990

84.	R. Broglia A. Winther	Heavy Ion Reactions, Parts I and II, 1990
85.	N. Goldenfeld	Lectures on Phase Transitions and the Renormalization Group, 1992
86.	R. D. Hazeltine J. D. Meiss	Plasma Confinement, 1992
87.	S. Ichimaru	Statistical Plasma Physics, Volume I: Basic Principles, 1992 (updated 2004)
88.	S. Ichimaru	Statistical Plasma Physics, Volume II: Condensed Plasmas, 1994 (updated 2004)
89.	G. Grüner	Density Waves in Solids, 1994
90.	S. Safran	Statistical Thermodynamics of Surfaces, Interfaces, and Membranes, 1994 (updated 2003)
91.	B. d'Espagnat	Veiled Reality: An Analysis of Present Day Quantum Mechanical Concepts, 1994 (updated 2003)
92.	J. Bahcall R. Davis, Jr. P. Parker A. Smirnov R. Ulrich	Solar Neutrinos: The First Thirty Years, 1994 (updated 2002)
93.	R. Feynman F. Morinigo W. Wagner	Feynman Lectures on Gravitation, 1995 (reprinted 2003)
94.	M. Peskin D. Schroeder	An Introduction to Quantum Field Theory, 1995
95.	R. Feynman	Feynman Lectures on Computation, 1996 (reprinted 1999)
96.	M. Brack R. Bhaduri	Semiclassical Physics, 1997 (updated 2003)
97.	D. Cline	Weak Neutral Currents, 1997 (reprinted 2004)
98.	T. Tajima K. Shibata	Plasma Astrophysics, 1997 (updated 2002)
99.	J. Rammer	Quantum Transport Theory, 1998
100.	R. Hazeltine F. Waelbroeck	The Framework of Plasma Physics, 1998 (updated 2004)
101.	P. Ramond	Journeys Beyond the Standard Model, 1999 (updated 2004)
102.	Y. Nutku C. Saclioglu T. Turgut	Conformal Field Theory: New Non-Perturbative Methods in String and Field Theory, 2000 (reprinted 2004)
103.	P. Philips	Advanced Solid State Physics, 2003

EDITOR'S FOREWORD

The problem of communicating in a coherent fashion recent developments in the most exciting and active fields of physics continues to be with us. The enormous growth in the number of physicists has tended to make the familiar channels of communication considerably less effective. It has become increasingly difficult for experts in a given field to keep up with the current literature; the novice can only be confused. What is needed is both a consistent account of a field and the presentation of a definite "point of view" concerning it. Formal monographs cannot meet such a need in a rapidly developing field, while the review article seems to have fallen into disfavor. Indeed, it would seem that the people most actively engaged in developing a given field are the people least likely to write at length about it.

FRONTIERS IN PHYSICS was conceived in 1961 in an effort to improve the situation in several ways. Leading physicists frequently give a series of lectures, a graduate seminar, or a graduate course in their special fields of interest. Such lectures serve to summarize the present status of a rapidly developing field and may well constitute the only coherent account available at the time. Often, notes on lectures exist (prepared by the lecturer himself, by graduate students, or by postdoctoral fellows) and are distributed in mimeographed form on a limited basis. One of the principal purposes of the FRONTIERS IN PHYSICS Series is to make such notes available to a wider audience of physicists.

It should be emphasized that lecture notes are necessarily rough and informal, both in style and content; and those in the series will prove no exception. This is as it should be. One point of the series is to offer new, rapid, more informal, and, it is hoped, more effective ways for physicists to teach one another. The point is lost if only elegant notes qualify.

During the past decade computational physics — the use of modern high speed computers to carry out numerical modeling of physical situations which cannot be addressed by either analytic theory or by experiments — has emerged as a distinct, and rapidly growing, field of physics. Toshiki Tajima has played a leading role in developing this approach to the physics of plasmas. As John Dawson, one of the founders of computational plasma physics, notes in his introductory remarks, computer simulations are properly regarded as numerical experiments and as such possess certain advantages over conventional experiments, in helping gain physical understanding.

Tajima's book not only provides a lucid introduction to computational plasma physics, but also offers the reader many examples of the way numerical modeling, properly handled, can provide valuable physical understanding of the complexity so often encountered in both laboratory and astrophysical plasmas. It should therefore prove of great value to the novice and experienced researcher alike, and it gives me pleasure to welcome Prof. Tajima to the "Frontiers" series.

David Pines
Urbana, Illinois
June, 1988

FOREWORD

The physics of plasmas is an extremely rich and complex subject as the variety of topics addressed by Tajima in this book clearly demonstrates. This richness and complexity demands new and powerful techniques for investigating plasma physics.

The richness of plasma physics arises because of the long range nature of the electromagnetic interactions between the charged particles. These give rise to collective modes, modes in which large numbers of particles move in unison. Furthermore, the various collective modes often interact nonlinearly; thus compounding the difficulties for understanding plasma behavior, but adding another layer of richness to the physics of plasmas. Finally, the long mean free paths for two body scattering of the energetic particles of which a plasma is made means that techniques of conventional fluid approaches also often fail. Particle wave (collective mode) interactions can come to dominate dissipative processes. These further enrich the physics of plasmas, but one pays the price of additional complexity. However, often if one understands which are the dominant effects, one can understand these processes. Thus, one is faced with the problem of eliminating from one's theories the unimportant phenomenon to simplify problems to manageable proportions.

For centuries, the classic method for obtaining a physical understanding of nature was a two pronged experimental-theoretical approach. Experiments asked questions of nature; the results were interpreted in terms of analytic theories applying well-established physics laws. This approach has been extremely successful. However, it has its limitations when one has many degrees of freedom interacting simultaneously and nonlinearly. Some classic examples are the problems of turbulence, understanding the weather, the dynamics of internal flow within the earth, and in the dynamics of stellar gases. There are also many situations of interest which experiments cannot address. Examples here are situations involving very large space and/or time scales, situations where we want to measure processes on a finer scale and in more detail than our instruments allow.

For these situations we believe we understand the underlying laws but are simply incapable of handling the complexity, a new and very powerful third method of attack has recently become possible; this is the method of numerical modeling using modern high speed computers. These devices are able to follow the dynamics of millions of interacting degrees of freedom and tell us the behavior of the system in as much detail as we desire.

It should be mentioned that computer simulation contains many advantages over conventional experiments. It is possible to obtain as complete information about the systems dynamics as we want; we can turn on and off various physical effects (waves of a certain wavelength, use reduced dimensional models, etc.) and thus determine their importance to the physics under study. We can even perform numerical experiments on almost identical systems; for example, one system might contain an extra particle or have a certain wave excited; but in all other respects the systems are identical. By subtracting the results we can find out exactly the effect of the imposed perturbation.

The goal of numerical modeling is to gain physics understanding; not to generate masses of computer output. Unless the results of the modeling can be condensed into a relatively simple physical picture or theory, they are of very limited value. Fortunately, often once the answer is known, we are able to see those things which are important and incorporate these into a relatively simple model which predicts the essential behavior. Thus, this approach puts a premium on physics understanding and insight; it increases the physics work and in no way substitutes for it. Tajima's book contains many examples of this.

This approach has another value; it develops our physical intuition. Using such an intuition, we can often guess physical situations which are of interest. These can be tested much more quickly and cheaply than by experiment. If our intuition is correct, we discover something new; if it is wrong, we must modify it, but in any case we learn something.

Numerical models, just as theories, are constructions of man; they involve simplification and approximations. There are no God-given numerical methods. There are only bad, good, and better methods. Thus, numerical models provide a wide range of challenges for physicists and model builders. In the wealth of techniques presented in this book, one glimpses the richness of the options that are available to the computer modeler. This is not a static or closed subject; it is very much a dynamic one with new approaches continually arising. Young scientists will find that this is an exciting field to which they can make real contributions.

Of course, numerical modeling has its limitations. Even the most powerful present computer can at best handle a few million degrees of freedom. The conditions investigated are usually much simpler than those encountered in nature. Nevertheless, the power of these machines has been increasing at an exponential rate (about a factor of 2 every 1.5 years) and this trend shows no sign of changing. As already mentioned, the techniques for handling problems have also seen explosive growth. Thus, we can expect that this will become a more and more powerful tool in the future.

J.M. Dawson
Los Angeles
April 1988

PREFACE

The torch illuminating the behavior of plasmas was carried by early pioneers such as Langmuir, Landau, Alfvén, and Chapman. It was not until the late 1950's, though, that a torrent of plasma research began and the study of plasmas assumed the form of what is now called plasma physics. In 1956 the Soviet Union and the West simultaneously declassified controlled thermonuclear research for the peaceful purpose of energy production, after realizing that controlling high temperature plasmas for fusion is a far more complex task than had been previously expected on the basis of the initially rapid success obtained in thermonuclear weapons research. Scientists realized that in order to achieve controlled thermonuclear fusion they had to learn much more about the very complex behavior of plasmas. On the other hand, in 1957, the first man-made satellite, Sputnik, was launched and it was soon discovered that the earth is surrounded by active plasmas, such as the Van Allen belt (1958). The International Geophysical Years (1957-1958), which were highlighted by the first man-made satellites, marked another milestone of man's realization of his outer environment surrounded by plasmas. The behavior of these plasmas is as enigmatic as that found in a laboratory for fusion research.

Since the late 1950's, fusion research and space (and, in a wider sense, astrophysical) research have furthered the knowledge of plasma physics and are presently the main areas of plasma research. Although various phenomena and approaches are different in these two areas, many fundamentals are common. This may be appreciated by the simple and ideal example of a nonmagnetized uniform electron plasma in which there exists only one fast time scale, the plasma period (a slow time is a collision time). Any time scale is measured relative to the plasma period, although it can vary from a femto second in a solid state plasma to a second in intergalactic medium, and yet the behavior can be described by the same set of equations.

Unlike solid state physics, there exists only one kind of interaction in plasma physics: that is, the Coulombic interaction (and its electromagnetic counterpart). In solid state physics, half the fun is in determining the Hamiltonian of the system, whereas in plasma physics the Hamiltonian of the system is always uniquely fixed. The fun of plasma physics is not in finding the Hamiltonian, but in unlocking the code of behavior of the plasma Hamiltonian. The main interest and difficulty in this arises from the *long range* interaction of the Coulombic force acting among *many bodies*. This is a notoriously difficult problem, well epitomized by D. Pines in "The Many-body Problem."[1]

The development of highly sophisticated computational plasma physics is largely due to this difficulty of resolving many-body problems of a highly nonlinear nature by means of traditional analytical methods. Although the use of computational plasma physics may be traced back to thermonuclear weapons research, a systematic and large scale effort in computational plasma physics seems to have started in the 1960's in fusion research and in the 1970's in space and astrophysical plasma research. Main line fusion physics is called magnetic fusion because of its employment of magnetic fields to confine particles. Moreover, because almost all astrophysical plasmas contain magnetic fields (there exists an approximate cosmical equipartition among the magnetic, kinetic, and cosmic ray energies), the physics of *magnetized plasmas* is of paramount importance in plasma physics. Equally important is the presence of *inhomogeneity* in the plasma. A filamentary (string-like) structure of cosmical plasma has been pointed out by many authors, including Alfvén; and a laboratory fusion plasma is necessarily inhomogeneous because it cannot be infinitely large.

As we shall see in the Introduction and in the other chapters, the presence of magnetic fields and inhomogeneities in a plasma introduces a rich and wide variety of phenomena, whose time and spatial scales vary over many orders of magnitude. One of the intricacies of plasma physics is this hierarchical structure of plasma behavior and interaction. Different levels in the hierarchy of plasma interaction exhibit quite distinct behavior and yet each level of the hierarchy interacts with the other levels. This forces us to develop many layers of theoretical models and thus many different kinds of computational modeling.

In this book we intend to discuss computational plasma physics in a way that reflects this diverse nature of plasma physics. Our motivations for this book are: (i) We present *physics-driven modeling* in contrast to computationally motivated thinking. The various computer models we discuss are devised and structured in order to solve outstanding problems in plasma physics and modeling. The consideration of physics, as opposed to computation, drives the modeling. (ii) We emphasize the richness and complexity of various algorithms for *magnetized plasmas*. This is a reflection of the hierarchical nature of plasmas. This necessitates that we discuss many levels of computer simulation techniques of varied sophistication starting from the simplest Cartesian non-

[1] D. Pines: "The Many-body Problem" (Benjamin, Reading, 1962)

magnetized electrostatic simulation to the general geometry of magnetized plasma simulation. (iii) Since the computational models are driven by the physics of interest, we try to emphasize the interwoven relation between the computational methodologies and their *applications* to actual physics problems.

The various simulation models and their richness of variety are a consequence of the hierarchical nature of magnetized plasmas in spite of the simplicity of the basic interaction of plasma particles, as we already mentioned. We will present various numerical techniques according to this hierarchical order, i.e., the most primitive (or fundamental) method first and the increasingly sophisticated methods follow in the successive order of chapters. The most primitive level corresponds to a nonmagnetized electrostatic Cartesian plasma in the high frequency domain, which is discussed in Chapters 2-4. As magnetic fields are added and longer time scales are considered, models to describe these phenomena must coarse-grain some of the high frequency phenomena in order to achieve meaningful computational efficiency. As such, in a general sense, the historical development of numerical techniques also follows the hierarchical structure of plasma physics.

After Chapter 1, the Introduction, we order Chapters 2 through 10 in the order of ascending sophistication and longer time scale models. Within this overall organization of Chapters 2-10, we can identify a substructure made up of two distinct approaches, the particle method and the fluid method. Phenomena that are less coarse-grained are typically more appropriately treated by the former, whereas phenomena that manifest themselves after sufficient coarse-graining by the latter. Chapters 2-5 discuss the particle approach, while Chapter 6 goes into the fluid approach. Later chapters treat a hybrid of the two or more sophisticated approaches. Corresponding applications are presented in a similar manner beginning in Chapter 12 through Chapter 15, from problems with the shortest time scales to ones with the longest time scales. Again the techniques employed gradually vary from the more predominantly particle-like to the more predominantly fluid-like in Chapters 12-15. Chapter 11 serves to connect the computational techniques discussed in Chapters 2-10 with their applications to physics in Chapters 12-15 by reviewing computation in general. Chapters 1-15 constitute building blocks for our numerical laboratory. Synthesis of these techniques and elements developed in these chapters is discussed in the epilogue "Numerical Laboratory."

In this book we try to emphasize symbiotic relationships in computational physics between numerics and physics. We often encounter theoretical physicists claiming that the only important results come from analysis (although they perhaps grudgingly admit results that come from experiments). These "purists" either dismiss the computational approach entirely or regard it as a peripheral tool. Applications of our computational techniques (in Chapters 12-15) to make new discoveries in nonlinear physics should be a testimony against the "purists'" bias. Many of these results prompted theoretical investigations, which would not have started without the insight into the crux of the matter that was gained from computation. Computational physicists would like to have their Magna Carta, or their bill of rights, stating their emergent important role in nonlinear physics investigation, a role that often has been neglected by their theoretical (and perhaps experimental) colleagues. On the other hand, we also often encounter computational physicists who are content with the development of codes or the investigation of numerics alone. We have to realize that the inherent importance of a particular code only arises from the results that an application of the code can uncover. Thus, this book does not intend to be a "how-to" recipe book. Since our major interest and emphasis are to discuss the intricate relation between the numerical methods and their actual application to physics, we do not intend to make a necessarily thorough survey of the numerics itself. These methods are presented only to the extent that the reader is provided an exposition of their basic ideas and fundamental techniques. The reader's actual contact with the codes may be done by following the instruction in the epilogue.

Wherever opportunities exist in this book, we try to make contact with the general context of physics as derived from (computational) plasma physics. We realize the danger is that such broadening of the context may be perceived as being a bit cocky, but we believe that it is important to point out and to emphasize some aspects of plasma physics that are relevant or useful to wider fields of physics in general. Certainly the computations in plasma physics are some of the most advanced in physics in general. The plasma physicists's approach to a nonlinear problem, to a problem of long time scales, etc., could be enlightening to problem solving in other branches of physics. In fact, I hope that this book proves to be useful also to readers outside of plasma physics.

This book is an outgrowth of my lecture notes for "Computational Physics I and II" (Physics 391T) given at the Department of Physics of The University of Texas at Austin, since the Spring of 1981. The course has been offered in several semesters in the past, in which many graduate students have participated and some of them have worked under my auspices till the end of their Ph.D. dissertation or beyond. Typically, the introductory "Computational Physics I" would cover the basic ideas of computational (plasma) physics and their fundamental techniques and simpler applications. This corresponds to most of Chapters 1-6 and 12 and part of 14. The advanced "Computational Physics II" would cover more advanced concepts and techniques, which are often those that have been developed in our group at the Institute for Fusion Studies of The University of Texas. This roughly corresponds to Chapters 7-11 and Chapter 13 and 15 and some portions of other chapters. Within the numerically technical first half (Chapters 1-10), Chapters 1-6 constitute a generic core (basics), while Chapters 7-10 present more advanced topics. Since there are many excellent books and reviews on the generic core, including Hockney and Eastwood's textbook,[2] Birdsall and Langdon's textbook,[3] and Dawson's review article,[4] the description of many computational technical questions is left to previous literature and we are content only with a preparation for what follows in later chapters. The discussion of advanced computational techniques is focused on those that have been frequently employed and/or developed by our group, to understand the complex labyrinth of magnetized plasma. Therefore, we must skip or neglect many important recent worthwhile developments. I would like to apologize, therefore, for any flagrant omissions of any important works due to my oversight or to the scope of the book. In this sense this book takes the flavor of a personal monograph, in which, beyond the generic portions of the chapters, the materials have been chosen and organized around my own principal involvements and research interests and work. This continues increasingly to be the case toward the later chapters.

The development of thoughts and techniques presented here is largely due to the encouragement, education, criticisms, and support of my mentors, who guided me and to whom I am deeply indebted: In my years at the University of Tokyo Professor R. Kubo taught me the fascination of the physics of many-body problems, while Professor S. Ichimaru cast my thoughts on plasma physics, as documented.[5] I learned from Professor Norman Rostoker how to approach a real problem, and he shaped my philosophical outlook on plasma physics during my Ph.D. years. Professor John Dawson taught me the joy of discovery, some of which we shared together during my staff scientist years. It is Professor Dawson from whom I learned everything from nuts-and-bolts to broader issues of computational plasma physics. Naturally, for these reasons, I imagine this book should look like a book compiling his thoughts and techniques. Professor Marshall Rosenbluth brought me to Texas and taught me rigor and critical evaluation of ideas. To these mentors who nurtured me I dedicate this book.

Many of my colleagues taught me and contributed indispensably to the development of various thoughts, methods, and specifics of this book. There are so many that it is impossible to name them all. Here I just mention only a few. Collaborations and associations with Dr. J.N. Leboeuf, Dr. F. Brunel, Dr. D.C. Barnes, Dr. T. Kamimura, and Dr. J.S. Wagner have been extremely

[2] R.W. Hockney and J.W. Eastwood: "Computer Simulation Using Particles" (McGraw-Hill, New York, 1981);

[3] C.K. Birdsall and A.B. Langdon: "Plasma Physics Via Computer Simulation" (McGraw-Hill, New York, 1985);

[4] J.M. Dawson: Rev. Mod. Phys. **55**, 403 (1983);

[5] S. Ichimaru: "Basic Principles of Plasma Physics – A Statistical Approach" (Benjamin, Reading, 1973)

fruitful and inspirational. Dr. H. Okuda, Dr. A.T. Lin, Dr. V. Decyk, Dr. P.L. Pritchett, Dr. A.Y. Aydemir, and Dr. D.D. Schnack taught me many important "trade secrets." Helpful comments and suggestions from Dr. J.U. Brackbill, Dr. R.S. Steinolfson, Dr. W. Horton, Dr. J. Meiss, Dr. H.L. Berk, Dr. W. Miner, Dr. N. Bekki, Dr. K. Shibata, Dr. J. Van Dam, Dr. D. Papadopoulos, and Dr. J. Scalo are greatly appreciated. Countless students contributed to this project by taking notes of my incomprehensible mumblings. In particular I appreciate the enthusiasm, sharing, and suffering together with my former students, Dr. R.D. Sydora, Dr. J.L. Geary, Dr. E.G. Zaidman, and Dr. M.J. Lebrun. Without Mrs. Suzy Crumley's dedication, typesetting skills, and patience, however, not one page of this book would have been printed. My deep appreciation goes to her. In addition, I thank Ms. Mindy Walker, Mr. S. Banerjee, and Ms. L. Williams for their editorial assistance.

Finally, I appreciate Professor S.-I. Akasofu who first suggested that I prepare my lecture notes for publication; Professor David Pines for his encouragement; and Mr. R. Minxter and more recently, Mr. A. Wylde for professionally guiding me in finishing the project. This project was made possible by the support of The University of Texas at Austin, the U.S. Department of Energy, the National Science Foundation, and the National Aeronautical and Space Administration, to whom I am grateful.

T. Tajima
Austin, Texas
June 1988

CONTENTS

COMPUTATIONAL PLASMA PHYSICS

WITH APPLICATIONS TO FUSION AND ASTROPHYSICS

1 INTRODUCTION

1.1 Computer and Computer Simulation

Recently, the growth in computer technology has been explosive. Its speed, cost^{-1}, capacity, or their product can be described as increasing explosively as a function of time. The trend of speed can be seen in Fig. 1.1. A set of orbit calculations which would take a human brain and hand seven hours to finish was carried out in a matter of three seconds on the first computer ENIAC in 1945. This is a reduction of computing time by a factor of 10^{-4}. A story of the first computer is described in Ref. 1. The latest model of high power scientific computers such as the CRAY-1 executes calculations in approximately 10^{-8} (of 7 hours) a human would take. We can look at this rapid development of scientific computing capabilities in a concrete example. The Magnetic Fusion Energy Computer Center (MFECC) CDC 7600 computer (1975), which revolutionized the way the plasma physicists compute, was installed only a decade ago, and the MFECC CRAY was introduced only several years ago. Correspondingly, many new disciplines of computational science burst into existence recently. Some examples are computational geophysics, computational astrophysics, computational solid state physics, and computational high energy physics. It all started only two or three decades ago.

Ever since Galileo Galilei dropped a metal ball from the Pisa tower and shattered Aristotle's theory of gravity, experiment and theory became the two essential ingredients of physics and (natural) sciences in general. Thus

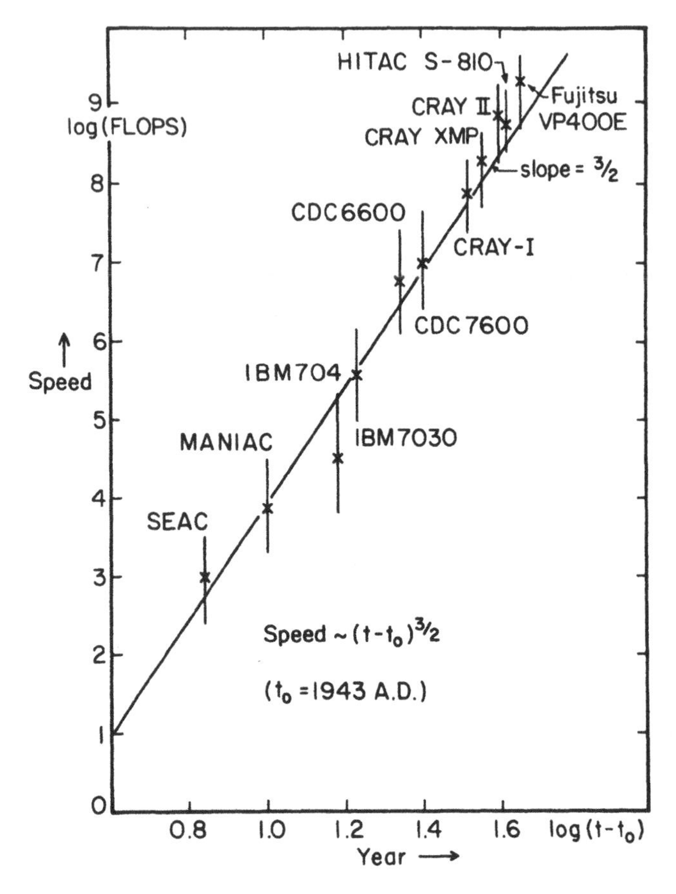

FIGURE 1.1 Rapid progress in the speed of computer over the years. (Flop = floating point operations per second)

the traditional means of investigating physical phenomena are through laboratory experiments and through the analytic application of well-established (or proposed) physical laws. In the case of large-scale natural phenomenon, one must often substitute observations of what is taking place for controlled experiments. In fact, this is almost always the case for astrophysical phenomena. The traditional methods have their limitations; often the complexity of the phenomenon and the simultaneous interaction of many effects make a complete analysis impossible. On the experimental side, one is limited to measurements of only a small fraction of the quantities of interest in a process and even these may be sampled only at a few times and positions and with a limited degree of accuracy. This is particularly true for observations of natural phenomena such as those that are encountered in fusion, space, and astrophysical plasmas. Thus, one is then faced with the task of interpreting limited observations with theories that are incomplete; and often, many different theories that can explain the observations exist. For the case of fusion research, the sustained and controlled thermonuclear plasma has yet to be realized. Even for plasmas with lesser parameters, it is very costly or time-consuming to perform experiments. For space and astrophysical plasmas, it is difficult or impossible to do experiments, and the observational data available are sparse and sporadic. It is very difficult to perform experiments only for instance, a gravitational system such as interacting galaxies. Yet another example is quark confinement. Here, because of the perfect confinement, no physicist has observed quarks, which interact strongly with each other.

Recently, a powerful tool has been added to these two traditional methods; it is that of computer simulation, a kind of Gedanken experiment of the physical system with the help of a modern computer. This was initiated by such visionaries as J. von Neumann, E. Fermi, and S. Ulam. This field is sometimes called computational physics (in a narrower sense). Particularly, it becomes powerful when the system is inherently many-bodied or interacting with itself strongly. Modern computers are extremely fast and do not complain of boredom when repeating the same procedure billions of times. Analytical methods have been plagued with this problem. A classic and typical example is that of classical turbulence, which has not yet seen a definitive resolution even after 100 years of scientific inquiry. Modern computers are able to follow the time evolution of systems containing many millions of degrees of freedom, all of which are simultaneously interacting with each other. One constructs mathematical models of the system of interest and then carries out numerical experiments based on these models. One starts the system out in some configuration of interest and observes and analyzes its behavior. Detail about the motion as complete as one desires can be obtained. One can often test theoretical predictions and the ansatzs and approximations which go into them in ways not possible with real experiments. Thus the computational physicist holds a power seldom allowed by the real world, the power to control the computer experiment. This approach complements the other two. The models are simpler and more idealized than the actual physical system; however, they are far more complete and realistic than we can hope to handle analytically for many-body systems. The method of nonlinear 'theoretical' mathematics is in an infantile stage of development. In addition, it is esoteric, extremely restrictive, and difficult to employ in an actual physical system. In fact, sometimes nonlinear mathematics is equivalent to experimental mathematics, meaning computational mathematics. The computer simulations reproduce both linear and nonlinear behavior. One can compare the results of such calculations with the behavior of real physical systems and with theory; the results can be used to predict the behavior, test theoretical predictions, and to gain understanding of the phenomenon involved. For some problems, the experimental method is impossible or very difficult. In these cases, the simulation method becomes not only important, but even indispensable to understand the physical systems better. This book discusses computational physics mainly in this sense.

Computational physics in a broader sense, however, may include not only the above mentioned modelling of nature, the computer simulation, but also the following two areas: (i) synergetics of the computer and the natural physical world and (ii) physics of computation. In his books,[2] Nobert Wiener first suggested a synergesis between man and computer. But this was primarily for the purpose of a (computing) machine. We can envisage an emerging field of physics that is qualitatively different with the presence of a computer in that physics is interfaced by or interacted with a computer. In this context the synergetic computation shares similar problems with quantum theory of observation.[3] Examples of this are: the feedback stabilization of plasma instabilities and the stochastic cooling of particle beams.[4] In the latter, for instance, the phase space volume of a beam (or the beam emittance) which otherwise would be conserved or increased in time can be contracted due to cooling with a feedback with the external "circuit" that is programmed in such a way. The other area that may be called computational physics is that of physics of computation, i.e., dynamics (or "physics") of thought and mind, perhaps a field overlapped with psychology. This may pertain to the science of computation and, moreover, intelligence. One of the prominent subjects that falls into this category is the field of artificial intelligence. The way information is processed for intelligence may be studied by the method of physics. This includes studies of how knowledge increases, is transported, associated, preserved, and destroyed via the method of nonlinear physical dynamics of information and information processing. This field has been in the mind of many scientists,[5] but has gained increased attention recently, particularly with regard to modelling of neural processing.[6] We shall discuss this subject again in a later chapter (Chapter 11). Until then, we shall concentrate on a discussion of computational physics in the narrower sense.

It may be said that the real power of numerical simulation does not lie in reproducing complex physical phenomena themselves. This would only replace a set of complex experimental data with equally complex sets of computer printouts with less detail. To a certain field of applications, this may still be an acceptable goal, where the prognosis of actual states of matter is a prime concern, such as in meteorological forecasts, financial forecasts, etc. Even in such cases, however, it is very important to understand the fundamental characteristics of the system in order to make accurate and long-ranged predictions. We shall touch on this again later. Physicists, on the other hand, wish to *understand* the working of the natural system and in fact the world.

If this is the goal, it is clear that piles of computer outputs are of little value unless one is able to construct a theoretical framework which condenses their essence into relatively simple models. The results of such computations often show us which are the important effects among many possibilities. These results thus let the physicist construct models which he would be hard pressed to justify *a priori* and provide insight which he might not attain otherwise. Often the system is too complicated even to begin to simulate. Thus he may have to extract the most important few ingredients to enter into the computer. Here, too, the ingenuity of the physicist counts. Computer calculations do not reduce the amount of physics and thought processes but rather increase the physicist's energy spent on physics and give physical intuition, inspiration, and imagination more room.

To construct a model for simulation, one must confront the fact that the modern computer is digital and treats events by digital words. To appreciate the significance of the physical models and the manner in which they are formulated for solution on the computer, it is necessary to consider the essential structure of digital computers. Although it is not well-understood and therefore it is premature to categorize, it seems that the function of human thought is associative and conceptual as well as analytical and synthetical. On the other hand, the computer of today handles essentially only digital numbers, not functions or abstract formulae. Many efforts have been made to utilize the computer in an analogous way and in a way of symbolic manipulation, but the basic function of the computer still lies in discrete number crunching. All computers have two features: a logical unit called the central processing unit (CPU) or the central processor in which simple individual computational operations are performed and a central memory where information may be stored.

Information in the central memory is stored in units called bits, where each bit has two modes: say, yes or no, or zero or one. However, the memory is conventionally organized by grouping a number of bits (typically 16, 32, 48, 60 or 64 bits) into a unit of information called a 'word.' In practice, information is dealt with in words. For example, a 'word' in a CRAY-1 machine is a unit of 64 bit information. A word may be assigned to represent an integer number, a real number, a name (that is, a set of alphabetic characters), a vector, the variables describing one particle, etc. Typically, on large contemporary computers, the core memory contains on the order of a few million words or variable locations and, although it is obviously capable of storing a very large number of variables, it is, nevertheless, still finite. It is true that the storage capacity of computers can be expanded greatly by the use of external (or peripheral) memories such as magnetic disks or tapes. But the input/output time associated with these additional memories is considerably longer than that of the central memory, and in many problems we are frequently limited to the main central memory. This is because the operation of drums, disks, or tapes is often mechanical and therefore limited by the speed of sound, while the operation of the central memory is electronic and therefore limited only by the speed of light. Recently, fast solid-state peripheral memory has been developed such as RAM512's (random access memory of 512 kilobytes) and beyond (1M and 4M bytes) and the peripheral memory speed has a tendency to increase considerably.

Both the finiteness of the capacity of computer memory and the finiteness of operation speed, force the physical systems that are described by computational physics to be represented by models of discrete finite mathematics. Thus a typical path for computer simulation is to develop a mathematical model, perhaps in a series of differential or integral equations and then to reduce it to a discrete form that can be numerically "integrated" or "differentiated" by a computer.

As a pertinent example of the discrete mathematical representations, we shall consider the case of the descriptions of galactic structures. In many galaxies, astronomy has revealed the apparent existence of spiral arms, as well as of the central core, the analysis of which can provide considerable information on the rotational properties of the galaxy, or the history of the evolution of particular galaxies. This problem can be studied in two obvious ways computationally: by the method of particles or fluids. In the first method, the galaxy may be thought of as a finite set of interacting stars or 'particles' in which each star moves under the self-consistent gravitational field arising from all the other stars in the assembly. However, the number of stars in a galaxy (typically 10^{10}) is considerably greater than the number which may be represented in the computer.

An alternative approach is to employ the methods of statistical mechanics, where it is assumed that the galaxy is composed of such a large number of stars that, effectively, it may be described as a continuous fluid. The discreteness parameter is put to zero in the modelling. Hence, equations which define the evolution in time of the distribution, or density of the galactic fluid may be derived to describe the system. However, to satisfy the finite requirements of the computer, the conceptually continuous fluid is now divided into a set of elementary cells to provide difference equations which can again describe the motion and structure of the galaxy.

Another simple example to illustrate the discrete requirements of computational mechanics occurs in the vibrations of ion lattices. Using the Debye theory, or the Einstein theory,[7] the solid may be described as a continuum, but the resulting differential equations must be reformulated as difference equations and consequently the properties of the elements or cells of the solid are described. Alternatively, the particle method could again be employed by directly following the motions of a limited number of interacting ions on the lattice.

Similarly, there are a large variety of models for simulating plasmas and, basically, they are of two types, particle models and fluid models. In a particle model, we attempt to emulate nature by following the motion of a large number of charged particles in their self-consistent electric and magnetic fields. Such models follow the motion of the plasma on the finest space scale and on the most rapid time scale at which things happen. They, therefore, are limited to looking at the phenomenon in a relatively small sample of plasma and over relatively short periods of time. However, the phenomena at this scale influence the macroscopic behavior of the plasmas and are fundamental to understanding plasma.

The second type of model is a fluid model. Here one adopts a set of fluid equations to describe the plasma. One of the justifications for this model is that in a typical nearly collisionless model there are so many particles in a given Debye sphere that the individual nature of particles is not as important as the collective nature of plasma particles as the discreteness parameter of the plasma becomes small. Here the discreteness parameter of the plasma is defined as the inverse of a number of particles in the Debye sphere.[7] These generally can describe the plasma behavior on large space and time scales. However, one must often insert *ad hoc* transport coefficients and other assumptions related to microscopic processes into such models.

1.2 Dynamical Systems of Many Degrees of Freedom

We may break down the method of computer simulation into two overall categories: the static simulation and the dynamical simulation. In what follows, we shall focus primarily on the dynamical simulation. The method of the Monte-Carlo simulation[9–11] prominently belongs to the static simulation, conceived by pioneers such as von Neumann, Ulam, and Metropolis. This method utilizes the random number generation and probability distribution to calculate integrals of various moments of the probability distribution function, since it is a good method to evaluate complicated multi-dimensional integrals. For statistical mechanics the integrals are often for the partition function. The Feynman path integral method introduces a ready possibility of doing quantum mechanical simulations. Here, for example, the canonical ensemble's temperature T is replaced by $i\,t$ with t being time. Since this method does not rely on expansion by any smallness parameter of interaction, it can be applied to strongly interactive systems. The study of strong interaction via this method has been quite an active research area of gauge field theory.[12,13]

The dynamical simulation may be classified into the phase volume conserving cases and the nonconserving cases. It may be argued that the former is more fundamental and certainly more easily handled. We will, again, concentrate only on the former. We shall touch on the latter in a section of Chapter 7. The quantum mechanical method for dynamical simulations is not well developed and only the classical method is described here. However, an interesting example of hydrodynamical simulation of quantum chromodynamics does exist.[14]

An example of classical phase volume conserving systems is Klimontovich's system.[15] For one ensemble (i.e., for a single particle) the dynamical equations may be given by

$$\dot{p}_j = F_j , \tag{1.1}$$

$$\dot{q}_j = G_j , \tag{1.2}$$

where p_j and q_j are particle j's (generalized) momentum and coordinate. One can construct Klimontovich's equation

$$\frac{\partial \mathcal{N}}{\partial t} + \nabla \cdot (u\mathcal{N}) \equiv \frac{d\mathcal{N}}{dt} = 0, \tag{1.3}$$

where $\mathcal{N} = \delta(q-q_j)\delta(p-p_j)$, $u = (\dot{q}_j, \dot{p}_j)$, $\nabla = (\partial_q, \partial_p)$. This is simply a statement of phase volume conservation. Consider a system of N particles with N being kept constant. The phase volume conservation means that the volume of $2 \times 3 \times N$ dimensional phase space (Γ-space) is constant.[7] The mathematical manifestation of this statement is called the Liouville equation. On the other hand, this system for one ensemble can be described by Klimontovich's equation Eq. (1.3) except that now

$$\mathcal{N}(p, q) = \sum_j \delta(q - q_j)\delta(p - p_j) . \tag{1.4}$$

If the system is Hamiltonian, we have a constraint

$$\frac{\partial \dot{q}_i}{\partial q_i} + \frac{\partial \dot{p}_i}{\partial p_i} = 0 . \tag{1.5}$$

The Liouville equation[16] makes an ensemble average of the spiky δ-functions in Klimontovich's equation for statistical description. In order to obtain statistical quantities, we make the ensemble average of various moments. For example, the one-point distribution function is given by

$$f_1(q_1, p_1, t) = \int dq_2 dp_2 \ldots dq_N dp_N D(q_1, p_1, \ldots) , \tag{1.6}$$

and likewise the two-point distribution function is

$$f_2(q_1, p_1, q_2, p_2, t) = \int dq_3 dp_3 \ldots dq_N dp_N D(q_1, p_1, \ldots, t) , \tag{1.7}$$

where D is the Liouville distribution.[16] The equation for f_1 is given by

$$\begin{aligned}\frac{\partial f_1}{\partial t} &= -\frac{\partial}{\partial q_1}(G(1,1)f_1) - \frac{\partial}{\partial p_1}(F(1,1)f_1) \\ &\quad -(N-1)\frac{\partial}{\partial q_1}\int dp_2 dq_2 G(1,2) f_2(q_1,p_1,q_2,p_2,t) \\ &\quad -(N-1)\partial/\partial p_1 \int dp_2 dq_2 F(1,2) f_2(q_1,p_1,q_2,p_2,t) \ , \end{aligned} \tag{1.8}$$

with

$$\begin{aligned}\dot{q}_i &= \sum_j G(i,j),\ \dot{p}_i = \sum_j F(i,j) \text{ and } f_2(q_1,p_1,q_2,p_2,t) \\ &\equiv f_1(q_1,p_1)f_1(p_2,q_2) + g(q_1,p_1,q_2,p_2) \ .\end{aligned}$$

The convention of $q_i \rightarrow i$ was used. Equation (1.8) is rewritten as

$$\begin{aligned}\frac{\partial f_1}{\partial t} &= -\frac{\partial}{\partial q_1}(Gf_1) - \frac{\partial}{\partial p_1}(Ff_1) \\ &\quad -(N-1)\frac{\partial f_1}{\partial p_1}\int dq_2 dp_2 F(1,2) f_1(q_2,p_2) \\ &\quad -(N-1)\frac{\partial f_1}{\partial q_1}\int dq_2 dp_2 G(1,2) f_1(q_2,p_2) \\ &\quad -(N-1)\int dq_2 dp_2 F(1,2)\frac{\partial}{\partial p_1} g(q_1,p_1,q_2,p_2) \\ &\quad -(N-1)\int dq_2 dp_2 G(1,2)\frac{\partial}{\partial q_1} g(q_1,p_1,q_2,p_2) \ . \end{aligned} \tag{1.9}$$

This is the first chain of the BBGKY hierarchy.[16,17] The equation which describes the dynamics of the one-point distribution function f_1 couples with the (intrinsic) two-point distribution of g. We shall touch on this again later. Higher order chains can be derived similarly. It is shown[17,8] that when the parameter called the plasma parameter $(n\lambda_D^3)^{-1}$ is small, the last two terms are small and we arrive at the Vlasov equation which describes collisionless plasmas. The terms involving the two-point distribution function are sometimes called the collision terms, since these terms give rise to collisional effects. These terms also express the highly correlated phenomena such as high density plasmas,[18] liquids,[19] and turbulence.[20] In gravitational systems, these terms play an essential role, because the gravitational system is intrinsically hierarchical.[21] The dynamical physics near the phase transition is similar to this.[22] In the physics of accelerators, the Vlasov equation is often used and the (often invariant) phase space volume is called emittance[23] (Chapter 15). When the short-range interaction is complicated or important, and thus, the higher order terms become crucial, one often resorts to the molecular dynamic calculations.[24]

For systems where the total number of particles is conserved, it is quite natural to introduce a particle model of the system. In the majority of this book we will discuss this model. If there is a sink and/or a source of particles, however, this technique becomes less powerful, although a technique to handle this situation will be discussed in relation to the gyrokinetic equation in Chapter 7. A different approach should be used if the creation and annihilation of particles are as essential as in high energy physics. The Monte-Carlo simulation has been perhaps the most successful in such situations.

Boltzmann's premise that all possible states (or phase space volumes) are occupied with equal probability leads to the basis of calculating the partition function in statistical mechanics. The partition function Z is given by

$$Z = \sum_{E'} e^{i\beta E'} \ , \tag{1.10}$$

for discrete energy levels $\{E'\}$ or

$$Z = \int \cdots \int d^N p d^N q e^{-\beta E(p,q)} \ , \tag{1.11}$$

for continuous energy levels.[7] The quantum mechanical calculations can be done through a similar functional summation/integral. The probability for each "energy level" E' in Eq. (1.10) is represented by the prescribed random number generator. This method has been quite successful in calculating static properties of a complicated many-body interaction system.[10]

In the method of Feynman's path integral formulation of quantum mechanics[25] the integral along all possible paths must be computed

$$I = \int \cdots \int D(q) e^{+i\beta \int \hat{L}dt} \ , \tag{1.12}$$

where $D(q)$ is along all possible paths and $\int \hat{L}_{dt}$ is the action integral. This is because the quantum mechanical uncertainty principle allows all possible paths. Here the "weight" function, the exponential function, has an imaginary quantity in the argument. The classical mechanics is deterministic in contrast to the quantum mechanics whose indeterministic characteristic is manifested in the integral over *all* possible paths. On the other hand, the classical dynamical equation is determined by *the* path along which the action integral is minimized: no integral over all possible paths is taken. The deterministic Liouville equation $\partial_t D + LD = 0$ with D being the Liouville distribution function in Γ-space is contrasted with the equation in quantum mechanics:

$$D(\{q_ip_i\};t) = e^{\int Ldt} D(\{q_ip_i\};t)\} \quad \text{vs.}$$

$$\Psi(t) = \int dq_1 \ldots dp_N e^{i\int \hat{L}dt}\Psi(0) \,.$$

One more important difference between the quantum mechanical system and the classical dynamical system is that while the quantum mechanical operator $\hat{L}$ consists of the free particle part $\hat{L}_0$ (i.e. the unperturbed part) which is a quadratic operator and the smaller perturbative part $\delta\hat{L}$, the classical counterpart does not in general possess this property. That is, the "unperturbed" operator is not definable in the classical system in general. In quantum mechanics one can develop an efficient integral method via Wick's contraction theorem because of this property. On the other hand, the classical case L_0 and δL neither commute nor truncate:

$$e_0^L e^{\delta L}$$

is not contracted, since with $e^X \equiv e^X e^Y$, Z becomes

$$X + Y + \frac{1}{2}[X,Y] + \frac{1}{12}[[X,Y],X] + \frac{1}{12}[[X,Y],Y] \ldots \quad .$$

This is due to the Campbell-Baker-Haussdorff theorem. Because of these difficulties in formulating calculational procedures similar to quantum mechanical systems, classical dynamical systems are usually not appropriately or efficiently treated by the Monte-Carlo simulation method. In this book we shall therefore concentrate on the dynamical simulation methods: (i) the particle simulation and (ii) the fluid simulation.

It may, however, be plausible[26] to apply the path integral method to the dynamical systems if the system is Gaussian. It turns out that the Gaussianness corresponds to having a quadratic form for the "unperturbed" ground state, corresponding to the "free particle part" $\hat{L}_0$. Suppose one wants to calculate the Green function of the following equation

$$\left(\frac{\partial}{\partial t} + \mathbf{v}\cdot\frac{\partial}{\partial \mathbf{x}} + \frac{\partial}{\partial \mathbf{v}}\cdot\mathbf{D}\cdot\frac{\partial}{\partial \mathbf{v}}\right) G(\mathbf{r},\mathbf{v},t;\mathbf{r}_0,\mathbf{v}_0,t_0) = \delta(t-t_0)\delta(\mathbf{r}-\mathbf{r}_0)\delta(\mathbf{v}-\mathbf{v}_0)\,. \tag{1.13}$$

The causal solution of Eq. (1.13) that is often called the retarded Green function is expressed as

$$G(\mathbf{r},\mathbf{v},t;\mathbf{r}_0,\mathbf{v}_0,t_0) = \theta(t-t_0)\left\langle \mathbf{r}\ ,\mathbf{v}\left|e^{\hat{L}(t-t_0)}\right|\mathbf{r}_0\mathbf{v}_0\right\rangle\ , \tag{1.14}$$

where $\theta(t-t_0)$ is the Heavyside function and $\hat{L} = -\mathbf{v}_{\hat{p}} + \hat{q}\cdot\mathbf{D}\cdot\hat{q}$ with $\hat{p}$ and $\hat{q}$ are operators conjugate to $\mathbf{x}$ and $\mathbf{v}$: $[\hat{p}_\mu, r_\nu] = \delta_{\mu\nu}$, $[\hat{q}_\mu, v_\nu] = \delta_{\mu\nu}$. The Dirac representation[27] $|r>(\equiv \Psi_r)$ and $<r|(\equiv \Psi_r^+)$ are such that

$$|\mathbf{r}, \mathbf{v}\rangle = e^{\hat{L}(t-t_0)} |\mathbf{r}_0(t_0), \mathbf{v}_0(t_0)\rangle \ . \tag{1.15}$$

The evaluation of Eq. (1.15) may be carried out by Feynman's method. Let us evaluate the factor on the right-hand side of Eq. (1.15)

$$\begin{aligned}
&\left\langle \mathbf{r}, \mathbf{v} \left| e^{(-\mathbf{v}\hat{\mathbf{p}}+\hat{\mathbf{q}}\cdot\mathbf{D}\cdot\hat{\mathbf{q}})(t-t_0)} \right| \mathbf{r}_0, \mathbf{v}_0 \right\rangle \\
&= \int \prod_{i=1}^{n} d\mathbf{r}_i d\mathbf{v}_i \left\langle \mathbf{r}, \mathbf{v} \left| e^{(-\mathbf{v}\cdot\hat{\mathbf{p}}+\hat{\mathbf{q}}\cdot\mathbf{D}\cdot\hat{\mathbf{q}})\epsilon} \right| \mathbf{r}_n, \mathbf{v}_n \right\rangle \cdots \\
&\left\langle \mathbf{r}_{k+1}, \mathbf{v}_{k+1} \left| e^{(-\mathbf{v}\cdot\hat{\mathbf{p}}+\hat{\mathbf{q}}\cdot\mathbf{D}\hat{\mathbf{q}})\epsilon} \right| \mathbf{r}_k, \mathbf{v}_k \right\rangle \cdots \\
&\left\langle \mathbf{r}_1, \mathbf{v}_1 \left| e^{(-\mathbf{v}\cdot\hat{\mathbf{p}}+\hat{\mathbf{q}}\cdot\mathbf{D}\cdot\hat{\mathbf{q}})\epsilon} \right| \mathbf{r}_0, \mathbf{v}_0 \right\rangle \ .
\end{aligned} \tag{1.16}$$

We now have to evaluate individual factors $\langle\cdots\rangle$ in Eq. (1.16).

$$\begin{aligned}
&\left\langle \mathbf{r}_{k+1}, \mathbf{v}_{k+1} \left| e^{(-\mathbf{v}\cdot\hat{\mathbf{p}}+\hat{\mathbf{q}}\cdot\mathbf{D}\cdot\hat{\mathbf{q}})\epsilon} \right| \mathbf{r}_k, \mathbf{v}_k \right\rangle \\
&= \int d\mathbf{p}d\mathbf{v}d\mathbf{r}d\mathbf{q} \left\langle \mathbf{r}_{k+1}, \mathbf{v}_{k+1} | e^{-\mathbf{v}\cdot\hat{\mathbf{p}}\epsilon} | \mathbf{p}, \mathbf{v} \right\rangle \\
&\cdot \left\langle \mathbf{p}, \mathbf{v} \left| e^{\hat{\mathbf{q}}\cdot\mathbf{D}\cdot\hat{\mathbf{q}}\epsilon} \right| \mathbf{r}, \mathbf{q} \right\rangle \langle \mathbf{r}, \mathbf{q} | \mathbf{r}_k, \mathbf{v}_k \rangle \\
&= \int d\mathbf{p}d\mathbf{v}d\mathbf{r}d\mathbf{q} e^{-i\mathbf{v}\cdot\mathbf{p}\epsilon} \delta(\mathbf{v}_{k+1} - \mathbf{v}) \langle \mathbf{r}_{k+1} | \mathbf{p} \rangle e^{-q^2 D\epsilon} \langle \mathbf{p} | \mathbf{r} \rangle \\
&\cdot \langle \mathbf{v} | \mathbf{q} \rangle \delta(\mathbf{r} - \mathbf{r}_k) \langle \mathbf{q} | \mathbf{v}_k \rangle \ .
\end{aligned} \tag{1.17}$$

Here

$$\begin{aligned}
\langle \mathbf{r}_{k+1} | \mathbf{p} \rangle &= (2\pi)^{-3} \exp(i\mathbf{p} \cdot \mathbf{r}_{k+1}) \ , \\
\langle \mathbf{p}\mathbf{r} \rangle &= (2\pi)^{-3} \exp(-i\mathbf{p} \cdot \mathbf{r}) \ , \\
\langle \mathbf{v} | \mathbf{q} \rangle &= (2\pi)^{-3} \exp(i\mathbf{v} \cdot \mathbf{q}) \ ,
\end{aligned}$$

and

$$\langle \mathbf{q} | \mathbf{v_k} \rangle = (2\pi)^{-3} \exp(-i\mathbf{v}_k \cdot \mathbf{q}) \ .$$

Thus Eq. (1.17) becomes

$$
\begin{aligned}
(\text{LHS}) &= \int d\mathbf{p}d\mathbf{q} \exp\left[i\mathbf{p}\cdot\left(\frac{\mathbf{r}_{k+1}-\mathbf{r}_k}{\epsilon}-\mathbf{v}_{k+1}\right)\epsilon \right. \\
&\quad \left. + \left(i\frac{\mathbf{v}_{k+1}-\mathbf{v}_k}{\epsilon}\cdot\mathbf{q}-q^2D\right)\epsilon\right] \\
&= \frac{1}{(2\pi)^3(4\pi D\epsilon)^{3/2}} \exp\left[-\frac{(\dot{\mathbf{v}}_{k+1}\epsilon)^2}{4D\epsilon}\right] \\
&\quad \times \int d\mathbf{p}_{k+1} \exp\left[i\mathbf{p}_{k+1}\left(\dot{\mathbf{r}}_{k+1}-\mathbf{v}_{k+1}\right)\epsilon\right] . \qquad (1.18)
\end{aligned}
$$

As we can see from these calculations, the factor in Eq. (1.17) and eventually Eq. (1.16) are similar to Eqs. (1.11) and (1.12). Thus the dynamical equation such as Eq. (1.13) may also be put into the framework of a Monte-Carlo method, if one wants, in principle for Gaussian systems. If the system is not Gaussian, the calculation becomes too complicated. With this diversion aside, we return to more well-trodden paths of the dynamical simulations: the particle simulation method and the fluid simulation method.

1.3 Particle Simulation and Finite-Size Particles

The older of the two methods, the fluid simulation method[28] is conceptually straightforward as long as we can regard the many-body system as being composed of so many particles that we can smear out the individual particles into a continuous fluid. Some of the applications to the magnetohydrodynamic aspects of plasmas (and other systems) are discussed in Chapters 6 and others. Here, we focus on the newer of the two, the particle simulation method. The simplest possible system for the particle simulation may be the electrostatic particle simulation . At the same time, this illustrates difficulties involved in the particle simulation and the central idea of how to cope with these difficulties. Therefore, we look at this problem a little closer now.

Figure 1.2 shows some examples of electrostatic particle models. They come in one, two and three dimensional versions. The one dimensional version may be thought of as a large number of charge sheets perpendicular to the x axis. The two dimensional model is a series of charged rods parallel to, say, the z axis. A three dimensional model, of course, consists of particles in three dimensions. The one dimensional model can be generalized to what we call a one and a half dimensional model by allowing the charge sheets to also move in

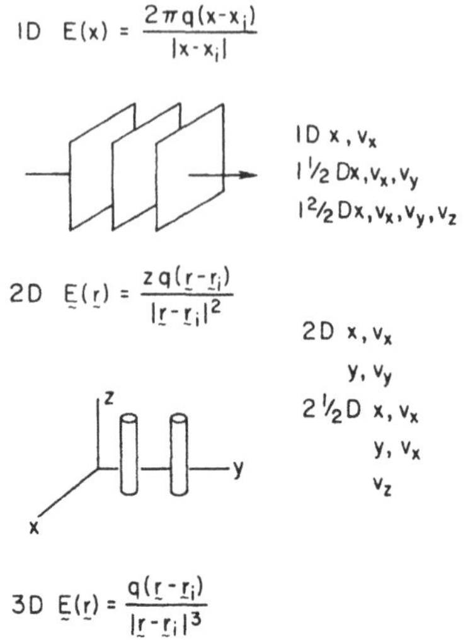

1D $E(x) = \frac{2\pi q(x-x_i)}{|x-x_i|}$

1D x, v_x
$1\frac{1}{2}$D x, v_x, v_y
$1\frac{2}{2}$D x, v_x, v_y, v_z

2D $\underline{E}(\underline{r}) = \frac{2q(\underline{r}-\underline{r}_i)}{|\underline{r}-\underline{r}_i|^2}$

2D x, v_x
y, v_y
$2\frac{1}{2}$D x, v_x
y, v_x
v_z

z
y
x

3D $\underline{E}(\underline{r}) = \frac{q(\underline{r}-\underline{r}_i)}{|\underline{r}-\underline{r}_i|^3}$

FIGURE 1.2 1D and 2D models vs. 3D force.

the y direction. There is then motion and current in the y direction, but none of the quantities such as charge density, potential, electric field, magnetic field, etc. vary in that direction. We would say that x is a full dimension, but that y is a half dimension. The dynamics in the y direction is, so to speak, fractal. We can further generalize the model by allowing the sheets to also move in the z direction. Then both the y and z directions would be considered half dimensions and the model is one and two halves dimensional. Similarly, the two dimensional model can be generalized by allowing the rods to move in the z direction. Again, there would be z directed currents but no quantities would vary in the z direction. The x and y directions would be considered full dimensions and the z direction a half dimension and the model would be considered a two and one half dimensional model. The $1-\frac{1}{2}$, $1-\frac{2}{2}$, and $2-\frac{1}{2}$ dimensional models are important for treating electromagnetic effects which involve currents perpendicular to wave vector $\mathbf{k}$. These fractal dimensions can also be important for magnetized plasmas where plasmas behave anisotropically. Some bearings on isotropic phenomena (e.g., turbulence) by the fractal dimensions may also become significant in certain cases. Figure 1.2 illustrates the electrostatic versions of these models.

Basically, such particle models seem simple; for the electrostatic case we wish to solve the equations:

$$\ddot{\mathbf{r}}_i = \frac{q_i}{M_i} \sum_j \mathbf{E}_j(\mathbf{r}_i) = \frac{(2\pi)^{\left(1-\frac{n}{2}\right)} q_i}{M_i} \sum_j \frac{q_j(\mathbf{r}_i - \mathbf{r}_j)}{|\mathbf{r}_i - \mathbf{r}_j|^n}, \tag{1.19}$$

for a large number of particles. Here i refers to the ith particle and q_i and M_i are its charge and mass; n is still the dimensionality. Typically, one wishes to solve these equations for $10^4 \sim 10^6$ particles; a simple straightforward estimate shows that it is not possible even on foreseeable computers. Suppose we have N particles, then for each particle the sum over j contains N terms and we must evaluate it for all N particles so that the number of operations is proportional to N^2. The evaluation of each term in the sum involves a number of arithmetic operations. For the purpose of evaluation let us assume that $10N^2$ operations are involved per time step δt.

$$N_{\text{op}} \sim 10N^2 \quad \text{per time step} . \tag{1.20}$$

If we used 10^5 particles and on the average the machine carried out one arithmetic operation in 10^{-7} sec, just evaluating the forces would require 10^4 sec per time step. Simulations involving 10^3 to 10^4 time steps would require from one month to one year to do. Clearly such calculations are not feasible unless the results are extremely valuable. If the force is short-ranged, one can perhaps afford to calculate the interaction in such a way. In molecular dynamics calculations, in fact, it is done this way.[24] We must find a faster, more efficient method for evaluating the *long-range forces* (see Problem 9); fortunately, these exist, i.e., the method of finite-sized particles and grid calculation of these quantities. The advent of such a method was a great breakthrough in particle simulation.

A second important consideration is that of particle collisions. In fact, this consideration sets the requirement on the number of particles that we must use. For a particle model there are collisions between particles just as there are in a real plasma. The kinetic theory of one, two and three dimensional plasmas can be carried out. The collision times, τ_{coll}, obtained are roughly given by the following expression:

$$\omega_p \tau_{\text{coll}} \cong \quad 2n\lambda_D, \quad (1D) \tag{1.21}$$

$$2\pi^2 n\lambda_D^2, \quad (2D) \tag{1.22}$$

$$2\pi n\lambda_D^3/\ell n\Lambda, \quad (3D) \tag{1.23}$$

where ω_p is the plasma frequency, τ_{coll}^{-1} is the collision frequency, n is the charge density, λ_D is the Debye length and $\ell n\Lambda$ is the Coulomb logarithm:

$$\omega_p^2 = \frac{4\pi n e^2}{m} \tag{1.24}$$

$$\lambda_D^2 = \frac{\kappa T}{4\pi n e^2} \tag{1.25}$$

$$\ell n\Lambda = \ell n(n\lambda_D^3) , \tag{1.26}$$

T is the temperature, e and m are the electronic charge and mass. We see from these formulas the collision times are proportional to the number of particles in a Debye sphere in 1, 2 and 3D.

The long range nature of the Coulomb force gives rise to collective modes – modes involving coherent motions of many particles. These are the modes of interest. If the short range forces such as collisions are important, the analysis or molecular dynamics may be employed. The collective modes are modes with wavelengths greater than the Debye length. Also for plasmas of interest, $\omega_p \tau_{\text{coll}}$ should be much larger than unity; collisional effects should be small on the time scale of plasma oscillations. This means that a large number of particles exist in the Debye sphere which is typical for the plasma of interest; thus we want:

$$N_D \equiv g^{-1} \gg 1 \tag{1.27}$$

where

$$N_D = n\lambda_D, \quad (1D) \tag{1.28}$$

$$2\pi n\lambda_D^2, \quad (2D) \tag{1.29}$$

$$\tfrac{4\pi}{3} n\lambda_D^3. \quad (3D) . \tag{1.30}$$

Further, for a significant simulation there should be many collective modes in the system. For the two dimensional case a significant simulation might involve a system of 100 Debye lengths on a side. Thus, the total number of particles must be large compared to 10^4 (hence, the number of particles used in the above estimate).

Before we introduce the method of finite-size particles, we shall revisit a simplified version of the BBGKY hierarchy. Let us examine the pioneering work by Landau[29] and by Lennard and Balescu[30] on separating two spatial scales of relevant plasma phenomena. The microscopic Klimontovich equation[15] describes a system of interacting N-bodies with the electric force:

$$\frac{\partial}{\partial t} F(\mathbf{x}, \mathbf{v}; t) + \mathbf{v} \cdot \frac{\partial}{\partial \mathbf{x}} F(\mathbf{x}; \mathbf{t}) + \frac{q}{m} \mathbf{E}(\mathbf{x}) \cdot \frac{\partial}{\partial \mathbf{v}} F(\mathbf{x}, \mathbf{v}; t) = 0 , \tag{1.31}$$

where

$$F(\mathbf{x}, \mathbf{v}; t) = \sum_j \delta[\mathbf{x} - \mathbf{x}_j(t)] \cdot \delta[\mathbf{v} - \mathbf{v}_j(t)] , \tag{1.32}$$

and $\mathbf{x}_j(t)$ and $\mathbf{v}_j(t)$ are the j-th particle's position and velocity at time t. This equation (1.31) can be transformed into the Vlasov[31]-like equation,[29,30,17] which describes the evolution of the Vlasov distribution $f(\mathbf{x}, \mathbf{v}; t)$ $[\equiv < F(\mathbf{x}, \mathbf{v}; t) >$, a coarse-grained distribution], except for the right-hand side:

$$\frac{\partial}{\partial t} f(\mathbf{x}, \mathbf{v}; t) + \mathbf{v} \cdot \frac{\partial}{\partial \mathbf{x}} f(\mathbf{x}, \mathbf{v}; t) + \frac{q}{m} \langle \mathbf{E}(\mathbf{x}) \rangle \cdot \frac{\partial}{\partial \mathbf{v}} f(\mathbf{x}, \mathbf{v}; t)$$

$$= \frac{\pi \omega_p^2}{n} \sum_k \omega_p^2 \frac{\mathbf{k}}{k^2} \cdot \frac{\partial}{\partial \mathbf{v}} \int d\mathbf{v}' \left[\left(\frac{\partial}{\partial \mathbf{v}} - \frac{\partial}{\partial \mathbf{v}'} \right) f(v) f(v') \right]$$

$$\times \frac{\delta(\mathbf{k} \cdot \mathbf{v} - \mathbf{k} \cdot \mathbf{v}')}{|\epsilon(\mathbf{k}, \mathbf{k} \cdot \mathbf{v})|^2}, \tag{1.33}$$

with $\epsilon(\mathbf{k}, \omega) = 1 + \frac{\omega_p^2}{k^2} \int dv \frac{1}{\omega - \mathbf{k} \cdot \mathbf{v}} \mathbf{k} \cdot \frac{\partial}{\partial \mathbf{v}} f(v)$, where the angular bracket originally meant the ensemble average and may be replaced by the spatial average over the short-range distance that is incorporated in the right-hand side so-called collision integral contribution. This equation (1.33) is a truncated version of Eq. (1.9). The Landau collision integral is recovered when the dielectric constant ϵ is taken as unity. The main message in this treatment is that the short-range collisional effects can be isolated from the (averaged) Vlasov-like equation which describes only the long-range interaction. For collisionless plasmas the right-hand side must be small. If the number of particles in a Debye sphere (or cylinder, etc.) is large, the discreteness parameter is small and the right-hand side of Eq. (1.33) is small. If the number of particles in a Debye sphere (or cylinder, etc.) is large, the discreteness parameter is small and the right-hand side of Eq. (1.33) is not small. For a computationally represented plasma that cannot contain too large a number of particles due to practical computational restraints, the right-hand side is bound to be large. This undesirable deviation, from what we wish to describe due to the small number of particles in a Debye sphere which we can realize, may be avoided by changing the behavior of the dielectric constant in the short-range domain in the integral of the right-hand side of Eq. (1.35). That is, we should reduce the strength of the short-range interaction to accomplish it. Since this manipulation involves only the short-range interaction represented by the right-hand side collision operation, it does not change the left-hand side long-range collective nature of interaction and describes it with satisfactory accuracy.

To carry out a task in the simulation that is equivalent to reducing the collision effects, the finite-size particle method was introduced.[32] Now the force between two charged particles has the general form shown at the top of Fig. 1.3 (except for one dimension where it is $x/|x|$). Collisions come about due to the rapid variation of the force as particles pass close to each other. If we could eliminate the strong variation of the force for close encounters, we could greatly reduce the collision rate. This can be done as shown in Fig. 1.3. Instead of point charges, we consider particles made of finite size charge clouds of radius a_0. For separations which are large compared to a_0,

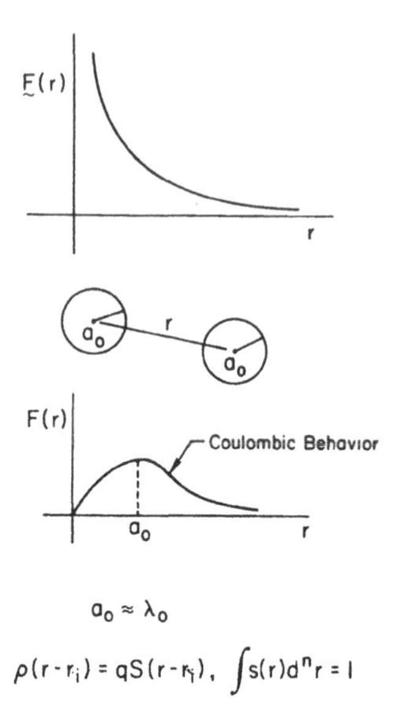

FIGURE 1.3 Coulombic forces by a point charge and by a finite-sized charge

the force between the particles is Coulombic. However, when the particles overlap (we allow them to pass freely through each other) the force falls to zero as r goes to zero, as illustrated in the lower drawing of Fig. 1.3.

Since we desire to model modes with scale lengths greater than the Debye length accurately, we generally take $a_0 \sim \lambda_D$ in the explicit simulation (see Chapter 3 for the definition of explicit simulation). [A significant deviation from this will be discussed in the implicit simulation (Chapter 9).] We can represent the charge density ρ of such a particle by

$$\rho(\mathbf{r} - \mathbf{r}_i) = qS(\mathbf{r} - \mathbf{r}_i) \ , \tag{1.34}$$

instead of

$$\rho(\mathbf{r} - \mathbf{r}_i) = q\delta(\mathbf{r} - \mathbf{r}_i) \ , \tag{1.35}$$

where r_i is the center of the particle, q is its total charge and S is a shape factor

$$\int S(\mathbf{r} - \mathbf{r}_i) d^n r = 1 \ . \tag{1.36}$$

The choice of shape and size of the charge cloud is at the discretion of the physicist performing the calculations. Two possible examples of shapes, a uniformly charged sphere and Gaussian charge distribution are shown in Fig. 1.4. In the simplest and canonical code, we generally choose the particle size to

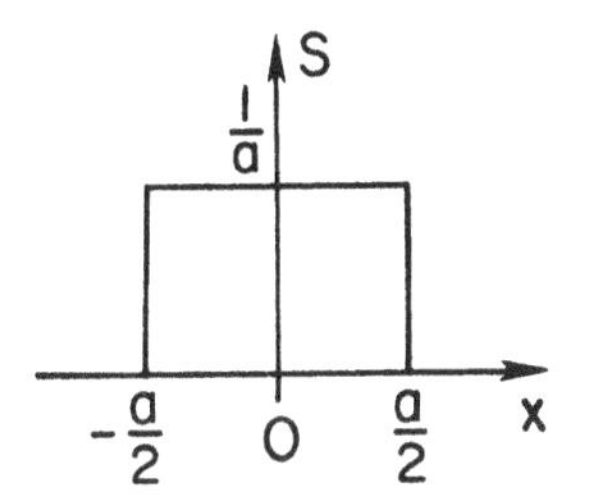

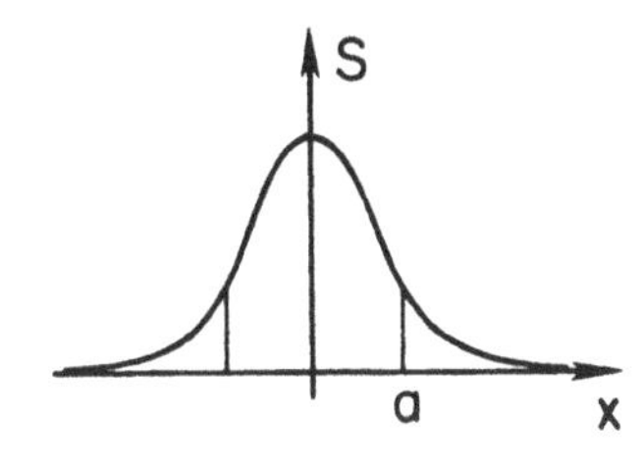

FIGURE 1.4 Shapes of finite-sized particles

be on the order of a Debye length. More sophisticated methods, however, are discussed in Chapters 4 and 9. If we choose it smaller than this, the reduction in collision rate is not so great (in addition the code becomes numerically unstable as we shall see later); on the other hand, if we choose it larger than this, we cease to accurately model modes with wavelengths a few Debye lengths long. As we shall see later, the size of the particle determines the collision rate to great extent.

The use of finite size particles meshes is another important method of rapid calculations. This is the method of using a grid to compute the force on a particle. Consider a force F on the i-th particle

$$\mathbf{F}_i = q_i \mathbf{E}(\mathbf{r}_i) \ . \tag{1.37}$$

For our finite sized particles this must be replaced by:

$$\mathbf{F}_i = q_i \int S(\mathbf{r} - \mathbf{r}_i) \mathbf{E}(\mathbf{r}_i) d^n r \ . \tag{1.38}$$

The electric field can be determined from Poisson's equation:

$$\boldsymbol{\nabla} \cdot \mathbf{E} = -\nabla^2 \phi = 4\pi \rho(r) \ . \tag{1.39}$$

If the charge density, $\rho(\mathbf{r})$, is that given by point particles, then calculating the field is equivalent to computing the sum in Eq. (1.9) and we have gained nothing. However, if the particles have finite size, we cannot resolve charge density variations smaller than the size of the particle. Therefore, we can divide the space by a uniform grid with grid spacing about equal to a particle radius as shown in Fig. 1.5. This coarse-graining of space which was originally continuous is needed in order to avoid spurious physics creeping in by going from continuous space to discrete space. A similar requirement becomes necessary in the simulation of lattice gauge theory where space is divided into

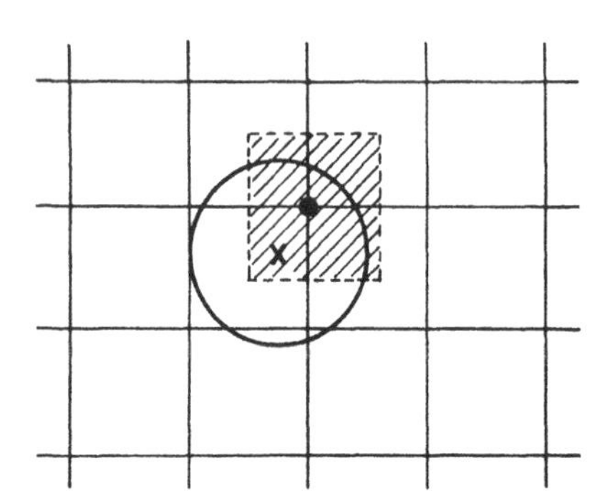

FIGURE 1.5 Finite-sized particle, grid, and grid assignment

a lattice to evaluate path integrals. In the present case, the finiteness of particle size plays the role of natural coarse-graining.[33] A second grid (grid 2) connecting the centers of the squares formed by the first grid (grid 1) is also introduced as shown by the dashed grid. Around each grid point of grid 1 is a square of grid 2. We may associate the charge of particles whose centers lie in this square with its associated grid 1 point; i.e., we can place all particles within a cell of grid 2 at its center. If the particle size is a grid spacing or greater, this is a reasonably accurate approximation and is called the nearest grid point approximation (NGP). We can improve on this approximation by putting a dipole at the grid point which is equal to the dipole moment of the particle with respect to the grid 1 point. We then get the following charge distribution:

$$\rho(\mathbf{r}) = \sum_g [Q(r_g)S(\mathbf{r}-\mathbf{r}_g) + \mathbf{D}(\mathbf{r}_g)\cdot\nabla_g S(\mathbf{r}-\mathbf{r}_g)] , \tag{1.40}$$

$$Q(\mathbf{r}_g) = \sum_{i\epsilon g} q_i \quad ; \quad \mathbf{D}(\mathbf{r}_g) = \sum_{i\epsilon g} q_i(\mathbf{r}_i - \mathbf{r}_g) . \tag{1.41}$$

A modification is to put fractional charges on the surrounding grid points so that that configuration has the same charge and dipole moment (charge sharing). For either approximation the result is a set of finite size charged particles located on a regularly spaced grid system.

Fast Poisson solvers and fast Fourier Transforms (FFT) are examples of fast methods for solving for the fields on such a regularly spaced grid. Once we obtain the field, we compute the forces on the particles; if the nearest grid point method is used, this is done by evaluating the field at the nearest grid point; if the dipole approximation is used, we either interpolate the fields at the position of the particles or in the charge sharing scheme we take an appropriate weighting of the forces on the nearby grid points.

The speed of this method of computing forces is much faster than the direct calculation method used in Eq. (1.19). The time to calculate the fields from the charges is proportional to $M\ell nM$ where M is the number of grid points. On the other hand, the time required to put the charges on the grids, compute the forces on the particles from the fields and to advance the particles according to the equations of motion is proportional to N. So the total time for the present fast method is given by

$$\tau = M\ell nM + bN \ , \tag{1.42}$$

as compared to N^2 for direct two-body force calculations. Here a and b are constant. The $M\ell nM$ term is much smaller than N^2 because it is not quadratic and M is much smaller than N since there are many particles per cell. The cells must simply give an adequate representation of the field while the particles must represent both the spatial charge variation and the velocity distribution function.

The above described method of simulation allows us a greatly enhanced computational power. It enables us to solve problems involving nonlinear dynamics of systems with many degrees of freedom such as microscopic dynamics of plasmas. The discussion of this method and its recent sophisticated development, along with its vivid applications to actual physics problems will be the main topic of this book.

1.4 Limitations on Simulation — Future Directions

As powerful as it can be, the computer simulation has limitations. Even if the computers are of the same make and, moreover, if the codes are of the same origin, the result and wealth one can obtain from simulations may vary disparately depending upon how the simulation is handled.

The first point is that the computer does not handle the analytical formulae theoretical physicists are fond of manipulating, but crunches many bits of numbers. This way we get only an "event" instead of a physical law out of the computer. In order to make use of the "event" and learn the general behavior and laws of nature, we have to interpret the computer results and then analyze, make a model out of it, hypothesize, and synthesize. This leads to another difficulty. That is, the computer produces only one ensemble per run in general, and for most statistical arguments, quantities that matter are statistically averaged quantities. Since we cannot typically afford to have so many runs to make ensemble-average, we have to replace ensemble-averaging by something else. If the system is in a (quasi-) stationary state, the time average may play that role. By averaging over many particles, we may also achieve a similar effect. In most cases (to date) the simulation system is isolated and thus the total energy is conserved (or should at least be approximately conserved). In this sense, we are talking about the microcanonical simulation in contrast to the canonical or grand canonical simulations. Even if the total energy given is exactly the same from run to run, each sample (run) would evolve differently because $E_{\text{tot}} = \sum_i p_i^2/2m=$ constant allows all kinds of combination $\{p_i\}$ possible. However, the local phase space average at $\vec{r}_1$ and $\vec{r}_2$ are about the same for thermal start cases. One would also expect that the overall average is approximately the same for a similar but slightly different run (sample). For example, an average velocity squared over a certain (small) volume of phase space (i.e., over a certain number of particles in phase space volume $\Delta\mathbf{x}_1\Delta\mathbf{p}_1$)

$$\overline{v_1^2} = \sum_{j\epsilon\Delta\mathbf{x}_1\Delta\mathbf{p}_1} v_j^2 \Bigg/ \left(\sum_{j\epsilon\Delta\mathbf{x}_1\Delta\mathbf{p}_1} 1\right)$$

is approximately equal to an average over a volume of phase space in a different position ($\Delta\mathbf{x}_2\Delta\mathbf{p}_2$)

$$\overline{v_2^2} = \sum_{j\epsilon\Delta\mathbf{x}_2\Delta\mathbf{p}_2} v_j^2 \Bigg/ \left(\sum_{j\epsilon\Delta\mathbf{x}_2\Delta\mathbf{p}_2} 1\right) .$$

Thus we may argue that the (expected) ensemble average $\langle v^2 \rangle$ is approximately given by

$$\langle v^2 \rangle \simeq \overline{v_1^2} = \overline{v_2^2} \ . \tag{1.43}$$

In this way we may avoid running many similar samples to make necessary ensemble averages. However, if the mode is global, this is not perhaps accurate. Often, by taking correlation function we can avoid quirks of choice of the initial values and pick out "phase-independent" information such as the energy spectrum. On the other hand, phases often depend on the individual event. In any case, it takes intelligence and experience to interpret only several runs to make a ensemble-averaged physical picture.

A third difficulty is due to the complexity of nature itself. When we try to simulate a natural phenomenon, there can be many factors that influence it. If we then try to incorporate all these factors, it is too complicated to reduce them into tractable differential (or integral) equations and, furthermore, too expensive to solve these equations. The physicist has to choose the most essential factors and mold them into a tractable form. This difficulty is more or less experienced by all physicists.

A fourth difficulty is that of uniquely simulating nature. But at the same time in most of the interesting problems we consider, this is often deeply coupled with the structure of nature itself. It is a matter of accuracy: how close we are approaching the true description of nature by discretizing all the physical processes for the computer. The near or perfectly continuous function is approximated by discrete grid or particle quantities. The grid quantity, for example, is only an approximation of the actual value. Moreover, as degrees of freedom increase, the number of interaction connections increases even faster ($\propto {}_N C_2$ or even $N!$ for some interaction), which may contribute to an excessive number of operations. It is, however, hopeful if the accuracy increases linearly (or even stronger than a linear fashion) with the number of grid points or particles. This is because we know that the capacity of modern computers is rapidly advancing (almost exponentially).

Some notoriously difficult problems resist such an increase of sheer capacity of the computer. These intractable problems involve mathematically ill-behaved nature and these include the so-called stiff differential equations. In these cases input and output of mathematical information are not proportional; systems are not easily scaled up or down. All scales simultaneously participate in their own right in these problems. The cluster of galaxies, the fluid turbulence and plasma turbulence are but three of such examples. The turbulence essentially involves only classically well-known interactions. Nevertheless, the fluid turbulence has resisted a solution for more than a century and the plasma turbulence close to a half century. Some of the difficulties with these problems lie in the fact that a larger eddy involves a smaller one, which in turn contains an even smaller one, and so on (structures within structures within structures··· *ad infinitum*). Such a structure of nature may be called hierarchical. Different layers of phenomena not only interact with their equals but also with their superiors and inferiors in a hierarchy of physical interaction. Therefore, to approximate or truncate the system at a certain layer or level may lead to a gross distortion or sometimes to a completely wrong description. This poses a tremendous challenge to physicists. In fact it may be said that one of the main and most challenging trends of present-day physics (and perhaps in the future) is to tackle the interaction *self-consistently* instead of to treat the system with all or most conditions fixed (or given) but a few. Only in this way can we approach some of the most challenging and intriguing problems that have resisted solution trials by the previous physicists' approaches. Physicists of today (and perhaps of the future) have come to pay more attention to interdependence, strong interaction, self-consistent physical pictures, and ever-evolving systems in the hierarchical world of physics.

Let us look at the problem of the Navier-Stokes turbulence.[34,35] A stationary homogeneous isotropic fully-developed turbulence with energy injected at a large wavelength ($2\pi/k_0$) was considered by Kolmogorov.[36] The energies of eddies with wavenumbers larger than k_0 are determined by an equilibrium between the injected energy, the viscous dissipation loss, and the energy flux to larger wavenumbers. In the inertial domain of wavenumbers[34-36] where the viscous damping and energy injection are negligible, there can be a quasi-stationary self-similar hierarchy of eddies. The largest eddies that receive energy from outside break up into smaller eddies, which in turn break up into even smaller ones, and so on, until the size of eddies is so small that the viscous dissipation is not any more negligible. This is the process of eddy energy cascade from large scales to small scales.

By choosing eddy wavenumbers $k_n = 2^n k_0$, the energy of eddies $E(k)$ and the energy flux $\Pi(k)$, we can relate $E(k)$ and $\Pi(k)$. Let

$$k_n = 2^n k_0 , \tag{1.44}$$

$$E_n = \int_{k_n}^{k_n+1} E(k)dk , \tag{1.45}$$

$$\Pi_n = \Pi(k_n) , \tag{1.46}$$

where Π_n measures the rate (flux) at which energy is being transferred out of the wavenumber interval $k_n \leq k \leq 2k_n$ into the interval $2k_n \leq k \leq 4k_n$. This rate is given by the amount of energy E_n divided by the eddy turn-over time τ_n

$$\Pi_n = E_n/\tau_n \tag{1.47}$$

$$\tau_n = \ell_n/v_n \tag{1.48}$$

$$v_n = \left\langle |u(x) - u(x+\ell_n)|^2 \right\rangle^{1/2} , \tag{1.49}$$

where $\ell_n = 2\pi/k_n$ and v_n is a typical velocity $(u(x))$ difference across an eddy of size ℓ_n. In the inertial range the flux $\Pi(k_n)$ is independent of k_n. Using $E_n \propto v_n^2$, we obtain

$$v_n^3/\ell_n \propto \epsilon = \text{ constant} , \tag{1.50}$$

and thus

$$E_n \propto \epsilon^{2/3}\ell_n^{2/3} \tag{1.51}$$

or

$$E(k) = C\epsilon^{2/3}k^{-5/3} . \tag{1.52}$$

Equation (1.52) is the Kolmogorov spectrum. As is clear from the definition of the inertial range, the range to which the hierarchical relation Eq. (1.52) applies increases when the only dimensionless parameter in the the Navier-Stokes turbulence, the Reynolds number, $R = \ell_0 v_0/\nu$, increases. This is because the viscosity damping is νk_n^2 and as ν decreases, the wavenumber limit k_d where the viscosity dissipation becomes negligible (i.e., the beginning of the dissipation zone) increases: $k_d \propto R^{3/4}$. Thus as $R \to \infty$, the fully-developed

	Turbulence	Phase Transition
toward critical point	$R_0^{-1} \to 0$	$\Delta T = T - T - c \to 0$
fundamental scale	largest, external	smallest, characteristic
	$\ell_0 (\equiv 2\pi/k_0)$	ξ_0
cascade	wavenumber space	real space
(critical opalescence)	$k \to \infty$	$r \to \infty$
dissipation wavenumber	$k_d = k_0 R_0^{\nu}$	$\nu = \xi_0 (\Delta T/T_c)^{-\nu}$
(correlation length)		
order parameter	vorticity	magnetization, density fluctuation, etc.
	$\omega(\mathbf{k})$	$M(\mathbf{r}), \delta n(r) \cdots$
asymptotic universal	$\omega(k)$: self-similar	$M(\mathbf{r})$: self-similar
self-similarity	for $k \to \infty$	for $r \to \infty$
	as $R^{-1} \to 0$	as $\Delta T \to 0$
correlation function	vorticity correlation	spin correlation
	$\langle \lVert \omega(k) \rVert^2 \rangle \neq 0$	$g(r) \neq 0$
	as $R^{-1} \to 0$	as $\Delta T \to 0$
(susceptibility)	$\int_0^{\infty} \omega(k) dk \propto R_0^{\gamma} a$	$\int_0^{\infty} r^2 g(r) dr \propto \Delta T^{-\gamma}$
universal spectrum	$E(k) \sim k^{-5/3} f\left(\frac{k}{k_d}\right)$	$g(r) \sim r^{1+\eta_f} \left(\frac{r}{\xi}\right)$

TABLE 1.1 Comparison of fully developed turbulence and phase transition in their wide ranges of scales (Ref. 34)

turbulence exhibits a wider and wider range of hierarchy of eddies: asymptotically universal self-similar behavior. This, of course, leads to a nightmare for computational physicists, because they now need more and more modes to represent the system's hierarchy correctly.

When the hierarchy becomes infinite-fold, we would need an infinite number of modes, which is clearly beyond hope for simulation. The present example of the accuracy problem is in a sense intrinsic, since the nature of interaction precisely demands infinite degrees of freedom, which contradicts our starting point of view of discretization. If the system is asymptotically self-similar, however, it remains invariant after magnifying the wavenumber k_n by a factor of two to k_{n+1}. This property of self-similarity is reminiscent[37] of that in phase transition critical phenomena.[34] Nelkin[37] and others compared the fully-developed three dimensional Navier-Stokes turbulence with critical phenomena[38] (see Table 1.1). Taking the asymptotic universality of $R^{-1} \to 0$ is contrasted with the limit of the temperature deviation from the critical temperature $|T - T_c| \to 0$. Such an analogy leads us to suspect that instead of worrying about infinite degrees of freedom, we could employ the renormalization group transformation[38,39] to reduce the freedom, if the system is asymptotically self-similar. If we can reduce the system by some method relying on the renormalization group transformation (or some other clever method), we might be able to handle a less formidable set of equations numerically. Research on this front is just beginning. One possible way is the subgrid scale modeling.[40] The existence of a long range of quickly evolving small eddies implies the possibility of a numerical method which only explicitly refers to the large eddies because the properties of the small eddies are universal and need not be calculated explicitly. Examples that may have some relevance to this point are in Refs. 41 and 42.

Even if the domain of applicability of self-similar structure of the fully-developed turbulence is wide, direct applications of the above-mentioned renormalization group transformation technique and the subgrid scale method remain only a possible approach in the problem of turbulence. The classic success of the renormalization group theory[39] relies on the existence of a smallness parameter with which one can expand the true function around the unperturbed value. For example, if the intrinsic correlation functions of orders higher than two (e.g., the skewness for the third order and the kurtosis for the fourth order) are much smaller than the correlation function of the second order, then we can expand the true functions around the Gaussian theory utilizing the cumulant expansion (Wick's theorem). This is the case for quantum electrodynamics (QED) and most quantum mechanical problems. Wilson's method is powerful when such an expansion can be found. One of the intriguing and fascinating aspects of turbulence is that, in spite of its nice correspondences with the phase transition problem, (Table 1.1), there exists some experimental and, in fact, computational evidence that the higher the order of correlations are, the more pronounced the deviations from the Gaussian are.[43,44] This is the phenomenon of intermittence in the fluid turbulence. Because of such problems, no straightforward transformation based on the expansion technique is known. The reader may recall a similar situation encountered in Section 1.2 in the Monte Carlo formulation of dynamical simulation that is effective only when a Gaussianness is realized. This situation puts the role of computational physics of turbulence in a very demanding position and, at the same time, a premier position for hope for resolution for a very high Reynolds number, since easier ways are not really open.

Because the aspect of hierarchical structures of nature is one of the most important, difficult, interesting, and unsolved problems, we not only have mentioned it here, but also will give detailed discussions of a few examples later. One is the problem of drift wave turbulence. Another is that of tearing instabilities and their nonlinear development. Another is the hierarchical structure of the universe. In these problems, the main difficulties are that the problems involve multilayer time scales and spatial scales that are interrelated. The drift wave problem involves the tiniest electrons microscopically resonant with waves that in turn cause an overall slow scale, large spatial scale particle and heat transport. Here the (fast) electron thermal motion couples with much slower drift waves, while the slow drift waves trigger even slower scale

	Electron Cyclotron Plasma Wave	RF Heating Lower Hybrid Wave	MHD	Driftwaves	Resistive MHD Trapped Particles	Transport Confinement Collisions
Timescale	$\lesssim$ psec	$\sim$ nsec	$\sim \mu$ sec	$\sim 10_\mu$ sec	sec	
Frequency	$\Omega_e = eB/mc$	$\Omega_i = eB/mc$	$\omega_A = kv_A$	$\omega_* = \frac{cT_e\kappa k_y}{eB}$	$\omega_A^\alpha \nu^{1-\alpha}$	
	$\omega_p = \left(\frac{4\pi ne^2}{m}\right)^{1/2}$	$\omega_{pi} = \left(\frac{4\pi ne^2}{m}\right)^{1/2}$				
Simulation Methods			MHD Codes (Explicit		Resistive MHD Codes Implicit)	
	Conventional Explicit Particle Codes	Ion-Scale Particle Codes (Quasi-Neutral Code etc.)				
						MHD Transport Equilibrium Codes Transport Codes
			Hybrid Codes Guiding Center Gyrokinetic Codes Implicit Particle Codes			

TABLE 1.2 Hierarchy of time scales and corresponding physics and numerics

transport. Spatially, the resonance phenomenon takes place microscopically, while the drift waves encompass multi-resonance layers, which induces overall spatial diffusion. The tearing problem involves again microscopic electron resonance with the wave (or magnetic island structure) that causes a (infinitesimally) small effective dissipation. This introduction of a small amount of dissipation triggers tearing and reconnection of the magnetic field lines and this in turn affects the overall structure of magnetohydrodynamic structure of magnetic field lines. This may give rise to an overall slow scale transport of the plasma and/or lead to an explosive disruption of magnetic fields and confined plasma. Again the tiniest scale of electrons leads to a larger structure of magnetic islands, which in turn couple to the overall magnetic structure of the confinement device. The problem of hierarchical structure of galactic and interstellar correlations is well known.[45,46] Not only stars form a galaxy, but galaxies form a cluster, and clusters make up super-clusters, etc. This superstructure comes about because of the long-range nature of the gravitational system.

Some of the most important topics for future investigations for computational progress may emerge from the discussion here. These may be summarized into three categories: (i) the fundamental spatial resolution size of simulation (mesh size) Δx as large as possible relative to a unit physical length such as the Debye length without compromising physics and its hierarchical structure; (ii) the fundamental temporal resolution size of simulation (time step) Δt as large as possible relative to a unit physical time such as the plasma period without a sacrifice in accuracy of physical description; (iii) the spatial and time grid as large as possible. The first is related to the spatial hierarchy problem of necessity of very many modes for turbulence, for example. To introduce an internal structure to describe short wavelengths may be one of them. Different-size mesh may be another. The second is related to the temporal hierarchy problem where many different time scales interact. Various hybrid models and implicit codes are some of the examples. The last is in a sense related to hardware. In terms of software, techniques of mapping as large mesh as possible on the working computer disc are examples here. To make these symbolic,

$$\text{(i)} \qquad \Delta x \rightarrow \infty \; ; \tag{1.53}$$

$$\text{(ii)} \qquad \Delta t \rightarrow \infty \; ; \tag{1.54}$$

$$\text{(iii)} \qquad L_x \cdot L_y \cdot L_z \cdot N_t \rightarrow \infty \; , \tag{1.55}$$

where L_x, L_y, L_z, and N_t are the numbers of mesh in the $x-$, $y-$, and z-directions and the number of time steps respectively.

The effort in the direction of Eq. (1.53) is picked up again for discussion in Chapters. 4, 6, 7, 8, 9 and 10, starting from the methods of smaller Δt at earlier chapters to those of larger Δt at later ones. The physics behind Eq. (1.54) is closely related and discussed in Chapters 3, 5, 6, 7, 8, and 9, starting from the methods of smaller Δt at earlier chapters to those of larger Δt at later ones. Finally, the effort in the direction of Eq. (1.55) is touched on in Chapter 11.

1.5 Hierarchical Nature and Simulation Methods

As we discussed, many unresolved problems involve the hierarchical nature of physical structures, i.e., a kind of structure made up of many layers of different yet intimately interrelated physical aspects. A pertinent example is plasmas encountered in laboratories for fusion experiments. The hierarchy of structures in fusion plasmas is both of multiple time scales and of multiple spatial scales. In terms of time scales the duration or maintaining time of a typical tokamak[47] fusion plasma is of the order of 1-10 seconds, whereas the electron cyclotron period and the electron plasma period are of the order of 1 psec.

There are various other time scales and they generally fall between these two extreme time scales. If there are high frequency electromagnetic waves (light), these frequencies are even higher than the plasma frequency. Next to the plasma oscillation period, there exists the ion plasma period (see Table 1.2). While the electron plasma frequency corresponds to the characteristic frequency associated with the (lighter) electrons, a frequency that is associated with characteristic ion oscillations is usually the ion acoustic frequency $\omega \sim kc_s(\ll \omega_{pi})$ with k and c_s being the wavenumber of oscillations and the sound speed, respectively. Ion oscillations are largely screened by electrons and their repulsive interaction is considerably mitigated, and thus the much lower frequency. In a magnetized plasma, however, the frequency of ion oscillations can be much higher than kc_s and in fact on the order of ω_{pi}, the characteristic ion plasma frequency, if the wave propagates nearly perpendicular to the magnetic field. This is because electrons are magnetized and have to follow the magnetic field line to respond to the charge separation. Thus ions determine the characteristic frequency. The ion plasma period is the square-root of the mass ratio $[(M/m)^{1/2}]$ times longer than the electron plasma period. This period is typical of radio-frequency heating of plasmas.

A still longer period is the ion cyclotron period ($2\pi/\Omega_i$, where $\Omega_i = eB/Mc$ with M being the ion mass). This is characteristic of ion gyration. Electrostatic ion cyclotron waves (ion Bernstein waves) and electromagnetic ion cyclotron waves are two examples of phenomena in this regime. In this frequency regime the electron motion is so fast that we can treat the electrons as a smeared-out fluid in certain cases: the coarse-graining of the electron time scale results in a smooth, continuous, fluid-like electron response. The magnetohydrodynamic (MHD) or hydrodynamic phenomena are even slower than the ion gyration. In fact, when both electron and ion motions are coarse-grained, both electrons and ions are described as fluids, or a fluid composed of electronic and ionic components. This is a typical fluid description of a plasma (although a two-fluid description can describe electron time scale problems to a certain degree). A variety of global waves and modulations of plasma configurations are often well described by the MHD equations. In such a low frequency regime ($\omega \ll \Omega_i$), both electrons and ions respond to electric fields by $\mathbf{v} = c\mathbf{E} \times \mathbf{B}/B^2$, i.e., the velocity of electrons and ions are perpendicular to the electric field as well as the (external) magnetic field: the motion is called drift or $\mathbf{E} \times \mathbf{B}$ drift. Also in this low frequency the displacement current in the Maxwell equation is negligibly small and at the same time fields are dominated by magnetic fields: $B = \frac{kc}{\omega}E \gg E$ (Faraday's law).

Even longer periods involve subtler effects. An inhomogeneous magnetized plasma has a characteristic frequency called the drift frequency $\omega_*(\omega_* = \frac{cT_e}{eB}k_y\kappa$, where k_y is the wavenumber of the drift wave, κ the inverse of the density gradient length, T_e the electron temperature, and B the external magnetic field). The drift frequency, a characteristic of the drift waves,[48,17] is typically a few hundred times smaller than the ion cyclotron frequency. The drift wave can be driven to instability by the presence of the plasma density gradient; the nature of the instability is kinetic, i.e., the electron resonance with the drift wave feeds energy into the wave and thus the growth of the wave. One of the subtleties of this wave is that although the frequency of the wave is much smaller than Ω_i and it might be possible to coarse-grain particle motion, the kinetic effect remains essential, while the slow period of the wave and slow evolution make the wave easily couple with transport processes that are generally slower than the drift period. Another subtlety is the effect of collisions which have not come into play in faster processes.

Collisions give rise to non-vanishing resistivity, which in turn, brings in fundamental changes in (ideal) MHD. In the ideal (or non-collisional) MHD the fluid is stuck to the magnetic field line,[49] whereas with resistivity (or other "non-ideal" effects) the fluid slips away from the field line. One of the most important consequences of this is the possibility of the reconnection of magnetic field lines.[50] An instability which grows due to the finite resistivity, tearing the field lines (the tearing instability), was discussed by Furth, Killeen, and Rosenbluth.[51] They found the growth rate of the instability between the Alfvén frequency ω_A (the ideal MHD frequency) and the collision frequency ν (Note that $\omega_A \gg \nu$ for a typical case). The resistive MHD phenomena are much faster than the collisional phenomena such as the collisional transport process.[52] The disruption of the tokamak confinement magnetic fields may take place as a nonlinear evolution of these resistive MHD processes.

The ultimate transport process and thus the plasma confinement time scale are related to the collisional process. In contrast to the resistive MHD where the typical frequency is $\omega_A^{\alpha}\nu^{1-\alpha}$ $\left(\frac{1}{3} \lesssim \alpha \lesssim \frac{2}{3}\right)$, the transport process has a frequency scale of ν (or time scale of ν^{-1}). One of the interesting and challenging experimental indications is that the confinement time scale seems to be determined by processes, other than the collisional process, that involve much faster time scales. For example, the transport may be determined by the drift wave process and its associated time scale ω_*^{-1}. The study of this problem thus involves several different time scales and layers of hierarchy.

We have discussed the multiplicity of the temporal scales in fusion plasmas so far. In terms of spatial scales, plasmas exhibit a multiplicity of spatial structures as well. The size of a plasma device obviously decides the maximum possible length, whereas the Debye length characterizes the minimum possible collective interaction length which marks the smallest spatial size of relevance. There are smaller scale lengths, such as the collisional impact parameter and the deBroglie length. However, in most plasmas and in most cases it is acceptable to drop the spatial lengths less than the Debye length. The spatial hierarchy will be discussed in detail in later occasions. In short, the plasma exhibits a variety of instabilities[48] and phenomena of various time scales and spatial scales that are intimately related to each other. It is as if the plasma forms a subtle mosaic pattern in which various sizes of different colored patches connect with each other.

It is naturally quite difficult to resolve all these aspects of plasmas in a single code. Although we wish to include as many layers of phenomena as possible, a typical approach is to coarse-grain some of the more detailed spatial and temporal structures and at the same time to freeze some of the grander spatial and temporal structures in favor of retaining and "pushing" the dynamics of plasmas between. This leads to a variety of types of computer codes, depending upon the main interest of the physicist. In Table 1.2 we outline some of the typical time scales and corresponding codes. Each code guards its own territory of responsibility corresponding to each layer of timescale. Each code takes a certain physics fully represented, while it coarse-grains others and freezes still other effects. If the higher frequency codes can be run long enough, the next layer of slower physics may be represented. It is, however, either too expensive or too inaccurate to cross and reach several layers in this manner.

Astrophysics also encounters many layers of interrelated phenomena. Typical time scales and relevant numerical techniques associated with them are considered in works of Refs. 55-62. By far the majority of the work has been done with traditional low order hydrodynamics techniques (e.g., Richtmyer and Morton Lagrangian codes,[28] donor cell Eulerian,[53] etc.), or N-body codes[32] which compute the direct particle-particle forces. Currently, higher order accuracy hydrodynamics algorithms, such as FCT (discussed in Chapter 6), are being used more frequently. Hockney and his collaborators[54] have used FFT particle techniques since the late sixties, and work using these methods is also becoming more common in the astrophysical literature. In some astrophysical applications, simple radiative transfer approximations have been coupled to hydrodynamics. A small amount of numerical MHD work has also been published.

As in plasma physics, no single simulation can resolve all of these scales. Fortunately, many of these problems are relatively uncoupled, e.g., the evolution of a solar mass star and the formation of the galaxy. However, gravity couples scales ranging in size from an individual star cluster to the whole universe. The evolution of a globular star cluster, for example, is dominated by the tidal field of the whole galaxy containing it. Similarly, the evolution of an individual galaxy may be strongly influenced by its interactions with neighboring galaxies. Direct encounters between galaxies are not uncommon at present, and they may have been even more frequent in earlier times.

In what follows, we shall discuss the basic principles of dynamic many-body simulation as well as various codes corresponding to the above-mentioned various levels of hierarchy of the physics we try to understand. In the first half of this book (Chapters 2-10) we present a technical discussion of computational principles and methods. Generally, we present these techniques which cover physics of corresponding time scales starting from the leftmost column of Table 1.2 to the rightmost as we go along. After discussing computational techniques from Chapters 2 to 10 we arrive at the discussion of computational environment (the hardware-software of the computer itself) in Chapter 11. In the second half of this book (Chapters 12-15) we discuss applications of the dynamic plasma simulation method to fusion plasma problems as well as to astrophysical plasma problems. Once again, we generally start from phenomena with short time scales and proceed to those with longer time scales.

Problems

1. What are the computational methods which circumvent the brute force calculations of the type of solving Eq. (1.19)? List the major methods.
2. What are the missions and advantages of the method of computer simulation? What are the limitations of this method? What do we have to be careful about in order to avoid the possible pitfalls in this method?
3. What are the major differences between the analytical expressions and the computational modelling?
4. What are the consequences of the finite-size particle method?
5. What is the advantage of employing the grid? Can you think of any disadvantage of employing the grid?
6. Besides the particle method and the fluid method, there are other simulation methods. One of the important ones is the Monte-Carlo method, often employed in the statistical physics in the lattice gauge theory, etc. Can you sketch the strategy of this method?
7. In which category of dynamical simulation does the finite element method fall. Sketch the strategy of this method. See Ref. 63.
8. Discuss the difficulty of pursuing a dynamical simulation of a quantum system. Cite a possible approach (e.g. Ref. 64). In this regard, simulations are categorized into the deterministic simulation and the probabilistic simulation with the quantum dynamical simulation belonging to the latter.
9. Study the recent development of direct algorithm of two-body force calculations without resorting to grid (e.g., Ref. 65).
10. Discuss hierarchical interplays between fast and slow time scales in the problems of (i) cosmic evolution, (ii) evolution of life, and (iii) information processing and memory formation (learning).

11. Why does the experimental plasma physicist typically have to confront more difficulty or less accuracy than the experimental solid state physicist? List some of the major physical reasons, one of which pertains to the densities of plasmas and solid states. This may explain why computation is so important in plasma physics.

References

1. H.H. Goldstone, *The Computer-from Pascal to von Neumann* (Princeton University Press, Princeton, 1972).
2. N. Wiener, *Cybernetics* (John Wiley & Sons, New York, 1948); *Human Use of Human Beings* (Houghton Mifflin, New York, 1950).
3. J.A. Wheeler, Int. J. Theor. Phys. **21**, 557 (1982).
4. D. Möhl, G. Petrucci, L. Thorndahl, and S. van der Meer, Phys. Rep. **58**, 73 (1980).
5. L. Brillouin, *Science and Information Theory* (Academic, New York, 1956).
6. J.J. Hopfield, Proc. Natl. Acad. Sci. USA **79** 2554 (1982).
7. R. Kubo, *Statistical Mechanics* (Syokabo, Tokyo, 1961); also see J.M. Ziman, (Cambridge Univ. Press, London, 1963).
8. N. Rostoker and M.N. Rosenbluth, Phys. Fluids **3**, 1 (1960).
9. N. Metropolis, A.W. Rosenbluth, M.N. Rosenbluth, A.H. Teller, and E. Teller, J. Chem. Phys. **21**, 1087 (1953).
10. S.G. Brush, H.L. Sahlin, and E. Teller, J. Chem. Phys. **45**, 2102 (1966).
11. D.M. Ceperly and M.H. Kalos, in *Monte Carlo Methods in Statistical Physics*, ed. by K. Binder (Springer-Verlag, New York), p. 145 (1979).
12. K. Wilson, Phys. Rev. D **10**, 2445 (1974).
13. M. Creutz, Phys. Rev. Lett. **50**, 1411 (1983); M. Okawa, Phys. Rev. Lett. **49**, 353 (1982).
14. M. Gyulassy and T. Matsui, Phys. Rev. D. **29**, 418 (1984).
15. Yu.L. Klimontovich, *The Statistical Theory of Non-Equilibrium Processes in a Plasma* (H.S.H. Massey and O.U. Blunn, Transl., D. ter Haar, eds.) (MIT Press, Cambridge, Mass., 1967).
16. B. Bogoliubov, in *Studies in Statistical Mechanics*, ed. by J. deBoer and G.E. Uhlenback (North-Holland, Amsterdam, 1962) Vol. I, p. 1.
17. S. Ichimaru, *Basic Principles of Plasma Physics* (W.A. Benjamin, Reading, Mass., 1973).

18. S. Ichimaru, Rev. Mod. Phys. **54**, 1017 (1982).

19. P.A. Egelstaff, *An Introduction to the Liquid State* (Academic, New York, 1967).

20. T.H. Dupree, Phys. Fluids **9**, 1773 (1966); Phys. Fluids **15**, 334 (1972); Phys. Fluids **25**, 277 (1982).

21. M. Lebrun, T. Tajima, and W. Schieve, U.T. Physics Department preprint (1983).

22. K. Kawasaki, in *Phase Transitions and Critical Phenomena*, ed. by C.Domb and M.S. Green (Academic, N.Y., 1976). p. 166.

23. J .D. Lawson, *The Physics of Charged Particle Beams* (Oxford Univ.Press, London, 1977); A.W. Chao, in *Physics of High Energy Particle Accelerators* (American Institute of Physics, New York, 1983) p. 353.

24. B.J. Alder and T.E. Wainwright, J. Chem. Phys. **31**, 459 (1959).

25. R.P. Feynman and A.R. Hibbs,*Quantum Mechanics and Path Integrals* (McGraw-Hill, New York, 1965).

26. Y.Z. Zhang, private discussion.

27. P.M. Dirac, *The Principles of Quantum Mechanics* (Clarendon, Oxford, 1958).

28. R.D. Richtmyer, *Difference Methods for Initial-Value Problems* (Interscience Publisher, Inc., New York, 1957).

29. L.D. Landau, Phys. Z. Sowjetunion **10**, 154 (1936); Zhur Eksptl. Teoret. Fix. **7**, 203 (1937).

30. A. Lennard, Ann. Phys. (NY) **10**, 390 (1960); R. Balescu, Phys. Fluids 3, 52 (1960).

31. A.A. Vlasov, Zhur, Eksptl. Teoret. Fiz. **8**, 291 (1938).

32. C.K. Birdsall, A.B. Langdon, and H. Okuda, in *Methods Comput. Phys.*, Vol. 9, p. 241 (1970).

33. J.M. Dawson, Rev. Mod. Phys. **55**, 403 (1983).

34. H.A. Rose and P.L. Sulem, J. Phys., (Paris) **39**, 441 (1978).

35. T. Nakano, Ann. Phys. (N.Y.) **73**, 326 (1972).

36. A.N. Kolmogorov, Dokl. Akad. Nauk SSSR **30**, 299 (1941).

37. M. Nelkin, Phys. Rev. **A11**, 1737 (1975).

38. L. Kadanoff et al., Rev. Mod. Phys. **39**, 395 (1967).

39. K.G. Wilson, Phys. Rev. **B4**, 3174 (1971).

40. H.A. Rose, J. Fluid Mech. **81**, 719 (1977).

41. J.W. Deardorff, J. Atm. Sci. **29**, 91 (1972).

42. R.H. Kraichnan, J. Atm. Sci. **33**, 1521 (1976).

43. E. Hopfinger, F. Browand, and Y. Gagne, J. Fluid Mech. **125**, 505 (1982); J.C. McWilliams et al. in *Eddies and Marine Science*, ed. by A. Robinson (Springer-Verlag, Berlin, 1983).

44. J.C. McWilliams, J. Fluid Mech. **146**, 21 (1984).

45. H. Totsuji and T. Kihara, Publ. Astr. Soc. Jpn. **21**, 221 (1969).

46. J.M. Scalo, in *Protostars and Planets II* ed. D.C. Black and M.S. Matthews (Univ. Arizona Press, Tucson, 1985) p. 201.

47. K. Miyamoto, *Plasma Physics for Nuclear Fusion* (MIT Press, Cambridge, Mass., 1976).

48. A.B. Mikhailovskii, *Theory of Plasma Instabilities* Vols. 1 and 2 (Consultant Bureau, New York, 1974).

49. L.D. Landau and E.M. Lifshitz, *Electrodynamics of Continuous Media* (Pergamon, Oxford, 1959) p. 216.

50. E.N. Parker, Ap. J. Suppl. **77**, 177 (1963); P.A. Sweet, Proc. Internat. Astr. Union Symp. on Electromagnetic Phenomena in Cosmic Physics No. 6 (Stockholm, 1958) p. 123; H.E. Petschek, Proc. of the AAS-NASA Symposium on the Physics of Solar Flares (NASA, Washington, D.C., 1984).

51. H.P. Furth, J. Killeen, and M.N. Rosenbluth, Phys. Fluids **6**, 459 (1963).

52. F.L. Hinton and R.D. Hazeltine, Rev. Mod. Phys. **48**, 239 (1976).

53. P.J. Roache, *Computational Fluid Dynamics*, (Hermosa Publ., Albuquerque, 1972).

54. R.W. Hockney and J.W. Eastwood, *Computer Simulation Using Particles* (McGraw-Hill, New York, 1981).

55. M. Schwarzschild, *Structure and Evolution of the Stars* (Princeton U. Press, Princeton, 1958) .

56. P. Woodward, Ann. Rev. Astronomy and Aphys. **16**, 555 (1978).

57. M.J. Duncan and S.L. Shapiro, in I.A.U. Colloquium 68 *Astrophysical Parameters for Globular Clusters*, A.G.D. Philip and D.S. Hayes, eds.

58. R.B. Larson, in *Galaxies: Sixth Advanced Course of the Soc. of Astron. and Aphys.*, S.M. Fal and D. Lyndon-Bell, eds. (1976).

59. S. Tremaine, in *The Structure and Evolution of Normal Galaxies*, p. 67, S.M. Fall and A. Lyndon Bell, eds. (Cambridge Univ. Press, Cambridge, 1981).

60. C.F. McKee and D.J. Hollenback, Ann. Rev. Astronomy and Aphys. **18**, 219 (1980).

61. D.S. Young, Ann. Rev. Astronomy and Aphys. **14**, 447 (1976).

62. A. Toomre, Ann. Rev. Astronomy and Aphys. **15**, 437 (1977).

63. G. Strang and G.I. Fix, *An Analysis of the Finite Element Method* (Prentice-Hall, Englewood Cliffs, N.J., 1973).

64. R.P. Feynman, Inter. J. Theor. Phys. **21**, 467 (1982).

65. L. Greengard and V. Rokhlin, J. Comp. Phys. **73**, 325 (1987).

2
FINITE SIZE PARTICLE METHOD

A computer simulation would have obvious difficulty processing a large number of ions and electrons for a realistic plasma. However, as long as we are interested in the collective behavior of the plasma, we can introduce a computational method, the Finite Size Particle (FSP) method,[1] which filters out the effects of the short range force, thus allowing the use of a much smaller number of particles, and still correctly describes the effects of the long range force with a greatly reduced number of particles.

A finite size particle has a radius Δx and a distribution $S(r)$ inside, i.e., shape factor. By introducing the FSP concept, we see that the divergence of the force potential at short range is removed as desired. This chapter presents the effects of the FSP method and studies how the real physics is modified by introducing the concept.

2.1 Gridless Theory of a Finite-Size Particle System

The charge densities for a point particle system and a FSP system can be written as

$$\rho_p(\mathbf{x}) = \sum_j q_j \delta(\mathbf{x} - \mathbf{x}_j) \tag{2.1}$$

$$\rho_f(\mathbf{x}) = \sum_j q_j S(\mathbf{x} - \mathbf{x}_j) . \tag{2.2}$$

Here the subscripts p and f refer to the quantities associated with the point particle case and the finite size particle case. If we take the Fourier transform, we obtain

$$\rho_p(\mathbf{k}) = \sum_j q_j e^{-i\mathbf{k}\cdot\mathbf{x}_j} \tag{2.3}$$

$$\rho_f(\mathbf{k}) = \sum_j q_j e^{-\mathbf{k}\cdot\mathbf{x}_j} S(\mathbf{k}) \tag{2.4}$$

$$= \rho_p(\mathbf{k}) S(\mathbf{k}) \tag{2.5}$$

where the Fourier transform is defined as

$$f(\mathbf{k}) = \int_{-\infty}^{\infty} f(\mathbf{x}) e^{-i\mathbf{k}\cdot\mathbf{x}} d\mathbf{x} . \tag{2.6}$$

Its inverse Fourier transform is an integral over an entire range of $\mathbf{k}$. More discussion will ensue on the Fourier transforms in Chapter 4.

From Eq. (2.5) the charge density for a FSP system simply equals the charge density for a point particle system multiplied by the shape factor in $\mathbf{k}$ space. And one can now rewrite most of the plasma theory for the FSP system by replacing the charge q by $qS(\mathbf{k})$, although some care must be taken.

As Eq. (2.5) is simple multiplication, it becomes convoluted if we take the inverse Fourier transform.

$$\rho_f(\mathbf{x}) = \int_{-\infty}^{\infty} d\mathbf{x}' S(x - \mathbf{x}') \rho_p(\mathbf{x}') . \tag{2.7}$$

Now, let us compare Gaussian $S(\mathbf{x})$, which is the usual assumption for shape factor, with $\delta(\mathbf{x})$. Notice that $S(\mathbf{x})$ becomes a delta function as $\Delta x \to 0$.

$$S(\mathbf{x}) = \frac{1}{\sqrt{2\pi}} \Delta x e^{-\left[\frac{(\mathbf{X} - x_0)^2}{2\Delta x^2}\right]} .$$

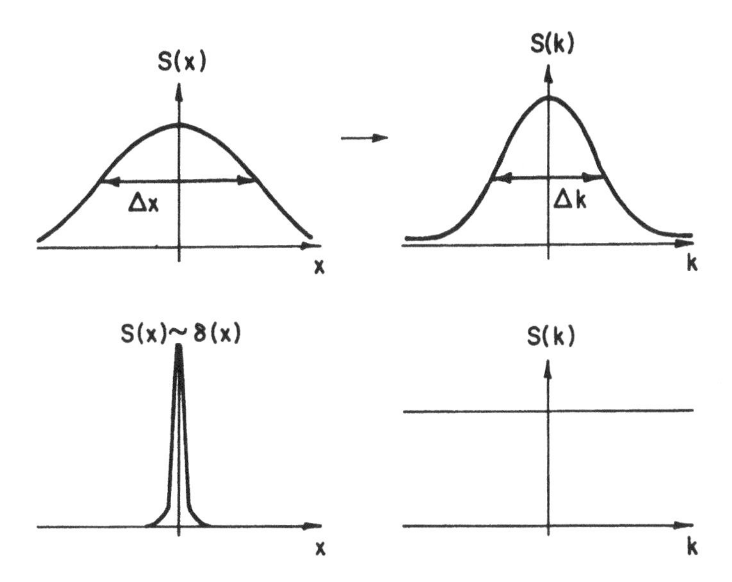

FIGURE 2.1 Finite-sized particle (and its form factor) vs. a point particle

Then

$$S(\mathbf{k}) = \sqrt{2\pi} e^{-i\mathbf{k}\cdot\mathbf{x}_0} e^{-k^2 \Delta x^2/2} ,$$

where $e^{-i\mathbf{k}\cdot\mathbf{x}_0}$ is the phase factor due to the shift of the origin. Notice that the Fourier transform of the Gaussian is another Gaussian, so the narrower the shape in x, the broader the shape in k. The widths Δx and Δk are roughly inverses of each other, $\Delta x \Delta k \simeq 2\pi$. One can see from Fig. 2.1 that the Gaussian shape factor reduces short wavelength effects considerably.

We now present some applications to clarify the transition from the point-charge plasma to FSP plasma.

A. Poisson's Equation

Poisson's equation with finite size particles is written as

$$\nabla^2 \phi(\mathbf{x}) = -4\pi \sum_j q_j S(\mathbf{x} - \mathbf{x}_j) . \tag{2.8}$$

We take the Fourier transform with respect to $\mathbf{x}$

$$-k^2 \phi(\mathbf{k}) = -4\pi \sum_j q_j e^{-i\mathbf{k}\cdot\mathbf{x}_j} S(\mathbf{k}) .$$

Hence

$$\phi(\mathbf{k}) = \frac{4\pi S(\mathbf{k})}{k^2} \sum_j q_j e^{-i\mathbf{k}\cdot\mathbf{x}_j} = \phi_p(\mathbf{k}) S(\mathbf{k}) . \tag{2.9}$$

And the electric field is

$$\begin{aligned} \mathbf{E}(\mathbf{k}) &= -i\mathbf{k}\phi(\mathbf{k}) = -4\pi i S(\mathbf{k}) \frac{\mathbf{k}}{k^2} \sum_j q_j e^{-i\mathbf{k}\,\mathbf{x}_j} \\ &= -i\mathbf{k}\phi_p(\mathbf{k}) S(\mathbf{k}) = E_p(\mathbf{k}) S(\mathbf{k}) . \end{aligned} \tag{2.10}$$

The electric force exerted on the jth FSP is

$$\begin{aligned} \mathbf{F}_f(\mathbf{x}_j) &= \int d\mathbf{x}' qj S(\mathbf{x}' - \mathbf{x}_j) \mathbf{E}(x') = q_j \int d\mathbf{k} \mathbf{E}(\mathbf{k}) S(-\mathbf{k}) e^{i\mathbf{k}\,\mathbf{x}_j} \\ &= q_j \int d\mathbf{k} e^{i\mathbf{k}\cdot\mathbf{x}_j} \frac{4\pi i}{k^2} S(\mathbf{k}) S(-\mathbf{k}) \mathbf{k} \sum_\ell q_\ell e^{i\mathbf{k}\cdot\mathbf{x}_\ell} \\ &= q_j \int d\mathbf{k} |S(\mathbf{k})|^2 \mathbf{E}_p(\mathbf{k}) e^{i\mathbf{k}\cdot\mathbf{x}_j} . \end{aligned} \tag{2.11}$$

Hence

$$\mathbf{F}_f(\mathbf{k}) = |S(\mathbf{k})|^2 q_j \mathbf{E}_p(\mathbf{k}) = |S(\mathbf{k})|^2 \mathbf{F}_p(\mathbf{k}) . \tag{2.12}$$

As we have shown, the electrostatic potential, the electric field strength and electric force for the FSP system can be obtained simply by replacing q by $qS(\mathbf{k})$ for those of the point particle system.

B. Lorentz Force

The Lorentz force on a FSP at position $\mathbf{x}$ with velocity $\mathbf{v}$ is

$$\mathbf{F}(\mathbf{x}, \mathbf{v}, t) = q \int d\mathbf{x}' S(\mathbf{x}' - \mathbf{x}) \left[\mathbf{E}(\mathbf{x}', t) + \frac{1}{c} \mathbf{v} \times \mathbf{B}(\mathbf{x}', t) \right] \tag{2.13}$$

and its Fourier transform is

$$\mathbf{F}(\mathbf{k}, \mathbf{v}, t) = qS(-\mathbf{k}) \left[\mathbf{E}(\mathbf{k}, t) + \frac{1}{c} \mathbf{v} \times \mathbf{B}(\mathbf{k}, t) \right] . \tag{2.14}$$

If we choose the symmetric shape factor, $S(\mathbf{k})$ equals $S(-\mathbf{k})$. Thus, again, the Lorentz force for a FSP system can be obtained by replacing q as $qS(\mathbf{k})$.

Hence

$$F(\mathbf{k},\omega) = -4\pi n_0 q^2 \frac{i\mathbf{k}}{k^2}|S(\mathbf{k})|^2 \int d\mathbf{v} f(\mathbf{k},\mathbf{v},\omega) \ . \tag{2.23}$$

Using Eq. (2.23), we can rewrite Eq. (2.20) as

$$f(\mathbf{k},\mathbf{v},\omega) = -i\frac{4\pi n_0 q^2}{m}\frac{|S(\mathbf{k})|^2}{k^2}\mathbf{k}\frac{\int d\mathbf{v} f(\mathbf{k},\mathbf{v},\omega)\frac{\partial f_0}{\partial \mathbf{v}}}{[i(\mathbf{k}\cdot\mathbf{v}-\omega)+\nu]} \ . \tag{2.24}$$

If we integrate both sides of Eq. (2.24) with respect to $\mathbf{v}$ and normalize, we obtain the plasma dispersion relation for a FSP system

$$1 = \frac{4\pi n_0 q^2}{m}\frac{|S(\mathbf{k})|^2}{k^2}\mathbf{k}\int d\mathbf{v}\frac{\partial f_0}{\partial \mathbf{v}}[(\omega - \mathbf{k}\cdot\mathbf{v}) + i\nu]^{-1} \ ,$$

and the dispersion function $D(\mathbf{k},\omega)$ is

$$D_f(\mathbf{k},\omega) = 1 - \frac{\omega_p^2|S(\mathbf{k})|^2}{k^2}\mathbf{k}\int\frac{d\mathbf{v}\frac{\partial f_0}{\partial \mathbf{v}}}{(\omega-\mathbf{k}\cdot\mathbf{v})+i\nu} \ , \tag{2.25}$$

where ω_p is plasma frequency $\omega_p = \left(\frac{4\pi n_0 q^2}{m}\right)^{1/2}$. Hence the dispersion relation for a FSP system can be obtained by replacing q by $qS(\mathbf{k})$.

Now let's solve Eq. (2.25) for 1-D Maxwellian

$$f_0(v) = \left(\frac{m}{2\pi T}\right)^{1/2} e^{-\frac{mv^2}{2T}} \ . \tag{2.26}$$

Then

$$\begin{aligned} D_f(k,\omega) &= 1 + \frac{\omega_p^2}{k^2}|S(k)|^2\frac{1}{\sqrt{2\pi}v_e^3}\int\frac{dv\, v e^{-\frac{mv^2}{2T}}}{v-\omega/k - i\nu/k} \\ &= 1 - \frac{1}{2}\left(\frac{\omega_p}{kv_e}\right)^2|S(k)|^2 Z'\left(\frac{\omega}{\sqrt{2}kv_e}\right) \ , \end{aligned} \tag{2.27}$$

where $v_e = \sqrt{T/m}$ and $Z(w)$ is defined as

$$Z(w) = \frac{1}{\sqrt{\pi}}\int_{-\infty}^{\infty}\frac{e^{-x^2}}{x-w-i\eta}dx . (\eta \to 0) \ . \tag{2.28}$$

If we expand Z function for $\frac{\omega}{kv_e} \gg 1$, Eq. (27) becomes

$$D_f(k,\omega) = 1 + \frac{1}{2}\left(\frac{\omega_p}{kv_e}\right)^2|S(k)|^2\left[i\sqrt{2\pi}\frac{\omega}{kv_e}e^{-\frac{\omega^2}{2k^2v_e^2}} - \frac{2k^2v_e^2}{\omega^2} - 6\frac{k^4v_e^4}{\omega^4}\right] \ . \tag{2.29}$$

We seek a solution to this equation in the form of a complex frequency

$$\omega = \omega_k + i\gamma_k \ . \tag{2.30}$$

Assuming $|\gamma_k/\omega_k| \ll 1$, we find that

$$\omega^2 \simeq \omega_k^2 = \omega_p^2 S^2(k) + 3(kv_e)^2 \tag{2.31}$$

and

$$\gamma_k = -\left(\frac{\pi}{8}\right)^{1/2} S(k)\omega_p \left(\frac{\omega_p}{k\lambda_D}\right)^3 e^{-1/2\left[\frac{S(k)}{k\lambda_D}\right]^2 - 3/2} \ , \tag{2.32}$$

where

$$\lambda_D = \left(\frac{T}{4\pi n q^2}\right)^{1/2} = \frac{v_e}{\omega_p} \ .$$

Notice that we have the frequency of collective mode and damping rate for a FSP system which is equivalent to those of a point particle system again by replacing q by $qS(k)$. Notice also that for large k, ω_k and γ_k are reduced considerably.

If $S(k)$ is Gaussian,

$$\omega^2 = \omega_p^2 e^{-k^2a^2} + 3k^2 v_e^2 \ . \tag{2.33}$$

For $k\lambda_D \ll 1$, we can approximate ω^2 as

$$\omega^2 \simeq \omega_p^2 + 3k^2\left(v_e^2 - \frac{1}{3}\omega_p^2 a^2\right) \ . \tag{2.34}$$

Thus, as long as the radius of a FSP is less han $\sqrt{3}\lambda_D$, the group velocity of plasma waves remains positive.

2.3 Collisional Effects Due to Finite-Size Particles

In a fully ionized plasma, the dynamics of collisions between any pair of charged particles are governed by the charges q_1, q_2, the masses m_1, m_e, the relative velocity v and the impact parameter b of the two particles. See Fig. 2.2. The Rutherford scattering cross-section can be calculated by the following integration

$$\sigma = 2\pi \int (1 - \cos\theta) b\, db \ . \tag{2.35}$$

This integral has the usual divergence which arises from the fact that the Coulomb potential has such a long range effect. In a plasma this divergence is removed with the reasonable approximation of taking

$$b_{\max} = \lambda_D \ .$$

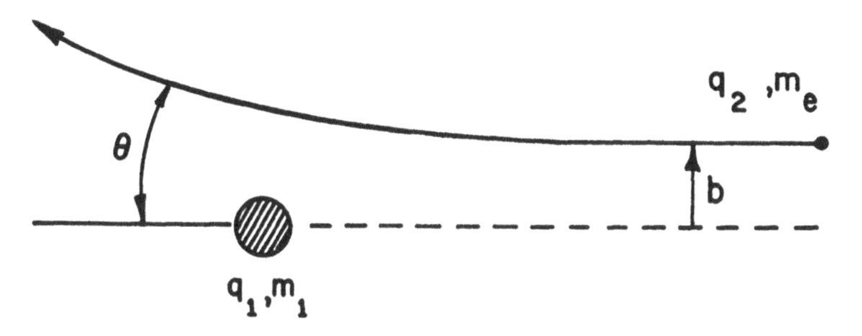

FIGURE 2.2 Collision and the impact parameter b

The integral then gives the total cross-section for point charges

$$\sigma_p = \frac{4\pi q^4}{(m_c v_c^2)^2} \ell n \left(\frac{m_c v_c^2 \lambda_D}{q^2} \right) , \tag{2.36}$$

where

$$m_c \quad : \quad \text{reduced mass} \tag{2.37}$$

$$v_c \quad : \quad \text{center of mass velocity} \tag{2.38}$$

For finite size particles, the Coulomb potential is a well-behaved function, but it is no longer a simple function of r.

The result of the cross section, σ_f, for various dimensionalities is plotted[1] in the graph, Fig. 2.3. With these cross-sections, σ_p and σ_f, we can determine collision frequencies ν_p and ν_f,

$$\nu = n\overline{\sigma v} \tag{2.39}$$

where $\overline{\sigma v}$ is the averaged collisional rate over the Maxwellian background.

A. Point Charge Case

$$1-D \quad \nu_p \cong \quad \tfrac{1}{2N_D}\omega_p \quad \text{with} \quad N_D = n\lambda_D \tag{2.40}$$

$$2-D \quad \nu_p \cong \quad n v_e \sigma \simeq \tfrac{\omega_p}{16N_D} \quad \text{with} \quad N_D = n\lambda_D^2 \tag{2.41}$$

$$3-D \quad \nu_p \simeq \quad \tfrac{1}{27}\tfrac{\omega_{pe}}{N_D}\ell n(9N_D) \quad \text{with} \quad N_D = n\lambda_D^3 \ . \tag{2.42}$$

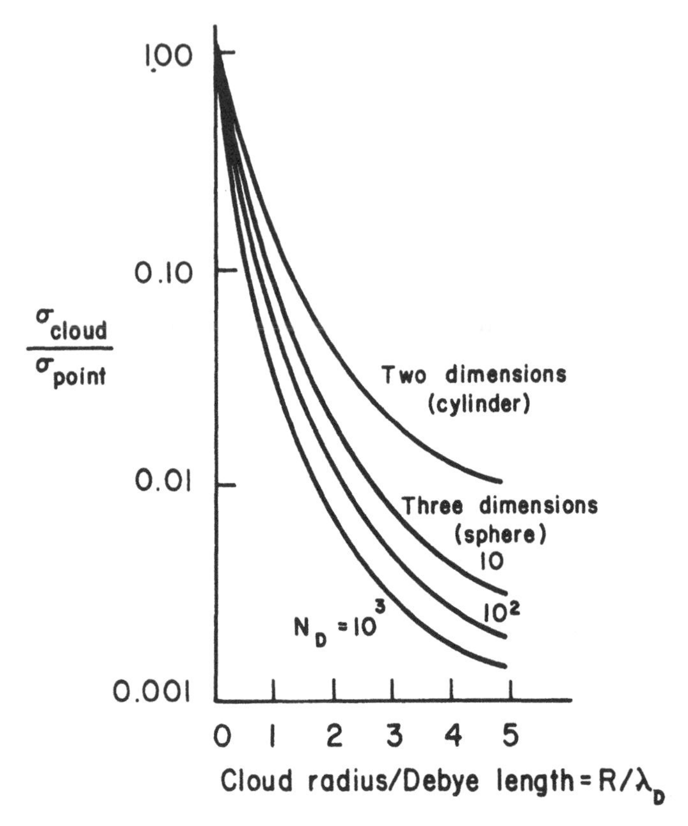

FIGURE 2.3 The ratio of cross-section of the finite-sized particle to that of the point particle as a function of the radius of the particle size (Ref. 1)

B. Finite-size Particle Case

$$1 - D \qquad \nu_f \simeq \frac{1}{2n\lambda_D} F\omega_p \tag{2.43}$$

$$2 - D \qquad \nu_f \simeq \frac{1}{16n\lambda_D^2} F\omega_p \tag{2.44}$$

$$3 - D \qquad \nu_f \cong \frac{1}{27n\lambda_D^3} F\omega_p \ell n(9N_D) \tag{2.45}$$

where $F = \frac{\sigma_f}{\sigma_p}$: reduction factor found in Fig. 2.3.

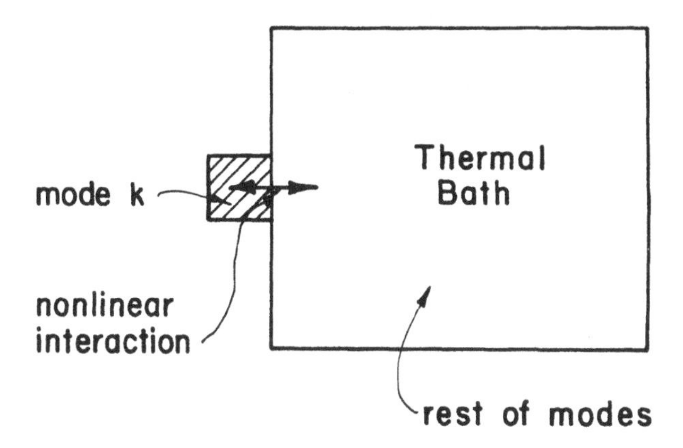

FIGURE 2.4 The canonical ensemble for the mode **k**

2.4 Fluctuations

In thermal equilibrium a plasma sustains fluctuations of various collective modes of (electrostatic) oscillations. This situation is depicted in Fig. 2.4. As we shall see, the property of fluctuation-dissipation relation manifests itself only in a fully nonlinear particle model. In a linearized model, for example, there is no coupling with other modes. According to statistical mechanics, the probability of finding a certain energy level in a canonical ensemble is given by the Boltzmann distribution.[2] We have

$$P(E_k)dE_k \propto \exp\left[-\Psi_k(E_k)/T\right] dE_k \ , \tag{2.46}$$

where $P(E_k)$ is the probability of finding a system in any one particular mode with wavenumber k in dE_k and E_k is electrostatic field of mode k=0, $\frac{2\pi}{L}, \ldots, \pi$. $\Psi_k(E_k)$ is the energy required to create fluctuation of mode E_k. Then $\Psi_k(E_k)$ is given by

$$\Psi_k(E_k) = \frac{E_k^2 L}{8\pi}\left(1 + k^2\lambda_D^2 e^{k^2a^2}\right) \ , \tag{2.47}$$

with a term proportional to $k^2\lambda_D^2$ due to the Debye screening effect and a expresses the size of the shape factor of finite size particles.

With proper normalization of this probability distribution,

$$\int_{-\infty}^{\infty} P(E_k)dE_k = 1 \ ,$$

2.2 Dispersion Relation

We have shown how the charge density, field strength, force and potential are modified by introducing the FSP concept. Using these relations, we can derive the dispersion relation for homogeneous unmagnetized plasma in thermodynamic equilibrium.

The Boltzmann transport equation with the Krook collision operator is

$$\frac{\partial f}{\partial t}(\mathbf{r},\mathbf{v},t)+\mathbf{v}\cdot\frac{\partial f}{\partial \mathbf{r}}(\mathbf{r},\mathbf{v},t)+\frac{F_f}{m}\cdot\frac{\partial f}{\partial \mathbf{v}}(\mathbf{r},\mathbf{v},t)=-\nu(f-f_0)\ , \tag{2.15}$$

where f_0 is the distribution function in thermal equilibrium. Since we are concerned with a collisionless plasma, the collision frequency ν is small and the equation reduces to Vlasov's equation. However, we keep the collision term for convenience and can let ν be small later.

The linearized Vlasov equation is derived by letting the distribution function $f(\mathbf{r},\mathbf{v},t)$ and the force F be expressed in terms of the equilibrium value plus a perturbation, i.e.,

$$f(\mathbf{r},\mathbf{v},t) = f_0(\mathbf{v})+\delta f(\mathbf{r},\mathbf{v},t)\ , \tag{2.16}$$

$$F(\mathbf{r},\mathbf{v},t) = 0+F(\mathbf{r},\mathbf{v},t)\ . \tag{2.17}$$

By neglecting second order nonlinear terms, one can obtain

$$\frac{\partial \delta f}{\partial t}+\mathbf{v}\cdot\frac{\partial \delta f}{\partial \mathbf{r}}+\frac{F}{m}\cdot\frac{\partial f_0}{\partial \mathbf{v}}=-\nu\delta f\ . \tag{2.18}$$

We assume that the ions are massive and fixed in space, and that the waves are plane waves. That is

$$\delta f \propto f(\mathbf{k},\omega)e^{i(\mathbf{k}\cdot\mathbf{r}-\omega t)},\quad \text{and} \quad F\propto F(\mathbf{k},\omega)e^{i(\mathbf{k}\cdot\mathbf{r}-\omega t)}\ .$$

Then

$$i(\mathbf{k}\cdot\mathbf{v}-\omega)f+\frac{F}{m}\cdot\frac{\partial f_0}{\partial \mathbf{v}}=-\nu f\ . \tag{2.19}$$

Thus

$$f(\mathbf{k},\mathbf{v},\omega)=-\frac{F(\mathbf{k},\omega)\cdot\frac{\partial f_0}{\partial \mathbf{v}}}{m[-(i\omega-\mathbf{k}\cdot\mathbf{v})+\nu]}\ . \tag{2.20}$$

The electrostatic force $F(\mathbf{k},\omega)$ is obtained from Poisson's equation

$$\mathbf{E}(\mathbf{k},\omega)=-4\pi q i\frac{S(\mathbf{k})n(\mathbf{k},\omega)}{k^2}\mathbf{k}=-4\pi i n_0 q\frac{S(\mathbf{k})}{k^2}\mathbf{k}\int d\mathbf{v} f(\mathbf{k},\mathbf{v},\omega)\ , \tag{2.21}$$

where

$$n(\mathbf{r},t)=n_0\int d\mathbf{v} f(\mathbf{r};\mathbf{v},t)\ . \tag{2.22}$$

one can obtain

$$P(E_k) = \frac{1}{2\sqrt{\pi}} \left[\frac{L}{4\pi T} \left(1 + k^2\lambda_D^2 e^{k^2a^2}\right)\right]^{1/2} \times \exp\left[-\frac{E_k^2 L}{8\pi T}\left(1 + k^2\lambda_D^2 e^{k^2a^2}\right)\right] .$$

With this probability distribution, the averaged electric field energy is

$$\left\langle \frac{E_k^2 L}{8\pi} \right\rangle = \int_{-\infty}^{\infty} \frac{E_k^2 L}{8\pi} P(E_k) dE_k = \frac{1}{2} \frac{T}{1 + k^2\lambda_D^2 e^{k^2a^2}} . \tag{2.48}$$

From the above relation one can immediately recognize the strong reduction of field energy for $k^2a^2 \gg 1$.

These results show that the finite size particle method describes long-wavelength fluctuations correctly. However, there is some deviation for E_k with very small wavenumber k due to the long relaxation time for these modes. In Fig. 2.5, the theoretical prediction and simulation results are plotted.

We can derive Eq. (2.48) in another way. According to the fluctuation-dissipation theorem,[3,4] the spectral intensity[5] of the fluctuations in thermal equilibrium is given as

$$S_{\substack{p\\f}}(k, \omega) = -\frac{N}{\pi\omega} \frac{k^2}{k_D^2} \,\mathrm{Im} \frac{1}{D_{\substack{p\\f}}(k,\omega)} . \tag{2.49}$$

Using Kramers-Krönig's relation,[3,6,7,8] we obtain

$$S_{\substack{p\\f}}(k) \equiv -\frac{k^2}{\pi k_{D^2}} \int_{-\infty}^{\infty} \frac{d\omega}{\omega} \,\mathrm{Im} \frac{1}{D_{\substack{p\\f}}(k,\omega)} = \frac{k^2}{k_D^2}\left[1 - \,\mathrm{Re}\frac{1}{D_{\substack{p\\f}}(k,0)}\right] . \tag{2.50}$$

From the expression of $D_p(k,0) = 1 + \frac{k_D^2}{k^2}$, we obtain

$$S_p(k) = \frac{k^2}{k^2 + k_D^2} . \tag{2.51}$$

It is important to notice that the result of Eq. (2.51) holds even for the magnetized plasma system. The dielectric constant for magnetized plasma is[5,9]

$$D_p(k,\omega) = 1 + \frac{k_{De}^2}{k^2} + \sum_{n=-\infty}^{\infty} e^{-k^2\rho_i^2} I_n\left(k^2\rho_i^2\right) \frac{\omega}{\omega - n\omega_i}\left[W\left(\frac{\omega - n\Omega_i}{k_{Di}v_i}\right) - 1\right] . \tag{2.52}$$

Since the Kramers-Krönig relation says that only the zero frequency (time integration) contribution matters, the sum over various ion cyclotron poles in Eq. (2.52) does not contribute to the result of Eq. (2.51), which leads to the result obtained earlier.[9]

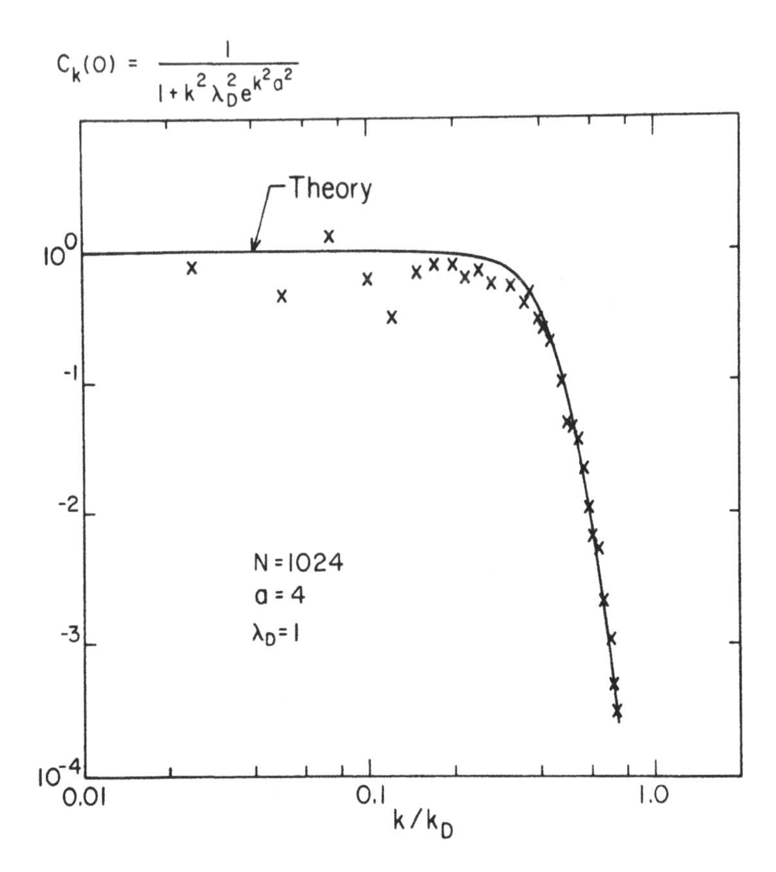

FIGURE 2.5 The value of Eq. (2.48) (with the curve) vs. the wave number obtained in a particle simulation with finite-size particles (with crosses)

On the other hand,

$$S_{\substack{p\\f}}(k) = \frac{1}{N}\left\langle |\delta\rho_k|^2 \right\rangle .$$

Therefore,

$$\langle \rho_k^2 \rangle_{\substack{p\\f}} = \frac{k^2}{4\pi} T \left[1 - \frac{1}{D_{\substack{p\\f}}(k,0)} \right] ,$$

and at the same time

$$\langle \rho_k^2 \rangle = \frac{k^2}{16\pi^2} \langle E_k^2 \rangle \cdot L \, .$$

Thus

$$\frac{\langle E_k^2 L \rangle_p}{8\pi} = \frac{T}{2} \frac{1}{1 + k^2 \lambda_D^2} \, ,$$

and

$$\frac{\langle E_k^2 L \rangle_f}{8\pi} = \frac{T}{2} \frac{1}{1 + k^2 \lambda_D^2 |S(k)|^2} \, .$$

Thus we reproduced Eq. (2.48) this time through the fluctuation dissipation theorem.

We have observed in this chapter that the finite size particle method reproduces correct linear (the dispersion relation) and nonlinear (the fluctuation-dissipation relation) properties of a plasma for a long wavelength regime of interest ($k < k_D$ and $k_D \Delta x \sim 1$). It also introduces a reduction in the collision frequency as desired. Thus the method allows us to reduce collisional effects without sacrificing collective effects within a manageable number of simulation particles. More discussion can be found in Chapters 7 and 9.

Problems

1. In your particle code measure the slowdown times when the test particle α is very fast ($v_\alpha \gg v_{th}$) and when the test particle α is very slow ($v_\alpha \ll v_{th}$), where v_α is the test particle initial velocity and v_{th} the field particles' thermal velocity. How about the particle scattering? Try this for the test particle being an electron and an ion. Examine the difference when you fix ions as immobile. Dawson's article (Ref. 10) may be a good guide.

2. Using your electrostatic particle code, try to reproduce the result of Fig. 2.5.

3. Derive Eq. (2.50).

4. When we need collisional effects in the collisionless particle model, we often introduce collisions artificially by the Monte-Carlo method. The Lorentz gas model has been tried by Shanny, Dawson, and Greene[11] as well as by Matsuda and Okuda.[12] Describe their method.

References

1. C.K. Birdsall, A.B. Langdon, and H. Okuda, in *Methods of Computational Physics*, ed. by B. Alder, S. Fernbach, and M Rotenberg (Academics, New York, 1970) vol. 9, p. 241.
2. J.M. Dawson, in *Methods of Computational Physics*, *ibid.*, vol. 9, p. 1.
3. R. Kubo, J. Phys. Soc. Jpn. **12**, 570 (1957).
4. A.G. Sitenko, *Electromagnetic Fluctuations in Plasma*, (Academic, New York, 1967) p. 61.
5. S. Ichimaru, *Basic Principles of Plasma Physics* (Benjamin, Reading, 1973) Chapters 1 and 9.
6. L.D. Landau and E.M. Lifshitz, *Electrodynamics of Continuous Media*, translated by J.B. Sykes and J.S. Bell (Pergamon, Oxford, 1960) p. 259.
7. N. Rostoker and M.N. Rosenbluth, Phys. Fluids **3**, 1 (1960).
8. A.B. Langdon, Phys. Fluids **22**, 163 (1979).
9. C.Z. Cheng and H. Okuda, J. Comp. Phys. **25**, 133 (1977).
10. J.M. Dawson, Rev. Mod. Phys. **55**, 403 (1983).
11. R. Shanny, J.M. Dawson, and J.M. Greene, Phys. Fluids **10**, 1281 (1967).
12. Y. Matsuda and H. Okuda, Phys. Fluids **18**, 1740 (1975).

3 TIME INTEGRATION

We try to solve (i.e., numerically integrate) differential equations for $u(t)$ of the type

$$\frac{du}{dt} = f(u,t) \ . \tag{3.1}$$

In order to do so, we discretize this differential equation in time. We must approximate the differentiation by finite differences. For example, the first time derivative of u, du/dt, becomes

$$\frac{du}{dt} \to \frac{u^{n+1} - u^n}{\Delta t} \ , \tag{3.2}$$

and the second derivative of u, d^2u/dt^2 becomes

$$\frac{d^2u}{dt^2} \to \frac{u^{n+1} - 2u^n + u^{n-1}}{\Delta t^2} \ , \tag{3.3}$$

where the superscripts n denote the time step. These expressions reduce to the original differentiation when Δt tends to zero. This statement can be verified by Taylor expanding the function u^n etc., on the right-hand side of Eqs. (3.2) and (3.3). For example, the right-hand side of Eq. (3.3) can be expanded around $t = n\Delta t$ as

$$\frac{1}{\Delta t^2}\left(u^{n+1} - 2u^n + u^{n-1}\right) = \frac{1}{\Delta t^2}\left[\left(u^n + \Delta t\frac{\partial u^n}{\partial t} + \frac{1}{2}\Delta t^2\frac{\partial^2 u^n}{\partial t^2} + \dots\right)\right.$$

$$- \quad 2u^n + \left(u^n - \Delta t \frac{\partial u^n}{\partial t} + \frac{1}{2}\Delta t^2 \frac{\partial^2 u^n}{\partial t^2} - \ldots\right)\Bigg] = \frac{\partial^2 u^n}{\partial t^2} + \mathcal{O}(\Delta t^4) \,.$$

There can be many schemes to represent the differential operators. This means that we should choose a finite difference scheme which reduces to the original differentiation in the limit of small Δt.[1,2] In addition, when we choose a finite difference scheme to describe the differential operator, we should consider accuracy and economy in computation.

Perhaps even more important is the question of the stability of the finite difference scheme in time. This is because errors due to inaccuracy of the algorithm accumulate randomly, while errors due to numerical instability geometrically magnify themselves. Let us consider the example of a case where the finite difference in time is numerically unstable. A system of a harmonic oscillator may be described by the equations

$$\frac{dv}{dt} = -\omega^2 x \tag{3.4}$$

$$\frac{dx}{dt} = v \,. \tag{3.5}$$

We now replace derivatives by finite difference:

$$\frac{dv}{dt} \quad \longrightarrow \quad \frac{v^{n+1} - v^n}{\Delta t} = -\omega^2 x^n (n = 0, 1, \ldots) \tag{3.6}$$

$$\frac{dx}{dt} \quad \longrightarrow \quad \frac{x^{n+1} - x^n}{\Delta t} = v^n \,. \tag{3.7}$$

This seemingly benign finite differencing is called the forward differencing, which turns out to be numerically unstable as we will see. In order to analyze the numerical properties of the system of Eqs. (3.6) and (3.7), we look for linear stability with an amplification factor g over the adjacent time steps.

Let us assume:

$$x^{n+1} = g x^n \,, \tag{3.8}$$

$$x^n \propto (g)^n = \left(e^{-i\alpha\Delta t}\right)^n \,, \tag{3.9}$$

$$v^n \propto g^n v^0 \,. \tag{3.10}$$

Equations (3.6) and (3.7) now become:

$$g^{n+1} v^0 - g^n v^0 = -\omega^2 \Delta t g^n x^0 \,, \tag{3.11}$$

$$g^{n+1} x^0 - g^n x^0 = \Delta t g^n v^0 \,. \tag{3.12}$$

We can obtain g by solving the determinant equation for Eqs. (3.11) and (3.12). This matrix is called the amplification matrix:

$$\begin{pmatrix} g-1 & \omega^2\Delta t \\ -\Delta t & g-1 \end{pmatrix} .$$

By setting the determinant of the amplification matrix equal to zero we obtain

$$(g-1)^2 + \omega^2\Delta t^2 = 0 , \tag{3.13}$$

or

$$g = 1 \pm i\omega\Delta t . \tag{3.14}$$

That is, if $|g|^2 > 1$, the solution blows up in time. Here $|g|^2 = 1 + \omega^2\Delta t^2 > 1$. Therefore, no matter how small Δt is taken to be, $|g|^2 > 1$, and the forward differencing scheme is thus numerically unstable.

Let us first make some general remarks on time integration. Suppose

$$\frac{du}{dt} + f(u,t) = 0 \tag{3.15}$$

with the initial value given, $u(0) = u^0$. How does one integrate in time? Theoretically, Eq. (3.15) is integrated as

$$u(t) = -\int_0^t f(t)dt ,$$

and over the time period (t^n, t^{n+1}) as

$$u^{n+1} = u^n - \int_{t^n}^{t^{n+1}} f(t)dt . \tag{3.16}$$

We discuss how to integrate Eq. (3.16) numerically in the following.

3.1 Euler's First-Order Scheme

Euler's first-order scheme[2] integrates as

$$u^{n+1} = u^n - f(u^n, t)(t^{n+1} - t^n) . \tag{3.17}$$

See Fig. 3.1. Let the true value be u^n and the computed value be $u^n + \varepsilon^n$. From Eq. (3.17) we obtain this relation for ε^n:

$$u^{n+1} + \varepsilon^{n+1} = u^n + \varepsilon^n - f(u^n + \varepsilon^n, t^n)\Delta t . \tag{3.18}$$

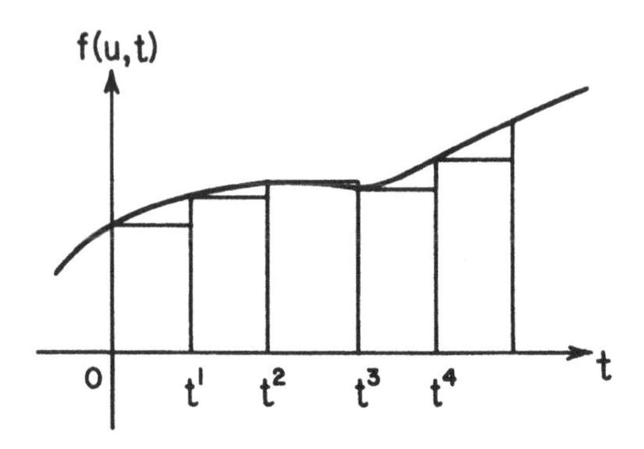

FIGURE 3.1 The function $f(u,t)$ and its finite-summation

We Taylor expand around the true value

$$u^{n+1} + \varepsilon^{n+1} = u^n + \varepsilon^n - \left[f(u^n, t^n)\Delta t + \left.\frac{\partial f}{\partial u}\right|^n \varepsilon^n \Delta t + \ldots \right] . \tag{3.19}$$

Then $\varepsilon^{n+1} = \varepsilon^n - \left.\frac{\partial f}{\partial u}\right|^n \Delta t \cdot \varepsilon^n$. The amplification factor is obtained as

$$g = 1 - \left.\frac{\partial f}{\partial u}\right|^n \Delta t . \tag{3.20}$$

If $2 > \left.\frac{\partial f}{\partial u}\right|^n \Delta t > 0$, $|g| < 1$ and the Euler scheme is stable.

3.2 Leapfrog Scheme

Let us consider the electric field acceleration:

$$\frac{dv_j}{dt} = \frac{q_j}{m} E(r) = a(t); \quad \frac{dx}{dt} = v . \tag{3.21}$$

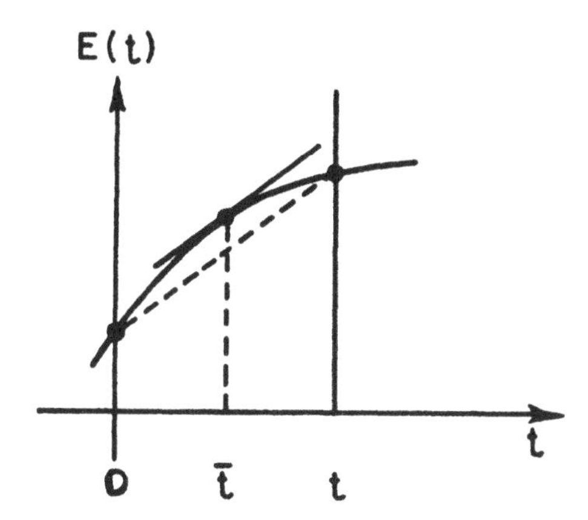

FIGURE 3.2 Finite differencing and the mean value theorem

From the mean value theorem, we can write Eq. (3.21) as

$$v(t+\Delta t) - v(t) = \Delta t \frac{q_j}{m} E(\bar{t}) \,, \tag{3.22}$$

where $\bar{t}$ is graphically given in Fig. 3.2. Notice, however, that we cannot utilize this theorem unless we know $\bar{t}$ whereas in the previous scheme (Euler's) we used $\bar{t} = t$. Here we consider the leapfrog which uses $\bar{t} \cong t + \frac{1}{2}\Delta t$:

$$v(t+\Delta t) - v(t) = a\left(t + \frac{1}{2}\Delta t\right) \Delta t \,, \tag{3.23}$$

$$r\left(t + \frac{3}{2}\Delta t\right) - r\left(t + \frac{1}{2}\Delta t\right) = v(t+\Delta t)\Delta t \,. \tag{3.24}$$

See Fig. 3.3 for this time-step arrangement.

To demonstrate the stability of this numerical scheme, we shall take the harmonic oscillator as an example. Equation (3.22) in a plasma admits the solution of plasma oscillations. Let the position of the electron $X = x_0 + x(x_0, t)$ with x_0 being the initial position. $x(x_0, t)$ is called the Lagrangian coordinate of the electron. Following the argument of Ref. 3, Gauss' law yields the electric field at the electron position X as

$$E = 4\pi e n_0 x(x_0, t) \,. \tag{3.25}$$

We can thus write the electric field as

$$\frac{q_j}{m} E(x) \cong -\omega_p^2 x \,. \tag{3.26}$$

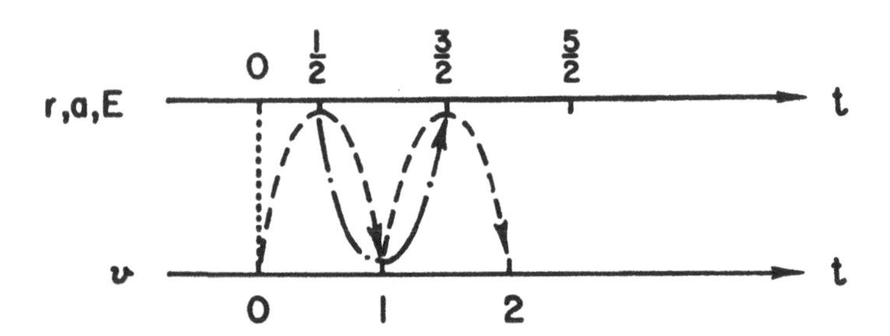

FIGURE 3.3 Leapfrog time-step assignment

Let $t = n\Delta t$. Then Eqs. (3.23) and (3.24) are

$$v^{n+1} - v^n = -\omega_p^2 \Delta t x^{n+1/2} \tag{3.27}$$

$$x^{n+3/2} - x^{n+1/2} = v^{n+1}\Delta t \ . \tag{3.28}$$

With the amplification factor g defined in $v^n \propto (g)^n = \left(e^{-i\alpha t}\right)^n$, we get the amplification matrix

$$\begin{pmatrix} g-1 & \omega_p^2 \Delta t g^{1/2} \\ -g\Delta t & g^{3/2} - g^{1/2} \end{pmatrix} .$$

By setting the determinant equal to zero, we obtain

$$g^2 - \left(2 - \omega_p^2 \Delta t^2\right) g + 1 = 0 \ . \tag{3.29}$$

This quadratic equation yields

$$\begin{aligned} g &= 1 - \frac{\omega_p^2 \Delta t^2}{2} \pm \sqrt{\left(1 - \frac{\omega_p^2 \Delta t^2}{2}\right)^2 - 1} \\ &= 1 - \frac{\omega_p^2 \Delta t^2}{2} \pm \sqrt{\left(\frac{\omega_p^2 \Delta t^2}{2}\right)^2 - \omega_p^2 \Delta t^2} \ . \end{aligned}$$

Finally, we obtain

$$g = 1 - \frac{\omega_p^2 \Delta t^2}{2} \pm \omega_p \Delta t \sqrt{\frac{\omega_p^2 \Delta t^2}{4} - 1} \ . \tag{3.30}$$

If

$$\frac{\omega_p^2 \Delta t^2}{4} - 1 < 0, \quad g = 1 - \frac{\omega_p^2 \Delta t^2}{2} \pm i(\omega_p \Delta t)\sqrt{1 - \left(\frac{\omega_p^2 \Delta t^2}{4}\right)} \ .$$

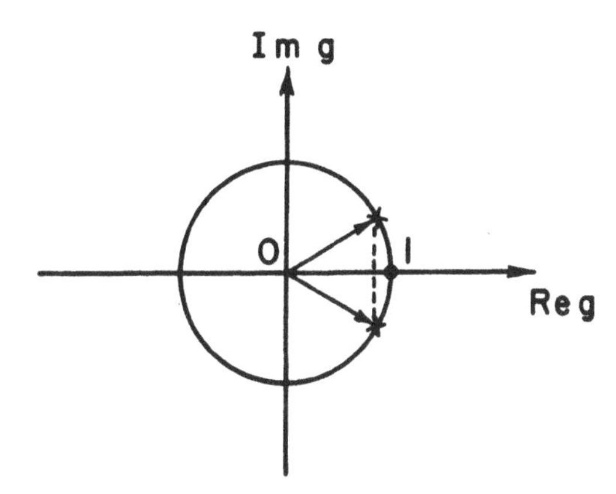

FIGURE 3.4 Complex amplification factor for the leapfrog

In this case, $|g|^2 = g\, g^* = 1$, and the system is therefore *stable*, and moreover *neutral*. See Fig. 3.4. And on the other hand, if

$$\frac{\omega_p^2 \Delta t^2}{4} - 1 > 0, \quad |g|^2 > 1 ,$$

then the system is numerically unstable. In summary, the stability condition requires

$$\Delta t \leq \frac{2}{\omega_p} \tag{3.31}$$

for the leapfrog scheme. In practice one needs a time step smaller than the value given by Eq. (3.31) in order to achieve desirable accuracy. More detailed discussions on temporal finite differencing and its consequences may be found in Refs. 4-6.

3.3 Biasing Scheme

Let us again consider Eqs. (3.4) and (3.5)

$$\begin{aligned} \dot{x} &= v \\ \dot{v} &= -\omega^2 x . \end{aligned}$$

We now introduce a non-integer time step, the "$n + \alpha$-th" time step to push the above equations. Here we use superscripts at the left shoulder for time step in order to distinguish it from power.

$$^{n+1}v - ^{n}v = -\omega^2 \Delta t^{n+\alpha}a \ , \tag{3.32}$$

$$^{n+1+\alpha}x - ^{n+\alpha}x = \Delta t^{n+1/2+\alpha}v \ , \tag{3.33}$$

where $\alpha = 1/2$ reduces these equations to the leapfrog scheme. We restrict $0 \leq \alpha \leq 1$, where we call $\alpha = 1/2$ the leapfrog scheme (explicit), $\alpha \geq 1/2$ the forward bias scheme (implicit), and $\alpha \leq 1/2$ the backward bias scheme (explicit). The implicit method gives the new future value in terms of the known old values as well as the now unknown value, whereas the explicit method gives the new value entirely in terms of the old values. The reason we call the forward bias implicit and the backward bias explicit is that, in general, when the time step of the right-hand side of Eqs. (3.32) and (3.33) is with $\alpha > 1/2$, the terms have to be expressed partially in terms of the future value. Since we do not know the future value yet, we cannot advance such an equation in an explicit straightforward way. We have to invert the equation. Thus, the name "implicit" arises. We will discuss implicit treatments further in Chapter 9.

We now study the stability analysis of the biasing scheme. Let

$$^{n}v \propto g^n$$

$$^{n}x \propto g^n .$$

Then the amplification matrix is

$$\begin{pmatrix} g-1 & \omega^2 \Delta t g^{\alpha} \\ -\Delta t g^{\alpha} & g-1 \end{pmatrix} .$$

By letting the determinant be 0, we obtain

$$(g-1)^2 + \omega^2 \Delta t^2 g^{2\alpha} = 0 \ . \tag{3.34}$$

For the fully implicit case ($\alpha = 1$), we obtain

$$(g-1)^2 + \omega^2 \Delta t^2 g^2 = 0 \ . \tag{3.35}$$

Then,

$$|g|^2 = \frac{1}{1+\omega^2 \Delta t^2} < 1 \ . \tag{3.36}$$

The scheme is always stable and the perturbation decays in time. Notice that the damping rate is frequency dependent and that it is less for smaller frequencies. When the frequency ω is small, we may want to use this, since damping is weak for small ω. On the other hand, if $\alpha = 0$, we obtain

$$|g|^2 = 1 + \omega^2 \Delta t^2 > 1 \ ,$$

and the scheme is absolutely unstable for any frequency.

Let us consider the particle motion in a static magnetic field due to the Lorentz force

$$\dot{v} = \frac{q}{mc} v \times B_0 \ . \tag{3.37}$$

Suppose $B_0 // \hat{z}$, then in component form we have

$$\dot{v}_x = \Omega v_y \tag{3.38}$$

$$\dot{v}_y = -\Omega v_x \tag{3.39}$$

where $\Omega = qB_0/mc$, the cyclotron frequency. By adding the components, we obtain

$$\dot{v}_x \mp i\dot{v}_y = \mp i\Omega v_x + \Omega v_y = \mp i\Omega(v_x \mp iv_y) \ . \tag{3.40}$$

Let $v^{\pm} = v_x \pm iv_y$, then $\dot{v}^{\pm} = \mp i\Omega v^{\pm}$ with solution

$$v^{\pm} = v_0^{\pm} e^{\mp i\Omega t} \ . \tag{3.41}$$

This is Larmor's theorem. Theoretically, Eq. (3.41) suggests $|v^{\pm}| = |v_0^{\pm}| =$ constant, correspond to having no spiraling in or out, therefore, it is important to have this property in the code. To push Eq. (3.37), we must discuss two schemes.

A. Time Centered Scheme

In the electrostatic case, the equation in the time-centered scheme was $v^{n+1} - v^n = E^{n+1/2}\Delta t$. Here we have

$$v_x^{n+1} - v_x^n = (v \times B)_x^{n+1/2} \Delta t = v_y^{n+1/2} B_0 \Delta\, t \ , \tag{3.42}$$

$$v_y^{n+1} - v_y^n = (v \times B)_y^{n+1/2} \Delta t = -v_x^{n+1/2} B_0 \Delta t \ . \tag{3.43}$$

But we don't know $v^{n+1/2}$ in the leapfrog scheme. We only have v^n and v^{n+1}, so we cannot proceed in exactly the same way as the electrostatic field case. We must approximate $v^{n+1/2}$ by

$$v_{\substack{y\\x}}^{n+1/2} = 1/2 \left[v_{\substack{y\\x}}^{n+1} + v_{\substack{y\\x}}^{n} \right] \ . \tag{3.44}$$

Then

$$v_x^{n+1} = v_x^n + \frac{1}{2}\frac{q}{mc} \left(v_y^{n+1} + v_y^n \right) B_0 \Delta t \ . \tag{3.45}$$

If we rewrite

$$v_x^{n+1} = v_x^n + \frac{\Delta t \Omega}{2} (v_y^{n+1} + v_y^n) \ , \tag{3.46}$$

and similarly

$$v_y^{n+1} = v_y^n - \frac{\Delta t \Omega}{2}\left(v_x^{n+1} + v_x^n\right) \tag{3.47}$$

are obtained. We have to solve Eqs. (3.46) and (3.47) by inverting the matrix in x and y. We shall now check this scheme for stability. If we let $g = \frac{v^{n+1}}{v^n}$ be the amplification factor, the amplification matrix is

$$\begin{pmatrix} (g-1) & -\frac{\Delta t \Omega}{2}(g+1) \\ \Delta t \frac{\Omega}{2}(g+1) & (g-1) \end{pmatrix} .$$

And the determinant equation of the amplification matrix yields

$$(g-1)^2 + \left(\frac{\Delta t \Omega}{2}\right)^2 (g+1)^2 = 0 . \tag{3.48}$$

The solutions to this quadratic equation satisfy (g_1, g_2 are two solutions)

$$|g_1|^2 = g_1 g_2 = g_1 g_1^* .$$

Then

$$g^2 \left[1 + \left(\frac{\Delta t \Omega}{2}\right)^2\right] - 2g\left[1 - \left(\frac{\Delta t \Omega}{2}\right)^2\right] + \left[1 + \left(\frac{\Delta t \Omega}{2}\right)^2\right] = 0 . \tag{3.49}$$

We now obtain

$$|g_1|^2 = \frac{1 + \left(\frac{\Delta t \Omega}{2}\right)^2}{1 + \left(\frac{\Delta t \Omega}{2}\right)^2} = 1 , \tag{3.50}$$

as long as $\Delta t \leq \frac{2}{\Omega}$. This result is similar to the leapfrog stability for the electrostatic field as shown in Eq. (3.30) and the discussion that follows. We found that such an algorithm is stable and, moreover, neutral.

B. Time-decentered Schemes

In order to advance the equation $\dot{\mathbf{v}} = -\boldsymbol{\Omega} \times \mathbf{v}$, we write

$$v_x^{n+1} - v_x^n = \Delta t \Omega v_y^{n+1/2+\alpha} , \tag{3.51}$$

$$v_y^{n+1} - v_y = -\Delta t \Omega v_x^{n+1/2+\alpha} , \tag{3.52}$$

where $-\frac{1}{2} \leq \alpha < \frac{1}{2}$ and $\alpha = 0$ corresponds to the time-centered scheme. One way to define $v^{n+1/2+\alpha}$ is to have

$$v_x^{n+1/2+\gamma} = (1/2 + \gamma) v^{n+1} + (1/2 - \gamma) x^n . \tag{3.53}$$

FIGURE 3.5 Particle orbit due to decentered Lorentz force

From Eq. (3.53), we calculate the amplification matrix as

$$\begin{pmatrix} (g-1) & -\Delta t\Omega\,[(1/2+\gamma)g+(1/2-\gamma)] \\ \Delta t\Omega\,[(1/2+\gamma)g+(1/2-\gamma)] & (g-1) \end{pmatrix} .$$

After some algebra, we obtain

$$|g|^2 = \frac{1-(\Delta t\Omega)^2(1/2+\gamma)^2}{1-(\Delta t\Omega)^2(1/2-\gamma)^2} \leq 1 \text{ (if } \alpha \geq 0) . \tag{3.54}$$

We obtain stability in the code if $\gamma \geq 0$ (i.e., implicit). That is, v^{n+1} is more weighted than v^n to the calculation of $v^{n+1/2+\gamma}$. This scheme, however, causes damping of the cyclotron motion. See Fig. 3.5.

We will now discuss the particle pushing in a magnetic field at greater length. Let us generalize by including electric fields. The equation of motion with electric and magnetic fields is

$$\dot{\mathbf{v}} = \frac{q}{m}\left(\mathbf{E}+\frac{1}{c}\mathbf{v}\times\mathbf{B}\right) = \frac{q}{m}\mathbf{E}+\mathbf{v}\times\boldsymbol{\Omega} , \tag{3.55}$$

$$\dot{\mathbf{r}} = \mathbf{v} , \tag{3.56}$$

where $\boldsymbol{\omega} \equiv \omega_c\hat{\mathbf{b}}$. How do we leapfrog these equations? The finite difference of the above equations becomes

$$\mathbf{v}(t+\Delta t) = \frac{q}{m}\mathbf{E}\left(t+\frac{\Delta t}{2}\right)\Delta t + \mathbf{v}\left(t+\frac{\Delta t}{2}\right)\times\boldsymbol{\Omega}\Delta t + \mathbf{v}(t) . \tag{3.57}$$

We replace the time-centered value of the velocity by an average of the previous and the future values as

$$\mathbf{v}(t+\Delta t) \cong \frac{q}{m}\mathbf{E}\left(t+\frac{\Delta t}{2}\right)\Delta t + \frac{\mathbf{v}(t)+\mathbf{v}(t+\Delta t)}{2}\times\boldsymbol{\Omega}\Delta t + \mathbf{v}(t) . \tag{3.58}$$

The velocity update is ordinary

$$\mathbf{r}(t+3/2\Delta t)=\mathbf{r}\left(t+\frac{1}{2}\Delta t\right)+\mathbf{v}(t+\Delta t)\Delta t\ . \tag{3.59}$$

Equation (3.58) is, of course, implicit in time step. In terms of matrices, we can write the above equations symbolically as

$$\left[\mathbf{I}-\mathbf{R}\frac{\Omega}{2}\Delta t\right]\cdot\mathbf{v}(t+\Delta t)=\left[\mathbf{I}+\mathbf{R}\frac{\Omega}{2}\Delta t\right]\mathbf{v}(t)+\frac{q}{m}\mathbf{E}\left(t+\frac{\Delta t}{2}\right)\Delta t \tag{3.60}$$

where

$$\begin{aligned}\mathbf{v}\times\boldsymbol{\Omega}&\equiv\mathbf{v}\times\mathbf{R}\Omega,\\ &\equiv\begin{pmatrix}v_yR_z-v_zR_y\\ v_zR_x-v_xR_z\\ v_xR_y-v_yR_x\end{pmatrix}\cdot\Omega\ .\end{aligned} \tag{3.61}$$

The explicit forms of the matrices are

$$\mathbf{I}\equiv\begin{pmatrix}1&&\\&1&\\&&1\end{pmatrix}$$

and

$$\mathbf{R}\equiv\begin{pmatrix}0&\frac{B_z}{B}&-\frac{B_y}{B}\\ -\frac{B_z}{B}&0&\frac{B_x}{B}\\ \frac{B_y}{B}&-\frac{B_x}{B}&0\end{pmatrix}=\frac{1}{B}\begin{pmatrix}0&B_z&-B_y\\ -B_z&0&B-x\\ B_y&-B_x&0\end{pmatrix}\ .$$

When we write $\varepsilon=\omega_c\Delta t/2$, we have the form

$$(\mathbf{I}-\mathbf{R}\varepsilon)=\begin{pmatrix}1&-\varepsilon b_z&\varepsilon b_y\\ \varepsilon b_z&1&-\varepsilon b_x\\ -\varepsilon b_y&\varepsilon b_x&1\end{pmatrix}\ .$$

The inverse matrix of $\mathbf{I}-\mathbf{R}\varepsilon$ is

$$[\mathbf{I}-\mathbf{R}\varepsilon]^{-1}=\frac{1}{\det(\mathbf{I}-\mathbf{R}\varepsilon)}\begin{pmatrix}1+\varepsilon^2b_x^2&\varepsilon b_z+\varepsilon^2b_xb_y&-\varepsilon b_y+\varepsilon^2b_xb_z\\ -\varepsilon b_z+\varepsilon^2b_xb_y&1+\varepsilon^2b_y^2&\varepsilon b_x+\varepsilon^2b_yb_z\\ \varepsilon b_y+\varepsilon^2b_xb_z&-\varepsilon b_x+\varepsilon^2b_yb_z&1+\varepsilon^2b_z^2\end{pmatrix}\ .$$

With this, the velocity in the next time step is solved by this inverted matrix as

$$\begin{aligned}\mathbf{v}(t+\Delta t)&=\left[\mathbf{I}-\mathbf{R}\frac{\Omega}{2}\Delta t\right]^{-1}\left[\mathbf{I}+\mathbf{R}\frac{\Omega}{2}\Delta t\right]\cdot\mathbf{v}(t)\\ &+\left[\mathbf{I}-\mathbf{R}\frac{\Omega}{2}\Delta t\right]^{-1}\cdot\mathbf{E}\left(t+\frac{\Delta t}{2}\right)\frac{q}{m}\Delta t\ .\end{aligned} \tag{3.62}$$

Since this finite difference is effectively time-centered, we expect that the scheme is energy conservative without the electric field effect. Let[7]

$$\mathbf{v}(t) = \mathbf{v}^- - \frac{q}{m}\frac{\Delta t}{2}\mathbf{E}\left(t + \frac{\Delta t}{2}\right)$$

$$\mathbf{v}(t+\Delta t) = \mathbf{v}^+ + \frac{q}{m}\frac{\Delta t}{2}\mathbf{E}\left(t + \frac{\Delta t}{2}\right) .$$

Then we substitute these into the original finite difference Eq. (3.58), obtaining

$$(\mathbf{v}^+ - \mathbf{v}^-) = \frac{q\Delta t}{2mc}(\mathbf{v} + \mathbf{v}^-) \times \mathbf{B} . \tag{3.63}$$

Multiplying $(\mathbf{v}^+ + \mathbf{v}^-)$ from left, we have

$$(v^+)^2 - (v^-)^2 = 0 . \tag{3.64}$$

If the electric field is zero, Eq. (3.64) reduces to

$$(v(t))^2 = (v(t+\Delta t))^2 .$$

That is, in this case, there is no kinetic energy gain for this particle by the finite difference scheme. In the explicit time integration, the time step has to be small enough to resolve the cyclotron motion, i.e., the Courant-Friedrichs-Lewy condition. Thus we consider the case where $\frac{\Omega}{2}\Delta t < 1$. Writing $\mathbf{R}\Omega\Delta t/2 \equiv \boldsymbol{\varepsilon}$, when $\varepsilon \ll 1$, we can use the expansion of the matrix

$$[\mathbf{I} - \boldsymbol{\varepsilon}]^{-1} \cong \mathbf{I} + \boldsymbol{\varepsilon} + \ldots . \tag{3.65}$$

Thus we obtain

$$\begin{aligned}
\mathbf{v}(t+\Delta t) &\cong \left[\mathbf{I} + \mathbf{R}\frac{\Omega\Delta t}{2}\right] \cdot \left[\mathbf{I} + \mathbf{R}\frac{\Omega}{2}\Delta t\right] \cdot v(t) \\
&\quad + \left[\mathbf{I} + \mathbf{R}\frac{\Omega}{2}\Delta t\right] \cdot \mathbf{E}\left(t + \frac{\Delta t}{2}\right)\frac{q}{m}\Delta t \\
&= [\mathbf{I} + \mathbf{R}\Omega\Delta t] \cdot \mathbf{v}(t) + \left[\mathbf{I} + \mathbf{R}\frac{\Omega\Delta t}{2}\right] \cdot \mathbf{E}\left(t + \frac{\Delta t}{2}\right)\frac{q}{m}\Delta t \\
&= [\mathbf{I} + \mathbf{R}\Omega\Delta t] \cdot \mathbf{v}(t) + \frac{q}{m}\mathbf{E}\left(t + \frac{\Delta t}{2}\right)\frac{\Delta t}{2} \\
&\quad + [\mathbf{I} + \mathbf{R}\Omega\Delta t] \cdot \mathbf{E}\left(t + \frac{\Delta t}{2}\right)\frac{q}{m}\frac{\Delta t}{2} .
\end{aligned} \tag{3.66}$$

This formulation leads the velocity pushing into two parts: (i) the electric field acceleration term over half time step and (ii) the full time-step rotation term after half time-step acceleration and (iii) $E\frac{\Delta t}{2}$

$$\mathbf{v}(t+\Delta t) = \frac{q}{m}\mathbf{E}\left(t+\frac{\Delta t}{2}\right)\frac{\Delta t}{2} + (\mathbf{I}+\mathbf{R}\Omega\Delta t)\cdot\left[\mathbf{v}(t) + \frac{q}{m}\frac{\Delta t}{2}\mathbf{E}\left(t+\frac{\Delta t}{2}\right)\right] . \tag{3.67}$$

If we use a time-decentered scheme, i.e., $1 \geq \alpha \geq 0$, the finite differencing becomes

$$\begin{aligned}\mathbf{v}(t+\Delta t) &= \frac{q}{m}\mathbf{E}\left(t+\frac{\Delta t}{2}\right)\Delta t + \frac{q}{m}\mathbf{v}(t+\alpha\Delta t)\times\mathbf{B}\Delta t \\ &= \frac{q}{m}\mathbf{E}\left(t+\frac{\Delta t}{2}\right)\Delta t + \frac{q}{mc}[(1-\alpha)\mathbf{v}(t)+\alpha\mathbf{v}(t+\Delta t)]\times\mathbf{B}\Delta t .\end{aligned}$$

In terms of matrix form we obtain

$$[\mathbf{I}-\mathbf{R}\alpha\Omega\Delta t]\cdot\mathbf{v}(t+\Delta t) = [\mathbf{I}+\mathbf{R}(1-\alpha)\Omega\Delta t]\cdot\mathbf{v}(t) + \frac{q}{m}\mathbf{E}\left(t+\frac{\Delta t}{2}\right)\Delta t . \tag{3.68}$$

Therefore,

$$\begin{aligned}\mathbf{v}(t+\Delta t) &= [\mathbf{I}-\mathbf{R}\alpha\Omega\Delta t]^{-1}[\mathbf{I}+\mathbf{R}(1-\alpha)\Omega\Delta t]\cdot\mathbf{v}(t) + [\mathbf{I}-\mathbf{R}\alpha\Omega\Delta t]^{-1} \\ &\quad \cdot\mathbf{E}\left(t+\frac{\Delta t}{2}\right)\frac{q\Delta t}{m} .\end{aligned}$$

This equation is often employed to advance velocities in magnetic fields. When $\alpha = \frac{1}{2}$, it is the time-centered scheme pusher.

3.4 Runge-Kutta Method

We compare the leapfrog and Runge-Kutta methods in detail in the following. The characteristics of the leapfrog method are: (a) It is an accurate second-order scheme for the acceleration equation, i.e., the error is proportional to $(\Delta t)^2$; (b) It is explicit and very simple (does not require any extra information); and (c) It is stable and neutral within the threshold Δt value. The Runge-Kutta method[8–10] is a popular and powerful method for integrating nonlinear differential equations. As we shall see, it is more complex and takes more memory than the previous leapfrog method, but it is more accurate.

To demonstrate the first point (a) for the leapfrog method, we combine the acceleration equation and the velocity definition for the standard pushing equation to obtain:

$$\frac{x^{n+1} - 2x^n + x^{n-1}}{(\Delta t)^2} = a(x^n) .$$

This finite difference equation corresponds to the differential equation —

$$\frac{d^2 x}{dt^2} = a(x) .$$

We evaluate the finite difference equation error for the leapfrog scheme with a Taylor expansion:

$$x^{n+1} = x^n + \left(x^{n+1} - x^n\right) \simeq x^n + \Delta t \frac{\partial x}{\partial t}\bigg|^n + 1/2(\Delta t)^2 \frac{\partial^2 x}{\partial t^2}\bigg|^n$$

$$+1/6(\Delta t)^3 \frac{\partial^3 x}{\partial t^3}\bigg|^n + \frac{1}{24}(\Delta t)^4 \frac{\partial^4 x}{\partial t^4 x}\bigg|^n .$$

Similarly,

$$x^{n-1} = x^n + \left(x^{n-1} - x^n\right) \simeq x^n - \Delta t \frac{\partial x}{\partial t}\bigg|^n + 1/2(\Delta t)^2 \frac{\partial^2 x}{\partial t^2}\bigg|^n$$

$$-1/6(\Delta t)^3 \frac{\partial^3 x}{\partial t^3}\bigg|^n + 1/24(\Delta t)^4 \frac{\partial^4 x}{\partial t^4}\bigg|^n .$$

Therefore, substitution into the above expression for $a(x^n)$ gives,

$$x^{n+1} - 2x^n + x^{n-1} \simeq (\Delta t)^2 \frac{\partial^2 x}{\partial t^2}\bigg|^n + \frac{1}{12}(\Delta t)^4 \frac{\partial^4 x}{\partial t^4}\bigg|^n + 0(\Delta t^6)$$

and

$$\frac{x^{n+1} - 2x^n + x^{n-1}}{(\Delta t)^2} \simeq \frac{\partial^2 x}{\partial t^2} + (\Delta t)^2 12 \frac{\partial^4 x}{\partial t^4} .$$

The error for the leapfrog scheme is on the order of $(\Delta t)^2 \frac{\partial^4 x}{\partial t^4}$, i.e., second order in Δt.

The characteristics of the Runge Kutta method, on the other hand, are: (a) It is more accurate than the leapfrog—fourth order accuracy,[10] i.e. errors $(\Delta t)^4$ or higher. (b) It is explicit (or implicit[11,12]). However, it does require some additional information at the intermediate time steps. (extra storage needed, about 3-4 times that needed for the leapfrog). (c) It is stable.

Let us consider the Runge-Kutta method. Let us consider an ordinary differential equation

$$\frac{dx}{dt} = f(x,t) . \tag{3.69}$$

We want to integrate this from x^n to x^{n+1}, when the superscripts refer to time steps. One example of the Runge-Kutta scheme[10,13] is

$$\Delta x_1 = \Delta t f(x^n, t_n) , \tag{3.70}$$

$$\Delta x_2 = \Delta t f\left(x^n + \frac{1}{2}\Delta x, t_n + \frac{1}{2}\Delta t\right) , \tag{3.71}$$

$$\Delta x_3 = \Delta t f\left(x^n + \frac{1}{4}\Delta x_1 + \frac{1}{4}\Delta x_2, t_n + \frac{1}{2}\Delta t\right) , \tag{3.72}$$

$$\Delta x_4 = \Delta t f\left(x^n - \Delta x_2 + 2\Delta x_3, t_n + \Delta t\right) , \tag{3.73}$$

$$x^{n+1} = x^n + \frac{1}{6}(\Delta x_1 + 4\Delta x_3 + \Delta x_4) . \tag{3.74}$$

The error estimate ε for a step is given by

$$\varepsilon \simeq \frac{1}{8}\Delta x_1 + \frac{2}{3}\Delta x_2 + \frac{1}{16}\Delta x_4 - \frac{27}{56}\Delta x_5 - \frac{125}{336}\Delta x_6 , \tag{3.75}$$

where

$$\Delta x_5 = \Delta t f\left(x^n + \frac{7}{27}\Delta x_1 + \frac{10}{27}\Delta x_2 + \frac{1}{27}\Delta x_4, t_n + \frac{2}{3}\Delta t\right) ,$$

$$\Delta x_6 = \Delta t f\left(x^n + \frac{28}{625}\Delta x_1 - \frac{1}{5}\Delta x_2 + \frac{546}{625}\Delta x_3 + \frac{54}{625}\Delta x_4 - \frac{378}{625}\Delta x_5, t_n + \frac{1}{5}\Delta t\right) .$$

The step size Δt may be increased until the error estimate approaches an allowed limit. The step size becomes smallest in regions where the equation becomes stiff, i.e., the solution rapidly changes its property.

It is instructive to compare these two popular methods.[14] Because the leapfrog scheme is time-reversible (for a non-dissipative system), it must be energy conserving as well. Thus leapfrog gives both stability and energy conservation, but because it is low order, it does not follow sudden changes in the exact solution very well for stiff equations.

One finds that for non-dissipative systems leapfrog trajectories close on themselves (to machine precision) whereas Runge-Kutta trajectories gain energy at a rate determined by the allowed error. This behavior is most easily verified by studying the pendulum near an unstable equilibrium,

$$m\ell^2\ddot{\theta} = -mg\sin\theta , \tag{3.76}$$

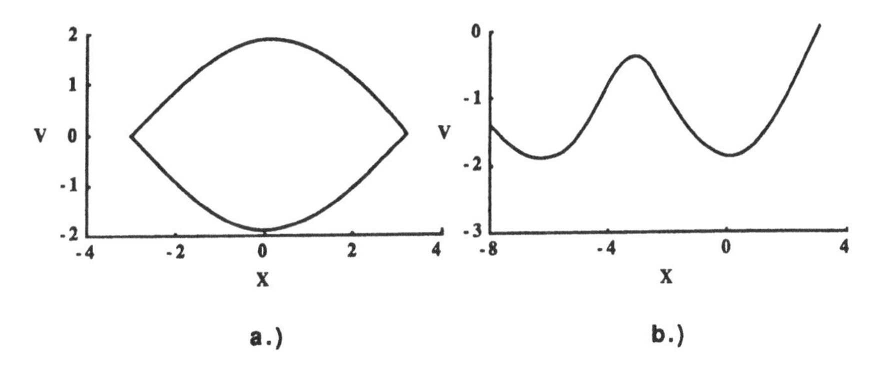

FIGURE 3.6 Phase space orbits by the leapfrog method (a) and by the Runge-Kutta (b) near the unstable point for a presumably closed orbit of the pendulum equation (Ref. 14)

where a small variation in the energy dramatically changes the character of the trajectory from periodic to non-periodic. In Fig. 3.6 the particle is supposed to be trapped but is very close to the unstable equilibrium. The leapfrog correctly shows (in Fig. 3.6a) a closed orbit, while the Runge-Kutta shows the orbit open up due to error (Fig. 3.6b). On the other hand, where sharp corners in the solution are expected, as is the case in the highly nonlinear regime for the Rayleigh oscillator

$$m\ddot{x} = -\mu\dot{x}(A + B\dot{x}^2 + C\dot{x}^4) - kx \ , \tag{3.77}$$

the Runge-Kutta trajectory follows the true trajectory much more closely than do low-order schemes. Here one can see the 'ringing' of Euler's solution as it repeatedly overshoots the actual trajectory, damping the error each time. (See Fig. 3.7.) In this example the ringing can be reduced by reducing the step size, but this is not always the case.

A particularly interesting contrast shows up for the van der Pol equation,

$$m\ddot{x} = -kx - \mu(x^2 - \ell^2)\dot{x} \ . \tag{3.78}$$

In the nearly linear regime, the solution as determined by the averaging method (see below) is given by an ellipse. Figure 3.8 shows the results of a simulation using leapfrog with a fairly large step size ($\Delta t = 0.8$). Clearly Runge-Kutta provides a better approximation. The leapfrog method is generally believed to lead to an approximate solution closer to the true solution

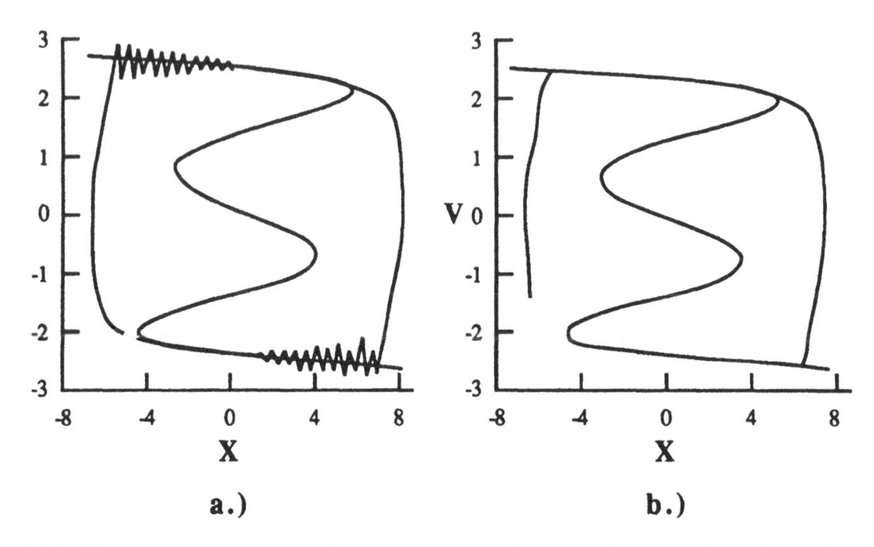

FIGURE 3.7 Phase space trajectories for a Rayleigh oscillator by the Euler method (a) and by the Runge-Kutta. The middle S-shaped curve is the characteristics to aid the eye (Ref. 14)

as the time step is chosen smaller. For this nonlinear oscillator, however, decreasing the step size for leapfrog does not yield better results; in fact we see an even more pronounced difference. Comparing this behavior with that of the Runge-Kutta solution as the nonlinearity (*not* the step size) is increased shows that the two sets of curves are quite similar. This strongly suggests that, in fact, the leapfrog method is unstable with respect to the particular type of nonlinearities involved in the van der Pol equation. In other words, numerical errors in the leapfrog approximation tend to exaggerate the nonlinearity of the problem.

3.5 Diffusion Equation

The transport phenomena in plasma physics is very important and involves the diffusion process. This process is described by a different type[15] of temporal differential equation than the wave-like ones we considered in the previous section. More discussion on types of differential equations are found in Secs. 5.6 and 5.7.

The diffusion equation is generally a parabolic differential equation that

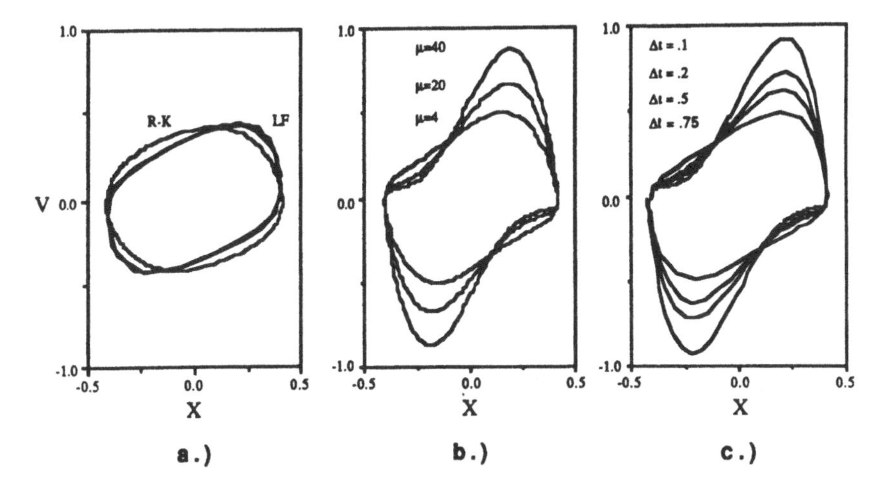

FIGURE 3.8 Phase space trajectories for a van der Pol oscillator by the leapfrog and the Runge-Kutta methods. (a) weakly nonlinear. (b) various nonlinearities by the Runge-Kutta. (c) various time-steps with the leapfrog method (Ref. 14)

assumes the form of

$$\frac{\partial u}{\partial t} - K\frac{\partial^2 u}{\partial x^2} = 0 \ . \tag{3.79}$$

The simplest way of solving the diffusion equation in time is to use a first-order explicit method, which is analogous to the Euler method for ordinary differential equations.

A. Explicit First-order Method

The diffusion equation in the explicit method of the first order is represented by the difference equation

$$u_j^{n+1} = u_j^n + \frac{K\Delta t}{\Delta x^2}\left(u_{j-1}^n - 2u_j^n + u_{j-1}^n\right) \ . \tag{3.80}$$

As has been discussed previously, there are errors associated with the time step Δt and the space step Δx. The general approach in considering the stability of the scheme is to derive the amplification factor for a Fourier mode $u_j^n \propto e^{ikj\Delta x - i\omega\Delta t \cdot n}$ and the amplification factor is defined as $g = e^{-i\omega\Delta t}$.

From the explicit method, the amplification factor is

$$g = 1 + \frac{2K\Delta t}{\Delta x^2}\left(\frac{1}{2}e^{ik\Delta x} + \frac{1}{2}e^{-ik\Delta x}\right) = 1 - \frac{4K\Delta t}{\Delta x^2}\sin^2\left(\frac{k\Delta x}{2}\right) \tag{3.81}$$

and stability requires that

$$|g|^2 \leq 1 .$$

This condition must hold for every wavenumber k. The requirement for stability is, therefore,

$$\Delta t \leq \frac{1}{2}\frac{\Delta x^2}{K}$$

or

$$K \leq \frac{1}{2}\left(\frac{\Delta x^2}{\Delta t}\right) . \tag{3.82}$$

From these results, one can expect that the maximum possible numerical diffusion is greater than the physical diffusion for stability.

Also this method requires:

- When one increases the spatial accuracy by a factor 2, the time step has to be reduced by a factor of 4.
- First order accuracy in time step and $(\Delta x)^2$ in the space step.

These limitations may be overcome by implementing an implicit approach.

B. The Crank-Nicholson Implicit Method

If we average the spatial diffusion term in time, then the equation becomes implicit:

$$u_i^{n+1} = u_i^n + \frac{K\Delta t}{2\Delta x^2}\left(u_{i+1}^{n+1} - 2u_i^{n+1} + u_{i-1}^{n+1}\right) + \frac{K\Delta t}{2\Delta x^2}\left(u_{i+1}^n - 2u_i^n + u_{i-1}^n\right) . \tag{3.83}$$

We analyze the method for a Fourier mode and the amplification factor satisfies the equation

$$g = 1 - \frac{K\Delta t}{\Delta x^2}[1 - \cos(k\Delta x)]g - \frac{K\Delta t}{\Delta x^2}[1 - \cos(k\Delta x)] . \tag{3.84}$$

Solving Eq. (3.84) for g, we have

$$g = \left[\frac{1 - \frac{2K\Delta t}{\Delta x^2}\sin^2\left(\frac{k\Delta x}{2}\right)}{1 + \frac{2K\Delta t}{\Delta x^2}\sin^2\left(\frac{k\Delta x}{2}\right)}\right] . \tag{3.85}$$

We learn from Eq. (3.85) that g is a real number and that the magnitude of g is always smaller than unity for all wavenumbers and all time steps. Therefore, the Crank-Nicholson method is unconditionally stable and accurate to the second order in both the time step and the space step. But because we are still left with the problem of inversion of the matrix at each time step due to the implicit nature of the algorithm, the method is expensive to use.

C. Leapfrog Method

The diffusion equation by the leapfrog method takes the form

$$u_i^{n+1} = u_i^{n-1} + 2\frac{K\Delta t}{\Delta x^2}\left(u_{i+1}^n - 2u_j^n + u_{i-1}^n\right) . \tag{3.86}$$

Analysis of the stability of the method yields the equation for the amplification factor,

$$g^2 = 1 - 4\frac{K\Delta t}{\Delta x^2}[1 - \cos(k\Delta x)]g = 1 - 8\frac{K\Delta t}{\Delta x^2}\sin^2\left(\frac{k\Delta x}{2}\right)g . \tag{3.87}$$

Solving this, we obtain

$$g = -\alpha \pm (\alpha^2 + 1)^{1/2} , \tag{3.88}$$

where

$$\alpha = 4\frac{K\Delta t}{\Delta x^2}\sin^2\left(\frac{k\Delta x}{2}\right) .$$

It turns out that the leapfrog method Eq. (3.86) is always unstable.

D. The Dufort-Frankel Method

By slightly altering the leapfrog method, a method for parabolic equations which is exceptional for being both explicit and unconditionally stable is obtained. It is a three-level formula in time-steps:

$$u_i^{n+1} = u_i^{n-1} + \frac{2K\Delta t}{\Delta x^2}[u_{i+1}^n - (u_i^{n+1} + u_i^{n-1}) + u_{i+1}^n] . \tag{3.89}$$

Note here that $2u_i^n$ is replaced by $u_i^{n+1} + u_i^{n-1}$. The dependent variable under the diffusion operator at the central time and space point has been replaced by a time average, so that the intermediate points need not be defined. We may solve explicitly for the new dependent variable u_i^{n+1} at each mesh point,

$$u_i^{n+1} = \left(\frac{1-\alpha}{1+\alpha}\right)u_i^{n-1} + \frac{\alpha}{1+\alpha}\left(u_{i+1}^n + u_{i-1}^n\right) , \tag{3.90}$$

where

$$\alpha = 2\frac{K\Delta t}{\Delta x^2} .$$

The amplification factor for the Dufort-Frankel method is obtained as

$$g^2 = \frac{1-\alpha}{1+\alpha} + \frac{2\alpha}{1+\alpha}\cos(k\Delta x)g ,$$

and therefore,

$$g = \frac{1}{1+\alpha}\left\{\alpha\cos(k\Delta x) \pm [1 - \alpha^2 \sin^2(k\Delta x)]^{1/2}\right\} . \tag{3.91}$$

To interpret this equation we note that there are two cases. If $\alpha^2 \sin^2(k\Delta x) \leq 1$ (i.e., small time step), the amplification factor, g, is real and always smaller than unity for both roots.

In the case where $\alpha^2 \sin^2(k\Delta x) > 1$ (i.e., large time step), the amplification factor, g, becomes complex and its magnitude is given by

$$|g| = \frac{1-\alpha}{1+\alpha} . \tag{3.92}$$

Thus, for all wavenumbers on the mesh, the magnitude of g is always smaller than unity, hence, the method is stable, but it gives oscillatory solutions. Although this scheme is stable, the method in this range is not accurate.

Problems

1. Discuss a slightly simpler version of the Runge-Kutta method for $\dot{x} = f(x)$:

$$\begin{aligned}
\Delta x_1 &= \Delta t f(x^n) , \\
\Delta x_2 &= \Delta f\left(x^n + \frac{1}{2}\Delta x_1\right) , \\
\Delta x_3 &= \Delta t f\left(x^n + \frac{1}{2}\Delta x_2\right) , \\
\Delta x_4 &= \Delta t f(x^n + \Delta x_3) , \\
x^{n+1} &= x^n + \frac{1}{6}(\Delta x_1 + 2\Delta x_2 + 2\Delta x_3 + \Delta x_4) .
\end{aligned}$$

2. Determine the phase error as a function of the timestep Δt in the leapfrog scheme.

3. Examine the situation when Eq. (3.26) breaks down.

4. Why would we want to use a temporally *damping difference scheme* in Section 3.3 when leapfrog differencing is both stable and neutral?

5. Let $\mathbf{v}^* = \mathbf{v}^n + \frac{q}{m}\mathbf{E}^{n+\alpha}\frac{\Delta t}{2}$. Then show that

$$\mathbf{v}^{n+1/2} = \frac{\mathbf{v}^* + \mathbf{v}^* \times \mathbf{R}\Omega\Delta t/2 + \mathbf{v}^* \cdot \mathbf{R}\Omega^2(\Delta t/2)^2}{1 + (\Omega\Delta t/2)^2},$$

and

$$\mathbf{v}^{n+1} = \mathbf{v}^n + 2\left(\mathbf{v}^{n+1/2} - \mathbf{v}^n\right).$$

This pushing recovers $\mathbf{E} \times \mathbf{B}$ drift in the limit $\Delta t \to \infty$ (Brackbill).

References

1. R. Hockney and J. Eastwood *Computer Simulation Using Particles*, (McGraw-Hill, New York, 1981) p. 94.
2. D.E. Potter, *Computational Physics* (Wiley, London, 1973), pgs. 18 and 37.
3. J.M. Dawson, Phys. Rev. **113**, 383 (1959).
4. A.B. Langdon, J. Comp. Phys. **30**, 202 (1979).
5. A.B. Langdon, Phys. Fluids **22**, 163 (1979).
6. R. Hockney and J. Eastwood *ibid*, p. 117.
7. A.B. Langdon and P.F. Lasinski, in *Methods in Computational Physics* (Academic, New York, 1976), Vol. 16, p. 327.
8. C. Runge, Math. Ann. (Ger.) **46**, 167 (1895).
9. W. Kutta, Z. Math. Phys. **46**, 435 (1901).
10. For example, A. Ralston and P. Rabinowitz, *A First Course in Numerical Analysis* (McGraw Hill, New York, 1978).
11. P. Henrici, *Discrete Variable Methods in Ordinary Differential Equations* (John Wiley and Sons, New York, 1962) Chaps. 1,3.
12. J.C. Butcher, Math. Comp. **18**, 50 (1964).
13. L.F. Shamine, H.A. Watts, and S. Davenport, SIAM Rev. **18**, 376 (1976).
14. S. Eubank, W. Miner, T. Tajima, and J. Wiley, to be published in Am. J. Phys.
15. R.D. Richtmyer and K.W. Morton, *Difference Methods for Initial-value Problems*, 2nd Ed. (1967), p. 185.

4

GRID METHOD

This chapter discusses the grid methods which relate the individual discrete spatial coordinates for computational efficiency. Spatial grids are commonly used in computer simulation models because the methods for calculating interaction forces through spatial grids are much more efficient than summing the direct interaction forces of all particles in the simulation. We will study how the physical properties of a simulation plasma are modified by the presence of a spatial grid.[1] Unfortunately some information loss occurs when a spatial grid is imposed. This happens when the current density and charge distribution of particles in the simulation are determined only at the grid points and when the particle force is interpolated from electric and magnetic fields that are also defined only at the grid points. We will see that interactions through the grids couple a perturbation to perturbations at other wavelengths called aliases. The significance of this coupling is limited if finite size particles are used, and if the spacing between two adjacent grid points is about the Debye length. This is an additional benefit of the finite-sized particle method discussed in Chapter 2.

4.1 Grid Method and the Dipole Expansion

We assume a uniform, constant grid. In this chapter we will use this type of grid unless otherwise specified. In Chapter 10 nonuniform, noncartesian grids will be discussed. Suppose we have N particles: $i = 1, \ldots, N$ and L grid points: $n = 1, \ldots, L$. If the system is periodic beyond the L grid points (i.e., repeats itself again), one can define (discrete) Fourier modes with wavenumber $k_n = 2\pi n/L\Delta$, where Δ is the grid separation (see Fig. 4.1).

The relation between the discrete and nondiscrete Fourier transforms is considered. The Fourier transform and its inverse nondiscrete points in infinite space are defined as

$$\tilde{f}(\mathbf{k}) \equiv \int_{-\infty}^{\infty} f(\mathbf{x}) e^{-i\mathbf{k}\cdot\mathbf{x}} \tag{4.1}$$

$$\longleftrightarrow f(\mathbf{x}) \equiv \frac{1}{2\pi} \int_{-\infty}^{\infty} d\mathbf{k} \tilde{f}(\mathbf{k}) e^{+i\mathbf{k}\cdot\mathbf{x}} . \tag{4.2}$$

The Fourier transform and its inverse at nondiscrete points in finite space with the periodic boundary condition (in two dimensions, for example) are defined as

$$\tilde{f}_{\ell m} \equiv \frac{1}{L_x L_y} \int_0^{L_x} dx \int_0^{L_y} dy \tilde{f}(\mathbf{x}) e^{-i\mathbf{k}\,\mathbf{x}} \tag{4.3}$$

$$\longleftrightarrow f(\mathbf{x}) \equiv \sum_{\ell,m=-\infty}^{\infty} \tilde{f}_{\ell m} e^{+i\mathbf{k}\cdot\mathbf{x}} , \tag{4.4}$$

where $\mathbf{k} = \left(\frac{2\pi\ell}{L_x}, \frac{2\pi m}{L_y}\right)$. The quantity $f = f(r,s)$ that is only defined at the discrete points (r,s) and periodic in the directions of r and s can be transformed by the discrete Fourier transform defined by

$$\tilde{f}_{\ell m} \equiv \frac{1}{N_x N_y} \sum_{r,s=0}^{N_x-1,N_y-1} f_{rs} \exp\left(-\frac{2\pi i \ell r}{N_x}\right) \exp\left(-\frac{2\pi i m s}{N_y}\right) \tag{4.5}$$

$$\longleftrightarrow f_{rs} \equiv \sum_{\ell,m=0}^{N_x-1,N_y-1} f_{\ell m} \exp\left(\frac{2\pi i r \ell}{N_x}\right) \exp\left(\frac{2\pi i s m}{N_y}\right) . \tag{4.6}$$

The grid accumulates charge and current from particles on the grid, converts the change and current into forces on the grid, and applies these forces at the grid points on the individual particles. Each particle exerts a force on each other particle. One could compute this *direct interaction* among all of the pairs. The grid method, an alternate view, is that each particle feels a force due to the *field* at its location. The grid method adopts the field philosophy:

Grid = "Field."

The advantage of the grid method is that *the number of operations is greatly reduced.*

The *charge* due to a finite size particle is,

$$\rho(x) = \sum_i S(x - x_i) \tag{4.7}$$

where S is the form factor. Write $x_i = x_g + (x_i - x_g)$, where x_g is the grid position and x_i the particle position. Then Taylor expanding in $(x_i - x_g)$ gives:

$$\begin{aligned}\rho(\mathbf{x}) &= \sum_i \Bigg[S(\mathbf{x} - \mathbf{x}_g) + (\mathbf{x} - \mathbf{x}_g) \cdot \nabla_g S(\mathbf{x} - \mathbf{x}) \\ &\quad + \frac{1}{2}(\mathbf{x}_i - \mathbf{x}_g)^2 : \nabla_g^2 S(\mathbf{x} - \mathbf{x}_g) + \ldots \Bigg] .\end{aligned} \tag{4.8}$$

The grid method procedure consists of three parts: 1) charge accumulation at grid. 2) From charge the force is calculated at grid. 3) Assign the force at grid to the particle.

If (as per (1)) we decide to look at $x = x'_g$ only, then we have

$$\rho(\mathbf{x}'_g) = \sum_i S(\mathbf{x}'_g - \mathbf{x}_i) = \sum_i [S(\mathbf{x}'_g - \mathbf{x}_g) + (\mathbf{x}_i - \mathbf{x}_g) \cdot \nabla_g S(\mathbf{x}'_g - \mathbf{x}_g) + \ldots] . \tag{4.9}$$

The Fourier transform of this equation becomes a *finite* set of discrete Fourier transforms instead of an infinite series of continuous ones (which would correspond to the continuous space of $\mathbf{x}$, instead of the grid points $\mathbf{x}_g$). For simplicity let us look at one-dimensional cases here.

$$\begin{aligned}\rho(k) &= \sum_i \Bigg[S(k)e^{-ikx_g} + (x_i - x_g)ikS(k)e^{-ikx_g} \\ &\quad + \frac{1}{2}(x_i - x_g)^2(ik)^2 S(k)e^{-ikx_g} + \ldots \Bigg]\end{aligned} \tag{4.10}$$

where $f(k) = \sum_{x_g} f(x_g)e^{-ikx_g}$ and $k = \frac{2\pi m}{\Delta L}$. The inverse of this is the discrete Fourier sum:

$$\rho(x_g) = \frac{1}{\Delta L} \sum_{k = \frac{2\pi m}{\Delta L}} \rho(k)e^{ikx_g} . \tag{4.11}$$

In general, we obtain the relation

$$\rho(k) = S(k) \sum_{n=0}^{\infty} \sum_g \left[\sum_{i \in g} \frac{(-ik\Delta x_i)^n}{n!} e^{-ikx_g} \right] , \tag{4.12}$$

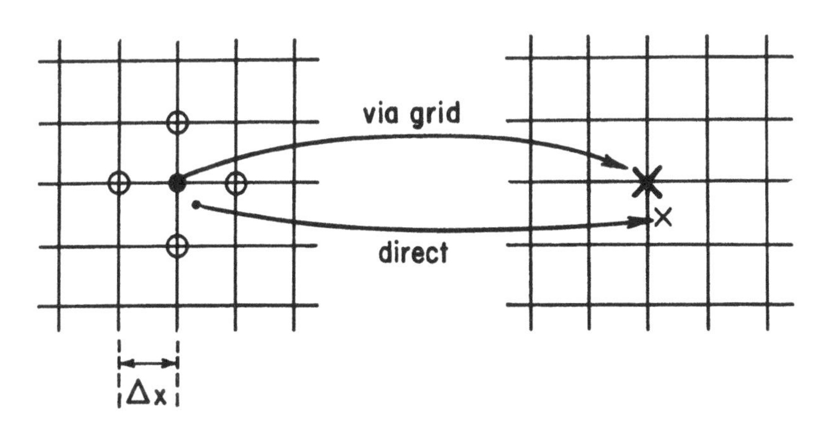

FIGURE 4.1 Interaction between particles and that via grid

where the outer summation is over the multipole orders, the middle sum is over the grid, and the inner summation is over all the particles that are closest to grid point g. The nearest grid point method sums over only the zeroth multipole term. In the *dipole approximation*, we truncate the expansion of the second ($n = 1$) term

$$\begin{aligned} \rho(x) &= \sum_k \left[\sum_{i \in g} S(x - x_g) + \sum_{i \in g} (x_i - x_g) \cdot \nabla_g S(x - x_g) \right] \\ &= \sum_k \left[\rho_{\mathrm{NGP}}(x_g) + D(x_g) \cdot \nabla_g \right] S(x - x_g) \ , \end{aligned} \tag{4.13}$$

where $\rho_{\mathrm{NGP}} = \sum_{i \in g} 1$, is the nearest grid point contribution and

$$D(x_g) \equiv \sum_{i \in g} (x_i - x_g) \ . \tag{4.14}$$

Similarly, the force acting on the ith particle is written,

$$F(x_i) = e \int_{-\infty}^{\infty} E(x) S(x_i - x) dx = e \int E(x_i + x) S(x) dx \ . \tag{4.15}$$

In the dipole approximation (again expanding x_i around x_g) ,

$$F(x_i) = \int S(x)[E(x + x_g) + (x_i - x_g) \cdot \nabla_g E(x + x_g)]dx \tag{4.16}$$

or

$$\simeq \sum_{x'_g} S(x'_g)[E(x'_g + x_g) + (x_i - x_g) \cdot \nabla_g E(x'_g + x_g)] , \tag{4.17}$$

where again we take $x = x'_g$ only.

In order to calculate Eq. (4.7), the scheme called the subtracted dipole scheme (SUD or SUDS) was introduced.[2] The subtracted dipole scheme works as follows. Using the grid as before, we calculate the effective field more precisely by using a dipole approximation to reduce the grid error

$$\rho(x) = \sum_g \sum_{i \in g} [S(x - x_g) + (x_i - x_g)S'(x - x_g) + O(x_i - x_g)^2] .$$

The derivative of the shape factor can be approximated as

$$S'(x - x_g) = -[S(x - x_{g+1}) - S(x - x_{g-1})]/2\Delta + O(\Delta^2) . \tag{4.18}$$

Then

$$\begin{aligned} \rho(x) &= \sum_g \sum_{i=g} S(x - x_g) - \sum_{i \in g} (x_i - x_g)[S(x - x_{g+1}) - S(x - x_{g-1})]/2\Delta \\ &= \sum_g S(x - x_g) \left[\sum_{i \in g} 1 - \sum_{i \in g} \frac{(x_i - x_{g-1})}{2\Delta} + \sum_{i \in g+1} \frac{(x_i - x_{g+1})}{2\Delta} \right] \\ &= \sum_g S(x - x_g)\rho_{\mathrm{NGP}}(x_g) + \frac{1}{2\Delta}[D(x_g + 1) - D(x_g - 1)] . \end{aligned} \tag{4.19}$$

The first term represents the nearest grid point charge (charge accumulated at its nearest grid point). The term $D(x_g + 1)$ in the second term in the curly bracket corresponds to a dipole moment over the distance $(x_i - x_{g+1})$ with charge of $\frac{1}{2}\Delta$. This is equivalent to having a dipole moment over a distance of 2Δ with charge $(x_i - x_{g+1})$. More discussion of this may be found in Ref. 3.

4.2 Area Weighting Scheme

The area weighting scheme assigns a fraction of charge of a particle to respective grid points according to the relative weight which the particle distributes

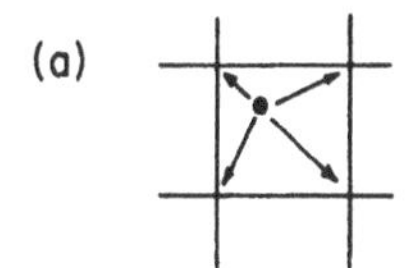

FIGURE 4.2 Area weighted interpolation of particle quantities to grid and vice versa

to the adjacent grid points. The simplest illustration may be given in the one-dimensional case. See Fig. 4.2. The jth particle at $x = x_j$ accumulates a fraction of charge $q(1 - dx)$ to the left grid point $x = i\Delta$ and charge $q_j dx$ to the right $x = (i+1)\Delta$, where q_j is the jth particle's charge and dx is the distance of the jth particle from the ith grid point $x_j - i\Delta$. After calculating the forces on the grid from the information gathered from the charge on the grid, we assign the force quantity to the particle in a similar way. The force given on $x = i\Delta$ and $x = (i+1)\Delta$ is interpolated in the same formula as the particle position $x = x_j$:

$$F(x_j) \cong (1 - dx)F_i + dx F_{i+1} \ , \tag{4.20}$$

where F_i and F_{i+1} are the forces given at the grid positions $x = i\Delta$ and $x = (i+1)\Delta$. Compare this with the nearest grid point method where dx is neglected, and with the subtracted dipole scheme where $F(x_j) \cong -\frac{dx}{2}F_{i-1} + F_i + \frac{dx}{2}F_{i+1}$.

In the two-dimensional case, this treatment can be generalized in a similar way. One way to calculate the weighting in two dimensions is to Taylor expand the function $f(x, y)$ to be assigned onto the grid:

$$\begin{aligned} f(x,y) &= f(x_i, y_i) + (x - x_i)\frac{\partial f}{\partial x} + (y - y_i)\frac{\partial f}{\partial y} \\ &+ \frac{1}{2!}\left[(x - x_i)\frac{2\partial^2 f}{\partial x^2} + 2(x - x_i)(y - y_i)\frac{\partial^2 f}{\partial x \partial y}\right. \\ &\left. +(y - y_i)^2\frac{\partial^2 f}{\partial y^2}\right] + \dots \ , \end{aligned} \tag{4.21}$$

where (x_i, y_i) is to the left lower corner of the cell in which the particle (x, y) is located, and we write $x = x_i + \Delta x$ and $y = y_i + \Delta y$. Then we have

$$f(x_i + \Delta x, y_i + \Delta y) \cong f(x_i, y_i) + \Delta x[f(x_{i+1}, y_i) - f(x_i, y_i)]$$

$$+\Delta y[f(x_i, y_{i+1}) - f(x_i, y_i)] + \Delta x \Delta y \frac{\partial}{\partial x}[f(x_i, y_{i+1}) - f(x_i, y_i)]$$

$$\cong f(x_i, y_i) + \Delta x[f(x_{i+1}, y_{i+1}) - f(x_i, y_{i+1}) - f(x_{i+1}, y_i)$$

$$+f(x_i, y_i)] \; . \tag{4.22}$$

Rearranging this, we obtain

$$f(x_i + \Delta x, y_i + \Delta y) = f(x_i, y_i)(1 - \Delta x - \Delta y + \Delta x \Delta y)$$

$$+f(x_{i+1}, y_i)\Delta x(1 - \Delta y)$$

$$+f(x_i, y_{i+1})\Delta y(1 - \Delta x) + f(x_{i+1}, y_{i+1})\Delta x \Delta y$$

$$= f(x_i, y_i)(1 - \Delta x)(1 - \Delta y) + f(x_{i+1}, y_i)\Delta x(1 - \Delta y)$$

$$+f(x_i, y_{i+1})\Delta y(1 - \Delta x) + f(x_{i+1}, y_{i+1})\Delta x \Delta y \; . \tag{4.23}$$

This result is essentially identical to Eq. (4.13) in one dimension if we neglect the y-dependence in Eq. (4.17).

A more general treatment of interpolation is derived by introducing the requirements for smoothness in the weighting function $w_j(x)$, where $w_j(x)$ represents the fraction of the charge of a particle at the position x assigned to the grid j. More discussion on this may be found in Ref. 4. If the nearest grid point is $j = 0$, the three-point interpolation assignment scheme may be described by quantities $w_{-1}(x)$, $w_0(x)$, and $w_1(x)$. The total charge should be constant and normalized:

$$w_{-1}(x) + w_0(x) + w_1(x) = 1 \; . \tag{4.24}$$

If we assume $w(x)$ is a polynomial and quadratic in x and consider its symmetry, we can generally write down

$$w_0(x) = ax^2 + b \tag{4.25}$$

and

$$w_{-1}(-x) = w_1(x) = cx^2 + dx + e \; . \tag{4.26}$$

The smoothness requirements demand that the amounts of charge assigned to each grid point $w(x)$ and their derivatives $w'(x)$, $w''(x)$ etc., should vary continuously as x varies. Since Eq. (4.25) and (4.26) are continuous as long as x belongs to one cell, we should only check the smoothness on the boundary $x = \pm\Delta/2$. For example,

$$w_{-1}(\Delta/2) = 0, \quad w_0(\Delta/2) = w_1(\Delta/2) \ , \tag{4.27}$$

and derivatives

$$w'_{-1}(\Delta/2) = 0, \quad w'_0(\Delta/2) = -w'_1(\Delta/2) \ . \tag{4.28}$$

Using these requirements, we determine the polynomials as

$$w_0(x) = \frac{3}{4} - \left(\frac{x}{\Delta}\right)^2 \ , \tag{4.29}$$

and

$$w_{-1}(-x) = w_1(x) = \frac{1}{2}\left(\frac{1}{2} + \frac{x}{\Delta}\right)^2 \ . \tag{4.30}$$

One can increase the order of polynomials (i.e., the number of freedoms) and impose more continuity in the higher order derivatives of these polynomials. It turns out, however, that more than third order interpolation does not bring much improvement in terms of the stability against the thermal instability when we reduce the number λ_{De}/Δ. This is cured only by resorting to the spline fitting in the exponential function.[5,6] This development is detailed in Section 4.6.

4.3 Examples of Electrostatic Codes

Let us look at the code which employs the Fourier transform in order to calculate the electric field or potential using Poisson's equation. Look at Table 4.1.

Starting from the collection of information of all the particle positions at time step 0, we accumulate the charge on the grid $\rho(x_g)$. We transform $\rho(x_g)$ by a discrete Fourier transform (see Section 4.1) into k-space as $\rho(k_m)$:

$$\rho(k_m) = \sum_{x_g} \rho(x_g) e^{-ik_m x_g} \ , \tag{4.31}$$

with $k_m = 2\pi m/L$ and $m = 0, 1, \ldots, L/2$. The inverse Fourier transform of Eq. (4.31) is

$$\rho(x_g) = \frac{1}{2\pi} L \sum_{k_{m=0}}^{L/2} \rho(k_m) e^{ik_m x_g} \ . \tag{4.32}$$

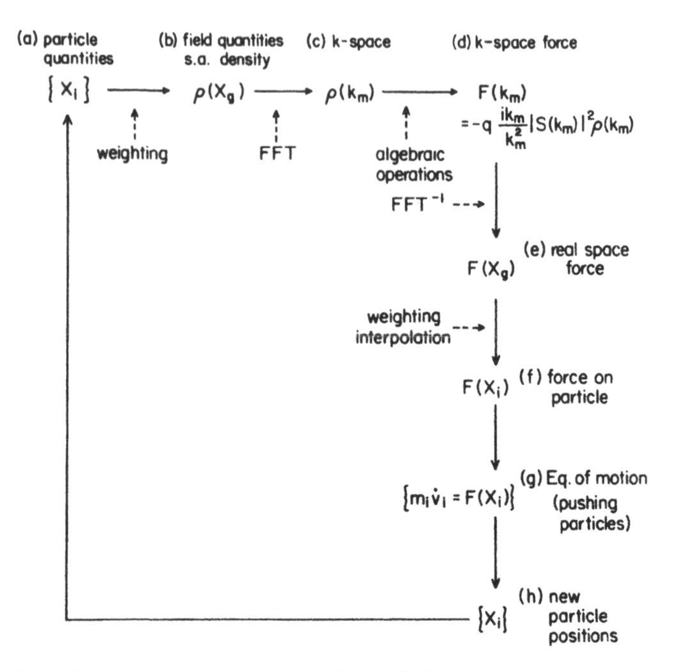

TABLE 4.1 A flow chart of an electrostatic particle code

Since the spatial points are discrete, Eq. (4.31) is a finite sum instead of an integral, and Eq. (4.32) is a finite sum instead of an infinite sum. Algebraic operations in k-space yield the force on the k-grid as

$$F(k_m) = -q\frac{i\mathbf{k}_m}{k_m^2}\left|S(k_m)\right|^2 \rho(k_m).$$

The Fourier inverse transform of this gives the force in real space on the grid. The force at the particle position is determined by the interpolation. The particle pushing is then carried out and new velocities and positions are found. This cycle is repeated many times. Also see Fig. 4.3.

If we want to determine the potential by finite differencing in real space instead of using the Fourier transform, we often employ the method of successive overrelaxation. The electric potential $\phi_{i,j}^{n+1}$ is the nth iterated value at the grid point (i, j) (two dimensional grid):

$$\phi_{i,j}^{n+1} = \phi_{i,j}^{n} + \frac{\alpha}{4}[\phi_{i+1,j}^{n} + \phi_{i,j+1}^{n} + \phi_{i-1,j}^{n+1} + \phi_{i,j}^{n+1} + \rho_{i,j}^{n} - 4\phi_{i,j}^{n}]\,. \quad (4.33)$$

When α is larger than 2, the scheme is unstable. The scheme $\alpha = 1$ corresponds to the Gauss-Seidel scheme. The optimum α is often determined

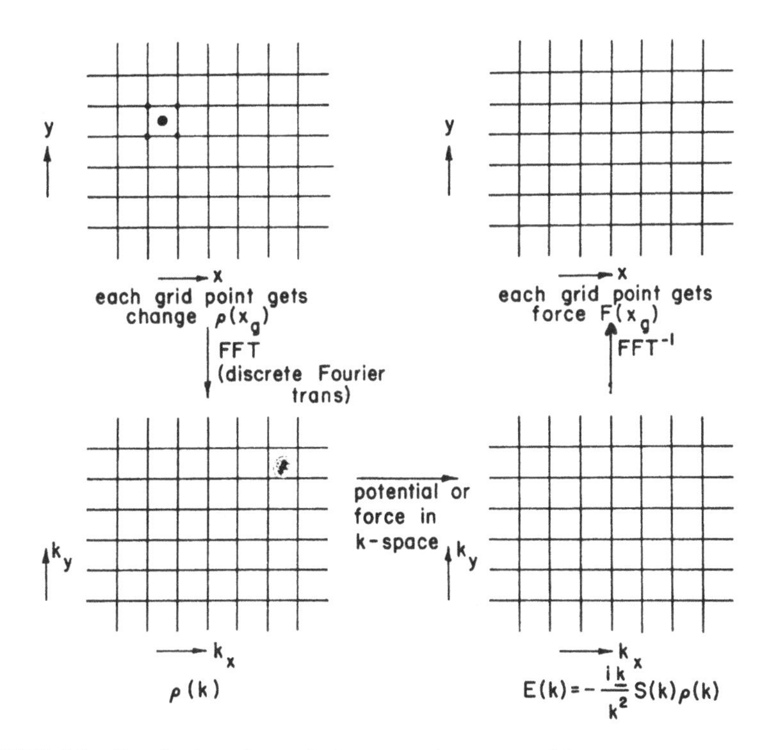

FIGURE 4.3 Fourier transforms between real space and **k**-space

empirically or $\alpha \simeq 2\left[1+\sin(2\pi/L_x)\right]^{-1}$.

4.4 Spatially Periodic Systems

A simulation plasma with a spatial grid is an example of a class of systems containing a periodic spatial nonuniformity. Another such example that bears resemblance to our problem is the behavior of an electron in a crystal lattice. The energy of a free electron in space can be written

$$\varepsilon = \frac{p^2}{2m} \propto k^2 \tag{4.34}$$

where

$$p = \hbar k \ .$$

A graph of ε as a function of k is shown in Fig. 4.4(a). This relation is changed if the electron is in the presence of a periodic potential such as a crystal lattice.[7] As can be seen in Fig. 4.4(b), the energy of a free electron cannot assume any value but is restricted to certain ranges that are known as conduction bands. The regions separating the conduction bands are known as forbidden bands. The discrete jumps in energy occur at wavenumbers k such that $k = \frac{(2n+1)\pi}{\Delta}$ where $n = 0, \pm 1, \pm 2 \ldots$, and where Δ is the period of the potential. The regions between the discontinuities are called Brillouin zones. The energy spectrum of an electron in a lattice has the same general shape as the energy spectrum of a free electron. However, for the spectrum of an electron in a crystal lattice, we notice that the curvature inverts at the edges of the Brillouin zones. Similar effects occur in our computational lattice, the spatial grid. Some of the representative pioneering works in this problem are Refs. 1 and 8.

Let us consider an interaction force $F(x, x')$ defined as the force on a particle at x due to a particle at x'. In normal physical systems the interaction force depends only on the distance between the particles $\delta x = x - x'$. In the presence of a spatial grid, however, $F(x, x')$ also depends on the average distance between the particles $\bar{x} = \frac{1}{2}(x + x')$ as well as δx. If we shift the grid while holding δx fixed as shown in Fig. 4.5, we generally obtain a difference value for $F(x, x')$. The translational invariance of the interaction force is destroyed. The dependence of $F(x, x')$ on the average positions of the two particles is the source of nonphysical mode coupling.

Since most simulations have a constant spacing between adjacent grid points, we will assume that the grid points are equally spaced by a distance Δ. In this case, $F(\bar{x} - \frac{1}{2}\delta x,\ \bar{x} + \frac{1}{2}\delta x)$ is periodic with period Δ if δx is held constant. For an infinite system, $F(\bar{x} - \frac{1}{2}\delta x,\ \bar{x} + \frac{1}{2}\delta x)$ can be expanded in a complex Fourier series in the following way

$$F\left(\bar{x} - \frac{1}{2}\delta x, \bar{x} + \frac{1}{2}\delta x\right) = \sum_{p=-\infty}^{\infty} F_p(\delta x) \exp(ipk_g \bar{x}) \ , \tag{4.35}$$

where p is an integer and where

$$F_p(\delta x) = \frac{1}{\Delta} \int_{\Delta} F\left(\bar{x} - \frac{1}{2}\delta x,\ \bar{x} + \frac{1}{2}\delta x\right) \exp(-ipk_g \bar{x}) dx \ . \tag{4.36}$$

If δx is allowed to vary, we can use a Fourier integral representation for $F(\bar{x} - \frac{1}{2}\delta x,\ \bar{x} + \frac{1}{2}\delta x)$ in terms of δx

$$F\left(\bar{x} - \frac{1}{2}\delta x, \bar{x} + \frac{1}{2}\delta x\right) = \int_{-\infty}^{\infty} \frac{dk}{2\pi} \exp(ik\delta x) F(\bar{x}, k) \ ,$$

$$F\left(\bar{x} - \frac{1}{2}\delta x, \bar{x} + \frac{1}{2}\delta x\right)$$

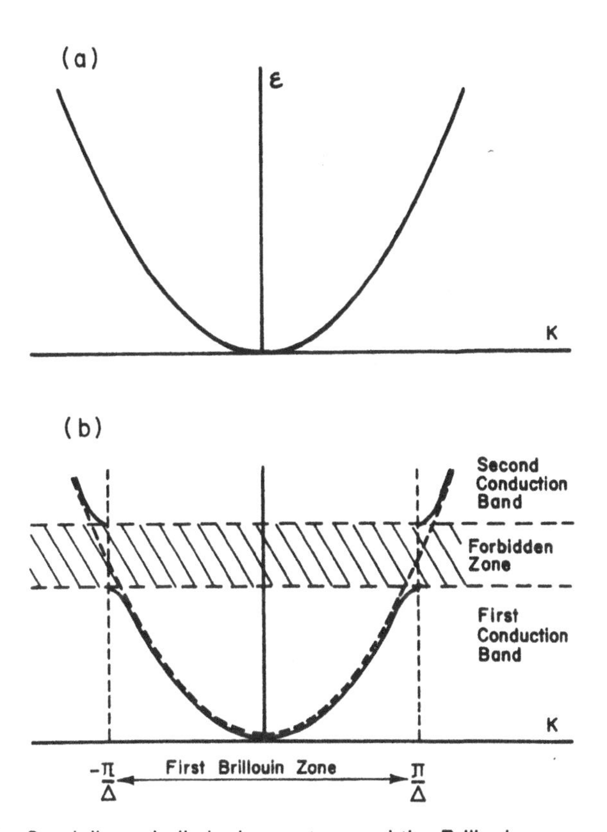

FIGURE 4.4 Spatially periodic lattice system and the Brillouin zone

$$= \int_{-\infty}^{\infty} \frac{dk}{2\pi} \exp(ik\delta x) \sum_{p=-\infty}^{\infty} \exp(ipk_g\bar{x})F_p(k) \ , \tag{4.37}$$

where

$$F_p(k) \equiv \int_{-\infty}^{\infty} d\delta x F_p(\delta x) \exp(-ik\delta x) \ . \tag{4.38}$$

The only Fourier component of Eq. (4.37) that exists in a gridless system is the $p = 0$ component. We then expect that the $p = 0$ component of the force should represent the two particle interaction force of the gridless system. The $p \neq 0$ components are expected to contribute a spurious force.

We can extend this concept from two point particles to a system with a continuous particle distribution.[1] For a particle density $n(x)$, the force $F(x)$

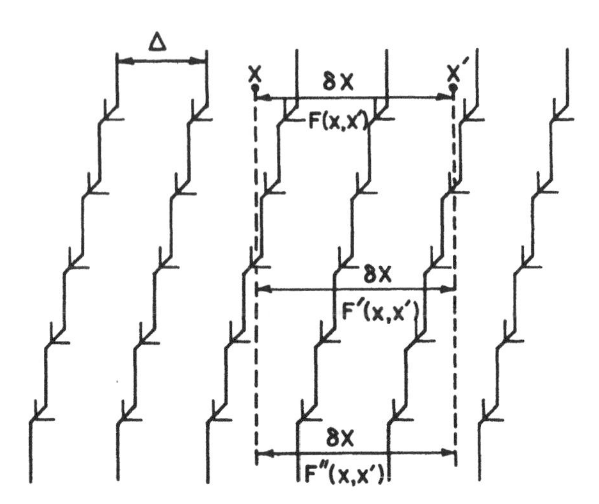

FIGURE 4.5 Translation of grid and interaction between two points

on a particle at x is given by

$$F(x) = \int_{-\infty}^{\infty} F(x, x')n(x')dx' \ . \tag{4.39}$$

The Fourier integral transform of $F(x)$ is given by

$$F(k) = \int dx e^{-ikx} F(x) = \int dx e^{-ikx} \int dx' F(x, x')n(x') \ . \tag{4.40}$$

If we use Eq. (4.37) for $F\left(\bar{x} - \frac{1}{2}\delta x, \bar{x} + \frac{1}{2}\delta x\right)$ and expand $n(x')$ in a Fourier integral, then the expression for $F(k)$ in Eq. (4.40) becomes

$$F(k) = 2 \int d\bar{x} \int d\delta x \exp\left[-ik\left(\bar{x} - \frac{1}{2}\delta x\right)\right] \int_{-\infty}^{\infty} \frac{dk'}{2\pi} \exp(ik'\delta x)$$

$$\times \sum_{p=-\infty}^{\infty} \exp(ipk_g\bar{x})F_p(k') \int_{-\infty}^{\infty} \frac{dk''}{2\pi} \exp\left[ik''\left(\bar{x} + \frac{1}{2}\delta x\right)\right] n(k'') \ . \tag{4.41}$$

The order of integration in Eq. (4.41) can be rearranged such that the integration over δx can be performed

$$F(k) = 2 \int_{-\infty}^{\infty} \frac{dk'}{2\pi} \int_{-\infty}^{\infty} \frac{dk''}{2\pi} \int d\bar{x} \exp(-ik\bar{x})n(k'') \exp(ik''\bar{x})$$

$$\times \sum_{p=-\infty} \exp(ipk_g\bar{x})F_p(k') \int_{-\infty}^{\infty} d\delta x \exp\left(\frac{1}{2}k\delta x\right) \exp(ik'\delta x) \exp\left(\frac{1}{2}k''\delta x\right)$$

$$F(k) = 2\int_{-\infty}^{\infty} \frac{dk'}{2\pi} \int_{-\infty}^{\infty} \frac{dk''}{2\pi} \int d\bar{x} \exp(-ik\bar{x}) n(k'') \exp(ik''\bar{x})$$

$$\times \sum_{p=-\infty}^{\infty} \exp(ipk_g\bar{x})F_p(k') 2\pi\delta\left(k' + \frac{k}{2} + \frac{k''}{2}\right) .$$

The integral over k'' is eliminated by the δ function such that

$$F(k) = 2\int_{-\infty}^{\infty} \frac{dk'}{2\pi} \int d\bar{x} \exp(-ikx) n(-2k' - k) \exp[-i(2k' + k)\bar{x}]$$

$$\times \sum_{p=-\infty}^{\infty} \exp(ipk_g\bar{x})F_p(k') . \tag{4.42}$$

We then repeat the procedure for the integral over $\bar{x}$ to obtain

$$F(k) = \sum_{p=-\infty}^{\infty} 2\int \frac{dk'}{2\pi} n(-2k' - k)F_p(k') 2\pi\delta\left(k + k' - \frac{p}{2}k_g\right) ,$$

$$F(k) = \sum_{p=-\infty}^{\infty} n(k - pk_g)F_p\left(\frac{p}{2}k_g - k\right) , \tag{4.43}$$

which can also be written

$$F(k) = \sum_{p=-\infty}^{\infty} n(k_p)F_p\left(k - pk_B\right) , \tag{4.44}$$

where $k_p = k - pk_g$ and $k_B = \frac{k_g}{2} = \frac{\pi}{\Delta}$. We can see that the effect of the grid is to couple density and force perturbations that differ by integral numbers of the grid wavenumber k_g. If the system were gridless, again only the $p = 0$ term of Eq. (4.44) would remain. For the gridless case, we would get

$$F(k) = n(k)F_{p=0}(k) . \tag{4.45}$$

The other $p \neq 0$ contributions to $F(k)$ in Eq. (4.44) are called aliases. The grid cannot properly resolve k-components whose wavenumbers are larger than the grid wavenumber. The grid is unable to distinguish between the principal harmonic k and its aliases, resulting in a nonphysical mode coupling between k and its aliases. An example of this coupling for the $p = 1$ alias is shown in Fig. 4.6. The shorter wavelength $p = 1$ alias is interpreted by the grid as the longer wavelength $p = 0$ mode. The problem of alias coupling will be mentioned again in Section 6.8.

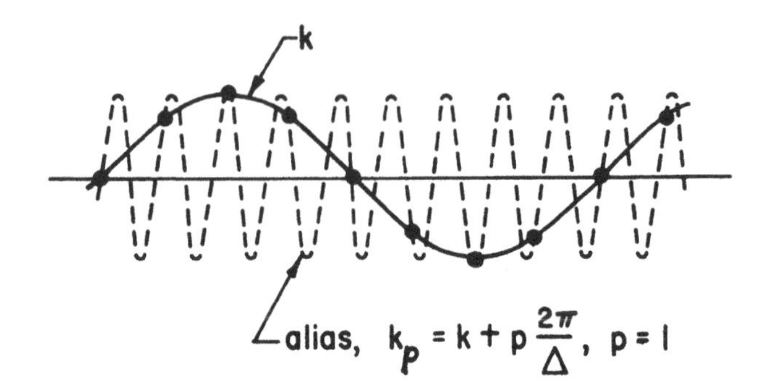

FIGURE 4.6 Indistinguishable aliases in a grid system

4.5 Consequences of the Grid for the Vlasov Theory of Plasmas

The importance of this spurious mode coupling in the linear regime of an electrostatic plasma model will be investigated in this section. For a system of finite size particles we found that the linearized Vlasov equation has the form

$$-i(\omega - kv)\delta f(k,\omega,v) + \frac{n_0 F(k,\omega)}{m}\frac{\partial f_0(v)}{\partial v} = 0 , \tag{4.46}$$

where δf is the perturbed distribution function, n_0 is the unperturbed density and f_0 is the unperturbed velocity distribution function. The perturbed density $n(k,\omega)$ is obtained by the integration of Eq. (4.46) over velocity space

$$n(k,\omega) = \int \delta f(k,\omega,v)dv,$$

which becomes

$$n(k,\omega) = F(k,\omega)\Phi(k,\omega) \tag{4.47}$$

where

$$\Phi(k,\omega) = \frac{n_0}{im}\int \frac{1}{\omega - kv}\frac{\partial f_0}{\partial v}dv . \tag{4.48}$$

We can combine Eqs. (4.44) and (4.47) to yield

$$n(k,\omega) \quad = \quad \Phi(k,\omega)\sum_{p=-\infty}^{\infty} F_p\left(k - \frac{1}{2}pk_g\right) n(k_p,\omega) ,$$

$$n(k,\omega) = \Phi(k,\omega) \sum_{p=-\infty}^{\infty} F_p\left(k - \frac{1}{2}pk_g\right) F(k_p,\omega)\Phi(k_p,\omega) , \tag{4.49}$$

or alternatively using Eq. (4.47),

$$F(k,\omega) = \sum_{p=-\infty}^{\infty} F_p\left(k - \frac{1}{2}pk_g\right) F(k_p,\omega)\Phi(k_p,\omega) . \tag{4.50}$$

The electric field $E(x_j) = E_j$ is calculated at the grid points, whose position is given by $x_j = j\Delta$. The Fourier transform of the electric field is

$$E(k) = \Delta \sum_{j=-\infty}^{\infty} E_j \exp(-ikx_j) , \tag{4.51}$$

with inverse transform

$$E(x) = \int_{k_g} \frac{dk}{2\pi} E(k) \exp(ikx_j) , \tag{4.52}$$

where the integral is over the range $k_g = \frac{2\pi}{\Delta}$. For any integer p we have $\exp(ik_p x_j) = \exp(ikx_j)$ such that $E(k_p) = E(k)$. This is another example of how the spatial grid cannot differentiate between the contributions of the principal harmonic and its aliases, as seen in Fig. 4.6.

The charge density ρ_j has the same type of Fourier expansion as the electric field as do all quantities that are defined at the grid points

$$\rho(k) = \Delta \sum_{j=-\infty}^{\infty} \rho_j \exp(-ikx_p) , \tag{4.53}$$

and

$$\rho_j = \int_{k_g} \frac{dk}{2\pi} \rho(k) \exp(ikx_j) . \tag{4.54}$$

During the discussion about finite size particle effects (Chapter 2), it was shown that the charge density is given by

$$\rho(x_p) = q \int n(x')S(x_p - x')dx' ,$$

where $S(x)$ is the particle shape for a finite size particle whose center is located at $x = 0$ and where q is the charge of a finite size particle. If the Fourier integrals of $n(x')$ and $S(x_p - x')$ are substituted into the expression above, we obtain

$$\begin{aligned} \rho(x_p) &= q \int_{-\infty}^{\infty} \int_{-\infty}^{\infty} n(k') \exp(ik'x') \frac{dk'}{2\pi} \\ &\times \int_{-\infty}^{\infty} S(k'') \exp[ik''(x_p - x')] \frac{dk''}{2\pi} dx' . \end{aligned}$$

Evaluation of the integral over x' yields

$$\rho(x_p) = q\int_{-\infty}^{\infty}\frac{dk'}{2\pi}\int_{-\infty}^{\infty}\frac{dk''}{2\pi}n(k')S(k'')\exp(ik''x_p)2\pi\delta(k'-k'') ,$$

or

$$\rho(x_p) = q\int_{-\infty}^{\infty}\frac{dk'}{2\pi}n(k')S(k')\exp(ik'x_p) .$$

From the periodicity of $\exp(ikx_p)$ we can re-express the integral as

$$\rho(x_p) = q\int_{k_g}\frac{dk'}{2\pi}\exp(ik'x_p)\sum_{n=-\infty}^{\infty}n(k'-nk_g)S(k'-nk_g) . \tag{4.55}$$

A comparison of Eqs. (4.54) and (4.55) shows that

$$\rho(k) = q\sum_{p=-\infty}^{\infty}n(k_p)S(k_p) . \tag{4.56}$$

The electric field is obtained from solving Poisson's equation. In the gridless case we have

$$\frac{-\partial^2\Phi}{\partial x^2} = 4\pi\rho , \tag{4.57}$$

$$E = -\frac{\partial\Phi}{\partial x} . \tag{4.58}$$

In the codes these equations take the form

$$K^2\Phi = 4\pi\rho , \tag{4.59}$$

$$E = -i\kappa\Phi , \tag{4.60}$$

where $K^2(k)$ and $\kappa(k)$ respectively represent the equivalent Laplacian and gradient operators in the grid system. If we solve Eqs. (4.51) and (4.52) in Fourier space, these operators are then $K^2 = k^2$ and $\kappa = k$, which is exact.

An alternative method for solving Eqs. (4.57) and (4.58) involves the use of a finite-difference scheme. There are three simple finite differencing schemes that may be applied to solving Eq. (4.58).

A. Centered Differencing

$$E_j = -\frac{\Phi_{j+1}-\Phi_{j-1}}{2\Delta} .$$

B. Forward Differencing

$$E_j = -\frac{\Phi_{j+1} - \Phi_j}{\Delta} .$$

C. Backward Differencing

$$E_j = -\frac{\Phi_j - \Phi_{j-1}}{2\Delta} .$$

When a differencing scheme is used, most simulations use the centered differencing. Application of the centered differencing scheme to Eqs. (4.57) and (4.58) yields

$$-4\pi\rho_j = \frac{\Phi_{j+1} - 2\Phi_j + \Phi_{j-1}}{\Delta^2} , \tag{4.61}$$

$$E_j = -\frac{\Phi_{j+1} - \Phi_{j-1}}{2\Delta} . \tag{4.62}$$

If we assume all quantities vary as $f_j \propto f_k \exp(ikx_j)$, then Eq. (4.61) becomes

$$-4\pi\rho_k \exp(ikx_j) = \Phi_k \frac{\exp[ik(x_j + \Delta)] - 2\exp(ikx_j) + \exp[ik(x_j - \Delta)]}{\Delta^2}$$

which reduces to

$$-4\pi\rho_k = \Phi_k k^2 \left(\frac{\sin\left(\frac{1}{2}k\Delta\right)}{\frac{1}{2}k\Delta} \right)^2 . \tag{4.63}$$

A similar analysis for Eq. (4.62) yields

$$E_k \exp(ikx_j) = -\Phi_k \frac{\exp[ik(x_j + \Delta)] - \exp[ik(x_j - \Delta)]}{2\Delta}$$

which reduces to

$$E_k = -ik \left[\frac{\sin(k\Delta)}{k\Delta} \right] \Phi_k . \tag{4.64}$$

A comparison of Eqs. (4.63) and (4.64) with Eqs. (4.59) and (4.60) shows that for the centered finite difference scheme

$$K^2(k) = k^2 \left[\frac{\sin\left(\frac{1}{2}k\Delta\right)}{\frac{1}{2}k\Delta} \right]^2 , \tag{4.65}$$

and

$$\kappa(k) = \frac{\sin(k\Delta)}{k\Delta} . \tag{4.66}$$

A plot comparing K^2 from the finite difference scheme with $K^2(k)$ from the Fourier space scheme is shown in Fig. 4.7. For small values of k, both schemes

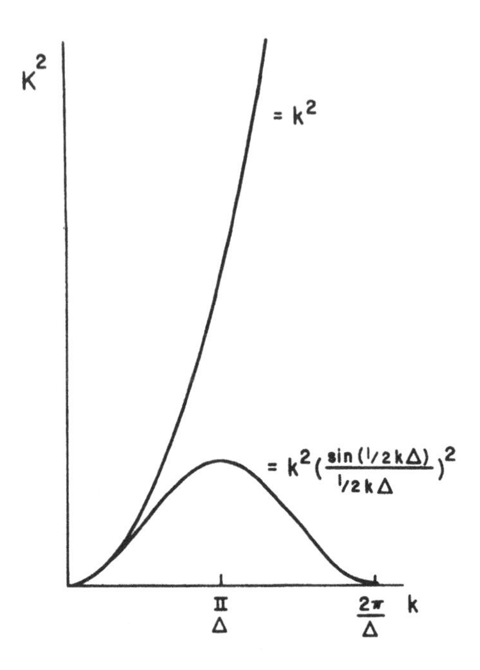

FIGURE 4.7 The finite differencing operator $K^2(k)$ vs. k^2

give approximately the same results. The finite difference scheme introduces additional attenuation for large values of k.

Either scheme can be used to derive the linearized electrostatic dispersion from Vlasov theory including both spatial grid and finite size particle effects. The combination of Eqs. (4.59) and (4.60) yields the following expression for $E(k)$

$$E(k) = -i\frac{4\pi q\kappa(k)}{K^2}\rho(k) . \tag{4.67}$$

If we assume the electric field and charge density $\propto \exp(-i\omega t)$, then using Eq. (4.56) for $\rho(k)$ we obtain

$$E(k,\omega) = -i\frac{4\pi q\kappa(k)}{K^2}\sum_{p=-\infty}^{\infty} S(k_p)n(k_p,\omega) .$$

The expression for the perturbed density was obtained from the Vlasov equation in Eq. (4.47). Usage of Eq. (4.47) in the above expression gives

$$E(k,\omega) = -i\frac{4\pi q\kappa(k)}{K^2}\sum_{p=-\infty}^{\infty} S(k_p)F(k_p,\omega)\Phi(k_p,\omega) . \tag{4.68}$$

It was shown that the force equation for finite size particles is

$$F(k) = qS(-k)E(k) \ . \tag{4.69}$$

When this force is substituted into Eq. (4.68), then

$$E(k,\omega) = -i\frac{4\pi q^2 \kappa(k)}{K^2} \sum_{p=-\infty}^{\infty} |S(k_p)|^2 E(k_p,\omega)\Phi(k_p,\omega) \ . \tag{4.70}$$

We proved earlier that $E(k,\omega)$ is periodic in k_g. The dispersion function is then obtained by dividing Eq. (4.70) by $E(k,\omega)$ to give

$$D(k,\omega) = 1 - i\frac{4\pi q^2 \kappa(k)}{K^2} \sum_{p=-\infty}^{\infty} |S(k_p)|^2 \Phi(k_p,\omega) \ . \tag{4.71}$$

The normal modes of this computational plasma system are determined from Eq. (4.71) when $D(k,\omega) = 0$. There are an infinite number of k-modes separated by integral multiples of k_g for a given $\omega(k)$. The grid imposes alias contributions to oscillations that otherwise would not be present.

For a Maxwellian distribution function in a plasma without equilibrium electric and magnetic fields, the evaluation of $\Phi(k_p,\omega)$ yields

$$D(k,\omega) = 1 - \frac{\omega_p^2 \kappa(k)}{2K^2(k)} \sum_{p=-\infty}^{\infty} |S(k_p)|^2 \frac{1}{k_p} Z'\left(\frac{\omega}{|k_p| v_e \sqrt{2}}\right) , \tag{4.72}$$

where Z is the Fried-Conte plasma dispersion function given by

$$Z(z) = \frac{1}{\sqrt{\pi}} \int_{-\infty}^{\infty} \frac{\exp(-x^2)}{x - z} dx \ .$$

In comparison the gridless case was shown to be

$$D(k,\omega) = 1 - \frac{\omega_p^2}{2k^2} |S(k)|^2 Z'\left(\frac{\omega}{|k| v_e \sqrt{2}}\right) \ . \tag{4.73}$$

The reason for the requirement that the width of the shape factor be about the same size as or larger than the grid spacing is apparent from Eq. (4.64) and from a typical plot of $S(k)$ in Fig. 4.8. The factor $|S(k)|^2$ greatly attenuates the aliasing terms in Eq. (4.72), so the finite size particle shape acts essentially like a low-pass wavenumber filter. In position space the charge of a finite size particle is smeared out such that density variations over regions smaller than the particle are difficult to resolve.

The reason for the requirement that the grid spacing be about the same size as, or larger than λ_D, can also be understood from Eq. (4.72). The derivative, Z', of the Fried-Conte dispersion function is much smaller for the $p \neq 0$

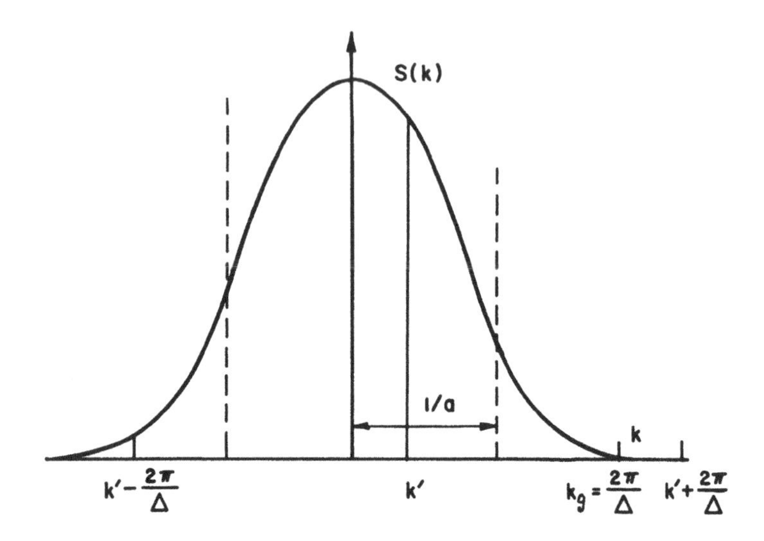

FIGURE 4.8 The particle form factor and the alias location

terms than for the $p = 0$ term if the argument is large i.e., $\lambda_D \geq \Delta$. If $\lambda_D \ll \Delta$, a weak numerical instability can develop. This instability has been observed and discussed by Okuda,[8] Langdon,[9] and others[10] to heat up a cold plasma until $\lambda_D \sim \mathcal{O}(\Delta)$.

In summary, the use of a spatial grid causes the loss of information that is present in the continuous system. This information loss manifests itself as the coupling of an oscillation of principal harmonic k to oscillations of harmonics that differ by an integer number p times the grid wavenumber $k_g = \frac{2\pi}{\Delta}$ from the principal harmonic. This will be revisited in Chapter 6 for the pseudo-spectral method. Fortunately, the aliasing effect is usually insignificant for linear plasma oscillations if the grid spacing is about the size of a finite size particle, and if the grid spacing is about the size of the Debye length for explicit time integration simulations. This statement does not apply to magnetized plasma simulation in the directions perpendicular to the magnetic field. We shall see a greatly relaxed condition on this by using different, more sophisticated techniques in Section 4.6 and Chapter 9.

4.6 Smoother Grid Assignment

A grid assignment smoother than ones discussed in Sections 4.1 and 4.2 could reduce the errors and greatly reduce the minimum Debye length that avoids the thermal instability mentioned in Section 4.5.

We revisit the discussion of the grid method and the interpolation of grid quantities to the particle and of the particle quantities to the grid already discussed in Sections 4.1 and 4.2. A typical equation we may want to solve is

$$\nabla \cdot \mathbf{E} = 4\pi q \sum_j \delta(\mathbf{x} - \mathbf{x}_j) \,, \tag{4.74}$$

where x_j is the jth particle position. In Fourier space, Eq. (4.74) takes the form

$$i\mathbf{k} \cdot \mathbf{E}_k = 4\pi q \sum_j e^{-ik\cdot x_j} \,. \tag{4.75}$$

The linear interpolation of function $f(x_j)$ is done as

$$f(x_j) = f(x_g)(1 - \Delta x_j) + f(x_g + 1)\Delta x_j \,, \tag{4.76}$$

where $\Delta x_j = x_j - x_g$. This can be derived by the Taylor expansion

$$\begin{aligned} f(x_j) &= f(x_g + \Delta x_j) = f\left(x_g + \frac{1}{2}\right) + \left(\Delta x_j - \frac{1}{2}\right) \frac{\partial}{\partial x} f(x)\bigg|_{x_g+\frac{1}{2}} + \ldots \\ &\cong \frac{1}{2}[f(x_g) + f(x_g+1)] + \left(\Delta x_j - \frac{1}{2}\right)[f(x_g+1) - f(x_g)] + \cdots \\ &= f(x_g)(1 - \Delta x_j) + f(x_g+1)\Delta x_j \,. \end{aligned}$$

The dipole expansion scheme discussed in Section 4.2 is also revisited here. Since the charge in Fourier space $\rho(k) = \sum_{j\varepsilon g} e^{-ik(x_g+\delta)}$, we consider the exponential function e^{-ikx_g}, where $\delta \equiv \Delta x_j$. The dipole expansion scheme relies on the expansion of the exponential phase factor in terms of powers of δ:

$$\begin{aligned} e^{-ikx_j} &= e^{-ik(x_g+\delta)} \cong e^{-ikx_g}\left[1 - ik\delta + \frac{1}{2}(-ik\delta)^2 + \ldots\right] \\ &\cong e^{-ikx_g}\left[1 + \frac{\delta}{2\Delta}\left(e^{ik\Delta} - e^{ik\Delta}\right)\right] \,, \end{aligned} \tag{4.77}$$

where we have retained only the first power of δ. Equation (4.77) is rewritten as

$$e^{-ikx_j} \cong e^{-ikx_g} + \frac{\delta}{2\Delta} e^{-ik(x_g+\Delta)} - \frac{\delta}{2\Delta} e^{-ik(x_g-\Delta)} \,. \tag{4.78}$$

This implies that we assign a unity charge to the grid x_g, $\frac{\delta}{2}$ charge (to include the grid separation length Δ, $\delta/2\Delta$) to the grid $x_g+\Delta$, and $-\frac{\delta}{2}$ charge to $x_g-\Delta$. This assignment was called the subtracted dipole scheme.

Instead of terminating the power series expansion at power of δ, a smoother interpolation scheme can collect terms up to powers of δ^2.

An expression equivalent to Eq. (4.77) is

$$\begin{aligned} e^{-ikx_j} &\cong e^{-ikx_g}\left(1-ik\delta-\frac{1}{2}k^2\delta^2\right) \simeq e^{-ikx_g}\left[1-\frac{\delta}{2\Delta}\left(e^{ik\Delta}-e^{-ik\Delta}\right)\right. \\ &\left. +\frac{1}{2}\frac{\delta^2}{\Delta^2}\left(e^{ik\Delta}+e^{-ik\Delta}-2\right)\right] . \end{aligned} \tag{4.79}$$

This corresponds to assigning charges to the grid as follows

$$\begin{array}{llll} 1 & -\dfrac{\delta^2}{\Delta^2} & \text{to grid} & x_g \\ \dfrac{1}{2}\dfrac{\delta}{\Delta} & +\dfrac{1}{2}\dfrac{\delta^2}{\Delta^2} & \text{to grid} & x_g+\Delta \\ -\dfrac{1}{2}\dfrac{\delta}{\Delta} & +\dfrac{1}{2}\dfrac{\delta^2}{\Delta^2} & \text{to grid} & x_g-\Delta . \end{array}$$

This is the quadrupole expansion scheme or subtracted quadrupole scheme.

Inversely, the force assignment can be written, for instance, with the subtracted dipole scheme as

$$\begin{aligned} F(x_j) &\simeq \left(1+\delta\frac{\partial}{\partial x}\right) F(x_g) = F(x_g)+\delta\sum_k ikF_k e^{ikx_g} \\ &= \sum_k F_k e^{ikx_g}(1+ik\delta) = \sum_k F_k e^{ikx_g}\left[1+\frac{\delta}{2\Delta}(ik\Delta+ik\Delta)\right] \\ &= \sum_k e^{ikx_g}\left\{F_k+F_k\frac{\delta}{2\Delta}[e^{ik(x_g+\Delta)}-e^{ik(x_g-\Delta)}]\right\} \\ &= F(x_g)+\frac{\delta}{2\Delta}[F(x_g+\Delta)-F(x_g-\Delta)] . \end{aligned} \tag{4.80}$$

As is pointed out in Ref. 5, the code becomes numerically unstable when the Debye length becomes much smaller than the grid length Δ. For the subtracted dipole scheme, the code becomes unstable when $v_{\text{th}} \lesssim 0.1\omega_p\Delta$ (i.e., $\lambda_{De} \lesssim 0.1\Delta$), while in the subtracted quadrupole scheme the limit is

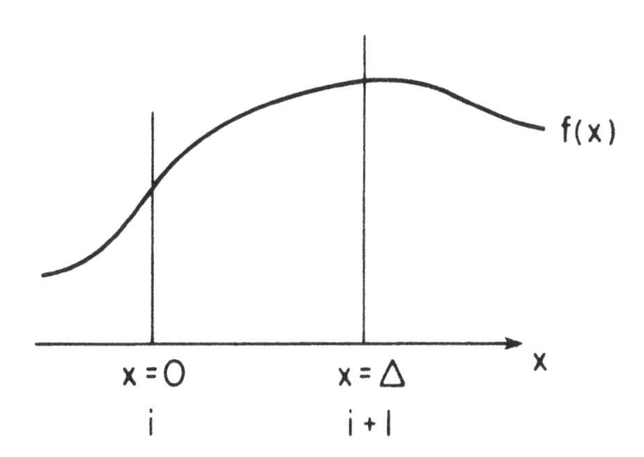

FIGURE 4.9 Function $f(x)$ and the grid

slightly improved. This numerical instability is due to aliasing. As the thermal velocity becomes smaller, higher order aliases begin to couple effectively with the fundamental $k_p(p = 0)$, as can be seen in Eq. (4.72). In order to have substantial coupling, $\omega/|k_p|v_e$ in the argument of Z', has to be finite. This implies that as v_e becomes smaller, $|k_p|$ can become larger to keep $\omega/|k_p|v_e$ finite. The smaller λ_{De} entails more coupling of aliases, and thus numerical instability results.

In order to overcome this difficulty, Okuda and Cheng have introduced[5] the spline method applied to the phase factor Eq. (4.75). With this method (the cubic spline case as we will discuss later) the code does not become numerically unstable until we reach machine accuracy ($\sim 10^{-7}$ on a VAX computer) due to the truncation errors.[6]

We now discuss the spline method. Let us start with the quadratic spline. The problem is to fit a function $f(x)$ by a curve (i.e., a "spline" function)

$$f(x) = a_i + b_i(x - x_i) + c_i(x - x_i)^2 , \tag{4.81}$$

where i designates the ith grid point. See Figure 4.9. Here the third term in Eq. (4.81) is quadratic. To determine the spline function, we demand:

$$\text{continuity at} \quad x = 0 \quad \rightarrow \quad a_i = f(0) , \tag{4.82}$$

$$\text{continuity at} \quad x = \Delta \quad \rightarrow \quad a_i + b_i\Delta + c_i\Delta^2 = a_{i+1} , \tag{4.83}$$

$$\text{derivative at} \quad x = 0 \quad \rightarrow \quad b_i = f'(0) . \tag{4.84}$$

Since we know $a_i = f(0)$ and $b_i = f'(0)$ already, all we need is the coefficient c_i. Therefore, from Eq. (4.83) we obtain

$$c_i = \frac{1}{\Delta^2}(a_{i+1} - a_i - b_i\Delta) . \tag{4.85}$$

Thus we have

$$\begin{aligned} f(x) &\cong a_i + b_i\delta + c_i\delta^2 = a_i + b_i\delta + \frac{\delta^2}{\Delta^2}(a_{i+1} - a_i - b_i\Delta) \\ &= a_i\left(1 - \frac{\delta^2}{\Delta^2}\right) + a_{i+1}\frac{\delta^2}{\Delta^2} + b_i\left(\delta - \frac{\delta^2}{\Delta^2}\right) . \end{aligned} \tag{4.86}$$

Using δ as a new definition for δ/Δ, we now have $0 < \delta < 1$. Equation (4.86) becomes

$$f(x) = a_i(1 - \delta^2) + a_{i+1}\delta^2 + b_i\Delta(\delta - \delta^2) . \tag{4.87}$$

Recall $a_i = f(0)$, $a_{i+1} = f(\Delta)$, and $b_i = f'(0)$.

We are interested in an equation of the type Eq. (4.77). So take $f(x) = e^{-ik(x_g+\delta)}$. Equation (4.87) yields

$$e^{-ik(x_g+\delta)} = e^{-ikx_g}(1 - \delta^2) + e^{-ik(x_g+\Delta)}\delta^2 - ik\Delta e^{-ikx_g}(\delta - \delta^2) . \tag{4.88}$$

If we expand $ik\Delta e^{-ikx_g}$ in terms of the finite difference form, i.e.,

$$- ike^{-ikx_g} = \frac{1}{2\Delta}[e^{-ik(x_g+\Delta)} - e^{-ik(x_g-\Delta)}] , \tag{4.89}$$

we obtain

$$\begin{aligned} e^{-ik(x_g+\delta)} &\cong e^{-ikx_g}\left(1 - \frac{\delta^2}{\Delta^2}\right) + e^{-ik(x_g+\Delta)}\frac{\delta^2}{\Delta^2} \\ &\quad + \frac{1}{2\Delta}[e^{-ik(x_g+\Delta)} - e^{-ik(x_g-\Delta)}]\left(\frac{\delta}{\Delta} - \frac{\delta^2}{\Delta^2}\right) \\ &= e^{-ikx_g}\left(1 - \frac{\delta^2}{\Delta^2}\right) + e^{-ik(x_g+\Delta)}\left(\frac{1}{2}\frac{\delta}{\Delta} + \frac{1}{2}\frac{\delta^2}{\Delta^2}\right) \\ &\quad + e^{-ik(x_g-\Delta)}\left(-\frac{1}{2}\frac{\delta}{\Delta} + \frac{1}{2}\frac{\delta^2}{\Delta^2}\right) . \end{aligned} \tag{4.90}$$

As we can see, Eq. (4.90) is identical to Eq. (4.79), which was obtained by the subtracted quadrupole expansion. We have shown that the quadratic spline with the finite differencing of the phase factor yields the identical result to the subtracted quadrupole expansion. It turns out, however, that it is advantageous not to expand the function $f'(0)$ using Eq. (4.89), but rather to keep the "exact" expression of derivative at $x = x_g$. Then the quadratic spline is far superior to the subtracted quadrupole expansion scheme, which will be shown in the following.

We can increase the accuracy of the spline function $f(x)$ by increasing the order of the polynomial of the spline. Let the spline function be a cubic polynomial, and this is called the cubic spline:

$$f(x) = a_i + b_i\delta + c_i\delta^2 + d_i\delta^3 , \tag{4.91}$$

where δ is again $\delta = x - x_g = x - x_i$. In order to determine the spline, we demand:

$$\text{continuity at} \quad x = 0 \quad : \quad a_i = f(0) , \tag{4.92}$$

$$\text{continuity at} \quad x = \Delta \quad : \quad a_{i+1} = f(\Delta) = a_i + b_i\Delta + c_i\Delta^2 + d_i\Delta^3 , \tag{4.93}$$

$$\text{derivative at} \quad x = 0 \quad : \quad b_i = f'(0) , \tag{4.94}$$

$$\text{derivative at} \quad x = \Delta \quad : \quad b_{i+1} = f'(\Delta) = b_i + 2c_i\Delta + 3d_i\Delta^2 . \tag{4.95}$$

From Eqs. (4.93) and (4.95), we obtain

$$c_i = (b_{i+1} - b_i - 3d_i\Delta^2)/2\Delta . \tag{4.96}$$

Solving for a_{i+1}, we get

$$a_{i+1} = a_i + b_{i\Delta} + \frac{\Delta}{2}(b_{i+1} - b_i) - \frac{\Delta^3}{2}d_i , \tag{4.97}$$

and using this, we get

$$d_i = \frac{2}{\Delta^3}\left(a_i - a_{i+1} + \frac{\Delta}{2}b_i + \frac{\Delta}{2}b_{i+1}\right) . \tag{4.98}$$

Substituting Eqs. (4.96) and (4.98) back into Eq. (4.91), we obtain

$$\begin{aligned} f(x) &= a_i\left(1 - 3\frac{\delta^2}{\Delta^2} + 2\frac{\delta^3}{\Delta^3}\right) + a_{i+1}\left(3\frac{\delta^2}{\Delta^2} - 2\frac{\delta^3}{\Delta^3}\right) \\ &\quad + b_i\left(\delta - 2\frac{\delta^2}{\Delta} + \frac{\delta^3}{\Delta^2}\right) + b_{i+1}\left(-\frac{\delta^2}{\Delta} + \frac{\delta^3}{\Delta^2}\right) . \end{aligned} \tag{4.99}$$

In the specific case where $f(x) = e^{-ik(x_g+\delta)}$, we obtain

$$\begin{aligned} e^{-ik(x_g+\delta)} &= e^{-ikx_g}(1 - 3\delta^2 + 2\delta^3) + e^{-ik(x_g+\Delta)}(3\delta^2 - 2\delta^3) \\ &\quad - ike^{-ikx_g}(\delta - 2\delta^2 + \delta^3) - ike^{-i(x_g+\Delta)}(-\delta^2 + \delta^3) , \end{aligned} \tag{4.100}$$

where δ is redefined to denote the old δ/Δ. Again much more accurate results can be obtained using the formula Eq. (4.100) itself, i.e., without expanding the terms proportional to $-ike^{-ikx_g}$ (the derivative terms). We may be able to say that our strategy is to interpolate the phase factor, not the exponent (i.e., the coordinate itself).

In contrast with the subtracted dipole scheme, let us write the force assignment:

$$\begin{aligned} F(x) &= \sum_k F(k)e^{ikx} = \sum_k F(k)e^{ik(x_g+\delta)} \\ &= \sum_k F(k)e^{ikx_g}e^{ik\delta} = \sum_k F(k)e^{ikx_g}(s_i + s_2e^{ik\Delta} + s_3ik + s_4ike^{ik\Delta}) \\ &= s_1\sum_k F(k)e^{ikx_g} + s_2\sum_k F(k)e^{ik(x_g+\Delta)} \\ &\quad +s_3\sum_k ikF(k)e^{ikx_g} + s_4\sum_k ikF(k)e^{ik(x_g+\Delta)} \\ &= s_1F(x_g) + s_2F(x_g+\Delta) + s_3\frac{\partial}{\partial x}F(x_g) + s_4\frac{\partial}{\partial x}F(x_g+\delta)\ , \end{aligned} \tag{4.101}$$

where s_1, s_2, s_3, and s_4 are the spline coefficients.

Following this strategy, let us write down the charge accumulation equation as

$$\rho(x) = \sum_j q_j e^{-(x-x_j)^2/2a^2}\ . \tag{4.102}$$

$$\begin{aligned} \rho(k) &= e^{-k^2a^2/2}\sum_j q_j e^{-ikx_j} = e^{-k^2a^2/2}\sum_j q_j e^{-ik(x_g+\delta_j)} \\ &= qe^{-k^2a^2/2}\sum_g e^{-ikx_g}\sum_{j\varepsilon g} e^{-ik\delta_j}\ . \end{aligned} \tag{4.103}$$

As discussed, we write

$$e^{-ik\delta_j} \simeq s_1 + s_2e^{ik\Delta} - iks_3 - ike^{ik\Delta}s_4\ , \tag{4.104}$$

where

$$\begin{aligned} s_1 &= (1-\delta_j)^2(1+2\delta_j) \\ s_2 &= \delta_j^2(3-2\delta_j) \\ s_3 &= \delta_j(1-\delta_j)^2\Delta \end{aligned}$$

$$s_4 = -(1-\delta_j)\delta_j^2\Delta \,. \tag{4.105}$$

Therefore, we have

$$\begin{aligned}\rho(k) &= qe^{-k^2a^2/2}\sum_g e^{-ikx_g}\sum_{j\varepsilon g}(s_1 + e^{-ik\Delta}s_2 - iks_3 - ike^{ik\Delta}s_4)\\ &= qe^{-ik^2a^2/2}\left[\sum_g e^{-ikx_g}\left(\sum_{j\varepsilon g}s_1 - ik\sum_{j\varepsilon g}s_3\right)\right.\\ &\quad \left.+\sum_g e^{-ik(x_g+\Delta)}\left(\sum_{j\varepsilon g}s_3 - ik\sum_{j\varepsilon g}s_4\right)\right] .\end{aligned} \tag{4.106}$$

Rewriting Eq. (4.106), we obtain

$$\begin{aligned}\rho(k) = qe^{-k^2a/2}&\left[\sum_g e^{-ikx_g}\left(\sum_{j\varepsilon g}s_1 + \sum_{j\varepsilon g-1}s_2\right)\right.\\ &\left.-ik\sum_g e^{-ikx_g}\left(\sum_{j\varepsilon g}s_3 + \sum_{j\varepsilon g-1}s_4\right)\right] .\end{aligned} \tag{4.107}$$

The electric field can be calculated by

$$E(k) = \frac{iqe^{-k^2a^2/2}}{k}[\mathrm{FFT(G1X)}\text{-}ik\,\mathrm{FFT(G2X)}] \,, \tag{4.108}$$

where FFT is the Fast Fourier Transform and G1X is the array containing the charge contribution arising from the first term in the square bracket in Eq. (4.107), while G2X contains the second term. Note that the derivative is not handled by finite differencing, but by the accurate Fourier representation ik here.

The force interpolation can be done as follows:

$$\begin{aligned}F(x) &= qE(x) = \frac{q}{2\pi L}\sum_k e^{-k^2a^2/2}E(k)e^{ik(x_g+\delta)}\\ &= \frac{q}{2\pi L}\sum_k e^{ikx_g}e^{-k^2a^2/2}E(k)\left(s_1 + s_2e^{ik\Delta} + iks_3 + ike^{+ik\Delta}s_4\right)\\ &= \frac{q}{2\pi L}\sum_k e^{-k^2a^2/2}\left[E(k)(s_1e^{ikx_g} + s_2e^{ik(x_g+\delta)})\right.\end{aligned}$$

$$+ikE(k)(s_3 e^{ikx_g} + s_4 e^{ik(x_g+\Delta)})\Big]$$

$$\begin{aligned} &= s_1(\delta)F_s(x_g) + s_2(\delta)F_s(x_g+\Delta) \\ &\quad + s_3(\delta)F_d(x_g) + s_4(\delta)F_d(x_\delta+\Delta) \ , \end{aligned} \tag{4.109}$$

where $F_d = \frac{q}{2\pi L}\mathrm{FFT}^{-1}[+ike^{-k^2a^2/2}E(k)]$. The present technique, often called the cubic spline interpolation,[5] allows us to reduce the Debye length λ_{De} as small as $10^{-6}\Delta$ and still to avoid the thermal instability.[6] This technique thus allows us to have effectively much larger system than less sophisticated techniques do. A typical wavelength is $\lambda = L\Delta/m$, where m is an integer and L a number of grid points. Then, the wavelength with respect to a physical length, the Debye length, is

$$\frac{\lambda}{\lambda_{\mathrm{De}}} = \left(\frac{L}{m}\right)\left(\frac{\Delta}{\lambda_{\mathrm{De}}}\right) , \tag{4.110}$$

where (L/m) is not less than unity but cannot be extremely large because the computer memory size and computer CPU will be the limit in particular for multidimensional simulation. The methods discussed in Sections 4.1 and 4.2 typically allow $(\Delta/\lambda_{\mathrm{De}}) \sim 10$, which means $\lambda/\lambda_{\mathrm{De}}$ is typically $< 10^3$ according to Eq. (4.100). On the other hand, the present method allows $(\Delta/\lambda_{\mathrm{De}}) \sim \mathcal{O}(10^6)$, leading to a much larger $\lambda/\lambda_{\mathrm{De}}$. An alternative method is to use exact positions instead of interpolation to grid as developed by Cheng and Okuda,[11] as will be discussed in Section 7.3. We shall also learn that the implicit method allows the removal of this restriction $\Delta/\lambda_{\mathrm{De}} \sim \mathcal{O}(1)$. We will come back to this point in Section 9.7.

If one elects, one can use higher order approximations (higher order spline). When the fifth order polynomial is used to fit the function $y(x)$, we have

$$y(x) = a + b\delta + c\delta^2 + d\delta^3 + e\delta^4 + f\delta^5 \ . \tag{4.111}$$

Our conditions are

$$\begin{aligned} y(0) &= a \ , \ y'(0) = b, \ \ y''(0) = 2c, \\ y(\Delta) &= a' \ , \ y'(\Delta) = b', y''(\Delta) = 2c' \ , \end{aligned} \tag{4.112}$$

where the quantities with a prime are at the grid point $x = (i+1)\Delta$ and ones without it are at $x = i\Delta$. We now abbreviate $a \to a$, $b\Delta \to b$, $c\Delta^2 \to c$, $d\Delta^3 \to d$, $e\Delta^4 \to e$, and $f\Delta^5 \to f$. From conditions (4.112), we obtain

$$\begin{aligned} d &= -10a + 10a' - 6b - 4b' - 3c + c' \ , \\ e &= 15a - 15a' + 8b + 7b' + 3c - 2c' \ , \end{aligned}$$

$$f = 6a' - 6a - 3b - 3b' - c + c' .$$

Eliminating d, e, and f in Eq. (4.111) and rearranging terms, we obtain

$$\begin{aligned} y(x) &= a(1 - 10\delta^2 + 15\delta^4 - 6\delta^5) + a'(10\delta^3 - 15\delta^4 + 7\delta^5) \\ &\quad + b(\delta - 6\Delta\delta^3 + 8\Delta\delta^4 - 3\Delta\delta^5) + b'(-4\Delta\delta^3 + 7\Delta\delta^4 - 3De\delta^5) \\ &\quad + c(\delta^2 - 3\Delta^2\delta^3 + 3\Delta^2\delta^4 - \Delta^2\delta^5) \\ &\quad + c'(\Delta^2\delta^3 - 2\Delta^2\delta^4 + \Delta^2\delta^5) . \end{aligned} \tag{4.113}$$

In this chapter, we have discussed the motivation to introduce grid and its benefits, its consequences, its deleterious effects, and methods to overcome these.

Problems

1. In Eq. (4.70) the summation should be double sums over p and p', to be rigorous. Write down a fuller expression using Eq. (4.44).
2. Discuss qualitatively why the numerical thermal instability in explicit particle simulation in a magnetized plasma does not necessarily arise under certain conditions, even when the conditions for the thermal stability in an unmagnetized plasma are not fulfilled (Section 4.5).
3. Derive Eq. (4.112). Then describe the algorithm of the 5-*th* order spline interpolation method.

References

1. A.B. Langdon, J. Comp. Phys. **6**, 247 (1970).
2. W.L. Kruer, J.M. Dawson, and B. Rosen, J. Comp. Phys. **13**, 114 (1973).
3. C.K. Birdsall and A.B. Langdon, *Plasma Physics via Computer Simulation* (McGraw-Hill, New York, 1985).
4. R.W. Hockney and J.W. Eastwood, *Computer Simulation Using Particles* (McGraw Hill, New York, 1981) p. 152-222.
5. H. Okuda and C.Z. Cheng, Comp. Phys. Comm. **14**, 169 (1978).
6. J.S. Wagner, private communication.

7. C. Kittel, *Introduction to Solid State Physics*, 4th ed. (John Wiley and Sons, Inc., New York, 1971) p. 296; also J. Ziman, *Principles of Solid State Physics* (Cambridge U.P., Cambridge, 1972).

8. H. Okuda, J. Comp. Phys. **10**, 475 (1972).

9. A.B. Langdon, J. Comp. Phys. **12**, 247 (1973).

10. C.K. Birdsall and N. Maron, J. Comp. Phys. **36**, 1 (1980).

11. C.Z. Cheng and H. Okuda, J. Comp. Phys. **25**, 133 (1977).

5 ELECTROMAGNETIC MODEL

Thus far we have primarily discussed the electrostatic model of plasmas for particle simulation. In this chapter we shall discuss the electromagnetic model of plasmas and the treatment of the Maxwell equations. It turns out that most of the previously discussed techniques can be employed here as well.

5.1 Electromagnetic Particle Simulation Code

The electromagnetic particle code is an extension of the electrostatic code to include the full electromagnetic effects. The electrostatic code is a special case in this case.

The Maxwell Equations read

$$\nabla \times \mathbf{E} = -\frac{1}{c}\frac{\partial \mathbf{B}}{\partial t}, \tag{5.1}$$

$$\nabla \times \mathbf{B} = \frac{1}{c}\frac{\partial \mathbf{E}}{\partial t} + \frac{4\pi}{c}\mathbf{J} = \frac{1}{c}\frac{\partial \mathbf{E}}{\partial t} + \frac{4\pi}{c}\sum_j q_j \dot{r}_j \delta(\mathbf{r} - \mathbf{r}_j) . \tag{5.2}$$

When the size of the particles is finite, we should replace the delta function by the shape factor $S(\mathbf{r}-\mathbf{r}_j)$. In addition, Poisson's equation employed in the electrostatic code reads

$$\nabla \cdot \mathbf{E} = 4\pi \sum_j q_j \delta(\mathbf{r}-\mathbf{r}_j) . \tag{5.3}$$

Again, when the size of particles is finite, we should replace $\delta(\mathbf{r}-\mathbf{r}_j)$ by $S(\mathbf{r}-\mathbf{r}_j)$. In addition we impose $\nabla \cdot \mathbf{B} = 0$.

It is convenient to separate the longitudinal and transverse components of electric fields. The magnetic field is all transverse. The longitudinal field is curl-free, $\nabla \times \mathbf{E}_L = 0$, and the transverse one is divergence-free, $\nabla \cdot \mathbf{E}_T = 0$.

$$\mathbf{E}(\mathbf{k}) = \hat{\mathbf{k}}\hat{\mathbf{k}} \cdot \mathbf{E}_L + \left(\mathbf{I} - \hat{\mathbf{k}}\hat{\mathbf{k}}\right) \cdot \mathbf{E}_T \tag{5.4}$$

and

$$\mathbf{E}(\mathbf{r}) = \sum_k \mathbf{E}(\mathbf{k}) e^{i\mathbf{k}\cdot\mathbf{r}} . \tag{5.5}$$

Here we have $\hat{\mathbf{k}} \equiv \frac{\mathbf{k}}{k}$. In Fourier space, Eq. (5.3) becomes

$$i\mathbf{k} \cdot \mathbf{E} = ikE_L = 4\pi \sum_j q_j S(\mathbf{k}) e^{-i\mathbf{k}\cdot\mathbf{x}_j} . \tag{5.6}$$

The Maxwell equations become

$$i\mathbf{k} \times \mathbf{B}(\mathbf{k}) = \frac{1}{c}\frac{\partial}{\partial t}\mathbf{E}_T(\mathbf{k}) + \frac{4\pi}{c}\mathbf{J}_T , \tag{5.7}$$

$$0 = \frac{1}{c}\frac{\partial}{\partial t}\mathbf{E}_L(\mathbf{k}) + \frac{4\pi}{c}\mathbf{J}_L , \tag{5.8}$$

$$i\mathbf{k} \times \mathbf{E}_T = -\frac{1}{c}\frac{\partial \mathbf{B}_T}{\partial t} , \tag{5.9}$$

where

$$\mathbf{J}_T(\mathbf{k}) = \left(\mathbf{I} - \hat{\mathbf{k}}\hat{\mathbf{k}}\right) \cdot \mathbf{J}(\mathbf{k}) , \tag{5.10}$$

and

$$\mathbf{J}(\mathbf{k}) = \sum_j q_j S(\mathbf{k}) \dot{\mathbf{r}}_j e^{-i\mathbf{k}\cdot\mathbf{r}_j} . \tag{5.11}$$

Writing the time derivative for $\mathbf{E}_T(\mathbf{k})$ and $\mathbf{B}(\mathbf{k})$ from the Fourier transformed Maxwell equations, we get

$$\frac{\partial \mathbf{E}_T}{\partial t}(\mathbf{k}) = ic\mathbf{k} \times \mathbf{B}(\mathbf{k}) - 4\pi \mathbf{J}_T(\mathbf{k}) , \tag{5.12}$$

and

$$\frac{\partial \mathbf{B}}{\partial t}(\mathbf{k}) = -ic\mathbf{k} \times \mathbf{E}_T . \tag{5.13}$$

Integrating in time in finite difference forms, we obtain

$$\frac{\mathbf{E}_T(\mathbf{k})^{t+\Delta t} - \mathbf{E}_T(\mathbf{k})^t}{\Delta t} = ic\mathbf{k} \times \mathbf{B}_T(\mathbf{k})^{t+\frac{\Delta t}{2}} - 4\pi \mathbf{J}_T(\mathbf{k})^{t+\frac{\Delta t}{2}} , \tag{5.14}$$

and

$$\frac{\mathbf{B}(\mathbf{k})^{t+\frac{3}{2}\Delta t} - \mathbf{B}(\mathbf{k})^{t+\frac{\Delta t}{2}}}{\Delta t} = -ic\mathbf{k} \times \mathbf{E}_T(\mathbf{k})^{t+\Delta t} . \tag{5.15}$$

Notice that Eqs. (5.14) and (5.15) are both time-centered. Therefore, the above time-stepping (the leapfrog) of the Maxwell equations can be numerically stable and neutral as long as the Courant-Friedrichs-Lewy condition is satisfied.[1-3] In this sense, the Maxwell equations are benign and easy to integrate numerically. This is a general property of the second-order hyperbolic (linear) differential equation.

There is a complication, however, when we try to advance particles with the fields obtained by Eqs. (5.14) and (5.15). The force equation is

$$\frac{d\mathbf{v}_j}{dt} = \frac{q}{m}\left[\mathbf{E}(\mathbf{r}_j) + \frac{\mathbf{v}_j}{c} \times \mathbf{B}(\mathbf{r}_j)\right] . \tag{5.16}$$

As seen in Eqs. (5.14) and (5.15), $\mathbf{E}(\mathbf{k})$ is evaluated at either t or $t + \Delta t$ and $\mathbf{B}(\mathbf{k})$ is evaluated at either $t + \frac{1}{2}\Delta t$ or $t + \frac{3}{2}\Delta t$. In the force equation, the term $(\mathbf{v} \times \mathbf{B})$ must be evaluated at $(t + \Delta t)$ in order to keep the time-step compatible with the electric field in Eq. (5.16). The force equation, therefore, should be

$$\frac{\mathbf{v}^{t+\frac{3}{2}\Delta t} - \mathbf{v}^{t+\frac{1}{2}\Delta t}}{\Delta t} = \frac{q}{m}\left\{\mathbf{E}^{t+\Delta t} + \frac{1}{c}(\mathbf{v} \times \mathbf{B})^{t+\Delta t}\right\} . \tag{5.17}$$

In order to find $\mathbf{v} \times \mathbf{B}$ at time step $t + \Delta t$, we approximate as follows:

$$\bar{\mathbf{B}} = \mathbf{B}^{t+\Delta t} = \frac{1}{2}\left(\mathbf{B}^{t+\frac{3}{2}\Delta t} + \mathbf{B}^{t+\frac{1}{2}\Delta t}\right) , \tag{5.18}$$

and

$$\bar{\mathbf{v}} = \mathbf{v}^{t+\Delta t} = \frac{1}{2}\left(\mathbf{v}^{t+\frac{3}{2}\Delta t} + \mathbf{v}^{t+\frac{1}{2}\Delta t}\right) . \tag{5.19}$$

These assignments of the time averaged $\bar{\mathbf{B}}$ and $\bar{\mathbf{v}}$ make Eq. (5.19) an implicit equation for $\mathbf{v}^{t+\frac{3}{2}\Delta t}$. This implicit equation has to be solved by matrix inversion as done in Chapter 3 for the electrostatic code with the external magnetic field. Because of these properties, the electromagnetic code is the most complete description of a plasma with stability, neutrality, and numerical resilience. At the same time, because this model describes "all" aspects of plasmas, the time-step used to satisfy the Courant-Friedrichs-Lewy condition has to be smaller than "all" the characteristic periods of the plasma. This includes $(kv)^{-1} \sim (\pi v/\Delta)^{-1}$, the electron cyclotron period ω_{ce}^{-1}, the electron plasma period ω_{pe}^{-1}, and the photon oscillation period $\omega^{-1} \cong (kc)^{-1}$. The major drawback of this model, therefore, is that very many time-steps must be run in order to simulate a slow phenomenon and, thus, the run is very expensive. Many models, which will be expounded upon in this and in later chapters, have been developed in order to circumvent this difficulty.

5.2 Analogy Between Electrodynamics and General Relativity

The Newtonian self-gravitational system can be described by "Poisson's equation" except for the difference in the attractive nature of the gravitational field.[3,4] A similar analogy exists in a more generalized gravitational system. An electromagnetic analogy[5,6] of the gravitational field around a black hole may be described by

"Gravitoelectric" field $\mathbf{g}$

"Gravitomagnetic" field $\mathbf{H}$.

In weak gravity (Newtonian limit) $\mathbf{g}$ reduces to the conventional gravitational field, and $\mathbf{H} \to 0$.

In the limit of weak gravity and slow motion $\left(\frac{v}{c} \ll 1\right)$, the Einstein equation becomes[5,6] (almost) identical to the Maxwell equation in such a way that: the equation of motion is

$$\frac{d\mathbf{v}}{dt} = \mathbf{g} + \frac{\mathbf{v}}{c} \times \mathbf{H} , \tag{5.20}$$

and the field equations are

$$\nabla \cdot \mathbf{g} = -4\pi G\rho , \tag{5.21}$$

$$\nabla \cdot \mathbf{H} = 0 , \tag{5.22}$$

$$\nabla \times \mathbf{g} = 0 , \tag{5.23}$$

$$\nabla \times \mathbf{H} = \frac{4\pi}{c}\left(-4G\rho\mathbf{v} + \frac{\partial \mathbf{g}}{\partial t}\right) . \tag{5.24}$$

This is almost identical to Maxwell's equations except for the minus sign in the density term in "Poisson's equation" Eq. (5.21) and in the current term in "Ampere-Maxwell's equation" Eq. (5.24). We may, therefore, apply an "electromagnetic" code as discussed in Section 5.1 to the case of the general relativistic equations (gravitoelectrodynamics) in this limit with only minor adjustments.

5.3 Absorbing Boundary for the Electromagnetic Model

Since the spatial extent of plasma simulation is necessarily finite and bounded, it is quite important to develop techniques which approximately handle the boundaries or extrapolate the system. The simplest to implement is the periodic boundary condition. Most of the basic particle codes have been implemented this way. For electrostatic codes, different and more sophisticated boundary condition techniques have been developed. One is the so-called capacity matrix technique[7,8] which adjusts the boundary condition by adding the boundary charge in real space. This technique can be extended to magnetostatic codes. Another[9] matches the vacuum solutions. Yet another[10] constructs a solution from a sum of the homogeneous solution and the inhomogeneous solution to Poisson's equation. An extension of this to magnetostatic codes is also possible.

On the other hand, the electromagnetic boundary treatment is less developed. Among the most important conditions beside the periodic one is the absorbing boundary condition. One technique,[11] originally by Lindman, Dawson and Langdon, utilizes the fact that the hyperbolic operator $\partial_t^2 - c^2\partial_x^2$ in the Maxwell equation can be factorized into the left- and right-ward propagation operators $\partial_t \pm c\partial_x$. This technique can be employed only in strictly one dimension. To date the most complete method for this problem has been developed by Lindman.[12] Lindman's method utilizes the projection operators (left- and right-ward), which allow oblique angle incidence to be taken, and approximates the operators by a numerically stable partial fraction expression (a Padé approximation). The technique requires solving several (three or six) finite difference equations (in order to restrict the amplitude error to less than 1%) to determine the "reflection" coefficients.

A naive approach by extrapolation cannot yield a complete absorption. Consider a three-point extrapolation for example,

$$E_{-1} - 3E_0 + 3E_1 - E_2 = 0 , \tag{5.25}$$

where subscripts indicate the grid position, i.e., 0 means on the boundary and -1 the extrapolated grid and so on. Let us express the solution at the n-th grid as

$$E_n = Ae^{inb} + Be^{-inb} , \tag{5.26}$$

where the coefficient A indicates the amplitude of the outgoing wave and B the incoming one. Substituting Eq. (5.26) into Eq. (5.25) yields[13]

$$B = Ae^{-2ib}\left(\frac{1-e^{ib}}{1-e^{-ib}}\right)^3 . \tag{5.27}$$

It follows from Eq. (5.27) that $|B| = |A|$. We therefore conclude that the extrapolation cannot make the incoming wave amplitude vanish ($|B| = 0$ or $|B| \ll |A|$) and that no absorbing condition is achieved. Equation (5.27), however, suggests that extrapolation with damping (complex b) may lead to $|B| \ll |A|$, but not $|B| = 0$. (A more sophisticated boundary treatment for absorption or open boundaries by extrapolation is discussed by Orlanski.[14]) Another simple approach to absorbing boundary conditions is the method of coordinate stretching or nonuniform grid (see Chapter 10). Beyond the physical volume where the grid is regular, the grid spacing in the ramp is exponentially stretched so that, for example, over 10 grid points, the actual distance is $\sum_{i=1}^{10} a^i \Delta_x$, where a is a ratio of stretching from one grid point to the next neighbor and Δ_x is the regular grid distance. If this distance is long enough, the information traveling over the distance of the ramp might never come back in the time span of a practical simulation, even if the wave is reflected at the last edge point. This method, however, also suffers from incomplete absorption. The finite difference of the distance in grid space is $x(j+1) - x(j) = a^j \Delta_x$ in the ramp. Consider the wave equation $\phi'' + k^2\phi = 0$:

$$\left(D_x^2 + k^2\right)\phi = \left[\partial_x^2 + p\partial_x + k^2\right]\phi = 0 \;, \tag{5.28}$$

where D_x is the finite difference operator and $p = 2(1-a)/(1+a)\Delta_x$. For $p^2 < 4k^2$, the stretching adds some effective imaginary part to the wavenumber k. On the other hand, too much stretching, $p^2 > 4k^2$, severely distorts the nature of the wave equation, leading to complete reflection of the incoming information. We have thus seen that the naive or simple techniques considered above cannot lead to the completely absorbing boundary condition. Next, we consider an algorithm of masking[13] in the Maxwell equations which provides a good absorbing boundary condition for electromagnetic waves in electromagnetic particle codes.

Let us introduce a masking function $f(x)$. A masking procedure of an electric field, for example, is

$$\tilde{E}(x) = f(x)E(x) \;, \tag{5.29}$$

where we consider the one-dimensional case for simplicity. We adopt the masking function as parabolically and smoothly matched at the plasma (or real system) boundary:

$$\begin{aligned} f(x) &= -d^{-2}x^2 + 2d^{-1}x && (0 \le x < d) \\ &= 1 && (d \le x \le L-d) \\ &= d^{-2}x^2 + 2d^{-2}(L-d)x + 1 + d^{-2}(L-d)^2 && (L-d < x \le L) \;, \end{aligned} \tag{5.30}$$

where d is the length of the ramps and L the total system length (see Fig. 5.1). Note that $f(x)$ approaches $x = x_0$ (x_0: 0 or L) linearly: $f(x) \propto (x - x_0)$. The

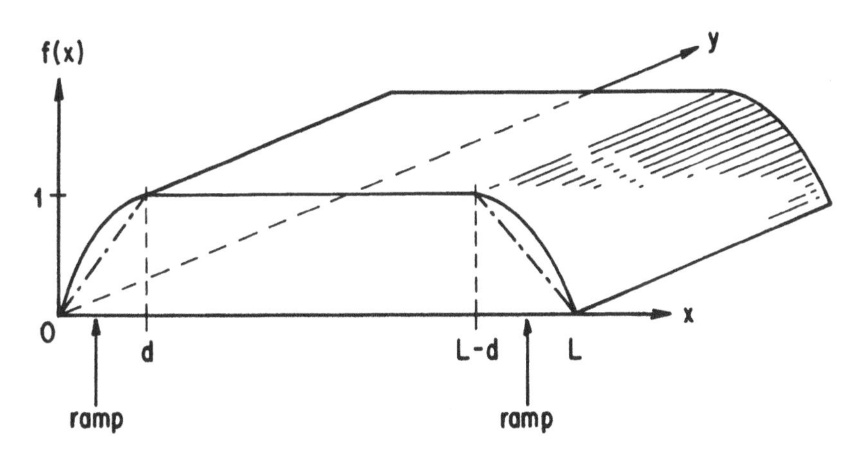

FIGURE 5.1 The masking function (after Ref. 13)

plasma is contained in $d \leq x \leq L-d$. We mask either transverse electric fields E_y and E_z or magnetic fields B_y and B_z, but not both. For each time step, the current (and charge) generates electromagnetic fields. The algorithm to calculate the Maxwell equations is otherwise exactly the same as the conventional code. (See Section 5.1.) Since particles are confined in $d \leq x \leq L-d$, either masked or unmasked fields make no difference to the fields which push particles in $d \leq x \leq L-d$. If the system is two-dimensional, the generalization of the masking procedure is straightforward: when the absorbing ramps are located parallel to the y-axis from $x = 0$ to $x = d$ and from $x = L - d$ to $x = L$, the masking is done for the transverse electric (or magnetic fields) whose directions are perpendicular to the normal to the boundary.

The singly masked Maxwell equations (electric field masked) in one dimension in the ramp area may be written as

$$\begin{aligned} -i\omega B &= -c\partial_x(fE), \\ -i\omega E &= c\partial_x B \,. \end{aligned} \tag{5.31}$$

If it is two dimensional with the ramps along the x-direction, we have

$$\begin{aligned} -i\omega E_x &= ik_y c B_z \,, \\ -i\omega E_y &= -c\partial_x B_z \,, \end{aligned}$$

$$\begin{aligned} -i\omega E_z &= -ik_y cB_x + c\partial_x B_y , \\ -i\omega B_x &= ik_y cE_z , \\ -i\omega B_y &= c\partial_x (fE_z) , \\ -i\omega B_z &= ik_y cE_x - c\partial_x (fE_y) . \end{aligned} \tag{5.32}$$

From Eq. (5.31) we obtain

$$(\partial_x^2 f + \omega^2/c^2)E = 0 . \tag{5.33}$$

Alternatively we obtain from Eq. (5.32),

$$\left(\partial_x^2 f - k_y^2 + \frac{\omega^2}{c^2}\right) E_y = 0 . \tag{5.34}$$

If f is linearly proportional to x at $x = 0$ or $x = L$, Eq. (5.33) [or Eq. (5.34)] can be written with $\mathcal{E} = fE$ as

$$\left(\partial_x^2 + \frac{\alpha}{x + i\eta}\right) \mathcal{E} = 0 , \tag{5.35}$$

where α is a real number and η is an infinitesimal real number. The differential equation (5.35) has a singular turning point at $x = 0$. The connection formula of asymptotic solutions for the singular turning point is[15,16]

$$\begin{aligned} &k_1^{-1/2} \{(A - iB) \exp [(\xi_1 + \pi/4)] + (A + iB) \exp [-i(\xi_1 + \pi/4)]\} \\ &\leftrightarrow |k_2|^{-1/2} [(A \pm iB) \exp(|\xi_2|) + 2B \exp(-|\xi_2|)] , \end{aligned} \tag{5.36}$$

where subscripts 1 and 2 refer to before and behind the turning point $x = 0$ (or $x = L$). ξ is a normalized coordinate of x. If the wave is coming from the left of the turning point, the term proportional to $\exp [i(\xi_1 + \pi/4)]$ is the incoming wave and the other $\exp [-i(\xi_1 + \pi/4)]$ is the reflected outgoing wave. When we demand that the solution be regular at $|\xi_2| \to \infty$, the Budden condition $A - iB = 0$ for $\eta > 0$ is required. As is seen from Eq. (5.36), the Budden condition leads to a vanishing coefficient for the reflected wave amplitude. Thus we find that the masking function $f(x)$, linearly proportional to $x - x_0$, (x_0 being either end of the system) results in a complete absorption of a plasma wave at $x = x_0$. This is the basic reason for absorption of electromagnetic waves by virtue of the above prescribed algorithm: the masking function (5.36) has an appropriate form of $(x - x_0)^{-1}$ near $x = x_0$. When we chose f as the one shown by broken lines in Fig. 5.1, we still observe high absorption but slightly worse than the case with f as given by Eq. (5.36). This may be due to lack of smooth joining of f at $x = d$ or $x = L - d$. This masking method may be looked upon physically as changing the electric (or magnetic) permeability primarily in a reactive way. Other works[17,18] have resorted to resistive methods, which theoretically cannot absorb 100% of waves over a finite length.

Although the absorption with this method[13] is worse than that with Ref. 12, the present method is simple, flexible, and versatile.

5.4 Magnetoinductive Particle Model

The fully electromagnetic particle code discussed in Section 5.1 is simple, neutral and resilient. It can be said that this model contains, in principle, all the information we want about the plasma. The problem is, on the other hand, that the fully electromagnetic model contains the high frequency radiation whose frequency is $\omega = kc$ in a vacuum and $\omega = (\omega_p^2 + k^2c^2)^{1/2}$ in a plasma. This light wave is ordinarily of such high frequency that the necessary time step $\Delta t \lesssim \omega^{-1}$ has to be very small. For example in a typical plasma the visible light frequency is $\sim 10^{15}$ Hz while the typical laboratory plasma frequency is $\sim 10^{12}$ Hz. These electromagnetic waves are mostly superluminous in a plasma, where the phase velocity is faster than the speed of light. In a magnetized plasma, however, there are many branches of subluminous nonevanescent electromagnetic waves which have phase velocities much less than the speed of light, such as the whistler wave and the Alfvén wave. It was Alfvén who first discovered a subluminous and nonevanescent wave in a magnetized plasma, the Alfvén wave. When one is interested in such lower phase velocities and lower frequencies (see Table 1.2), one would like to neglect the radiation phenomena but to retain the subluminous waves such as Alfvén waves. This can be done by dropping the displacement current term in the Ampere-Maxwell's equation. This treatment is sometimes called Darwin's model.[19,20] Here we call it the magnetoinductive model.

This model can become very efficient when other high frequencies in the system such as the plasma frequency ω_p and the electron cyclotron frequency Ω_e along with the radiation frequency $\omega = kc$ are also suppressed. We shall discuss this subject later.

Maxwell's equations for the transverse E and B fields (with radiation) are

$$\dot{\mathbf{E}} = c\nabla \times \mathbf{B} - 4\pi\mathbf{J} , \tag{5.37}$$

$$\dot{\mathbf{B}} = -c\nabla \times \mathbf{E} . \tag{5.38}$$

These equations are easily time centered, i.e.

$$E^{n+1} = E^n + \Delta t f\left(B^{n+1/2},\ J^{n+1/2}\right) ,$$

$$B^{n+3/2} = B^{n+1/2} + \Delta t f(E^{n+1}) ,$$

as we discussed in Section 5.1.

We consider the following three versions of the magnetoinductive model:

a. General model

b. Canonical momentum model

c. 3D model

A. General Model (Refs. 19,20)

In this model we drop the displacement current term $1/c\partial E_T/\partial t$ in Maxwell's equations above. We can show that this term is of order $(\omega/kc)^2$ times smaller than the $\nabla \times B$ term. From Faraday's law $\nabla \times E_T = -1/c(\partial B/\partial t)$ we have on the order of magnitude,

$$E \sim \left(\frac{\omega}{kc}\right) B .$$

Thus, on the order of magnitude,

$$\left(\frac{\partial E}{\partial t}\right) \Big/ (\nabla \times B) \sim \left(\frac{\omega}{kc}\right)^2 . \tag{5.39}$$

Since we are primarily interested in low frequency phenomena $\omega/kc \ll 1$, we are justified in neglecting the displacement current. For example, in the case of the Alfvén waves, $\omega \sim kv_A$ with $v_A^2 = B^2/4\pi nM$ and $(\omega/kc)^2 \simeq v_A^2/c^2 \ll 1$.

In neglecting the $\dot{E}$ term we gain by achieving a longer time-step, but now we have lost the symmetry in the equations that is needed to construct the time-centered leapfrog differencing. The differential equation is no longer hyperbolic. (See Section 5.6.) The method to calculate the transverse E field is as follows. First, take the curl of Faraday's law,

$$\mathbf{\nabla} \times \mathbf{\nabla} \times \mathbf{E}_T = -\frac{1}{c}\frac{\partial}{\partial t}(\mathbf{\nabla} \times \mathbf{B}) = -\frac{4\pi}{c^2}\frac{\partial}{\partial t}\mathbf{J}_T . \tag{5.40}$$

Here $\mathbf{J}_T$ is the transverse component of the current. With the aid of a vector identity this becomes

$$\mathbf{\nabla}^2 \mathbf{E}_T = \frac{4\pi}{c^2}\frac{\partial \mathbf{J}_T}{\partial t} , \tag{5.41}$$

where $\mathbf{J} = \sum_{j \epsilon q} q_j \mathbf{v}_j S(\mathbf{r} - \mathbf{r}_j)$.

The time derivative of the latter expression is tricky, since it involves both a Lagrangian and a convective part.

$$\frac{\partial \mathbf{J}}{\partial t} = \sum_j q_j \left[\frac{d\mathbf{v}_j}{dt} S(\mathbf{r} - \mathbf{r}_j) - \mathbf{v}_j (\mathbf{v}_j \cdot \boldsymbol{\nabla}) S(\mathbf{r} - \mathbf{r}_j) \right] \tag{5.42}$$

where

$$\frac{d\mathbf{v}_j}{dt} = \frac{q_j}{m} \int \left(\mathbf{E} + \frac{\mathbf{v}_j}{c} \times \mathbf{B} \right) S(\mathbf{r}' - \mathbf{r}_j) d\mathbf{r}' \ .$$

Thus,

$$\begin{aligned} \frac{\partial \mathbf{J}}{\partial t} &= \sum_j q_j \left\{ \frac{q_j}{m_j} S(\mathbf{r} - \mathbf{r}_j) \int S(\mathbf{r}' - \mathbf{r}_j) \left[\mathbf{E}(\mathbf{r}') + \frac{\mathbf{v}_j}{c} \times \mathbf{B}(\mathbf{r}') \right] d\mathbf{r}' \right. \\ & \left. -\mathbf{v}_j (\mathbf{v}_j \cdot \boldsymbol{\nabla}) S(\mathbf{r} - \mathbf{r}_j) \right\} \ . \end{aligned} \tag{5.43}$$

Substituting this expression Eq. (5.43) into Eq. (5.41) yields

$$\begin{aligned} & \nabla^2 \mathbf{E}_T - \frac{4\pi}{c^2} \left\{ \sum_j \frac{q_j^2}{m_j} S(\mathbf{r} - \mathbf{r}_j) \int S(\mathbf{r}' - \mathbf{r}_j) \mathbf{E}(\mathbf{r}') d\mathbf{r}' \right\}_T \\ & \frac{4\pi}{c^2} \left\{ \sum_j \frac{q_j^2}{mj} S(\mathbf{r} - \mathbf{r}_j) \int S(\mathbf{r}' - \mathbf{r}_j) \left[\mathbf{v}_j \times \frac{\mathbf{B}(\mathbf{r}')}{c} \right] d\mathbf{r}' \right. \\ & \left. -q_j \mathbf{v}_j (\mathbf{v}_j \cdot \boldsymbol{\nabla}) S(\mathbf{r} - \mathbf{r}_j) \right\}_T \ . \end{aligned} \tag{5.44}$$

Again the subscript T indicates the transverse components. This can be written symbolically as

$$\nabla^2 \mathbf{E}_T - \int G(\mathbf{r}, \mathbf{r}') \mathbf{E}_T(\mathbf{r}') d\mathbf{r}' = -\&_T(\mathbf{r}) \ . \tag{5.45}$$

If the particle shape factor $S(\mathbf{r})$ is the delta function, then $G(\mathbf{r}, \mathbf{r}') = G(\mathbf{r} - \mathbf{r}')$, and the integral is a simple convolution. The Fourier transformed equation is then,

$$\left[-k^2 - G(\mathbf{k}) \right] \mathbf{E}_T(\mathbf{k}) = -\&_T(\mathbf{k}) \ . \tag{5.46}$$

For general shape $S(\mathbf{r})$, the integral cannot be expressed as a convolution. The Fourier space representation is

$$k^2 \mathbf{E}_T(\mathbf{k}) + \frac{\omega_p^2}{n_0 c^2} \sum_{\mathbf{k}'} S(\mathbf{k}) \left[n_e(\mathbf{k}') + \frac{m}{M} n_i(\mathbf{k}') \right] \mathbf{E}_T(\mathbf{k} - \mathbf{k}') S(\mathbf{k} - \mathbf{k}') = \&_T(\mathbf{k}) \ . \tag{5.47}$$

Note that the source term $\mathcal{E}$ is a function of the known quantities $(\mathbf{v}_j, \mathbf{B}, n)$ coming from the terms on the right-hand side of Eq. (5.44).

If the plasma is relatively uniform and $S(\mathbf{r})$ is a relatively peaked function, then the major contribution in the k convolution in Eq. (5.47) arises from the $k' = 0$ term. If this is true then we renormalize Eq. (5.47). We keep only the $k' = 0$ terms on the left-hand side of Eq. (5.47) and get a Dyson-type equation

$$\left[k^2 + \frac{\omega_{pe}^2}{c^2}\left(1 + \frac{m}{M}\right) S(\mathbf{k})\right] \mathbf{E}_T(\mathbf{k})$$

$$= \mathcal{E}_T(\mathbf{k}) - \frac{\omega_{pe}^2}{n_0 c^2} \sum_{\mathbf{k}' \neq 0} S(\mathbf{k}) \left[n_e(\mathbf{k}') + \frac{m}{M} n_i(\mathbf{k}')\right]$$

$$\times \mathbf{E}_T(\mathbf{k} - \mathbf{k}') S(\mathbf{k} - \mathbf{k}') \ . \tag{5.48}$$

The second term in brackets on the left-hand side may be called the renormalized term.

We can solve Eq. (5.48) for $\mathbf{E}_T$ by perturbation theory i.e., by iteration numerically. In the first approximation we drop the terms in the sum on the right-hand side of Eq. (5.48). Define the function $g(\mathbf{k})$.

$$g(\mathbf{k}) \equiv k^2 + \frac{\omega_{pe}^2}{c^2}\left(1 + \frac{m}{M}\right) S(\mathbf{k}) \ . \tag{5.49}$$

Equation (5.48) can be written as

$$g(\mathbf{k}) \mathbf{E}_T(\mathbf{k}) = \mathcal{E}_T(\mathbf{k}) + \sum_{k' \neq 0} h(\mathbf{k}', \mathbf{k}) \mathbf{E}_T(\mathbf{k} - \mathbf{k}') \ . \tag{5.50}$$

In the first approximation

$$\mathbf{E}_T^0(\mathbf{k}) = \mathcal{E}_T(\mathbf{k}) / g(\mathbf{k}) \ . \tag{5.51}$$

The second approximation is

$$\mathbf{E}_T^1(\mathbf{k}) = \mathcal{E}_T(\mathbf{k}) / g(\mathbf{k}) + \frac{1}{g(\mathbf{k})} \sum_{\mathbf{k}'} h(\mathbf{k}', \mathbf{k}) \mathbf{E}_T^0(\mathbf{k} - \mathbf{k}') \ , \tag{5.52}$$

and similarly for higher order approximations until convergence. The convergence is more rapid if the operator $h(\mathbf{k}', \mathbf{k})$ is nearly diagonal. This is the case when the background plasma is nearly uniform and/or when the size of particles is large (more spatially spread).

Finally, we get $\mathbf{E}_T$ by transforming back to real space. We continue the calculation by advancing the velocities and positions of the electrons and ions as before with the equations

$$\frac{d\mathbf{v}_j}{dt} = \frac{q_j}{m_j}\left[(\mathbf{E}_L + \mathbf{E}_T) + \frac{\mathbf{v}_j}{c} \times \mathbf{B}\right], \frac{d\mathbf{x}_j}{dt} = \mathbf{v}_j \ ,$$

and so on.

B. Incompressible Two-dimensional Model (Ref. 21)

The above model in Section 5.4a keeps the full dynamics except for the displacement current. It is sometimes desirable to use a more simplified or abbreviated and, therefore, faster method. It is appropriate to use the incompressible two-dimensional model when the following two approximations are satisfied.

1. There is a strong "toroidal" external magnetic field in the z-direction and the perturbed B_z field is negligible compared to the external B_{z0}.
2. The z-direction dependence of B_x, B_y and the other variables is unimportant.

Thus, from 1 we need only δB_x and δB_y. The δB_z may be neglected in comparison with B_{z0}. From 2 $\partial_z = 0$ and therefore $k_z = 0$.

In terms of the vector potential $\mathbf{A}$ we have

$$\begin{aligned} \delta B_z &= \partial_x A_y - \partial_y A_x = 0 \\ \delta B_y &= \partial_z A_y - \partial_x A_z = -\partial_x A_z \\ \delta B_x &= \partial_y A_z - \partial_z A_y = \partial_y A_z \; . \end{aligned}$$

Now all electromagnetic fields may be described by A_z alone.

$$\mathbf{E} = -\nabla\phi + \dot{A}_z\hat{z} \; , \tag{5.53}$$

$$\mathbf{B} = \nabla \times \mathbf{A} = \nabla \times (A_z\hat{z}) \equiv \nabla \times (\psi\hat{z}) \; , \tag{5.54}$$

where we have defined a new (magnetic) flux function $\psi = A_z$. [(In the first equation there could also be $\dot{A}_x$ and $\dot{A}_y$ terms, but these are customarily smaller than the curl-free component of Eq. (5.53)].

We also have the equation

$$\nabla \times \mathbf{B} = \frac{4\pi}{c}\mathbf{J}_T \; ,$$

which can be written as an equation for the (poloidal) flux

$$\nabla^2 A_z = \nabla^2\psi = -\frac{4\pi}{c}J_z \; . \tag{5.55}$$

Since z-direction variations are assumed to be negligible, the z canonical momentum is constant

$$P_{zj} = mv_{zj} + \frac{q_j}{c}A_z(\mathbf{r}) = P_{0j} \; , \tag{5.56}$$

and the flux at a grid can be written as

$$\nabla^2 A_z = \frac{4\pi}{c}\left[\sum_j \frac{q_j^2}{m_j c} A_z(r_j) - \frac{q_j}{m_j} P_{0j}\right] S(\mathbf{r} - \mathbf{r}) \; . \tag{5.57}$$

Once again we can renormalize and solve this equation by iteration.

$$\nabla^2 A_z - \frac{4\pi}{c^2}\frac{q_e^2}{m}\left(1 + \frac{m}{M}\right) n_0 A_z$$

$$= \sum_j \left\{ \frac{4\pi}{c^2}\frac{q_e^2}{m}\left[(n_e - n_0) + \frac{m}{M}(n_i - n_0)\right] A_z - \frac{4\pi}{c}\frac{q_j P_{0j}}{m_j} \right\}$$

$$S(\mathbf{r} - \mathbf{r}_j) \; . \tag{5.58}$$

Because of the conservation of the canonical momentum the velocity in the z-direction does not have to be pushed [Eq. (5.56)]:

$$v_{zj} = p_{0j}/m_j - \frac{q_j}{m_j c} A_z(\mathbf{r}) \; .$$

The x, y velocities are pushed with

$$\frac{dv_{xj}}{dt} = \frac{q_j}{m_j} E_x + \frac{q_j}{m_j c}\left[v_{yj} B_0 - v_{zj} B_y\right] \; , \tag{5.59}$$

$$\frac{dv_{yj}}{dt} = \frac{q_j E_y}{m_j} + \frac{q_j}{m_j c}\left[v_{zj} B_x - v_{xj} B_y\right] \; , \tag{5.60}$$

thus completing the computational cycle.

C. Three-dimensional magnetoinductive model (Ref. 22)

Under conditions similar to those described above, the model becomes a little more complicated once it is three-dimensional, since we cannot invoke the z canonical momentum conservation anymore. This model is applicable under the following assumptions.

1. There is a strong external field B_0 in the z-direction and the perturbed field δB_z is negligible in comparison. When this assumption is satisfied the model is called incompressible, since compressions of the field lines in the z-direction have a negligible effect.
2. The plasma has a low $\beta = \frac{8\pi p}{B_0^2} \ll 1.$.
3. The transverse electric fields in the x-and y-directions are negligible.

$$E_{Tx}, E_{Ty} \ll E_{Lx}, E_{Ly} \; .$$

4. The length of the system is much greater in the z-direction: $L_z \gg L_x, L_y$, so that the wavenumbers are much smaller in the z-direction $k_z \ll k_x, k_y$ albeit that we do not assume $k_z = 0$ any more than we did in Section 5.4b.
5. As in Sections 5.4a and 5.4b, the displacement current $\dot{E}$ is neglected.

In this case Maxwell's equations are again equations (5.37) and (5.38) with $\dot{E} = 0$, and $\nabla \cdot E_L = -\nabla^2\phi = 4\pi\rho$ for a longitudinal field.

We begin by solving for B. Take the curl of the B equation.

$$- \nabla^2 \mathbf{B} = -\frac{4\pi}{c} \nabla \times \mathbf{J}_T \ . \tag{5.61}$$

Fourier transform this equation to get the component equations

$$-k^2 B_x(k) = -\frac{4\pi i}{c} \left[k_y J_{Tz}(k) - k_z J_{Ty}(k)\right] \ , \tag{5.62}$$

$$-k^2 B_y(k) = -\frac{4\pi i}{c} \left[k_z J_{Tx}(k) - k_x J_{Tz}(k)\right] \ . \tag{5.63}$$

According to assumption 4, the k_z terms are negligible compared to the k_x and k_y terms. Thus,

$$B_x(k) \simeq \frac{4\pi i}{c} \frac{k_y}{k^2} J_{Tz}(k) \ , \tag{5.64}$$

$$B_y(k) \simeq \frac{4\pi i}{c} \frac{k_x}{k^2} J_{Tz}(k) \ . \tag{5.65}$$

We have solved for B. Now we take the curl of the E equation.

$$\nabla^2 E_{T_z} = \frac{4\pi}{c^2} \frac{\partial}{\partial t} J_{T_z} \ . \tag{5.66}$$

This equation (5.66) can be cast into a form similar to Eq. (5.47). By assumption 3, we are only interested in the z component. This equation is solved by iteration as in models 5.4a and 5.4b.

5.5 Method of Relaxation

The magnetoinductive field equation (5.47) is solved by the iterative method, as discussed in Eq. (5.52). When there occurs a very nonlinear density-modulating effect or the original density $n(x)$ has a strong spatial variation, the iteration method above discussed ceases to converge.[20] The method of (over-)relaxation may relieve this trouble. For a general discussion of the relaxation method, see Ref. 23.

Suppose we have an elliptic equation of the type

$$\left[\nabla^2 + \mathcal{L}(x)\right] E(x) = \mathcal{S}(x) , \tag{5.67}$$

where $\mathcal{L}(x)$ is an operator in general. Equation (5.67) is recast into

$$\left[\nabla^2 + \bar{\mathcal{L}} + \delta\mathcal{L}(x)\right] E(x) = \mathcal{S}(x) , \tag{5.68}$$

where $\delta\mathcal{L}(x) = \mathcal{L}(x) - \bar{\mathcal{L}}$ and $\bar{\mathcal{L}}$ is independent of x. As before, the iterative equation is

$$\left(\nabla^2 + \bar{\mathcal{L}}\right) E^{m+1} = \mathcal{S} - \delta\mathcal{L}E^m , \tag{5.69}$$

where the superscripts refer to the step of iteration. Rewriting Eq. (5.69), we obtain

$$E^{m+1} = LE^m + S , \tag{5.70}$$

where $L = -\left(\nabla^2 + \bar{\mathcal{L}}\right)^{-1} \delta\mathcal{L}$ and $S = \left(\nabla^2 \bar{\mathcal{L}}\right)^{-1} \mathcal{S}$. By introducing a relaxation parameter $\lambda (0 \leq \lambda \leq 1)$, Eq. (5.70) may be modified to take the form

$$E^{m+1} = \lambda LE^m + \lambda S + (1 - \lambda)E^m . \tag{5.71}$$

By choosing an appropriate λ, Eq. (5.71) may be pursued. Subtracting $\left(\nabla^2 + \bar{\mathcal{L}}\right) E^m$ from both sides of Eq. (5.69), we obtain

$$\left(\nabla^2 + \bar{\mathcal{L}}\right) \xi^m = -S^m , \tag{5.72}$$

where $\xi^m \equiv E^{m+1} - E^m$ and $-S^m \equiv \mathcal{S} - \left(\nabla^2 + \mathcal{L}\right) E^m$. The equation for S^m is

$$S^{m+1} = S^m + \left(\nabla^2 + \mathcal{L}\right) \xi^m . \tag{5.73}$$

We can solve Eqs. (5.72) and (5.73) iteratively. These may be modified by the (over-)relaxation. By introducing an (over-)relaxation parameter α (α being typically larger than unity), we can derive the equation

$$\left(\nabla^2 + \alpha\bar{\mathcal{L}}\right) \xi^m = -S^m , \tag{5.74}$$

while the equation corresponding to Eq. (5.73) remains unchanged. This relaxation usually greatly increases the convergence for a steeply gradient background density. The equation of the same type as Eq. (5.67) will appear in the implicit particle code in Chapter 9. A similar treatment of field solvers and relaxation is employed there.

5.6 Transitions Among Hyperbolic, Parabolic, and Elliptic Equations

Although the original full electromagnetic model contains basically the structure of the hyperbolic partial differential equation (PDE), the magnetoinductive model has the transverse electric field determined by the elliptic partial differential equation. This transition of the nature of PDE from the hyperbolic to the elliptic type has taken place because the temporal evolution is frozen in the problems the magnetoinductive model addresses. The full electromagnetic particle code represents the bottom of the hierarchy of time scales (Table 1.2), containing all time scales including the fast radiation scale. The combined Maxwell equation for $\mathbf{E}_T$ reads

$$\nabla^2 \mathbf{E}_T - \frac{1}{c^2}\frac{\partial^2 \mathbf{E}_T}{\partial t^2} = \frac{4\pi}{c^2}\frac{\partial \mathbf{J}_T}{\partial t} . \tag{5.75}$$

The magnetoinductive code handles one layer higher hierarchy of time scales, so that the fastest time scale of radiation is averaged out. In the slower scale of this code, the time dependence of the displacement current is frozen. Thus the second term on the left-hand side of Eq. (5.75) is dropped. This is one manifestation of the occurrence of the change in the PDE nature due to the different level of hierarchy of the particular observation (or description). The elliptic equation has resulted from the hyperbolic Eq. (5.75) in the slower time scale, because the electric field instantaneously adjusts itself to satisfy Eq. (5.41) and the electric field in this time scale is instantaneously determined by Eq. (5.41).

This may be viewed in a slightly different way. The original Maxwell's equations are Lorentz invariant. A space-like disturbance $f(x,t)$ propagates along a space-like characteristic (see Fig. 5.2). Let the phase velocity of this disturbance $v_p = \omega/c$ (which is v_A if this is the Alfvén wave disturbance) $f(x,t) = f(x - v_p t)$. By the Lorentz transformation,

$$x' = \gamma(x - \beta ct) , \tag{5.76}$$

$$t' = \gamma(t - \beta x/c) , \tag{5.77}$$

$$\mathbf{E}'_\perp = \gamma(\mathbf{E}_\perp + \boldsymbol{\beta} \times \mathbf{B}_\perp) \tag{5.78}$$

$$\mathbf{B}'_\perp = \gamma(\mathbf{B}_\perp - \boldsymbol{\beta} \times \mathbf{E}_\perp) \tag{5.79}$$

where the quantities with primes are in the Lorentz transformed frame and $\beta = v_p/c$ and $\gamma = (1-\beta^2)^{-1/2}$ and $\perp$ means the quantities perpendicular to the transform direction. When we go onto the moving frame of the disturbance ($v = v_p = \beta c$), the disturbance becomes space-like $f(x - v_p t) = f(x')$ and

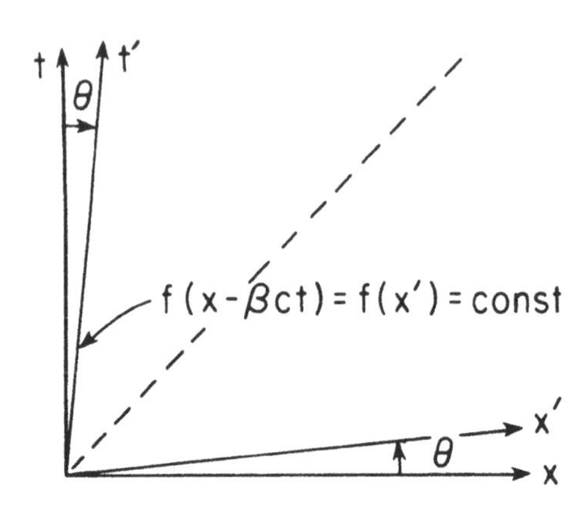

FIGURE 5.2 A space-like disturbance $f(x,t)$ and the Lorentz transform

time(t')-independent in this frame. If the disturbance is slow, as is the case of present consideration[19] ($v_p \ll c$, e.g. $v_p = v_A$), the angle θ between the characteristic and t-axis is very small and they are nearly parallel (Fig. 5.2). In this frame Eq. (5.75), where all the quantities are primed, becomes elliptic [Eq. (5.41)]. The Darwin model is not Lorentz invariant. In this model, instead of legitimately making Maxwell's equation space-like (and thus elliptic), we approximate θ by zero, or we neglect the second term on the left-hand side of Eq. (5.75) *without* carrying out the Lorentz transform. In the laboratory frame, the transverse electric field is not in general zero. However, it vanishes in the phase velocity frame [$\mathbf{E}'_L = 0$ in Eq. (5.78)]. In this frame all the fields are magnetic. As $\beta \ll 1$, we can expect that even in the laboratory frame the magnetic fields are much larger than the electric fields in such a region of slow time scales.

We now study the finite differenced forms of Maxwell's equations. A form of finite differenced equations, Eqs. (5.12) and (5.13) may be written as

$$B^{n+1/2} - B^{n-1/2} = -\Delta t cik \times E^n , \tag{5.80}$$

$$E^{n+1} - E^n = \Delta t \left(cik \times B^{n+1/2} - 4\pi J^{n+1/2} \right) , \tag{5.81}$$

$$J^{n+1/2} - J^{n-1/2} = \Delta t \frac{ne^2}{m} E^n . \tag{5.82}$$

Equation (5.82) arises from $\dot{v} = -eE/m$. The amplification analysis (see

Chapter 3) with $g = e^{-i\Delta t\omega}$ starts from

$$(g-1)B + \Delta tcikE = 0 , \tag{5.83}$$

$$(g-1)E + \Delta tcikBg + 4\pi\Delta tJg = 0 , \tag{5.84}$$

$$(g-1)J - \Delta t\frac{ne^2}{m}E = 0 . \tag{5.85}$$

The secular equation to Eqs. (5.83)-(5.85) is

$$(g-1)^3 + (g-1)\Delta t^2c^2k^2g + (g-1)\omega_p^2\Delta t^2 g = 0 ,$$

whose solutions are $g = 1$ or

$$g^2 - 2\left(1 - \frac{\omega_p^2 + k^2c^2}{2}\Delta t^2\right) + 1 = 0 . \tag{5.86}$$

This leads to

$$|g|^2 = 1 \quad \text{if} \quad 1 > (\omega_p^2 + k^2c^2)\Delta t^2/2 , \tag{5.87}$$

$$\text{(neutral)} ,$$

$$|g|^2 = \left(A + \sqrt{A^2-1}\right)^2 \quad \text{if} \quad 1 < (\omega_p^2 + k^2c^2)\Delta t^2/2 \tag{5.88}$$

$$\text{(unstable)}$$

where

$$A \equiv 1 - \frac{\omega_p^2 + k^2c^2}{2}\Delta t^2 .$$

We thus obtain the stability condition

$$\Delta t < \frac{\sqrt{2}}{\sqrt{\omega_p^2 + k^2c^2}} , \tag{5.89}$$

where the maximum possible wavenumber k is $k_{\max} = \frac{\pi}{2}$. The stability of the explicit scheme is determined by a sufficiently small time step to ensure that the fastest signal does not cross the unit cell in a time step. This is the Courant-Friedrichs-Lewy condition. In the explicit scheme, the hyperbolic equation keeps its essential nature as long as it is stable. Instead of Eqs. (5.80)-(5.82), we assign implicit finite-differencing time steps.

$$B^{n+1/2} - B^{n-1/2} = c\Delta t - k \times E^{n+1} , \tag{5.90}$$

and

$$E^{n+1} - E^n = \Delta t\left(cik \times B^{n+1/2} - 4\pi J^{n+1}\right) ,$$

and

$$J^{n+1} = J^n + \Delta t \frac{ne^2}{m} E^{n+1} . \quad (5.91)$$

Equation (5.90) yields

$$\left(1 + k^2c^2\Delta t^2 + \omega_p^2\Delta t^2\right) E^{n+1} = E^n + \Delta t cik \times B^{n-1/2} - 4\pi\Delta t J^n . \quad (5.92)$$

In large Δt limit, neglecting small $(\Delta t)^0$ terms in the above, Eq. (5.92) becomes

$$E^{n+1}(k^2c^2 + \omega_p^2)\Delta t = cik \times B^{n-1/2} - 4\pi J^{n-1} = \mathcal{S} \quad \text{(source)} . \quad (5.93)$$

If we drop the source terms, Eq. (5.93) becomes

$$c^2\nabla \times \nabla \times E + \omega_p^2 E = 0 , \quad (5.94)$$

which is a typical elliptic function in the P.D. equation. Recall Eq. (5.50), which has the form of P.D.E.

$$\nabla^2 E - F(E \text{ or } J) = 0 ,$$

where F is a function of E or J.

In this model we have frozen the radiative (displacement current) time scale. As a result, we obtain an elliptic P.D.E. out of the hyperbolic P.D.E. of the EM waves, as we discussed in the above.

The smooth transition from the (purely) hyperbolic P.D.E. nature to the elliptic P.D.E. nature of Maxwell's equations may be seen in the implicit finite differencing formulation of the P.D.E. of Maxwell's equations. Compare the implicit solution with the solution for the explicit scheme. The secular equation to Eqs. (5.90)-(5.92) is

$$(g-1)^3 + \frac{4\pi ne^2}{m}\Delta t^2 g^2(g-1) + \Delta t^2 c^2 k^2 (g-1) = 0 .$$

Besides $g - 1 = 0$, we obtain from the above

$$g^2\left[1 + \Delta t^2(\omega_p^2 + c^2k^2)\right] - 2g + 1 = 0 . \quad (5.95)$$

From Eq. (5.95) we obtain

$$|g|^2 = \frac{1}{1 + \Delta t^2(\omega_p^2 + k^2c^2)} < 1 , \quad (5.96)$$

indicating unconditional stability for all Δt. The solution is

$$e^{-i\omega t} = g = \frac{1 \pm i\Delta t\sqrt{\omega_p^2 + k^2c^2}}{1 + \Delta t^2(\omega_p^2 + k^2c^2)} \ . \tag{5.97}$$

When $\omega\Delta t \ll 1$, the numerical dispersion relation Eq. (5.97) gives rise to

$$e^{-i\omega\Delta t} \simeq 1 - i\omega\Delta t = \left[1 - \Delta t(\omega_p^2 + k^2c^2)\right] \pm i\Delta t\sqrt{\omega_p^2 + k^2c^2} + O(\Delta t^2\omega^2) \ . \tag{5.98}$$

This yields $\omega \simeq \pm\sqrt{\omega_p^2 + k^2c^2}$, as expected. This is a fully propagating wave solution, corresponding to the hyperbolic equation. When $\omega\Delta t \gg 1$, on the other hand, we have

$$e^{-i\omega\Delta t} \simeq i\frac{1}{\Delta t\sqrt{\omega_p^2 + k^2c^2}} \ , \tag{5.99}$$

leading to

$$\mathrm{Re}\omega = \pm\frac{\pi}{2}\frac{1}{\Delta t} \quad \text{(Nyquist frequency)} \ , \tag{5.100}$$

$$\mathrm{Im}\omega = -\frac{1}{\Delta t}\ell n\left(\Delta t\sqrt{\omega_p^2 + k^2c^2}\right) \ . \tag{5.101}$$

In the limit of large Δt, it reduces to a "static" solution, with the real frequency associated only with the time step. This corresponds to the magnetoinductive problem. Brackbill and Forslund discussed this problem in detail.[24] It is a manifestation of the properties of spectral compression and selective damping of the implicit finite differencing in time. More discussion ensues in Chapter 9.

The next example of the change in nature of the P.D.E. due to the finite differencing in time is the diffusion equation.[23] The diffusion equation is a parabolic equation:

$$\frac{\partial u}{\partial t} = -v\frac{\partial u}{\partial x}\kappa\frac{\partial^2 u}{\partial x^2} \ . \tag{5.102}$$

A typical finite differencing would be

$$\frac{u_j^{n+1} - u_j^n}{\Delta t} = -v\left(\frac{u_{j+1}^n - u_{j-1}^n}{2\Delta x}\right) + \kappa\frac{u_{j+1}^n - 2u_j^n + u_{j-1}^n}{\Delta x^2} \ . \tag{5.103}$$

We evaluate the equivalent differential equation by the Taylor expansion:

$$u_j^{n+1} = u_j^n + \Delta t\left.\frac{\partial u}{\partial t}\right|_j^n + \frac{1}{2}\left.\frac{\partial^2 u}{\partial t^2}\right|_j^n + O(\Delta t^3)$$

$$u_{j\pm 1}^n = u_j^n \pm \Delta x\left.\frac{\partial u}{\partial x}\right|_j^n + \frac{1}{2}\left.\frac{\partial^2 u}{\partial x^2}\right|_j^n \pm O(\Delta x^3) \ .$$

Using these expressions, we can express Eq. (5.103) as

$$\frac{1}{\Delta t}\left[\Delta t \frac{\partial u}{\partial t}\bigg|_j^n + \frac{1}{2}\Delta t^2 \frac{\partial^2 u}{\partial t^2}\bigg|_j^n + 0(\Delta t^3)\right]$$

$$= -\frac{v}{2\Delta x}\left[2\Delta x \frac{\partial u}{\partial x}\bigg|_j^n + O(\Delta x^3)\right] + \frac{\kappa}{\Delta x^2}\frac{\partial^2 u}{\partial x^2}\bigg|_j^n + O(\Delta x^4) .$$

This is written as an equivalent differential equation

$$\frac{\partial u}{\partial t} + \frac{\Delta t}{2}\frac{\partial^2 u}{\partial t^2} = -v\frac{\partial u}{\partial x} + \kappa\frac{\partial^2 u}{\partial x^2} + O(\Delta t^2, \Delta x^2) . \tag{5.104}$$

The second term on the left-hand side of Eq. (5.104) is a "hyperbolic term" appearing in the parabolic equation. This equation [Eq. (5.104)] now has a hyperbolic character. The hyperbolic equation exhibits a *domain of influence* of an arbitrary point (x, t) demarcated by the characteristic lines of slope $\pm\sqrt{\Delta t/2\kappa}$ through (x, t) as shown in Fig. 5.3.

A space-time diagram shows that each point has a domain of dependence (value at (x, t) depends on values in this zone), a domain of influence (value at (x, t) influences values in this zone), and a zone of silence. The slope of the characteristics that bound these zones is $\pm\sqrt{\Delta t/2\kappa}$.

In the finite differencing scheme, the point (j, n) depends on $(j \pm 1, n-1)$ and influences $(j \pm 1, n+1)$. The slopes of the characteristics are ± 1, which represents a physical quantity $\pm\Delta t/\Delta x$.

If we compare the domain of influence of the P.D.E. with that of the finite difference equation, we derive the CFL condition from the appropriate inequality of the slopes:

$$\frac{\Delta t}{\Delta x} \le \sqrt{\frac{\Delta t}{2\kappa}} \Longrightarrow \Delta t \le \frac{1}{2}\frac{\Delta x^2}{\kappa} . \tag{5.105}$$

This means we must make sure that we include the entire domain of influence of the P.D.E. within the domain of influence of the equation we actually calculate.

For the parabolic equation, the information has to transmit instantaneously, but to eliminate hyperbolic influences we say that influence propagates faster than the diffusive characteristic time.

5.7 Classification of Second-Order P.D.E.

We have encountered hyperbolic, parabolic, and elliptic partial differential equations and their crossovers. In this section we review the general prop-

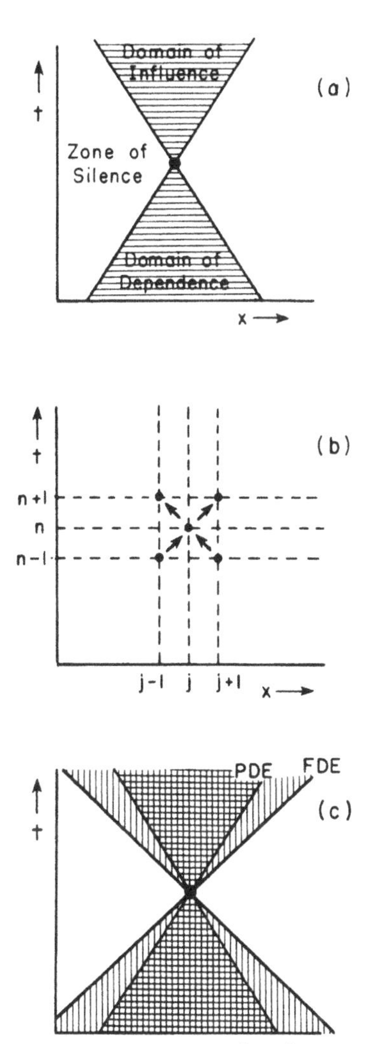

FIGURE 5.3 Hyperbolic differential equation and its domains of influence and silence. [(a) and (b)] The condition for finite differenced equation being parabolic (c)

erties of second order P.D.E.'s and their finite difference counterparts. Let A, B, C, D, E, F, G, H be functions of x, y. A general second order P.D.E. for

$f(x, y)$ may be written as[25]

$$A\frac{\partial^2 f}{\partial x^2} + B\frac{\partial^2 f}{2x\partial y} + C\frac{\partial^2 f}{2y^2} + D\frac{\partial f}{\partial x} + E\frac{\partial f}{\partial y} + Ff = G(x, y) \; , \tag{5.106}$$

where $G(x, y)$ is a source (inhomogeneous) term. Often it is advantageous to transform x, y to other variables in order to eliminate some of the functions. Consider the special case in which D, E, and F vanish, and also A, B, and C are constants for simplicity. One can always do this locally by an appropriate change of variables, if A, B, C, D, E, and F are well behaved locally. Choose

$$\xi = x + \lambda y \quad , \quad \eta = x + \mu y \; , \tag{5.107}$$

where

$$A + B\lambda + C\lambda^2 = 0 \quad , \quad A + B\mu + C\mu^2 = 0 \; ; \tag{5.108}$$

that is, λ and μ are roots of Eq. (5.108). Equations (5.107) are called characteristics.

Replace A, B, C, by $a, 2b, c$, respectively. We now get

$$(a + 2b\lambda + c\lambda^2)\frac{\partial^2 f}{\partial \xi^2} + 2\left[a + b(\lambda + \mu) + c\lambda\mu\right]\frac{\partial^2 f}{\partial \xi \partial \mu}$$

$$+(a + 2b\mu + c\mu^2)\frac{\partial^2 f}{\partial \eta^2} = G_1(\xi, \eta) \; .$$

A clever choice of the transformation makes the first and third term vanish, as we have specified λ and μ in Eq. (5.108). The solution is

$$f(\xi, \eta) = -\frac{c}{4(b^2 - ac)}\int d\xi d\eta G_1(\xi, \eta) + H(\eta) + I(\xi) \; , \tag{5.109}$$

where H and I are arbitrary functions of η and ξ, respectively. For example, if

$$\xi = x + ct \quad , \quad \eta = x - ct \; , \tag{5.110}$$

we have propagating wave solutions with phase velocity c.

Choose a different λ, μ such that

$$a + b(\lambda + \mu) + c\lambda\mu = 0 \; . \tag{5.111}$$

Now we see that

$$\lambda = -\frac{a + b\mu}{b + c\mu} \quad , \quad a + 2b\lambda + c\lambda^2 = -\frac{b^2 - ac}{(b + c\mu)^2}(a + 2b\mu + c\mu^2) \tag{5.112}$$

so

$$m^2\frac{\partial^2 f}{\partial \xi^2} - \frac{\partial^2 f}{\partial \eta^2} = -\frac{G_1(\xi, \eta)}{a + 2b\mu + c\mu^2} \quad , \quad m^2 \equiv \frac{b^2 - ac}{(b + c\mu)^2} \; . \tag{5.113}$$

This gives $1/m$ as the slope of the characteristics. On the characteristics the value of f is constant, if the source term G_1 does not exist.

For a parabolic equation, $b^2 - ac = 0$. There is a double root, λ. From Eq. (5.112) it follows that

$$a + 2b\mu + c\mu^2 = c(\lambda - \mu)^2 \; ; \tag{5.114}$$

thus,

$$0 \cdot \frac{\partial^2 f}{\partial \xi^2} + 0 \cdot \frac{\partial^2 f}{\partial \xi \partial \eta} + (a + 2b\mu + c\mu^2)\frac{\partial^2 f}{\partial \eta^2} = G_1(\xi, \eta) \; . \tag{5.115}$$

The first coefficient is zero by demand, while the second is zero by choice. The solution to Eq. (5.115) is

$$f(\xi, \eta) = \frac{1}{c(\lambda - \mu)^2} \int d\eta d\xi G_1(\xi, \eta) + \eta H(\xi) + I(\xi) \; . \tag{5.116}$$

Again H and I are arbitrary functions. Note, however, that H and I have the same argument.

For elliptic type equations, $b^2 - ac$ is negative. Choosing λ and μ according to Eq. (5.108), they are imaginary numbers. So we choose them instead according to Eq. (5.111). This gives us the relation,

$$a + 2b\lambda + c\lambda^2 = m^2(a + 2b\mu + c\mu^2) \quad , \quad m^2 \equiv \frac{ac - b^2}{(b + c\mu)^2} \; ; \tag{5.117}$$

so we obtain

$$m^2 \frac{\partial^2 f}{\partial \xi^2} + \frac{\partial^2 f}{\partial \eta^2} = \frac{G(\xi, \eta)}{a + 2b\mu + c\mu^2} \; . \tag{5.118}$$

This expression reduces to the standard form for m set equal to 1.

Let us discuss the method of characteristics.[25] Now let A, B, and C be functions of x, y such that

$$A(x, y)\frac{\partial^2 f}{\partial x^2} + B(x, y)\frac{\partial^2 f}{\partial x \partial y} + C(x, y)\frac{\partial^2 f}{\partial y^2} = G\left(x, y, f, \frac{\partial f}{\partial x}, \frac{\partial f}{\partial y}\right) \; . \tag{5.119}$$

Suppose we have f and its normal derivative on some strip $s = s(x_0, y_0)$; this is the Cauchy condition. See Fig. 5.4.

The method of characteristics yields[25] a solution if the determinant of the matrix

$$\Delta = \begin{vmatrix} \frac{dx_0}{ds} & \frac{dy_0}{ds} & 0 \\ 0 & \frac{dx_0}{ds} & \frac{dy_0}{ds} \\ A & B & C \end{vmatrix} = C\left(\frac{dx_0}{ds}\right)^2 - B\frac{dx_0}{ds}\frac{dy_0}{ds} + A\left(\frac{dy_0}{ds}\right)^2 \neq 0 \; , \tag{5.120}$$

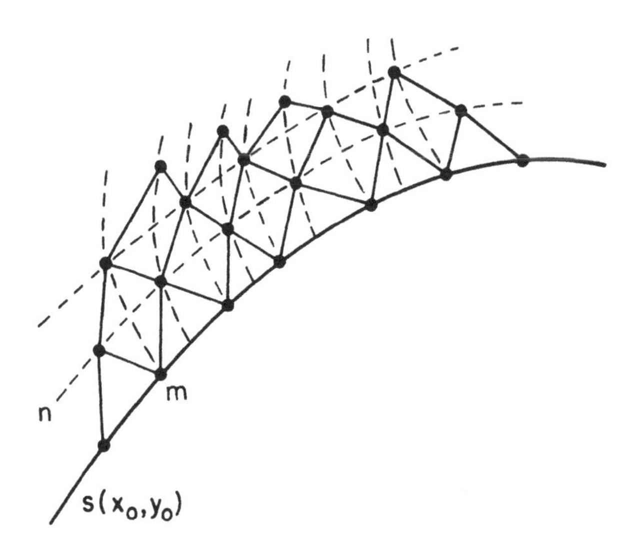

FIGURE 5.4 Boundary conditions and the nature of characteristics (Ref. 25)

where (x_0, y_0) is on the initial boundary $s(x_0, y_0)$. If the derivative vanishes, we obtain the characteristics

$$\Delta = 0 \Longrightarrow A dy = \frac{1}{2}\left(B \pm \sqrt{B^2 - 4AC}\right) dx \ . \tag{5.121}$$

That is, if the boundary coincides with one of the two characteristics, the Cauchy conditions do not uniquely specify the solutions (underspecification) [Fig. 5.5(a)]. For $B^2 - 4AC$ positive, the equation is hyperbolic. In order for this condition to mean anything physically, the two characteristics must be real. When the boundary crosses both families of characteristics, the Cauchy conditions uniquely determine solutions. On the other hand, Cauchy conditions on a closed boundary lead to overspecification [Fig. 5.5(c)].

The Cauchy condition requires defined values on two grid layers.[25] A hyperbolic equation relates $f(m \pm 1, n)$ to $f(m, n \pm 1)$. There is a conspicuous lack of participation of the middle point $f(m, n)$. The calculated values spread diagonally outward from the boundary. If the boundary is closed or badly

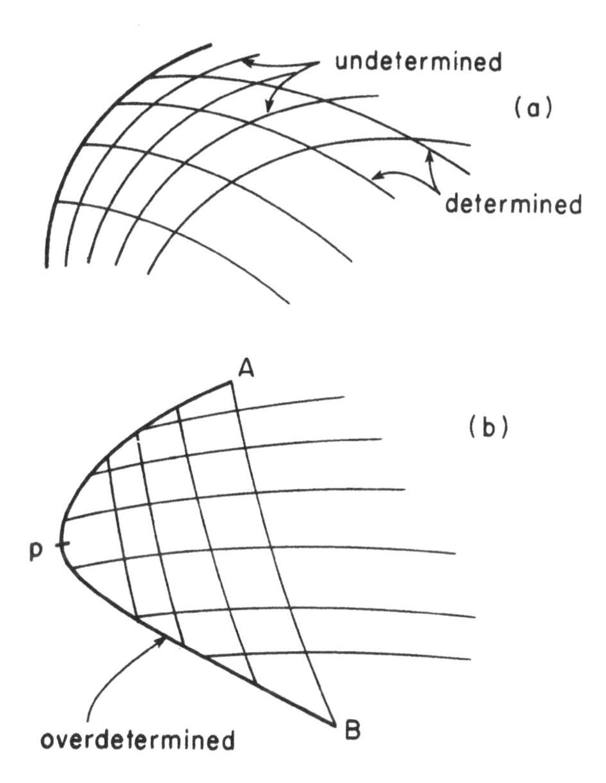

FIGURE 5.5 The boundary, numerical characteristics, and grids (Ref. 25)

warped, the calculated values propagate into the boundary, and the equation is overspecified.

For an elliptic equation, on the other hand, the value of f needs to be specified on a closed boundary. Suppose the values of f are given only for $n = 0$. See Fig. 5.6. To solve the Laplace equation at f_{21}, the first step is

$$f_{21} = \frac{1}{4}(f_{31} + f_{11} + f_{22} + f_{20}) = \frac{1}{4}(f_{31} + f_{11} + f_{22}) + \frac{1}{4}f_{20} \,. \qquad (5.122)$$

In the next step expand the non-boundary terms:

$$f_{21} = \frac{1}{13}(f_{41} + f_{23} + 2f_{32} + 2f_{12}) + \frac{1}{13}(f_{01} + f_{10} + f_{30} + 4f_{20}) \,. \qquad (5.123)$$

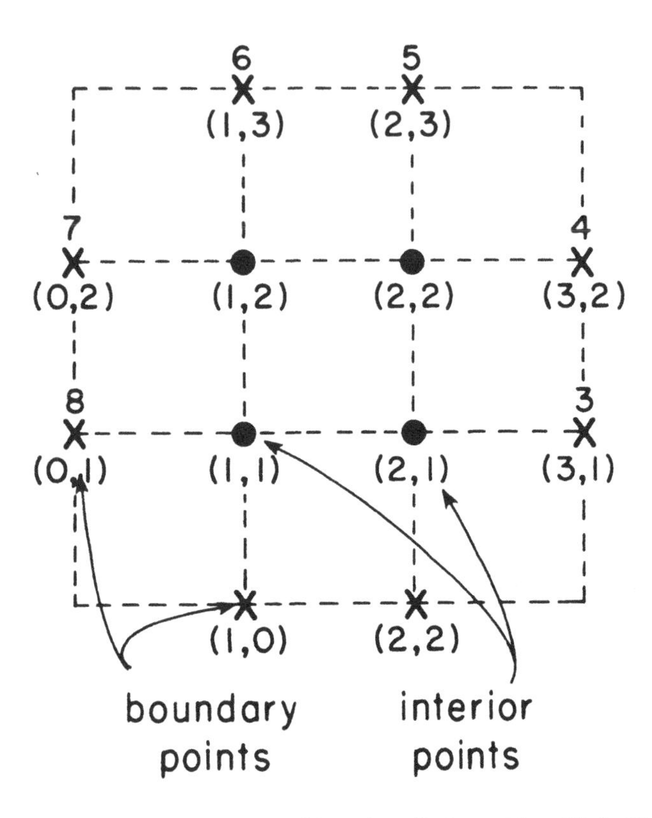

FIGURE 5.6 Grid and boundary conditions for elliptic problem (Ref. 15)

We can show[23] that all other steps are of the form (see Fig. 5.6).

$$f_{21} = \begin{pmatrix} \text{sum over} \\ \text{interior points} \end{pmatrix} + (\text{sum over boundary points}) \ . \qquad (5.124)$$

Only if there is a boundary on the other side, will all the interior points eventually become manageable.

Problems

1. If and when does the condition $v_A \ll c$ break down? Can you relate this to the condition for the magnetoinductive approximation?
2. The magnetoinductive field equation (5.48) is usually solved efficiently by the so-called pseudo-spectral method. That is, the convolution term on the right-hand side of Eq. (5.48) is computed as the Fourier transform of the product in real space. Describe this strategy. See also Section 6.8.
3. Show that Eq. (5.71) is equivalent to Eq. (5.70) for the converged solution.
4. Derive Eq. (5.74). Also by choosing a slightly different relaxation, derive the equation $\alpha\left(\nabla^2 + \bar{\mathcal{L}}\right)\xi^m = -S^m$.
5. Write down the dispersion relation of the quark-gluon plasma. References include B. Müller[26] and S.M. Mahajan and P. Valanju.[27] Consider a non-relativistic limit of "electromagnetic" particle code for the Yang-Mills fields of the quark-gluon system. Here a non-relativistic limit allows us to prohibit creation and annihilation of quarks.
6. Run the "EM" code for the gravitoelectrodynamics and observe any difference from the usual EM code result.
7. According to Feynman[28] the total angular momentum that is the sum of the particle angular momentum and the electromagnetic angular momentum conserves in a closed system. Define the electromagnetic angular momentum properly. Run the electromagnetic particle code in one dimension $\left(1\text{-}\frac{2}{2}D\right)$ and show the above.[29]
8. Show that the finite difference form of the diffusion equation Eq. (5.102) has the equivalent differential equation

$$\frac{\partial u}{\partial t} = \kappa' \frac{\partial^2 u}{\partial x^2} - v\frac{\partial u}{\partial x},$$

with $\kappa' = \kappa - v^2 \Delta t/2$, an effective diffusion coefficient. The finite difference equation or its equivalent differential equation becomes unstable when $\Delta t < 2\kappa/v^2$ or $\kappa' < 0$.

9. Obtain analytically a one-dimensional Green's function for Eq. (5.94) and compare it with numerical result from Eq. (5.93).

References

1. A.T. Lin, J.M. Dawson, and H. Okuda, Phys. Fluids **21**, 1995 (1974).

2. A.B. Langdon and B.F. Lasinski, *Methods in Computational Physics* vol. 16, p. 327 (1976).
3. R.W. Hockney and J.W. Eastwood, *Computer Simulation Using Particles* (McGraw Hill, New York, 1981) p. 409.
4. T. Tajima, A. Clark, G.G. Craddock, D.L. Gilden, W.W. Leung, Y.M. Li, J.A. Robertson, and B.J. Saltzman, Am. J. Phys. **53**, 305 (1985).
5. K.S. Thorne and R.D. Blandford, *Extragalactic Radio Sources* (International Astronomical Union Symposium No. 97) ed. by D.S. Heeshen and C.M. Wade (D. Reidel, Dordreich, Holland, 1982) p. 255.
6. C.W. Misner, K.S. Thorne, and J.A. Wheeler, *Gravitation* (W.H. Freeman, San Francisco, 1970) p. 486.
7. R.W. Hockney, *Methods in Computational Physics*, (Academic Press, New York, 1970) Vol. 9, p. 136.
8. J.P. Matte and G. Lafrance, J. Comput. Phys. 23, 86 (1977).
9. O. Buneman, J. Comput. Phys. **12**, 124 (1973).
10. V.K. Decyk and J.M. Dawson, J. Comput. Phys. **30**, 407 (1979).
11. A.B. Langdon and B.F. Lasinski, *Methods in Computational Physics*, (Academic Press, New York, 1976) Vol. 16, p. 327 .
12. E.L. Lindman, J. Comput. Phys. **18**, 66 (1975).
13. T. Tajima and Y.C. Lee, J. Comput. Phys. **42**, 406 (1981).
14. I. Orlanski, J. Comput. Phys. **21**, 215 (1976).
15. K.G. Budden, *Physics of the Ionosphere: Report of Phys. Soc. Conf. Cavendish Lab.*, (Physics Society, London, 1955) p. 320.
16. T.H. Stix, *The Theory of Plasma Waves*, Chapter 10, (McGraw-Hill, New York, 1962).
17. F.H. Bushby and M.S. Timpson, Q.J.R. Meteorol Soc. **93**, 1 (1967).
18. O. Buneman, Comp. Phys. Comm. **12**, 21 (1976).
19. C.W. Nielson and H.R. Lewis, *Methods in Computation Physics*, vol. 16, p. 367 (1976).
20. J. Busnardo-Neto, P.L. Pritchett, A.T. Lin, and J.M. Dawson, J. Comp. Phys. **23**, 300 (1977).
21. A.T. Lin, Phys. Fluids **21**, 1026 (1978).
22. H. Okuda, W.W. Lee, and C.Z. Cheng, Comp. Phys. Comm. **17**, 233 (1979).
23. G.D. Smith, *Numerical Solution of Partial Differential Equations*, 1st ed. (Clarendon, Oxford, 1965) p. 76; P.J. Roache, *Computational Fluid Dynamics* (Hermosa Publishers, Albuquerque, 1972) p. 46.

24. J.U. Brackbill and D.W. Forslund, in *Multiple Time Scales* eds. J. U. Brackbill and B.I. Cohen (Academic, New York, 1985) p. 272.

25. P.M. Morse and H. Feshbach, *Mathematical Method of Physics* (McGraw Hill, New York, 1953) p. 692.

26. B. Müller, *The Physics of the Quark-Gluon Plasma* (Springer-Verlag, Berlin, 1985).

27. S.M. Mahajan and P. Valanju, Phys. Rev. **D35**, 2543 (1987).

28. R.P. Feynman, R.B. Leighton, and M. Sands, *The Feynman Lectures on Physics,* vol. 2 (Addison-Wesley, Reading, Ma., 1965).

29. M. Thaker, J.N. Leboeuf, and T. Tajima, IFSR#201 (1981).

6
MAGNETOHYDRO-DYNAMIC MODEL OF PLASMAS

One of the most popular approaches that coarse-grain microscopic plasma processes is the fluid approach. In a magnetized plasma this description of the plasma is often called magnetohydrodynamics. If the magnetic field vanishes, it reduces to hydrodynamics. We shall derive the magnetohydrodynamic (MHD) model of a plasma. We start from the Vlasov equation for the distribution function $f(\mathbf{x}, \mathbf{v}, t)$.

$$0 = \frac{df}{dt} = \frac{\partial f}{\partial t} + \mathbf{v} \cdot \nabla f + \frac{q}{m}\left(\mathbf{E} + \frac{\mathbf{v}}{c} \times \mathbf{B}\right) \cdot \frac{\partial f}{\partial \mathbf{v}} . \tag{6.1}$$

If we have collisions, $\frac{df}{dt} = \left(\frac{\partial f}{\partial t}\right)_{\text{coll.}}$ See Chapter 1.

The velocity moment is defined as:

$$\bar{Q} = \int d\mathbf{v} Q(\mathbf{v}) f(x, \mathbf{v}, t) . \tag{6.2}$$

For example, when $Q(\mathbf{v})$ takes specific forms, the moment corresponds to some well-known macroscopic quantities

$$Q(\mathbf{v}) \begin{cases} 1 & \longrightarrow \quad \text{density} \\ \mathbf{v} & \longrightarrow \quad \text{momentum} \\ \\ \mathbf{v}^2 & \longrightarrow \quad \text{energy} \end{cases} \tag{6.3}$$

a) $Q = 1$ case

$$\frac{\partial n}{\partial t} = -\boldsymbol{\nabla} \cdot (n\mathbf{V}) \quad \text{continuity equation} \tag{6.4}$$

where $\mathbf{V} = \langle \mathbf{v} \rangle$, the average velocity

b) $Q = \mathbf{v}$ case

$$\frac{\partial}{\partial t}(n\mathbf{V}) = -\boldsymbol{\nabla} \cdot (n \langle \mathbf{vv} \rangle) - \frac{q}{m} n\mathbf{E} - \frac{q}{mn} \frac{\mathbf{V} \times \mathbf{B}}{c}$$
$$+ \frac{\partial}{\partial t}(n\mathbf{V})_{\text{coll.}} \tag{6.5}$$

c) $Q = v^2$ case

$$\frac{\partial}{\partial t} \left(n \left\langle v^2 \right\rangle \right) = -\nabla \cdot \left(n \left\langle v^2 \mathbf{v} \right\rangle \right) + \frac{2nq}{m} \mathbf{E} \cdot \mathbf{V}$$
$$+ \left(\frac{\partial}{\partial t} \left(n \left\langle v^2 \right\rangle \right) \right)_{\text{coll.}} \quad . \tag{6.6}$$

Very often terms beyond the second moment are neglected, or at best, approximated by some combination of lower moment quantities. One can write the above equation for both electron and ion species. Then, by summing, one obtains equations for one fluid description.

Let us write $\mathbf{v} = \mathbf{V} + \delta\mathbf{v}$ where $\delta\mathbf{v}$ is the random velocity. Expand $\mathbf{v}$ around the average noting that $\langle \mathbf{W} \rangle = 0$ and define the pressure tensor

$$\overleftrightarrow{\Pi}_\alpha \equiv m_\alpha \langle \delta\mathbf{v} \delta\mathbf{v} \rangle = m_\alpha \langle (\mathbf{v} - \mathbf{V})(\mathbf{v} - \mathbf{V}) \rangle \ .$$

The resulting momentum equations for electrons and ions are:

$$mn_e \left(\frac{\partial}{\partial t} \mathbf{V}_e + \mathbf{V}_e \cdot \boldsymbol{\nabla} \mathbf{V}_e \right) + \nabla \cdot \overleftrightarrow{\Pi}_e + n_e e \mathbf{E} + \frac{n_e e}{c} \mathbf{V}_e \times \mathbf{B}$$
$$= \left(\frac{\partial \mathbf{P}_e}{\partial t} \right)_{i \rightarrow e} , \tag{6.7}$$

$$Mn_i\left(\frac{\partial}{\partial t}\mathbf{V}_i + \mathbf{V}_i\cdot\nabla\mathbf{V}_i\right) + \nabla\cdot\overleftrightarrow{\Pi}_i - n_i Ze\mathbf{E}$$

$$-n_i Ze\frac{\mathbf{V}_i}{c}\times\mathbf{B} = \left(\frac{\partial\mathbf{P}_i}{\partial t}\right)_{e\to i} = -\left(\frac{\partial\mathbf{P}_e}{\partial t}\right)_{i\to e} \tag{6.8}$$

where the right-hand sides of Eqs. (6.7) and (6.8) represent the momentum exchange by collisions from ions to electrons ($i \to e$) and vice versa.

We define total quantities summed over electrons and ions

$\mathbf{V} =$	$\frac{n_i M\mathbf{V}_i + n_e m\mathbf{V}_e}{n_i M + n_e m}$	Center of mass velocity
$\mathbf{J} =$	$e\left(Zn_i\mathbf{V}_i - n_e\mathbf{V}_e\right)$	Current density
$\rho =$	$n_i M + n_e m$	Mass density
$\sigma =$	$e(n_i Z - n_e)$	Charge density .

Summing the appropriate equations, we get the following equations:

the continuity equation

$$\frac{\partial}{\partial t}\rho + \boldsymbol{\nabla}\cdot(\rho\mathbf{V}) = 0\ , \tag{6.9}$$

the momentum conservation equation:

$$\frac{\partial}{\partial t}(\rho\mathbf{V}) + \boldsymbol{\nabla}\cdot\left\{\rho\mathbf{V}\mathbf{V} + \frac{m}{n_e e^2}\mathbf{J}\mathbf{J}\left(1 + \frac{Zm_e}{M}\right) + \overleftrightarrow{\Pi}_e + \overleftrightarrow{\Pi}_i\right\}$$

$$-\sigma\mathbf{E} - \frac{\mathbf{J}\times\mathbf{B}}{c} = 0\ , \tag{6.10}$$

the equation for current:

$$\frac{m}{n_e e^2}\frac{\partial\mathbf{J}}{\partial t} + \boldsymbol{\nabla}\cdot\left\{-\frac{1}{n_e e^2}\mathbf{J}\mathbf{J} + (\mathbf{V}\mathbf{J} + \mathbf{J}\mathbf{V}) + \frac{e}{m}\left(\frac{Zm}{M}\overleftrightarrow{\Pi}_i - \overleftrightarrow{\Pi}_e\right)\right\}$$

$$-\left(\mathbf{E} + \frac{\mathbf{V}\times\mathbf{B}}{c}\right) + \frac{1}{n_e ec}\mathbf{J}\times\mathbf{B} = -\eta\mathbf{J}\ . \tag{6.11}$$

(η resistivity)

In order to arrive at more conventional MHD equations, we "linearize" (drop terms like **VV**, **JV**, **VJ**, etc.). We then obtain

$$\rho\frac{d\mathbf{V}}{dt} + \boldsymbol{\nabla}\cdot\overleftrightarrow{P} - \sigma\mathbf{E} - \frac{\mathbf{J}\times\mathbf{B}}{c} = 0\ , \tag{6.12}$$

where $\overleftrightarrow{P} = \overleftrightarrow{\Pi}_e + \overleftrightarrow{\Pi}_i$ and $d\mathbf{V}/dt \equiv \partial\mathbf{V}/\partial t + \mathbf{V}\cdot\nabla\mathbf{V}$. The third term may be dropped because $\sigma \simeq 0$ since $n_i \simeq n_e$ (quasineutrality).

$$\frac{4\pi}{\omega_p^2}\frac{\partial \mathbf{J}}{\partial t} + \frac{1}{en_e}\nabla\cdot\left(\frac{Zm}{M}\overleftrightarrow{\Pi}_i - \overleftrightarrow{\Pi}_e\right) - \left(\mathbf{E} + \frac{\mathbf{V}\times\mathbf{B}}{c}\right)$$

$$+\frac{\mathbf{J}\times\mathbf{B}}{n_e ec} = -\eta\mathbf{J}\ . \tag{6.13}$$

The first term may be dropped because the MHD frequency is small compared to ω_p. The last term on the left-hand side of Eq. (6.13) is called the Hall term. [See also Eq. (6.25))]. In the low frequency regime it may be neglected ($\omega \ll \Omega_i$). After all these approximations and assumptions, we recover the (conventional) MHD equations:

$$\frac{\partial\rho}{\partial t} + \nabla\cdot(\rho\mathbf{V}) = 0\ , \tag{6.14}$$

$$\rho\frac{d\mathbf{V}}{dt} = \frac{1}{c}\mathbf{J}\times\mathbf{B} - \nabla\cdot\overleftrightarrow{P}\ , \tag{6.15}$$

$$\mathbf{E} + \frac{\mathbf{V}\times\mathbf{B}}{c} - \frac{1}{en}\boldsymbol{\nabla}\overleftrightarrow{\Pi}_e = \eta\mathbf{J}\ . \tag{6.16}$$

And yet another "simpler" set of equations is arrived at by making more simplifications. This set of equations is the "Ideal MHD Equations."

$$\frac{\partial\rho}{\partial t} + \boldsymbol{\nabla}\cdot(\rho\mathbf{V}) = 0$$

$$\rho\frac{d\mathbf{V}}{dt} = \frac{1}{c}\mathbf{J}\times\mathbf{B} - \boldsymbol{\nabla}\cdot\overleftrightarrow{P} \tag{6.17}$$

$$\mathbf{E} + \frac{\mathbf{V}\times\mathbf{B}}{c} = 0\ , \tag{6.18}$$

where the resistivity is neglected (perfect conductor approximation) and the pressure term is neglected, as will be discussed in the paragraph after Eq. (6.26). Equation (6.18) is equivalent to saying that the plasma conductivity is infinite and no electric field exists in the plasma in the rest frame of the plasma.[1]

From Maxwell's equation, Faraday's law is

$$\boldsymbol{\nabla}\times\mathbf{E} = -\frac{1}{c}\frac{\partial\mathbf{B}}{\partial t}\ . \tag{6.19}$$

Equation (6.19) can be written using Eq. (6.18) as

$$\frac{\partial\mathbf{B}}{\partial t} = \boldsymbol{\nabla}\times(\mathbf{V}\times\mathbf{B})\ . \tag{6.20}$$

This equation is often called the magnetic induction equation.

An order of magnitude comparison of the terms in the preceding equations and a justification for the approximations made may be found also in Ref. 2.

An alternative way to derive the MHD equation, Eq. (6.17), also exists.[3] In this derivation we notice that in the MHD frequency range (low frequency compared to ω_p, for example) the electron oscillation motion is so fast that the time averaged electron acceleration can be largely cancelled. Once we time-average the equation of motion for electrons Eq. (6.7), the left-hand side is, therefore, negligibly small. This is sometimes called the massless electron (or force-free) approximation

$$0 = -m\frac{d\mathbf{V}_e}{dt} = e\left(\mathbf{E} + \frac{\mathbf{V}_e}{c} \times \mathbf{B}\right) - m\nu_{ei}(\mathbf{V}_i - \mathbf{V}_e)$$

$$+\frac{1}{n_e}\boldsymbol{\nabla} \cdot \overleftrightarrow{\Pi}_e \, . \tag{6.21}$$

The equation of motion for ions is

$$M\frac{d\mathbf{V}_i}{dt} = e\left(\mathbf{E} + \frac{\mathbf{V}_i}{c} \times \mathbf{B}\right) - \frac{1}{n_i}\boldsymbol{\nabla} \cdot \overleftrightarrow{\Pi}_i - M\nu_{ie}(\mathbf{V}_i - \mathbf{V}_e) \, . \tag{6.22}$$

Solving for $\mathbf{E}$ in Eq. (6.21) and substituting $\mathbf{E}$ into Eq. (6.22), we obtain

$$nM\frac{d\mathbf{V}_i}{dt} = \frac{e}{c}n(\mathbf{v}_i - \mathbf{v}_e) \times \mathbf{B} - \boldsymbol{\nabla} \cdot \left(\overleftrightarrow{\Pi}_i + \frac{n}{n_e}\overleftrightarrow{\Pi}_e\right) , \tag{6.23}$$

where the quasineutrality $n_i \approx n_e (n_i \equiv n)$ is used. Equation (6.23) can be cast in the form

$$\rho\frac{d\mathbf{v}_i}{dt} = \frac{1}{c}\mathbf{J} \times \mathbf{B} - \boldsymbol{\nabla} \cdot \overleftrightarrow{P} \, . \tag{6.24}$$

Using Eq. (6.21) the same way into Eq. (6.19), we obtain

$$\frac{\partial \mathbf{B}}{\partial t} = -c\boldsymbol{\nabla} \times \mathbf{E} = c\boldsymbol{\nabla} \times \left[\frac{\mathbf{v}_i}{c} \times \mathbf{B} + \frac{1}{4\pi n e^2}\mathbf{B} \times (\boldsymbol{\nabla} \times \mathbf{B})\right.$$

$$\left. - \frac{m e \nu_{ei}}{4\pi n e^2}\boldsymbol{\nabla} \times \mathbf{B} + \frac{1}{n_e}\boldsymbol{\nabla} \cdot \overleftrightarrow{\Pi}_e\right] , \tag{6.25}$$

where the relation from the Maxwell equation

$$\nabla \times \mathbf{B} = \frac{4\pi}{c}\mathbf{J} \tag{6.26}$$

was used. In Eq. (6.25) the Hall term appears in the second term in the braces in much the same way as in Eq. (6.13) (the last term on the left-hand side). In the ideal MHD approximation Eq. (6.25) reduces to Eq. (6.20). The reasons are as follows. (i) The Hall current term in Eq. (6.25) is much smaller than the first term if the electrons and ions move roughly together as is the case in a low frequency phenomenon such as MHD: the current $J/ne = v_i - v_e \ll v_i$ because $v_i \approx v_e$. Thus the Hall current term can be neglected in comparison with the first term. (ii) The third term in the braces vanishes as the collision frequency goes to zero. (iii) The fourth term can be dropped if the pressure tensor is diagonal. Then this term becomes a gradient of a scalar in the braces in Eq. (6.25). The curl of this gradient is zero, provided that ∇n is parallel to $\nabla \Pi_e$. In addition, the pressure is much smaller than the magnetic pressure (the second term) in low β plasma, which in turn is much smaller than the first term due to reason (i). Thus all the terms in the braces in Eq. (6.25) except for the first term can be dropped in MHD phenomena. Thus we arrive at the ideal MHD induction equation, Eq. (6.20), again.

Further reduction of the ideal MHD equations Eqs. (6.14), (6.17) and (6.20) leads to hydrodynamic equations in the absence of magnetic fields:

$$\frac{\partial \rho}{\partial t} + \nabla \cdot (\rho \mathbf{V}) = 0 , \tag{6.27}$$

$$\frac{d\mathbf{V}}{dt} = -\frac{1}{\rho} \nabla \cdot \overleftrightarrow{\Pi} . \tag{6.28}$$

In addition, we often assume that the pressure is solely a function of the density:

$$\overleftrightarrow{\Pi} = \overleftrightarrow{F}(\rho) , \tag{6.29}$$

or most often

$$\Pi = C\rho^{\gamma} , \tag{6.30}$$

where γ is the adiabatic gas constant. Equation (6.30) is called the adiabatic equation of state of a gas. In Eq. (6.30), the pressure tensor was assumed to be diagonal and isotropic.

As shown above, the set of MHD (or HD) equations are derived by truncating the higher order moment terms. This is based on the assumptions that (i) the quantities expressed by velocity moment Eq. (6.2) are reasonably representative of the state of the system; and (ii) the "intrinsic" higher order moments such as skewness and kurtosis are small or, even more strictly, the higher order moments are unimportant. The assumption that the velocity distribution function $f(v)$ is close to the Maxwellian underlies these assumptions. In cases where the velocity distribution cannot be regarded as Maxwellian, the MHD or HD equations can lead to an unphysical result (see Problem 1). The collisionless plasma simulation by the MHD model, therefore, needs special care in this point. This point will be revisited in Section 6.6 in the topic of multistreaming and the MHD particle model.

6.1 Difficulty with the Advective Term

The set of MHD Equations (6.14), (6.17), (6.20), and (6.26) are not amenable to straight forward time-integration. This is true even for the hydrodynamic equations (6.27) and (6.28). Both Eq. (6.14) and Eq. (6.17) have the convective derivative term (also called advective term) $\mathbf{v} \cdot \nabla f$, where f can be a function of space, time and $\mathbf{v}$. Let us take Eq. (6.14) as an example. The equation is a partial differential with respect to t and $\mathbf{x}$. It seems perfectly straightforward to discretize Eq. (6.14) in space and time as it appears. This approach is called an Eulerian approach, where the temporal variation is observed at the fixed spatial coordinates. It turns out that this is a difficult problem and consequently a large amount of research[4,5] has been and is still being done on the problem of integrating this equation. Since one of the problems results from the spatially-fixed time derivative (Eulerian picture), there have been different approaches, most notably the Lagrangian approach, where the temporal variation of a fluid quantity is observed on the frame on the moving fluid. Although the Lagrangian approach cures some of the problems in the Eulerian approach, it brings in its own new type of difficulties, most notably the grid entanglement as the fluid churns. Because of these numerical difficulties, as well as the cumbersome complexity of the Lagrangian algorithm, it seems instructive to start from the Eulerian approach and discuss it in detail.

Some difficulties associated with the Eulerian approach in solving Eq. (6.14) are: (i) the negative density difficulty, and (ii) the advective term difficulty. These are related. The first difficulty arises because Eq. (6.14) does not explicitly preclude negative density *per se*. Therefore, if for some (numerical) reason the density becomes negative, Eq. (6.17) behaves badly and usually leads to numerical instabilities. As for the second difficulty, discussion may be found in various papers,[4,6] and there have been huge areas of applications. In the following, we discuss some of the fundamental points only. The magnetic induction equation, Eq. (6.20), may give rise to a numerical instability if it is integrated in a naive way, because Eq. (6.20) contains the magnetic field that is to be pushed again on the right-hand side. Therefore, this equation, too, needs special care to be integrated numerically. From now on the mean fluid velocity $\mathbf{V}$ is written as a lower case $\mathbf{v}$.

The advective difficulty associated with the Eulerian algorithm may be looked at in a simple example. Let us consider a model equation

$$\frac{\partial u}{\partial t} = -v \frac{\partial u}{\partial x} , \tag{6.31}$$

where u is an arbitrary function of t and x. Equations (6.14), (6.15) and/or (6.27), (6.28) have such a structure. In Eqs. (6.14) and (6.27), u is ρ; while in Eqs. (6.17) and (6.28), u is $\mathbf{v}$. We have simplified the vectors v and x into one dimension in Eq. (6.31) to be constant for simplicity. We then readily obtain an analytical solution for $u(x,t)$ as

$$u(x,t) = f(x - vt) , \tag{6.32}$$

where f is an arbitrary function. On the other hand, a numerical approximation to Eq. (6.31) may be

$$\frac{u_j^{n+1} - u_j^n}{\Delta t} = -v \cdot \frac{u_{j+1}^n - u_{j-1}^n}{2\Delta x} , \tag{6.33}$$

where Δt and Δx are the time step and spatial grid step respectively. Equation (6.33) is rearranged into the form

$$u_j^{n+1} = u_j^n - \frac{v\Delta t}{2\Delta x}\left(u_{j+1}^n - u_{j-1}^n\right) . \tag{6.34}$$

Let u vary like $u_j^n \propto \exp\left[i(kxj - \omega n\Delta t)\right]$, where x_j is the j-th grid point coordinate $j\Delta x$. From Eq. (6.34) we obtain

$$e^{-i\omega\Delta t} = 1 - \frac{v\Delta t}{2\Delta x}\left(e^{-ik\Delta x} - e^{ik\Delta x}\right) , \tag{6.35}$$

which with $g \equiv e^{-i\omega\Delta t}$ leads to the amplification factor

$$g = 1 - i\frac{v\Delta t}{\Delta x}\sin(k\Delta x) . \tag{6.36}$$

The modulus squared of Eq. (6.36) is

$$|g|^2 = 1 + \left(\frac{v\Delta t}{\Delta x}\right)^2 \sin^2(k\Delta x) > 1 , \tag{6.37}$$

which suggests that such an integration scheme is always numerically unstable. This indicates that the algorithm such as Eq. (6.33) cannot be utilized.

6.2 Lax Scheme

The Lax scheme is introduced to overcome the above mentioned advective instability (see Ref. 7, for example). In solving Eq. (6.31), we follow a similar finite difference equation method like Eq. (6.34) except that we replace u_j^n in Eq. (6.34) by a spatially averaged term $\langle u_j^n \rangle$ in the Lax scheme.

$$\begin{aligned} u_j^{n+1} &= \langle u_j^n \rangle - \frac{v\Delta t}{2\Delta x}\left(u_{j+1}^n - u_{j-i}^n\right) \\ &= \frac{1}{2}\left(u_{j+1}^n + u_{j-1}^n\right) - \frac{v\Delta t}{2\Delta x}\left(u_{j+1}^n - u_{j-1}^n\right) . \end{aligned} \tag{6.38}$$

In higher dimensions, the average $\langle u_j \rangle$ involves an average over all the adjacent grid points. The amplification factor to Eq. (6.38) is

$$g = \cos(k\Delta x) - i\frac{v\Delta t}{\Delta x}\sin(k\Delta x) \ , \tag{6.39}$$

which yields $|g|^2$ as

$$|g|^2 = 1 - \sin^2(k\Delta x)\left[1 - \left(\frac{v\Delta t}{\Delta x}\right)^2\right] \ . \tag{6.40}$$

The stability of the Lax scheme is given by $|g|^2 \leq 1$ in Eq. (6.40). This condition reads

$$\Delta t \leq \Delta x/v \ . \tag{6.41}$$

This is a manifestation of the Courant-Friedrichs-Lewy condition.[5] Although this method is stable, the accuracy is first order in Δx. When $\Delta t > \Delta x/v$, an oscillating solution grows exponentially in time. Rearranging Eq. (6.38) we have

$$\frac{1}{2}\left(u_j^{n+1} - u_j^{n-1}\right) + \frac{1}{2}\left(u_j^{n+1} - 2u_j^n + u_j^{n-1}\right) = \frac{1}{2}\left(u_{j+1}^n - 2u_j^n + u_{j-1}^n\right)$$

$$-\frac{v\Delta t}{2\Delta x}\left(u_{j+1}^n - u_{j-1}^n\right) \ , \tag{6.42}$$

whose corresponding differential form is

$$\frac{\partial u}{\partial t} + \frac{\Delta t}{2}\frac{\partial^2 u}{\partial t^2} - \frac{(\Delta x)^2}{2N\Delta t}\frac{\partial^2 u}{\partial x^2} + v\frac{\partial u}{\partial x} = 0 \ , \tag{6.43}$$

where N is the dimensionality. An effective diffusion is incurred in this equation. With $\partial_t^2 \sim v^2\partial_x^2$ the spatial numerical diffusion incurred by this algorithm is

$$\text{“D”}\,\nabla^2 u = \left[(\Delta x)^2/2N\Delta t - \Delta t v^2/2\right]\nabla^2 u \ . \tag{6.44}$$

When $\Delta t < \Delta x/v$ (with $N = 1$), the numerical diffusion coefficient, D, is positive and the scheme is stable. When $\Delta t > \Delta x/v$, D is negative and the scheme is unstable, as suggested by the general theory of parabolic partial differential equations. When $\Delta t \sim \Delta x/v$, the numerical diffusion is minimized. In practice, the velocity v is a variable, so it is difficult to satisfy the exact condition $\Delta t = \Delta x/v$ across the system. In addition to this, the algorithm causes distortion in the dispersion relation of the eigenmodes. The analytic dispersion relation to Eq. (6.31) is $\omega = kv$, where ω is the frequency and k the wavenumber. In Eq. (6.38) with $g \equiv e^{-i\omega\Delta t}$, we equate the real part and imaginary part on both sides, where $\omega = \Omega - i\gamma$ with Ω and γ are both real

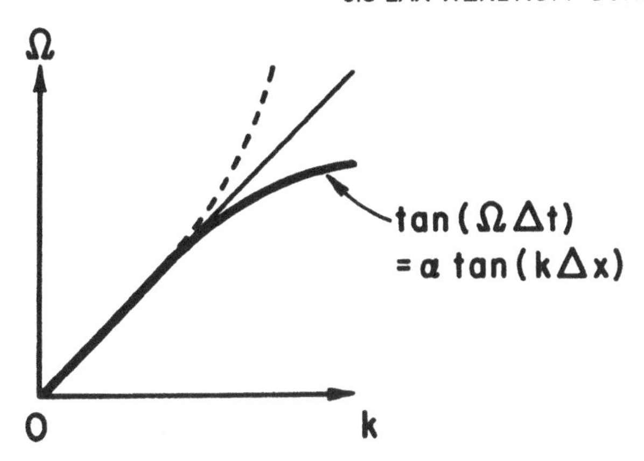

FIGURE 6.1 The finite difference dispersion relation Eq. (6.45)

numbers. Dividing, we obtain the dispersion relation of the finite $(\Delta x, \Delta t)$ difference system of Eq. (6.38) as

$$\tan(\Omega \Delta t) = \frac{v\Delta t}{\Delta x}\tan(k\Delta x) \ . \tag{6.45}$$

Since the stability condition Eq. (6.41) dictates $\alpha \equiv v\Delta t/\Delta x < 1$, the dispersion curve is bent as shown in Fig. 6.1. In the limit of $k\Delta x \to 0$, $|g|^2 \to 1$ and $\gamma \to 0$. Then $\Omega\Delta t \simeq kv\Delta t$, thus we have $\Omega = kv$ in $k\Delta x \to 0$ limit. The Lax scheme can be applied to the magnetic induction equation Eq. (6.20). The finite difference equation according to the Lax scheme is

$$B_j^{n+1} = \langle B_j^n \rangle + [D \times (v \times B)^n]_j \,\Delta t = \frac{1}{2}\left(B_{j+1}^n + B_{j-1}^n\right)$$

$$+\frac{\Delta t}{2}\left\{\frac{1}{\Delta y}\left[(v_x B_y - v_y B_x)_{j_y+1}^n - (v_x B_y - v_y B_x)_{j_y-1}^n\right]\right.$$

$$\left. -\frac{1}{\Delta z}\left[(v_z B_x - v_x B_z)_{j_z+1}^n - (v_z B_x - v_x B_z)_{j_z-1}^n\right]\right\} , \tag{6.46}$$

where only the x-component was shown and the notation D means the finite differencing equivalent to $\mathbf{\nabla}$.

6.3 Lax-Wendroff Scheme

To avoid excessive numerical diffusion and distortion of the dispersion relation by the Lax scheme of integration of Eq. (6.31), as seen in Section 6.2, the two-step Lax-Wendroff scheme has been introduced.[5,7] Let $F = uv$ and v be constant as before for simplicity. Then Eq. (6.31) reads

$$\frac{\partial u}{\partial t} + \frac{\partial F}{\partial x} = 0 .$$

The first step of this algorithm is the Lax step over a half time step:

$$u_j^{n+1/2} = 1/2\left(u_{j-1}^n + u_{j+1}^n\right) - \frac{\Delta t/2}{2\Delta x} \cdot \left(F_{j+1}^n - F_{j-1}^n\right) . \tag{6.47}$$

Sometimes instead of Eq. (6.47), the form

$$u_j^{n+1/2} = 1/2(u_{j-1/2}^n + u_{j+1/2}^n) - \frac{\Delta t/2}{\Delta x}(F_{j+1/2}^n - F_{j-1/2}^n) \tag{6.48}$$

is used. Here $F_j^{n+1/2} = F_j^{n+1/2}\left(u_j^{n+1/2}\right)$. The second step is a full time step:

$$u_j^{n+1} = u_j^n - \frac{\Delta t}{2\Delta x}\left(F_{j+1}^{n+1/2} - F_{j-1}^{n+1/2}\right) . \tag{6.49}$$

See Fig. 6.2. Again, alternatively, sometimes Eq. (6.49) becomes

$$u_j^{n+1} = u_j^n - \frac{\Delta t}{\Delta x}\left(F_{j+1/2}^{n+1/2} - F_{j-1/2}^{n+1/2}\right) .$$

The main features of this algorithm are (i) the first step serves just to provide the time-centered value $u_j^{n+1/2}$ albeit its value is relatively inaccurately calculated through the first order Lax scheme; and that (ii) the second step is virtually time-centered using the predicted value of Eq. (6.47). As a result, this provides a second order accurate, yet numerically stable algorithm. Again, let $u_j^n \propto \exp[i(kj\Delta x - \omega n\Delta t)]$. Equation (6.49) combined with Eq. (6.47) is written as

$$\begin{aligned} u_j^{n+1} = u_j^n - \frac{v\Delta t}{2\Delta x}\Bigg[&\frac{1}{2}(u_j^n + u_{j+2}^n) + \frac{\Delta t/2}{2\Delta x}(F_{j+2}^n - F_j^n) - \frac{1}{2}(u_j^n + u_{j-2}^n) \\ &- \frac{\Delta t/2}{2\Delta x}(F_j^n - F_{j-2}^n)\Bigg] \\ = u_j^n - \frac{v\Delta t}{2\Delta x}\Bigg[&\frac{1}{2}(u_{j+2}^n - u_{j-2}^n) + \frac{\Delta t/2}{2\Delta x}(F_{j+2}^n - 2F_j^n + F_{j-2}^n)\Bigg] . \end{aligned} \tag{6.50}$$

The second term in the braces in Eq. (6.50) is a diffusion term called the Lax-Wendroff diffusion. The amplification factor obtained from Eq. (6.50) is

$$g = 1 - i\alpha\sin(2k\Delta x) + \alpha^2\left[\cos(2k\Delta x) - 1\right] , \tag{6.51}$$

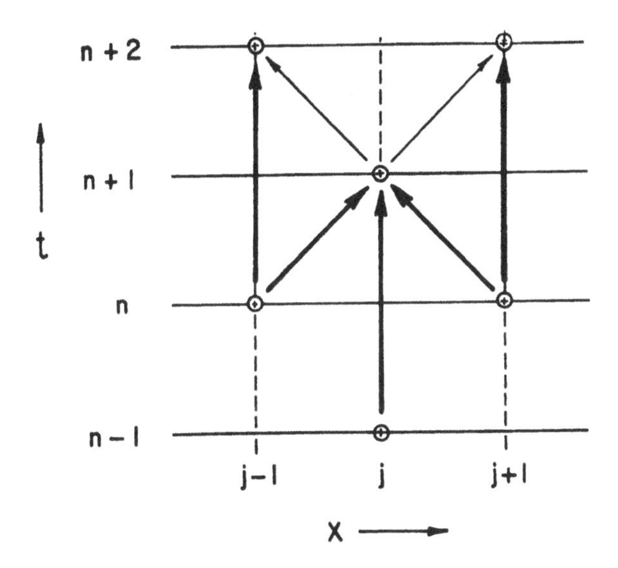

FIGURE 6.2 The temporal and spatial grid assignment for the Lax-Wendroff scheme

where $\alpha \equiv v\Delta t/2\Delta x$. The stability condition $|g|^2 \leq 1$ is calculated from Eq. (6.51) to yield

$$|g|^2 = 1 - \alpha^2(1-\alpha^2)\left[1-\cos(2k\Delta x)\right]^2 \leq 1 ,$$

which tells us the stability $\alpha^2 \leq 1$, i.e.,

$$\Delta t \leq \frac{\Delta x}{|v|} . \tag{6.52}$$

This is, again, a manifestation of the Courant-Friedrichs-Lewy condition.

In the small $k\Delta x$ limit, $|g|^2$ may be written as

$$|g|^2 \approx 1 - \alpha^2(1-\alpha^2)\frac{(2k\Delta x)^4}{4} + \mathcal{O}\left((k\Delta x)^6\right) . \tag{6.53}$$

The numerical diffusion of the Lax-Wendroff scheme is fourth order in $k\Delta x$. This scheme is only slightly affected by the numerical diffusion for long wavelengths.

The above discussion, of course, is confined to linear stability. It turns out,[6] unfortunately, that the system of equations under the algorithm is *nonlinearly* unstable, although it is stable linearly. Because of this, one needs a fictitious diffusion term to stabilize the equation:

$$\frac{\partial u}{\partial t} + \frac{\partial}{\partial x}(uv) = D\frac{\partial^2}{\partial x^2}u . \tag{6.54}$$

Although we have to introduce the numerical diffusion term, the Lax-Wendroff scheme is explicit, relatively simple, and straightforward for various geometrical and other requirements. For this reason this algorithm has been widely used, in spite of its numerical diffusion. The time step requirement is relatively stringent, as this is an explicit scheme.

6.4 Leapfrog Scheme

It is possible to make a leapfrog scheme for Eq. (6.17) if we introduce a double mesh.[7] Now Eq. (6.17) is pushed as follows:

$$u_j^{n+1} = u_j^{n-1} - \frac{\Delta t}{\Delta x}(F_{j+1}^n - F_{j-1}^n) \tag{6.55}$$

and then

$$u_{j+1}^{n+2} = u_{j+1}^n - \frac{\Delta t}{\Delta x}\left(F_{j+2}^{n+1} - F_j^{n+1}\right) , \tag{6.56}$$

where $F_{j+1}^n = F(u_{j+1}^n)$, etc. The linear stability of these equations is analyzed by calculating the amplification factor. Equation (6.56) yields

$$g^2 = 1 - i\frac{2\Delta t}{\Delta x}\sin(k\Delta x)\cdot g \ . \tag{6.57}$$

We then obtain

$$g = i\alpha \pm \sqrt{-\alpha^2} + 1 , \tag{6.58}$$

where $\alpha = \frac{v\Delta t}{\Delta x}\sin(k\Delta x)$.

Equation (6.58) gives two roots. If $|\alpha| \leq 1$ and $|g| = 1$, we have stability. If $|\alpha| > 1$ and $|g| > 1$, the algorithm is unstable. Thus the linear stability condition for all possible k demands

$$\Delta t \leq \frac{\Delta x}{|v|} , \tag{6.59}$$

which is again the Courant-Friedrichs-Lewy condition. In the case of $|\alpha| > 1$ where two separate roots result, these two roots correspond to the fact that the two spatial (and temporal) meshes drift apart. This happens because two meshes (the odd mesh and the even mesh) are not coupled. This is sometimes also called the odd-even instability.

In order to arrest this mesh drift, we usually try to couple two layers of meshes by implementing a fictitious diffusion term:

$$\frac{\partial u}{\partial t} + \frac{\partial}{\partial x}(uv) = D\frac{\partial^2}{\partial x^2}u \ .$$

6.5 Flux-Corrected Transport Method

Boris and Book[6,8] analyzed the advective difficulty in the Eulerian method in detail. The main point of their method may be characterized by the introduction of anti-diffusion after the equation is sufficiently stabilized by a large amount of fictitious diffusion as we have seen in Sections 6.3 and 6.4. [See Eq. (6.54)]. This is called the flux-corrected transport method. As we mentioned before, two of the Eulerian difficulties are the possible negative density and the advective instability.

The first step of the flux-corrected transport method is to ensure the non-negativity of density upon transporting (i.e., pushing) the density. The densities at grid points j and $j+1$ are transported according to the formulae

$$\rho_p = \rho_{j+1}^0 \Delta x / \left[\Delta x + \Delta t \left(v_{j+1}^{1/2} - v_j^{1/2}\right)\right] , \tag{6.60}$$

$$\rho_m = \rho_j^0 \Delta x / \left[\Delta x + \Delta t \left(v_{j+1}^{1/2} - v_j^{1/2}\right)\right] . \tag{6.61}$$

See Fig. 6.3. For example, Eq. (6.61) is nothing but the forward differencing of the continuity equation:

$$\rho_m = \rho_j^0 - \frac{\Delta t}{\Delta x} \rho_m \left(v_{j+1}^{1/2} - v_j^{1/2}\right) . \tag{6.62}$$

Equations (6.60) and (6.61) preserve the non-negativity of density ρ_p and ρ_m if $|v\Delta t/\Delta x| < \frac{1}{2}$ and ρ_j^0 is originally non-negative. Since the velocities are evaluated at the centered time but not at the centered space, the first step is not second order accurate in space.

To rectify this problem, we replace $v_j^{1/2}$ by

$$\bar{v}^{1/2} = (1 - \epsilon_j/2) v_j^{1/2} + (\epsilon_j/2) v_{j+1}^{1/2}, \left(\text{if} \quad v_j^{1/2} \geq 0\right) \tag{6.63}$$

or

$$\bar{v}^{1/2} = (1 - \epsilon_j/2) v_j^{1/2} + (\epsilon_j/2) v_{j-1}^{1/2}, \left(\text{if} \quad v_j^{1/2} < 0\right) , \tag{6.64}$$

where $\epsilon_j \equiv v_j^{1/2} \Delta t / \Delta x$. Thus the new density is defined as

$$\rho_j^{n+1} = \frac{1}{2} Q_-^2 \left(\rho_{j-1}^n - \rho_j^n\right) + \frac{1}{2} Q_+^2 \left(\rho_{j+1}^n - \rho_j^n\right) + (Q_+ + Q_-) \rho_j^n , \tag{6.65}$$

where

$$Q_\pm = \left(\frac{1}{2} \mp v_j^{1/2} \frac{\Delta t}{\Delta x}\right) \Big/ \left[1 \pm \left(v_{j\pm1}^{1/2} - v_j^{1/2}\right) \frac{\Delta t}{\Delta x}\right] .$$

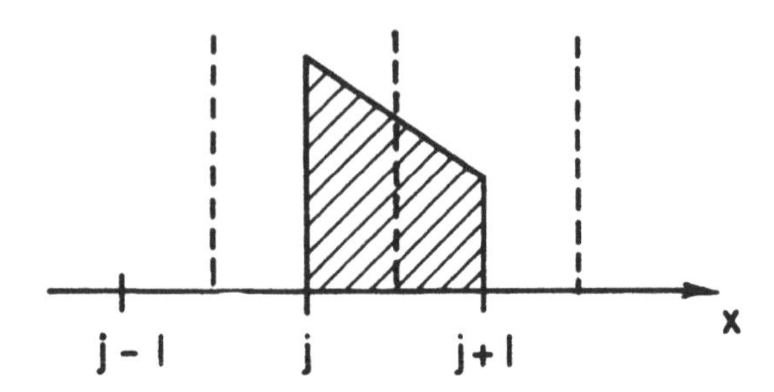

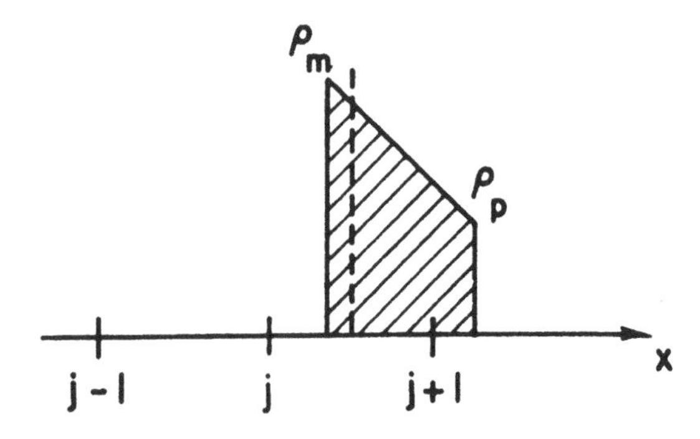

FIGURE 6.3 The initial density profile and the transported density (after Ref. 8)

In order to analyze the above algorithm, let us assume $v_j = v$, a constant, i.e., the velocity is uniform in space. Then Eq. (6.65) becomes

$$\rho_j^{n+1} = \rho_j^n - \frac{\epsilon}{2}\left(\rho_{j+1}^n - \rho_{j-1}^n\right) + \left(\frac{1}{8} + \frac{\epsilon^2}{2}\right)\left(\rho_{j+1}^n - 2\rho_j^n + \rho_{j-1}^n\right) , \qquad (6.66)$$

where $\epsilon = \Delta t v/\Delta x$. This step (transport step) preserves the mass and positivity. However, we have a large diffusion as seen in $\frac{1}{8}$ in the first factor of the third term on the right-hand side of Eq. (6.66). If $\epsilon^2 \ll 1$, the diffusion is $\sim \frac{1}{8}(\rho_{j+1} - 2\rho_j + \rho_{j+1})$. This result is very close to the form obtained by the two-step Lax-Wendroff scheme except for the factor $\frac{1}{8}$. Besides, the spatial scale was twice that in the Lax-Wendroff scheme. In short, in this positive diffusion process we have strong diffusion

$$\rho_j^{n+1} = \rho_j^n + \eta\left(\rho_{j+1}^n - 2\rho_j^n + \rho_{j-1}^n\right) \qquad (6.67)$$

with $\eta = \frac{1}{8} + \epsilon^2 \sim \frac{1}{8}$. To analyze the linear stability, let $\rho(x, t) \propto \exp(ikx -$

$ikvt$). Then we have

$$\rho_j^{n+1} = \rho_j^n \left[1 - \left(\frac{1}{4} + \epsilon^2\right)(1 - \cos k\Delta t) - i\epsilon \sin(k\Delta x)\right] . \tag{6.68}$$

Therefore, the $|g|^2$ (amplification factor modulus squared) is

$$|g|^2 = \left[1 - \frac{1}{4}(1 - \cos k\Delta x)\right]^2 - \frac{\epsilon^2}{2}(1 - 2\epsilon^2)(1 - \cos k\Delta x)^2 . \tag{6.69}$$

This suggests stability when $\epsilon < \frac{1}{2}$.

In order to counter the strong diffusion

$$\rho_j^{n+1} = \rho_j^n + \frac{1}{8}\left(\rho_{j+1}^n - 2\rho_j^n + \rho_{j-1}^n\right) , \tag{6.70}$$

the second step of this algorithm tries to bring in anti-diffusion. In the fully implicit form the density value $\bar{\rho}$ corrected by the anti-diffusion satisfies the following relation:

$$\rho_j^{n+1} = \bar{\rho}_j^{n+1} + \frac{1}{8}\left(\bar{\rho}_{j+1}^{n+1} - 2\bar{\rho}_j^{n+1} + \bar{\rho}_{j-1}^{n+1}\right) . \tag{6.71}$$

An explicit alternative is

$$\bar{\rho}_j^{n+1} = \rho_j^{n+1} - \frac{1}{8}\left(\rho_{j+1}^{n+1} - 2\rho_j^{n+1} + \rho_{j-1}^{n+1}\right) . \tag{6.72}$$

If we define the flux

$$f_{j\pm 1/2} = \pm\frac{1}{8}\left(\rho_{j\pm 1}^{n+1} - \rho_j^{n+1}\right) \tag{6.73}$$

or

$$F_{j\pm 1/2} = \pm\frac{1}{8}v\left(\rho_{j\pm 1}^{n+1} - \rho_j^{n+1}\right) ,$$

then we can write Eq. (6.72) as

$$\bar{\rho}_j^{n+1} = \rho_j^{n+1} - f_{j+1/2} + f_{j-1/2} . \tag{6.74}$$

As seen in Eq. (6.74), the anti-diffusion flux describes the transfer of mass. It can be proven that the anti-diffusion Eq. (6.74) is conservative (i.e., mass preserving). However, it is *not* non-negative (see Fig. 6.4).

To avoid this possible negativeness of the density, the flux-corrected transport is introduced; $f_{j+1/2}$ is corrected to be:

$$f_{j+1/2}^c = \operatorname{sgn}\Delta_{j+1/2} \max\left\{0, \min\left(8f_{j-1/2}\operatorname{sgn}f_{j+1/2}, f_{j+1/2} ,\right.\right.$$

$$\left.\left. 8f_{j+3/2}\operatorname{sgn}f_{j+1/2}\right)\right\} . \tag{6.75}$$

This is called the flux limiter to prohibit the density from becoming negative. Because of this flux limiter, the nonlinear odd-even grid instability, generally encountered in the Lax-Wendroff scheme is now cured, because the limiter binds the grid.

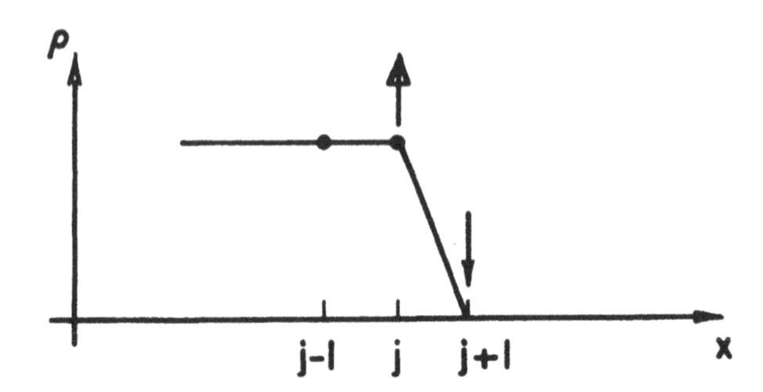

FIGURE 6.4 The tendency to create negative density without flux limiter (after Ref. 8)

6.6 Magnetohydrodynamic Particle Model

Two types of MHD plasma simulation codes, i.e., Eulerian code and Lagrangian code, have been mentioned already. But the Eulerian Method has difficulties such as the advective instability and negative density which we encountered before. On the other hand, the conventional Lagrangian method has its own difficulties such as non-conservativeness of mass, flux and momentum as well as an expensive running cost because of the complexity of the codes.

In order to avoid these difficulties, the MHD-particle model has been developed[9,10] by treating elements of the fluid as finite-size particles and the particle quantities are pushed in a Lagrangian way while meshes are fixed (Eulerian) and uniform. The tradition of pushing fluid Lagrangian elements may be traced back to such people as Pasta and Ulam.[11]

This MHD-particle model removes the advective instability and negative density difficulty and guarantees no mesh reconnection and the conservation of particle quantities.

A. Ideal-MHD Case

The equation of motion of the i-th finite-size particle can be written as

$$\frac{d\mathbf{v}_i}{dt} = -\frac{1}{\rho}\nabla p + \frac{1}{4\pi\rho}\mathbf{J}\times\mathbf{B} , \qquad (6.76)$$

$$\frac{d\mathbf{r}_i}{dt} = \mathbf{v}_i . \qquad (6.77)$$

This equation reduces to

$$\frac{d\mathbf{v}_i}{dt} = -\frac{1}{\rho}\boldsymbol{\nabla} p - \frac{1}{8\pi\rho}\boldsymbol{\nabla} B^2 + \frac{1}{4\pi\rho}\nabla \cdot (\mathbf{BB}) , \tag{6.78}$$

and if we include the collisional drag term

$$\frac{d\mathbf{v}_i}{dt} = -\frac{1}{\rho}\boldsymbol{\nabla} p - \frac{1}{8\pi\rho}\boldsymbol{\nabla} B^2 + \frac{1}{4\pi\rho}\boldsymbol{\nabla} \cdot (\mathbf{BB}) - \nu(\mathbf{v}_i - \langle\mathbf{v}\rangle) , \tag{6.79}$$

where $\langle\mathbf{v}\rangle$ is the averaged velocity of fluid elements in the cell. To close this equation, we may need the equation of state such as

$$P = Cn^\gamma(\gamma \geq 1) . \tag{6.80}$$

With these equations of motion, the relation between particle quantities and grid quantities may in turn be expressed as

$$n(\mathbf{r}) = \sum_{i \epsilon g} S(\mathbf{r} - \mathbf{r}_i) , \tag{6.81}$$

where the summation runs over all particles within the cell g.

The quantity $S(\mathbf{r} - \mathbf{r}_i)$ represents the shape factor of finite size particles. (See Chapter 2.) The particle positions $\mathbf{r}_i$ are determined in terms of the nearest grid point, for example, and the density then has the form

$$n(\mathbf{r}) = \sum_{g} S(\mathbf{r} - \mathbf{r}_g)\rho_{\mathrm{NGP}}(g) . \tag{6.82}$$

The summation is over all points g on the grid and

$$\rho_{\mathrm{NGP}}(g) = \sum_{i \epsilon g} 1 . \tag{6.83}$$

The the fluid velocity can be expressed as

$$\langle\mathbf{v}\rangle = \left[\sum_{i \epsilon g} \mathbf{v}_i S(\mathbf{r} - \mathbf{r}_i)\right] \left[\sum_{i \epsilon g} S(\mathbf{r} - \mathbf{r}_i)\right]^{-1} . \tag{6.84}$$

The induced magnetic field is determined from a differential equation in an Eulerian frame,

$$\frac{\partial \mathbf{B}}{\partial t} = \boldsymbol{\nabla} \times (\langle\mathbf{v}\rangle \times \mathbf{B}) . \tag{6.85}$$

In the above equation, $\langle\mathbf{v}\rangle$ stands for the fluid velocity, and the magnetic field $\mathbf{B}$ on the right-hand side is the total field. Now, we define magnetic flux

$$\Phi = \sum_{g} \mathbf{B}_g \cdot \hat{n} \tag{6.86}$$

and with Eqs. (6.85) and (6.86) one can show the conservation of flux by

$$\frac{\partial \Phi}{\partial t} = \sum_g \frac{\partial \mathbf{B}_g}{\partial t} \hat{n} = \sum_g [\nabla \times (\langle \mathbf{v} \rangle \times \mathbf{B})]_g = 0 , \tag{6.87}$$

where $\hat{n}$ is a unit vector normal to the surface across which flux goes through. The field pushing equation is readily integrated in time by the Lax-Wendroff method which we mentioned earlier and this method is known to be stable as long as the Courant-Friedrichs-Lewy condition is satisfied. It is known that the Lax-Wendroff method is unstable in the nonlinear range and we need an artificial diffusion term to avoid this difficulty in a typical MHD code. But this difficulty is removed automatically in the MHD-particle code, hence we do not need the artificial diffusion term.

The first approximate step (Lax step) is taken

$$\mathbf{B}^{n+1/2} = \langle \mathbf{B} \rangle^n + [\nabla \times \langle \mathbf{v} \rangle^n \times \mathbf{B}^n] \frac{\Delta t}{2} , \tag{6.88}$$

where $\langle \mathbf{B} \rangle$ is defined similar to Eq. (6.38) with particle quantity advancing equation

$$\mathbf{v}_i^n = \mathbf{v}_i^{n-1/2} + \mathbf{F} \frac{\Delta t}{2} , \tag{6.89}$$

followed by the full step

$$\mathbf{B}^{n+1} = \mathbf{B}^n + \nabla \times \left(\langle \mathbf{v} \rangle^{n+1/2} \times \mathbf{B}^{n+1/2} \right) \Delta t , \tag{6.90}$$

with

$$\mathbf{v}_i^{n+1/2} = \mathbf{v}_i^n + \mathbf{F}^n \frac{\Delta t}{2} . \tag{6.91}$$

The virtually time-centered difference can be achieved by this method. By doing similar analysis as before one can find the amplification factor for Lax-Wendroff method by

$$|g|^2 = 1 + 4k^2 \Delta x^2 \alpha \left(\alpha - \frac{1}{N} \right) , \tag{6.92}$$

where $\alpha = \left(\frac{v \Delta t}{2 \Delta x}\right)^2$ and N is the number of spatial dimensions. A linear stability criterion is obtained from Eq. (6.92) as

$$\Delta t \leq \sqrt{2} \Delta x / |v| . \tag{6.93}$$

with $|v|$ being the greatest velocity of the system.

B. The Resistive MHD Scheme

When the MHD fluid is resistive, some transport coefficients such as resistivity and heat conductivity must be included in our model. While the equations of motion are unchanged, the fluid equation becomes

$$\frac{\partial \mathbf{B}}{\partial t} = \nabla \times (\langle \mathbf{v} \rangle \times \mathbf{B}) - \frac{c^2}{4\pi} \nabla \times \overleftrightarrow{\eta} \cdot (\nabla \times \mathbf{B}) , \tag{6.94}$$

where the resistivity $\overleftrightarrow{\eta}$ is, in general, a tensor. The temperature equation, obtained from the second moment of the kinetic equation, can be written as

$$\frac{dT_i}{dt} = -\frac{1}{n} \left[(\overleftrightarrow{P} \cdot \nabla) \cdot \mathbf{v} - \mathbf{J} \cdot \left(\mathbf{E} + \frac{1}{c} \mathbf{v} \times \mathbf{B} \right) \right.$$

$$\left. + \rho_e (\mathbf{E} \cdot \mathbf{v}) + \nabla \cdot \mathbf{W} \right] , \tag{6.95}$$

where ρ_e is the charge density and $\mathbf{W}$ is the heat flux due to collisions, namely,

$$\mathbf{W} = -\overleftrightarrow{\kappa} \cdot \nabla T , \tag{6.96}$$

with the heat conductivity tensor $\overleftrightarrow{\kappa}$.

The electric field is expressed through Ohm's law as

$$\mathbf{E} = -\frac{\mathbf{v} \times \mathbf{B}}{c} + \overleftrightarrow{\eta} \cdot \mathbf{J} \tag{6.97}$$

where

$$\mathbf{J} = \frac{c}{4\pi} \nabla \times \mathbf{B} . \tag{6.98}$$

Combining these equations, Eq. (6.95) then yields

$$\frac{dT_i}{dt} = -\frac{1}{n} (\overleftrightarrow{P} \cdot \nabla) \cdot \mathbf{v} + \frac{c^2}{(4\pi)^2 n} \left[(\nabla \times \mathbf{B}) \cdot \overleftrightarrow{\eta} \cdot (\nabla \times \mathbf{B}) \right]$$

$$+ \frac{1}{n} (\nabla \cdot \overleftrightarrow{\kappa} \cdot \nabla) T , \tag{6.99}$$

where we may define T as

$$T \equiv \left[\sum_{i \epsilon g} T_i S(\vec{r} - \vec{r}_i) \right] \Big/ \left[\sum_{i \epsilon g} S(\vec{r} - \vec{r}_i) \right] \tag{6.100}$$

for finite size particles.

Properties of the MHD-Particle Code

Finite grid size effects in the MHD particle code are studied and dispersion relations for the sound and Alfvén waves are obtained.[12] For simplicity let us assume a functional variation in the x-direction only. Then the force acting on α-th particle is expressed as

$$F(x_\alpha) = \int dx f(x_\alpha - x) \sum_g w(x - x_g) F_g \ , \tag{6.101}$$

where f is the shape factor of the particle and w is a weighting function arising from the interpolation to the particle position of a grid quantity force.

The linearized force on the grid g is

$$\mathbf{F}_g = -\frac{1}{n}\nabla P + \frac{1}{nc}\mathbf{J} \times \mathbf{B} = -\frac{c_s^2}{n_0}\nabla n_g + \frac{1}{\rho_0}\frac{1}{4\pi}(\nabla \times \delta\mathbf{B}_g) \times \mathbf{B}_0 \ , \tag{6.102}$$

with the grid quantities n_g and $\delta\mathbf{B}_g$ being the density and magnetic field perturbations.

In Fourier space, Eqs. (6.101) and (6.102) become

$$\mathbf{F}(k) = S(k)\left[-\frac{c_s^2}{n_0}\kappa n(k)\right] + \frac{1}{4\pi\rho_0} i \left[\kappa \times \mathbf{B}(k)\right] \times \mathbf{B}_0 \ , \tag{6.103}$$

with $S(k) = f(k)w(k)$ and the term $\kappa(k)$ represents effective differentiation in k-space.

The perturbation of the distribution function for collisionless fluid model is given by

$$f_1 = -\delta\mathbf{v}_\alpha \cdot \frac{\partial f_0}{\partial \mathbf{v}} = -i\frac{F(k)}{\omega - \mathbf{k}\cdot\mathbf{v}} \cdot \frac{\partial f_0}{\partial \mathbf{v}} \ , \tag{6.104}$$

where $\delta\mathbf{v}_\alpha$ is the velocity perturbation due to the force $\mathbf{F}(k)$.

The density perturbation can be calculated by integration over velocity space of f_1 i.e.,

$$n(k) = i\frac{n_0}{\omega} \sum_{p=-\infty}^{\infty} S(k_p)\mathbf{F}(k_p) \cdot \psi(k_p, \omega) \ , \tag{6.105}$$

where

$$\psi(k_p, \omega) = -\frac{1}{n_0}\int \frac{d^3v}{1 - \mathbf{k}\cdot\mathbf{v}/\omega}\frac{\partial f_0}{\partial \mathbf{v}} \ , \tag{6.106}$$

and $k_p = k + 2\pi p$ with p being an integer. The terms with $p \neq 0$ in the summation represents the spatial aliases (see Chapter 4). If the unperturbed distribution function is Maxwellian, then

$$\psi(k,\omega) = \frac{\sqrt{\pi}}{n_0}\frac{\omega}{k} Z\left(\frac{\omega}{\sqrt{2}kv_t}\right) , \tag{6.107}$$

where Z is the plasma dispersion function.

Let us consider two examples.

The dispersion relation obtained for the acoustic wave is the following

$$F(k) = -if(k)w(k)\frac{c_s^2}{n_0}\kappa(k)n(k) , \tag{6.108}$$

and

$$n(k) = i\frac{n_0}{\omega}\sum_p S(k_p)F(k_p)\psi(k_p,\omega) . \tag{6.109}$$

Hence

$$\omega^2 = k\kappa(k)c_s^2\left[1 + 3\left(\frac{kvt}{\omega}\right)^2\right] f^2(k)w^2(k) , \tag{6.110}$$

for $\frac{kv_t}{\omega} \ll 1$ case, given by large argument expansion of the Z-function. The thermal control run is shown in Fig. 6.5 to confirm Eq. (6.110).

On the other hand the magnetic wave (Alfvén wave) is

$$-i\omega\mathbf{B} = i\mathbf{k} \times (\langle\mathbf{v}\rangle \times \mathbf{B}_0) , \tag{6.111}$$

where

$$\langle\mathbf{v}\rangle = i\frac{\mathbf{F}(k)}{\omega}\sum_p S(k_p)\psi_v(k_p) , \tag{6.112}$$

with

$$\psi_v(k) = \frac{1}{n_0}\int \alpha^3 v \frac{\mathbf{v}}{1 - \frac{kv_x}{\omega}} \cdot \frac{\partial f_0}{\partial \mathbf{v}} . \tag{6.113}$$

If $B_0 || k_x$, then

$$\omega^2 = \kappa^2(k)f(k)^2w^2(k)v_A^2 . \tag{6.114}$$

Here we have kept only the fundamental and have neglected all the aliases. The ion thermal speed, v_t, is assumed much smaller than c_s and v_A, where $c_s \simeq \sqrt{\frac{T_e}{M_i}}$ (Sound velocity), $v_A = \frac{B_O}{\sqrt{4\pi n M_i}}$ (Alfvén wave speed). Note that there are only ions in this model. The thermal control run to check the dispersion relation is shown in Fig. 6.6.

Because of its particle nature it has a noise problem. The numerical error introduced by the crude nearest grid point approximation in Eq. (6.84) adds to numerical diffusion and the computational overhead for particle quantities could also be substantial. The hydrodynamic particle code has its advantages in handling the non-negative density and the advective nonlinearities. This

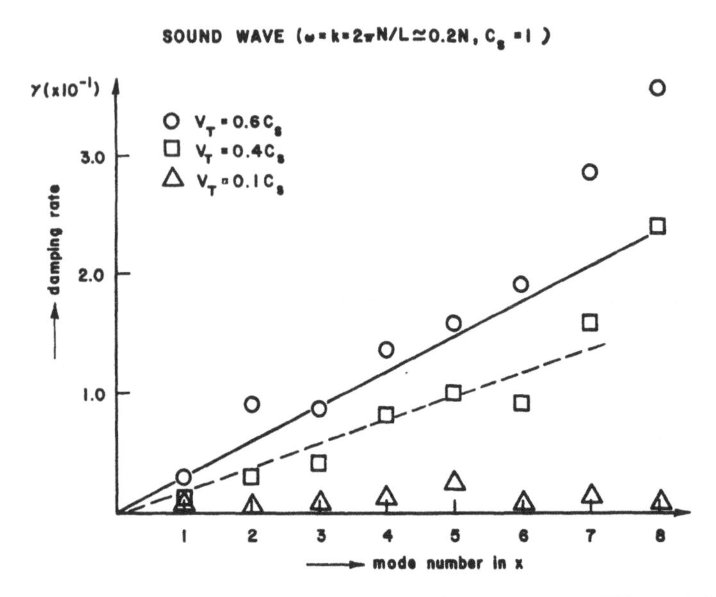

FIGURE 6.5 The ion Landau damping of sound wave in the MHD particle code. The real frequency shown in Ref. 12

leads to a number of applications unique to this model, including the simulation of the geomagnetosphere, astrophysical plasmas, laser fusion plasmas, shocks and reconnection. In these applications, plasmas behave quite violently so that a resilient code is necessary.

Yabe and his colleagues[13] developed a code which carries MHD particles, but with *moving* grid in the laser fusion application. In the laser implosion process, their code tries to relieve the heavy duty particles to treat nonlinearities by spreading a part of the burden onto the moving, and therefore, Lagrangian, mesh. If we were to let the moving Lagrangian mesh absorb all the nonlinearities, the mesh would go through the well-known numerical difficulties such as the mesh reconnection and numerical diffusion and/or stability. In Yabe's model the mesh behaves in a regular way representing the overall implosion, while particles carry most of the fluid nonlinearities. This method has been highly successful in simulating laser fusion plasmas. More recent work has been reported in Ref. 14 and 15.

In the Lagrangian mesh method[16] the advective terms have been treated by moving the mesh with velocity of the fluid $\mathbf{v}$. This gives rise to reconnection of meshes, convoluted meshes etc., when the fluid motion becomes vigorous.

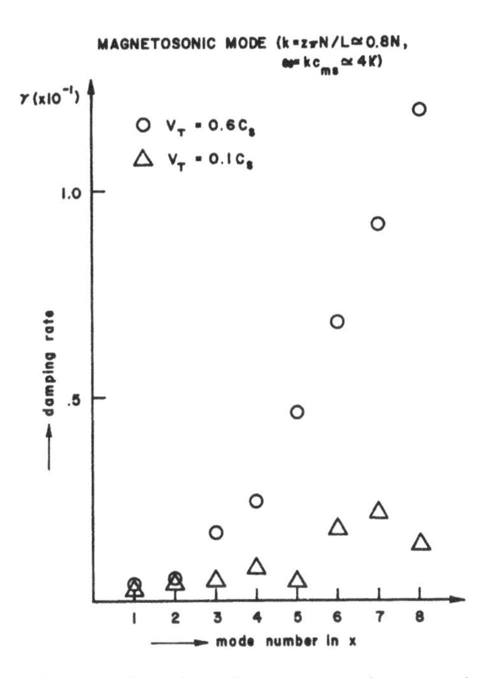

FIGURE 6.6 The ion Landau damping of magnetosonic waves in the MHD particle code. The real point of frequency shown in Ref. 12

In order to realize reflection of more fluid aspects of the fluid particle simulation, the following combination of the Eulerian and Lagrangian methods is considered.[13] For a set of hydrodynamic equations

$$\frac{\partial \rho}{\partial t} + \nabla \cdot (\rho \mathbf{v}) = 0 , \tag{6.115}$$

$$\frac{\partial (\rho \mathbf{v})}{\partial t} + \nabla \cdot (\rho \mathbf{v}\mathbf{v}) = -\nabla P , \tag{6.116}$$

$$\frac{\partial \varepsilon}{\partial t} + \nabla \cdot (\varepsilon \mathbf{v}) = -\nabla \cdot (P\mathbf{v}) . \tag{6.117}$$

Let us look at Eq. (6.115).

Phase 1: Eulerian Calculation

We advance v according to

$$\tilde{v}_{xi,j} = v^n_{x_{i,j}} - \frac{1}{\rho^n_{i,j}} \frac{\Delta t}{\Delta x} \left(P^n_{i+\frac{1}{2},j} - P^n_{i-\frac{1}{2},j} \right) , \tag{6.118}$$

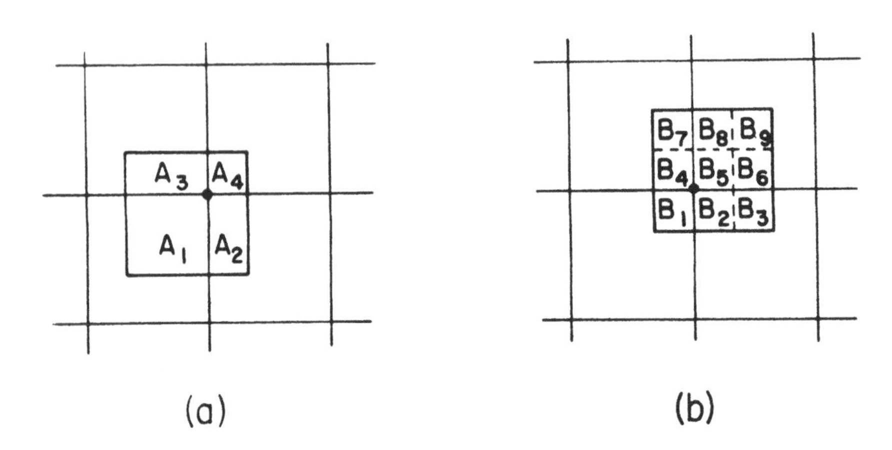

FIGURE 6.7 The assignment of the fluid-in-cell weighting (Ref. 13)

$$\tilde{v}_{yi,j} = v^n_{xi,j} - \frac{1}{\rho^n_{i,j}} \frac{\Delta t}{\Delta y} \left(P^n_{i,j+\frac{1}{2}} - P^n_{i,j-\frac{1}{2}} \right) . \tag{6.119}$$

Defining the internal energy $I = \varepsilon - \rho v^2/2$, it is advanced as

$$\tilde{I}_{i,j} = I^n_{i,j} - \frac{P^n_{i,j}}{\rho^n_{i,j}} \left[\frac{\Delta t}{\Delta x} \left(\bar{v}_{xi+\frac{1}{2},j} - \bar{v}_{xi-\frac{1}{2},j} \right) - \frac{\Delta t}{\Delta y} \left(\bar{v}_{yi,j+\frac{1}{2}} - \bar{v}_{yi,j-\frac{1}{2}} \right) \right] , \tag{6.120}$$

where $\bar{v} = (\tilde{v} + v^n)/2$.

Phase 2: Particle Push

The particle velocity is defined as

$$v_p = A_1 \tilde{v}_{i,j} + A_2 \tilde{v}_{i+1,j} + A_3 \tilde{v}_{i,j+1} + A_3 \tilde{v}_{i+1,j+1} ,$$

where the total area $A = \sum_{n=1}^{4} A_n = \Delta x \Delta y (= 1)$ and A_n is defined in Fig. 6.7.

The new particle position is given by

$$\mathbf{x}^{n+1}_p = x^n_p + \mathbf{v}_p \Delta t . \tag{6.121}$$

In order to reduce the numerical noise associated with particles, the area weighting is adopted.[12] Particles carry mass and internal energy, but not momentum.

After particle transport, the areas B_4 and B_7 in Fig. 6.7(b) are averaged out to give $(i, j+1)$th cell's quantity, for example. Physical quantity U is updated according to

$$U_{i,j}^{\text{new}} = B_1 U_{i,j}^{\text{old}} , \tag{6.122}$$

$$U_{i,j+1}^{\text{new}} = B_4 U_{i,j}^{\text{old}} + B_7 U_{i,j+1}^{\text{old}} , \tag{6.123}$$

$$U_{i+1,j}^{\text{new}} = B_2 U_{i,j}^{\text{old}} + B_3 U_{i+1,j}^{\text{old}} , \tag{6.124}$$

$$U_{i+1,j+1}^{\text{new}} = B_5 U_{i,j}^{\text{old}} + B_6 U_{i+1,j}^{\text{old}} + B_8 U_{i,j+1}^{\text{old}} + B_9 U_{i+1,j+1}^{\text{old}} , \tag{6.125}$$

where $\sum A_n = 1$, $\sum B_n = 1$. No zeroth order diffusion arises. This is a simple distribution of quantity to avoid the first order diffusion:

$$U(x,y) = [(U_{i+1,j+1} - U_{i+1,j})(2y - \delta y) + U_{i+1,j}](2x - \delta x)$$
$$- \{(U_{i,j+1} - U_{i,j})(2y - \delta y) + U_{i,j}\}(2x - \delta x - 1) , \tag{6.126}$$

where (x, y) is the particle position within the (i, j)th cell, $\Delta x, \Delta y$ are the particle center's deviation from the cell center, and a the size of particle. Then

$$\delta x = a + \left(\frac{\Delta x}{2} - \Delta x\right) , \tag{6.127}$$

$$\delta y = a + \left(\frac{\Delta y}{2} - \Delta y\right) , \tag{6.128}$$

where Δx and Δy are the cell sizes in the x- and y-directions.

6.7 Reduced Magnetohydrodynamic Equations

We first analyze the MHD equations and characterize the time scale hierarchy (see Table 1.2). We shall see that there are three distinct time scales in the resistive MHD physics. Accordingly, computational models can be constructed to be more efficient to follow longer time scales by reducing the original set of equations. These are generally called the reduced MHD equations. Precise forms of these vary, depending on the problems under consideration. Equations (6.14)-(6.16) or (6.18) along with the second order moment equation derived from Eq. (6.6) are linearized around the equilibrium of a uniform stationary magnetized plasma. We follow the development by Aydemir and Barnes[17,18] here. However, there are many other works in this field.[19,20] The equilibrium magnetic field is denoted by $\mathbf{B}_0$ and if no external fluid flow exists $\mathbf{u}_0 = 0$. All linearized (perturbed) quantities have a subscript 1: for example, $\mathbf{B} = \mathbf{B}_0 + \mathbf{B}_1$, $\mathbf{V} = 0 + \mathbf{u}_1$, $\rho = \rho_0 + \rho_1$ etc. All linearized quantities are assumed to vary as $\exp(i\mathbf{k}\cdot\mathbf{x} - i\omega t)$. Then we have

$$-i\omega\rho_1 + \rho_0(i\mathbf{k}\cdot\mathbf{u}_1) = 0 \quad (6.129)$$

$$-i\omega\mathbf{u}_1 = \frac{i(\mathbf{k}\cdot\mathbf{B}_0)}{4\pi_0\rho_0}\mathbf{B}_1 + i\mathbf{k}\left[\frac{p_1 + \mathbf{B}_0\cdot\mathbf{B}_1/4\pi}{\rho_0}\right] = 0\,, \quad (6.130)$$

$$-i\omega\mathbf{B}_1 - (i\mathbf{k}\cdot\mathbf{B}_0)\mathbf{u}_1 + (i\mathbf{k}\cdot\mathbf{u}_1)\mathbf{B}_0 = 0\,, \quad (6.131)$$

$$-i\omega p_1^* + p_0\left[\frac{\gamma p_0 + B_0^2/4\pi}{p_0}\right](i\mathbf{k}\cdot\mathbf{u}_1) - \frac{(i\mathbf{k}\cdot\mathbf{B}_0)(\mathbf{u}_1\cdot\mathbf{B}_0)}{4\pi} = 0\,. \quad (6.132)$$

Here $p_1^* = p_1 + \mathbf{B}_0\cdot\mathbf{B}_1/4\pi$.

a) We consider the wave propagating parallel to the external field ($\mathbf{k}||\mathbf{B}_0$). In this case, the wave satisfies $\mathbf{v}_1 \perp \mathbf{B}_0$, $\mathbf{B}_1 \perp \mathbf{B}_0$, $i\omega p_1 = 0$. From this we obtain

$$\begin{aligned} p_1^* &= p_1 + \mathbf{B}_0\cdot\mathbf{B}_1/\mu_0 = 0\,, \\ -i\omega\mathbf{B}_1 &- (i\mathbf{k}\cdot\mathbf{B}_0)\mu_1 = 0\,, \\ -i\omega\mathbf{u}_1 &- \left(\frac{i\mathbf{k}\cdot\mathbf{B}_0}{4\pi\rho_0}\right)\mathbf{B}_1 = 0\,. \end{aligned} \quad (6.133)$$

From the last two equations we obtain the dispersion relation for the shear Alfvén wave $\omega^2 = k_{||}^2 v_A^2$, where $v_A = (B_0^2/4\pi\rho_0)^{1/2}$, the Alfvén velocity.

b) We now consider the wave propagating perpendicular to $\mathbf{B}_0(\mathbf{k}\perp\mathbf{B}_0)$. In this case the wave satisfies $\mathbf{u}_1 \perp \mathbf{B}_0$, $\mathbf{B}_1||\mathbf{B}_0$. From this we obtain

$$\begin{aligned} -i\omega\rho_1 &+ \rho_0(i\mathbf{k}\cdot\mathbf{u}_1) = 0\,, \\ -i\omega\mathbf{u}_1 &+ i\mathbf{k}\left[\frac{p_1 + \mathbf{B}_0\cdot\mathbf{B}_1/4\pi}{\rho_0}\right] = 0\,, \\ -i\omega\mathbf{B}_1 &+ (i\mathbf{k}\cdot\mathbf{u}_1)\mathbf{B}_0 = 0\,, \\ -i\omega p_1^* &+ \rho_0\left[\frac{\gamma p_0 + B_0^2/4\pi}{\rho_0}\right](i\mathbf{k}\cdot\mathbf{u}_1) = 0\,. \end{aligned}$$

Let $v_s^2 = \frac{\gamma p_0 + B_0^2/4\pi}{\rho_0}$ (magnetosonic speed v_s) then $v_s \sim v_A$ for low β plasmas. The dispersion relation can be obtained from the secular equation for the

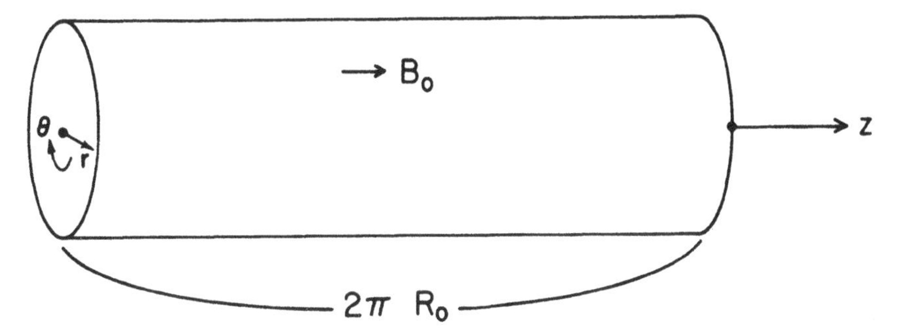

FIGURE 6.8 Cylindrical geometry for the reduced MHD

vector $(\rho_1, \mathbf{u}_1, \mathbf{B}_1, p_1^*)$.

$$\det \begin{bmatrix} -i\omega & ik\rho_0 & 0 & 0 \\ 0 & -i\omega & 0 & ik/\rho_0 \\ 0 & ikB_0 & -i\omega & 0 \\ 0 & ikv_s^2 & 0 & -i\omega \end{bmatrix} = -i\omega \begin{bmatrix} -i\omega & 0 & ik/\rho_0 \\ ikB_0 & -i\omega & 0 \\ ikv_s^2\rho_0 & 0 & -i\omega \end{bmatrix} = 0 \, , \tag{6.134}$$

or

$$i\omega \left\{ -i\omega^3 + (ik/\rho_0)\left[\rho_0 \omega k v_s^2\right] \right\} = 0 \, . \tag{6.135}$$

Equation (6.134) yields the dispersion relation for the magnetosonic wave $\omega^2 = k^2 v_s^2$, where k is the perpendicular wavenumber. This wave is also called the compressional Alfvén wave because the fluid is compressible $(\mathbf{k} \cdot \mathbf{u} \neq 0)$ and, also, the magnetic field is compressible by changing its strength $\mathbf{B}_1 || \mathbf{B}_0$.

Let us consider time scales of the resistive MHD. A typical tokamak,[21] or cylindrical equivalent (i.e., straightened tokamak in the toroidal direction) is depicted in Fig. 6.8. As we discussed above, the shear Alfvén wave propagates along external magnetic field lines and the compressional Alfvén wave propagates across field lines. The time scale that is associated with the compressional Alfvén wave in the tokamak geometry is measured by the transit time of the compressional Alfvén wave over the minor radius a, $\tau_{\text{comp}} \cong a/v_s \sim a/v_A$. The shear time scale that is related to the shear Alfvén wave is that of the transit time of the shear Alfvén wave over the major radius R_0 (or the periodicity length in that direction), $\tau_{\text{shear}} \cong R_0/v_A$. The third time scale that is associated with the diffusion of magnetic fields is that of the resistive time across the field, $\tau_{\text{res}} \simeq \left(\frac{4\pi}{c^2}\right) \frac{a^2}{\eta}$ (for cgs and without the quantity in parenthesis for MKS). Since $a/R_0 \equiv \epsilon \ll 1$ for a typical tokamak or toroidal machine, where ϵ is called the inverse aspect ratio, the compressional time is much shorter than the shear time. The Lundquist number (or the magnetic

Reynolds number) S is defined as

$$S = \frac{\tau_{\text{res}}}{\tau_{\text{comp}}} = \left(\frac{4\pi}{c^2}\right) \frac{a v_A}{\eta} . \tag{6.136}$$

This number is typically very large, $10^6 - 10^8$ for present day tokamaks and $10^{10} - 10^{12}$ for coronal loops. Thus the hierarchy of three time scales in the resistive MHD is

$$\tau_{\text{comp}} \ll \tau_{\text{shear}} \ll \tau_{\text{res.}} \cdot \tag{6.137}$$

Except for ideal MHD instabilities, the problems that one may wish to study have much longer time scales than τ_{comp}. How does one efficiently study such problems? This leads to the necessity of the removal of such fast time scales from the computational system. One way of accomplishing this goal is to eliminate the freedom of the compressional Alfvén waves and restrict the plasma so that it stays incompressible. This is sometimes called incompressible MHD.

Consider the continuity equation, Eq. (6.14), from which we have

$$\frac{d\rho}{dt} = \frac{\partial \rho}{\partial t} + (\mathbf{u} \cdot \nabla)\rho = -\rho \nabla \cdot \mathbf{u} . \tag{6.138}$$

If the incompressibility $\nabla \cdot \mathbf{u} = 0$ is imposed, Eq. (6.27) suggests that the fluid density ρ_1 at $\mathbf{x}_1$ stays equal to ρ_2 at $\mathbf{x}_2$ which has been convected from $\mathbf{x}_1$. Thus, if the density is originally uniform $\rho = \rho_0 = \text{const.}$, it stays constant. Then the equations can be written as

$$\frac{\partial \mathbf{u}}{\partial t} = -\nabla \cdot \mathbf{uu} \pm \mathbf{J} \times \mathbf{B} - \nabla p , \tag{6.139}$$

$$\frac{\partial \mathbf{B}}{\partial t} = -\nabla \times \mathbf{E} , \tag{6.140}$$

$$\mathbf{E} = -\mathbf{u} \times \mathbf{B} + \frac{1}{S}\mathbf{J} , \tag{6.141}$$

$$\mathbf{J} = \nabla \times \mathbf{B} , \tag{6.142}$$

$$\nabla \cdot \mathbf{u} = 0 \quad \text{and} \quad \nabla \cdot \mathbf{B} = 0 , \tag{6.143}$$

where the appropriate normalizations are now taken.[17] What about pressure? Let us take the divergence of the momentum equation (6.28).

$$\frac{\partial}{\partial t} \nabla \cdot \mathbf{u} = 0 = \nabla \cdot \{-\nabla \cdot \mathbf{uu} + \nabla \cdot \mathbf{J} \times \mathbf{B}\} - \nabla^2 p . \tag{6.144}$$

This suggests

$$\nabla^2 p = \nabla \cdot \{-\nabla \cdot \mathbf{uu} + \mathbf{J} \times \mathbf{B}\} - \frac{\partial}{\partial t}\nabla \cdot \mathbf{u} \,. \tag{6.145}$$

This is an elliptic equation for p that needs to be solved at each time step for pressure p.

See more discussion on the reduced MHD from Eqs. (6.190) and in Section 6.9.

6.8 Spectral Method

Many recent MHD simulations have been carried out based on the spectral method.[22,23] Sometimes this is called the Galerkin method. It computes the field variables as has been done in the particle simulation (Chapter 2). In particle simulation neither the field equations nor the pushing equation are explicitly nonlinear (Chapter 2). The nonlinearity in particle simulation emerges through the particle dynamics and the many degrees of freedom (many-bodies). Furthermore, products of quantities in these equations are mostly convolutions. Thus the spectral method has been the natural choice in the particle simulation almost from the beginning. In the fluid simulation the nonlinearity explicitly arises in equations. They are usually productive in real space, which becomes convolution in Fourier space. Nevertheless, the spectral method has become popular even in the fluid simulation. This is because of its accuracy in spite of its necessary cost and care for nonlinear terms. Its early advocate, S. Orszag,[22] states "Fourier equations involving N^d degrees of freedom, where d is spatial dimensions, give simulations of at least as accurate as finite difference simulations involving $(2N)^d$ degrees of freedom."

The spectral method's second advantage is its lack of phase errors. As before (Section 6.1), let us examine a problem of the scalar advection

$$\frac{\partial u}{\partial t} + v\frac{\partial u}{\partial x} = 0 \,, \tag{6.146}$$

with v being constant. The analytic solution can be obtained by the method of characteristics: $du/ds = 0$ along the characteristics $dt/ds = 1$ and $dx/ds = v$. This leads to the solution, $u = f(x - vt)$, with f being an arbitrary function as we saw before.

The spectral method expresses the field as

$$u = \sum_k \hat{u} e^{ikx} \,. \tag{6.147}$$

Then the differential equation (6.146) becomes

$$\frac{\partial \hat{u}}{\partial t} + ikv\hat{u} = 0 , \tag{6.148}$$

whose solution yields

$$\omega = kv , \tag{6.149}$$

and therefore the constant phase velocity $\omega/k = v$.

On the other hand, the finite difference equivalent to Eq. (6.148) is

$$\frac{\partial u_j}{\partial t} + v\frac{u_{j+1} - u_{j-1}}{2\Delta x} = 0 . \tag{6.150}$$

As discussed in Chapter 4 with $u_j = \hat{u}\exp(ik \cdot x)$, Eq. (6.150) gives rise to

$$\frac{\partial \hat{u}}{\partial t} + iv\frac{\sin k\Delta x}{\Delta x}\hat{u} = 0 , \tag{6.151}$$

whose solution yields

$$\omega = \frac{\sin k\Delta x}{\Delta x}v = g(k)kv . \tag{6.152}$$

The function $g(k)$ is (correctly) constant only when $k\Delta x \ll 1$. This produces spurious dispersion in the wave.

Interesting comparisons have been made by Boris and Book[6] and by Orszag.[22] Figure 6.9 shows a comparison[6,13] of numerical tests of the ballistic propagation of a square wave

$$\frac{d\rho}{dt} = \frac{\partial \rho}{\partial t} + \mathbf{v} \cdot \nabla \rho = 0 . \tag{6.153}$$

The leapfrog scheme (see Chapter 3 and Section 6.4), the Lax-Wendroff scheme (see Section 6.3), the donor cell method and the flux-corrected transport (Shasta) method (Section 6.5) are compared. The FCT's finesse is apparent, and so in Yabe et al.'s (CIP in Fig. 6.9) method.

Figure 6.10 shows comparison[22,23] of numerical tests of the scalar convection by a circular flow

$$\frac{\partial A}{\partial t} + (\mathbf{v} \cdot \nabla)A = 0 , \tag{6.154}$$

where the circular flow is given as $\mathbf{v} = \mathbf{c} \times \mathbf{r}$. The well-known finite difference scheme (Arakawa's scheme)[24] in various orders produces a fictitious wake trailing the peak of the function A initially given as a conic shape. Also, the originally sharp cone has been rounded. On the other hand, the spectral method based on a comparable sized grid produces a conic shape hardly modified from the original one, as it should be.

It is, however, the nonlinear terms that need careful treatment for the spectral method because the nonlinear terms in the MHD equations are products in real space, which give rise to convolution integrals (or sums) in Fourier

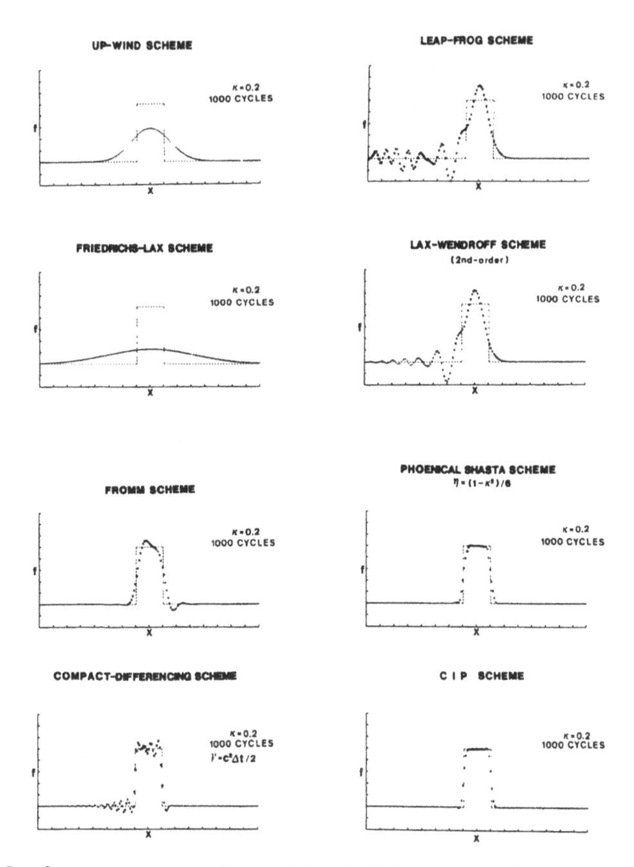

FIGURE 6.9 Accuracy comparison of the ballistic propagation of a square wave (after Ref. 13)

space. This causes two types of consequences: one is an increase of the number of computational operations for the nonlinear term and the other is a more serious numerical instability due to the convolution encountered in the derivative of the spectral method, called the pseudo-spectral method (see later). Let us examine a typical nonlinear term $\mathbf{E} = -\mathbf{u} \times \mathbf{B}$ in Eq. (6.140). One of the terms of the r-component is $E(r,\theta,\zeta) = u_z(r,\theta,\zeta)B_\theta(r,\theta,\zeta)$. Let

$$\begin{pmatrix} E_r \\ u_z \\ B_\theta \end{pmatrix} = \sum_{m=-M}^{M} \sum_{n=-N}^{N} \left\{ \begin{array}{c} E_r(r;mn) \\ u_z(r;mn) \\ B_\theta(r;mn) \end{array} \right\} e^{i(m\theta+n\zeta)} \tag{6.155}$$

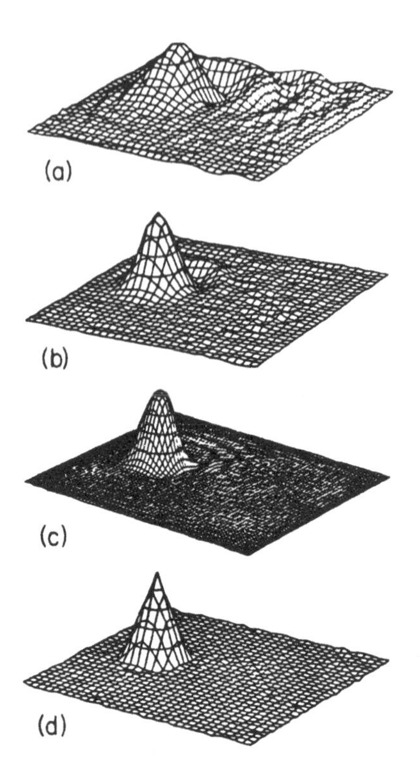

FIGURE 6.10 Accuracy comparison of passive scalar convection (after Ref. 22). (a) 2nd order Arakawa method (Ref. 24) 32 × 32 grid; (b) 4th order Arakawa method, 32 × 32 grid; (c) same, 64 × 64 grid; (d) spectral method, 32 × 32 grid

$$\sum^{'''} E_r(r; m'''n''')e^{i(m'''\theta + n'''\zeta)} = \left(\sum^{'} u_\zeta(r; m'n')e^{i(m'\theta + n'\zeta)} \right)$$

$$\times \quad \left(\sum^{''} B_\theta(r; m''n'')e^{m''\theta + n''\zeta} \right) . \tag{6.156}$$

From Eq. (6.156) we obtain $E_r(r; mn)$ on the left-hand side, and thus obtain

$$E_r(r; mn) = \sum_{m',n'} u_\zeta(r; m'n')B_\theta(r; m - m', n - n') . \tag{6.157}$$

The RHS of Eq. (6.157) has now been expressed in convolution form. For each mode (m, n), there are $O(MN)$ terms in the sum: thus the computation work for all these in the convolution scales as $O((MN)^2)$. The convolution sum involves a number of algebraic operations of $\sim M^2N^2$, while the real space product of the same quantity involves that of $\sim MN$, where M is the number of meshes in the one direction (θ-direction) and N in the other direction (ζ-direction) for the two-dimensional grid system. Thus as the grid system becomes larger, the spectral method becomes increasingly more expensive.

For the cylindrical problem[17] the radial direction is treated in real space. For the radial mesh we may use a staggered mesh for the following benefits.

i) to eliminate decoupling of odd and even mesh points.

ii) to satisfy certain vector identities identically: (in general not the case for an "arbitrary" mesh). e.g., $\nabla \cdot \nabla \times \mathbf{u} = 0$ in finite different form. Let us look at the numerical fulfillment of $\nabla \cdot \mathbf{B} = 0$.

$$\nabla \times \mathbf{E} = \hat{\mathbf{r}} \left\{ \frac{im}{r_i} E_{zi} - \frac{in}{R_0} E_{\theta_i} \right\} + \hat{\theta} \left\{ \frac{in}{R_0} E_{r_i+1/2} - \frac{E_{\zeta_i+1} - E_{\zeta_i}}{r_{i+1} - r_i} \right\}$$

$$+ \hat{\mathbf{z}} \left\{ \frac{r_{i+1} E_{\theta_i+1} - r_i E_{\theta_i}}{r_{i+1/2}(r_{i+1} - r_i)} - \frac{im}{r_{i+12}} E_{r_i+1/2} \right\} \tag{6.158}$$

$$r_{i+1/2} = (r_i + r_{i+1})/2 \tag{6.159}$$

$$\nabla \cdot \mathbf{B} = \frac{r_{i+1} Br_{i+1} - r_i Br_i}{r_{i+1/2}(r_{1+i} - r_i)} + \frac{im}{r_{1+1/2}} B_{\theta 1+1/2} + \frac{in}{R_0} B_\zeta^{1+1/2} \, . \tag{6.160}$$

By substitution, we see identically

$$\frac{\partial}{\partial t} \nabla \cdot \mathbf{B} = -\nabla \cdot \nabla \times \mathbf{E} = 0 \, .$$

If $\nabla \cdot \mathbf{B} = 0$ at $t = 0$, therefore, $\nabla \cdot \mathbf{B} = 0$ at any time.

For time advance differencing, Aydemir and Barnes,[17] for example, adopted a first-order predictor-corrector scheme.

Predictor

$$\mathbf{u}^* = \mathbf{u}^n + \Delta t \theta \left(-\nabla \mathbf{u}'' \mathbf{u}''' + \mathbf{J}^n \times \mathbf{B}^n - \nabla p \right) \, , \tag{6.161}$$

$$\mathbf{B}^* = \mathbf{B}^n + \Delta t \theta \left(\nabla \times \mathbf{u}^n \times \mathbf{B}^n \right) \, , \tag{6.162}$$

where

$$\nabla^2 p^* = \frac{1}{\theta \Delta t}\nabla \cdot \mathbf{u}^n + \nabla \cdot (-\nabla \cdot \mathbf{u}\mathbf{u}^n + \mathbf{J}^n \times \mathbf{B}^n) . \tag{6.163}$$

Corrector

$$\mathbf{u}^{n+1} = \mathbf{u}^n + \Delta t \left\{ -\nabla \cdot \mathbf{u}^* \mathbf{u}^* + \mathbf{J}^* \times \mathbf{B}^* - \nabla p^{n+1} \right\} , \tag{6.164}$$

$$\widetilde{\mathbf{B}}^{n+1} = \mathbf{B}^n + \Delta t \nabla \times (\mathbf{u}^* \times \mathbf{B}^*) , \tag{6.165}$$

where

$$\nabla^2 p^{n+1} = \frac{1}{\Delta t}\nabla \cdot \mathbf{u}^n + \nabla \cdot \left\{ -\nabla \cdot \mathbf{u}^* \mathbf{u}^* + \mathbf{J}^* \times \mathbf{B}^* \right\} , \tag{6.166}$$

$$\mathbf{B}^{n+1} = \mathbf{B}^n - \Delta t \nabla \times \frac{1}{S} \nabla \times \mathbf{B}^{n+1} , \tag{6.167}$$

where the resistive diffusion is treated in this step. Except for diffusion terms, the method is second-order accurate for $\theta = 1/2$; however, in the absence of any dissipation, $\theta = 1/2$ is unstable. Typically $\theta = \frac{1}{2} + \varepsilon$ is chosen.

Pseudo-spectral Method

We so far discussed purely spectral codes where we work in Fourier space only and the nonlinear terms become convolution sums.

$$\omega(x) = u(x)v(x) \rightarrow \omega(k) = \sum_{|k'| \leq K} u(k')v(k - k')$$

where the range of k' is $-K < k' \leq K$ with K being the largest K in the system. For large K, convolution sums become quite expensive; however, simple products in x-space are a lot less expensive. From this consideration, the strategy of pseudo-spectral method is:

— to evaluate nonlinear terms in x-space

— to avoid phase errors, evaluate derivatives in k-space.

Consider an example of $\frac{\partial u}{\partial t} + u\frac{\partial u}{\partial x} = 0$ with the discrete Fourier representation $u(x_j) = \sum_k u(k)e^{ikx_j}$ and its inverse $u(k) = \frac{1}{N}\sum_{j=1}^{N-1} u(x_j)e^{-ikx_j}$. Here the sum is over $k = 2\pi m/N$ with $-N/2 \leq m \leq N/2$ instead of $-\infty < m < \infty$ because the position is represented only at discrete positions. The position x is given only at a discrete point $x = x_j = 2\pi j/N(j = 0, 1, 2, k \ldots, N - 1)$. The linear terms are computed just as in the spectral code. The nonlinear term $u\partial u/\partial x$ is computed as follows. We Fourier transform $u(x_j)$ into k-space and multiply ik. We then make inverse Fourier transform $iku(k)$ into real space and multiply it by $u(x_j)$. The total number of operations in the pseudo-spectral method is: 2FFT's takes $2N \log N$ operations plus 2 products (in a k-space and real space) takes $2N$ operations. The purely spectral code, on the other hand, involves convolutions which require N^2 operations. For large N, $N \log_2 N \ll N^2$.

In the pseudo-spectral method, however, we have to pay the price for the speed of computation, as we mentioned earlier, in terms of the stability. This is because the aliases coupling in the pseudo-spectral method, which comes in coupling in because of the discrete real space representation of functions, rather than the continuous representation.

Examine the nonlinear terms $\omega(x_j) = u(x_j)v(x_j)$ more closely.

$$\begin{aligned}
\omega(k) &= \sum_{|k'|}\sum_{|k''|} u(k')v(k'')\frac{1}{N}\sum_{j=0}^{N-1} e^{i(k'+k''-k)x_j} \\
&= \sum_{|k'|<\frac{N}{2}}\ \sum_{|k''|<\frac{N}{2}} u(k')v(k'')\delta_{k'+k'',k+pN} \\
&= \sum\sum_{k'+k''=k} u(k')v(k'') + \sum\sum_{\substack{k'+k''=k+pN \\ p=\pm1,\pm2,k\ldots}} u(k')v(k'') \,. \qquad (6.168)
\end{aligned}$$

Thus $\omega(k)$ is expressed as

$$\omega(k) = \omega_e(k) + \omega_e(k \pm N) + \omega_e(k \pm 2N) + \cdots \,, \qquad (6.169)$$

where $\omega_e(k)$ is the first term on RHS of Eq. (6.168) and other terms are aliases (see Chapter 4). In practice, since $|k'| < \frac{N}{2}$ and $|k''| < \frac{N}{2}$, only $\omega_e(k \pm N)$ will be non-zero. Aliasing errors usually, but not always, lead to numerical instabilities; they, however, always lead to inaccuracies, especially for "high-k" modes.

The method of eliminating this aliasing error is called dealiasing. First, an obvious method is discussed. Let $\omega(k) = \omega_e(k) + \omega_e(k \pm N)$. We define regions I and II by $-M < k \le 0$ (I) and by $0 < k \le M$ (II) with $M = N/2$. (see Fig. 6.11). But we should require

$$\begin{aligned}
\omega(k_I - 2M) &= 0 \\
\omega(k_{II} + 2M) &= 0 \,. \qquad (6.170)
\end{aligned}$$

When k is in I , alias is in III, while when k is in II, aliasing is in IV. If we

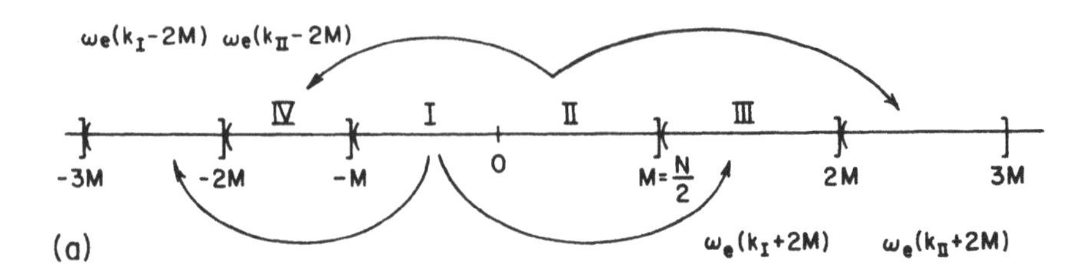

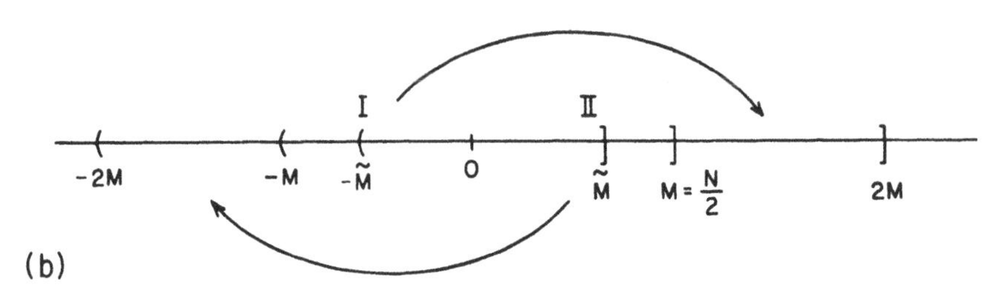

FIGURE 6.11 Aliases coupling and dealiasing

define the dealiased quantities $\tilde{u}$ and $\tilde{v}$ as

$$\begin{pmatrix} \tilde{u}(k) \\ \tilde{v}(k) \end{pmatrix} = \begin{cases} \begin{pmatrix} u(k) \\ v(k) \end{pmatrix} & \text{for } -\frac{M}{2} < k \leq \frac{M}{2} = \frac{N}{4} \\ 0 & \text{otherwise} \end{cases} , \tag{6.171}$$

then $\omega(k_I + 2M)$ and $\omega(k_{II} - 2M)$ are also zero, and thus we have no aliasing errors. This means that although we perform as N-point transform, we retain only $N/2$ modes.

We can somewhat improve the previous dealiasing. Now look at Fig. 6.11(b). In the previous method $\widetilde{M} = M/2 = N/4$. Consider Region II and $\omega(k - 2M)$. If the cutoff wavenumber is $\widetilde{M}$, the quadratic nonlinear term gives rise to $2\widetilde{M}$ and its alias is $2\widetilde{M} - 2M$. If the wavenumber of this alias is out of our allowed region, this system is free of nonphysical alias coupling. Thus we demand $2\widetilde{M} - 2M < -\widetilde{M}$. We thus have the condition

$$\widetilde{M} < \frac{2}{3}M \text{ , rather than } \frac{M}{2}$$

$$\begin{pmatrix} \tilde{u}(k) \\ \tilde{v}(k) \end{pmatrix} = \begin{cases} \begin{Bmatrix} u(k) \\ v(k) \end{Bmatrix} & \text{for } k < \frac{2}{3}M \\ 0 & \text{otherwise} \end{cases} . \tag{6.172}$$

Similarly, for $\omega(k)$ this will result in dealiased pseudo-spatial convolution sums. For general three-dimensional cases, see Patterson and Orszag.[23]

In particle simulation codes, we also often run into such alias coupling as discussed here. The transverse field solver Eq. (5.44) in the magnetoinductive algorithm is an example. Here the nonlinear term $n(x)E_T(x)$ has been treated by the pseudo-spectral method. A typical technique to dealias in this situation is the finite size particle form factor which reduces or cuts off the large wavenumber components of n and E_T or the cutoff of the modes as discussed here. A similar equation appears in the field solver for implicit particle simulation (Chapter 9). In Chapter 4, more discussion on aliases may be found.

6.9 Semi-Implicit Method

In order to describe the long-time scales of the MHD behavior such as the shear waves and resistive processes, there are basically two aproaches. One is to reduce the original MHD equations by eliminating the fast compressible effects by physical considerations.[19,20] This approach has already been partially examined in Section 6.7. The other possible approach is to solve the original (or the already reduced) MHD equations by the implicit time differencing method. As discussed in Chapter 3, the implicit advancement in time removes (or relaxes) the time step constraint, by selectively damping out high frequency components of the field response in the system. Typically, the larger the time step, the wider the spectrum of damping modes becomes. On the other hand, the obvious trade-off of this approach is to invert a large and cumbersome matrix every time step. By adopting the alternating direction implicit (ADI) method such as in Ref. 25, one can reduce the degree of complexity of this matrix inversion because of the reduction of the dimensionality of the "implicit direction." However, this method still requires a large block matrix inversion. The semi-implicit method[26–29] allows us to avoid a large matrix inversion in solving the full implicit equation. On the other hand, advancing the resistive term or viscous term implicitly does not usually pose great difficulties.

Let us consider a set of equations

$$\frac{\partial u}{\partial t} = a\frac{\partial v}{\partial x}, \tag{6.173}$$

$$\frac{\partial v}{\partial t} = a\frac{\partial u}{\partial x}. \tag{6.174}$$

If u is a magnetic field and v is a velocity of fluid for the MHD system in an appropriately normalized unit, a can be the Alfvén velocity. Combining

Eqs. (6.173) and (6.174), we can write down a hyperbolic equation (Chapter 5)

$$\frac{\partial^2 u}{\partial t^2} = a^2 \frac{\partial^2 u}{\partial x^2} , \tag{6.175}$$

describing a wave equation with wave phase speed a. Equation (6.175) may be written as

$$\frac{\partial^2 u}{\partial t^2} - a_0^2 \frac{\partial^2 u}{\partial x^2} = a^2 \frac{\partial^2 u}{\partial x^2} - a_0^2 \frac{\partial^2 u}{\partial x^2} , \tag{6.176}$$

where a_0 is a constant appropriately chosen for numerical stability considerations. In discretizing in time we make the new term on the LHS implicit. However, the same new term on the RHS is kept explicit because otherwise that value is not known and has to be inverted, which we try to avoid. Thus we have

$$u^{n+1} - a_0^2 \Delta t^2 \frac{\partial^2 u^{n+1}}{\partial x^2} = u^n + a\Delta t \left(\frac{\partial u}{\partial t} \right) + a^2 \Delta t^2 \frac{\partial^2 u^n}{\partial x^2} - a_0^2 \Delta t^2 \frac{\partial^2 u^n}{\partial x^2} . \tag{6.177}$$

This is equivalent to adding a linear term with a coefficient proportional to the time step Δt in Eq. (6.173) or Eq. (6.174):

$$\partial u / \partial t = a \partial v / \partial x + a \Delta t G \partial u / \partial t . \tag{6.178}$$

The analysis shows that the "unconditional" numerical stability is achieved provided that

$$aG\Delta t \geq \omega^2 \Delta t^2 / 4 - 1 , \tag{6.179}$$

where ω is the system's eigenfrequency such as $\omega = ka$. For the sake of accuracy, we furthermore choose

$$aG\Delta t \ll 1 . \tag{6.180}$$

A practical way to advance Eqs. (6.173) and (6.174) in order to obtain an algorithm equivalent to Eq. (6.177) is a two-step predictor-corrector scheme.

By writing $\partial u / \partial x$ as iku, the predicted values are

$$\begin{aligned} u^* &= u^n + \theta \Delta t (ika) v^n , \\ v^* &= v^n + \theta \Delta t (ika) u^n , \end{aligned} \tag{6.181}$$

The corrected values are

$$\begin{aligned} u^{n+1} &= u^n + i\Delta t a v^* = u^n + ik\Delta t a \left[v^n + \theta i \Delta t a u^n \right] , \\ v^{n+1} &= v^n + i\Delta t a u^* = v^n + ik\Delta t a \left[u^n + \theta i \Delta t a v^n \right] . \end{aligned} \tag{6.182}$$

Or Eqs. (6.181) and (6.182) are written as

$$\begin{bmatrix} u \\ v \end{bmatrix}^{n+1} = \begin{bmatrix} 1 - \theta \Delta t^2 a^2 & i\Delta t a \\ i\Delta t a & 1 - \theta \Delta t^2 a^2 \end{bmatrix} \begin{bmatrix} u \\ v \end{bmatrix}^n . \tag{6.183}$$

The stability of the system may be inspected by introducing the usual amplification factor g. We obtain

$$\left[\left(1-\theta\Delta t^2a^2\right)-g\right]^2+\Delta t^2a^2=0\ . \tag{6.184}$$

From Eq. (6.184) the square modulus of g is

$$|g|^2=1-\Delta t^2a^2(-1+2\theta)+\theta^2\Delta t^4a^4\ . \tag{6.185}$$

The minimum of $|g|^2$ in Eq. (6.185) takes place at $x=\frac{(2\theta-1)}{2\theta^2}$, where $x=\Delta ta$. The explicit system of the predictor-corrector equations is stable if

$$\Delta ta<(2\theta-1)/\theta^2\ . \tag{6.186}$$

For $\theta=\frac{1}{2}$, the system is unstable for all Δt. For $\theta=1$, $\Delta t\alpha<1$ is the stability condition.

Let us then try to modify the corrector-step according to the strategy of the semi-implicit technique as

$$u^{n+1}-\theta\Delta t^2a_0^2u^{n+1}=u^n+i\Delta tav^n+\theta\Delta t^2a^2u^n-\theta\Delta t^2a_0^2u^n\ , \tag{6.187}$$

where a_0 is some constant, as indicated in Eq. (6.176). For v we put

$$v^{n+1}=v^n+i\Delta ta\left(\frac{u^n+u^{n+1}}{2}\right)\ . \tag{6.188}$$

For the case

$$a_0^2<\frac{a^2}{16}\left(\frac{1+2\theta}{\theta}\right)^2\ , \tag{6.189}$$

this scheme is unconditionally stable as desired (stable for all Δt). For $\theta=1/2$, $a_0^2<a^2$. The advantages of this method are: a_0 is a constant. Trying to treat $\Delta t^2a^2u^n$ term implicitly would lead to large matrix inversion because a is a function instead of a constant.

As an application, let us examine semi-implicit reduced MHD calculations. The reduced MHD equations have been discussed in Section 6.7 and by Strauss[19] and others.[20] The assumption for the reduced MHD equations for a tokamak plasma is $\varepsilon=\frac{a}{R_0}\ll 1$, which is the small expansion parameter. The so-called tokamak scaling may be characterized by $\frac{\partial}{\partial z}\sim\varepsilon\nabla_\perp$, $B_\perp\sim\varepsilon B_z$, $B_z\sim 1$, $\rho\sim 1$ and $p\sim\varepsilon^2$. The reduced set of MHD equations may be written as

$$\frac{\partial U}{\partial t}+[\phi,U]+\nabla_\parallel J=0\ , \tag{6.190}$$

$$\frac{\partial\psi}{\partial t}+\nabla_\parallel\psi=\eta J\ , \tag{6.191}$$

where $U = \nabla_\perp^2 \phi$ vorticity, $J = \nabla_\perp^2 \psi$ "parallel current" (negative of $J_\parallel$ vector), ϕ is the electrostatic potential and ψ is the poloidal flux function. The usual predictor-corrector scheme leads to the usual stability condition $\Delta t |k_\parallel| v_A < 1$, ignoring diffusion and convection by $\mathbf{v}_\perp$.

Let us consider the semi-implicit treatment of the $\nabla_\parallel$ operator

$$\frac{\partial}{\partial t}\nabla_\perp^2 \phi + \nabla_\parallel \nabla_\perp^2 \psi = 0 , \tag{6.192}$$

$$\frac{\partial \psi}{\partial t} + \nabla_\parallel \phi = 0 , \tag{6.193}$$

which are similar to Eqs. (6.173) and (6.174). The predictor step yields

$$\nabla_\perp^2 \phi^* = \nabla_\perp^2 \phi^n - \theta \Delta t \nabla_\parallel^n \nabla_\perp^2 \psi^n , \tag{6.194}$$

$$\psi^* = \psi^n - \theta \Delta t \nabla_\parallel^n \phi^n , \tag{6.195}$$

where the operator $\nabla_\parallel$ has a time-step assignment because it involves projection to the magnetic field direction. The corrector produces

$$\nabla_\perp^2 \phi^{n+1} - \theta \Delta t^2 \nabla_{\parallel 0} \nabla_\perp^2 \nabla_{\parallel 0} \phi^{n+1} = \nabla_\perp^2 \phi^n - \Delta t \nabla_\parallel \nabla_\perp^2 \psi^n$$

$$+\theta \Delta t^2 \nabla_\parallel \nabla_\perp^2 \nabla_\parallel \phi^n - \theta \Delta t^2 \nabla_{\parallel 0} \nabla_\perp^2 \nabla_{\parallel 0} \phi^n , \tag{6.196}$$

where $\nabla_{\parallel 0} = \frac{\partial}{\partial z} - [\psi_0, \quad]$ and $\psi_0 = \psi_0(r)$ only. If ψ_0 is independent of θ and z, then $[\psi_0, \phi]$ does not lead to convolutions, i.e., it does not couple various harmonics. We have $\nabla_{\parallel 0} \rightarrow i k_{\parallel 0}(r)$ and $\nabla_\perp^2 \rightarrow \frac{1}{r}\frac{\partial}{\partial r} r \frac{\partial}{\partial r} - \left(\frac{m}{r}\right)^2$. Equation (6.196) may be cast into

$$\nabla_\perp^2 \phi^{n+1} + \theta \Delta t^2 k_{\parallel 0}(r) \nabla_\perp^2 k_{\parallel 0} \phi^{n+1}$$

$$= \nabla_\perp^2 \phi^n - \Delta t \nabla_\parallel \nabla_\perp^2 \psi^n + \theta \Delta t^2 \nabla_\parallel^n \nabla_\perp^2 \nabla_\parallel \phi^n + \theta \Delta t^2 k_{\parallel 0} \nabla_\perp^2 k_{\parallel 0} \phi^n \tag{6.197}$$

where $\phi = \phi(r; m, n)$. Tri-diagonal system for $\phi(r; m, n)$ can be inverted easily. Let us get

$$U^{n+1} = \nabla_\perp^2 \phi^{n+1} . \tag{6.198}$$

Finally, we have

$$\psi^{n+1} = \psi^n - \Delta t \nabla_\parallel^* \left(\frac{\phi^n + \phi^{n+1}}{2} \right) . \tag{6.199}$$

Unconditional stability is achieved if $|k_{\parallel 0}| > |\nabla_\parallel|$ (for $\theta = 1/2$).

When nonlinearities such as $v \cdot \partial v / \partial x$ become comparable to or larger than linear terms such as $v_A \cdot \partial v / \partial x$, however, the above prescribed semi-implicit method does not facilitate the production of large time steps. This is because the subtracted terms are based on (or related to) the terms in the real physical characteristics.

6.10 Upwind Differencing

Turbulence is one of the most fascinating and difficult problems in modern physics. This presents us with the challenge of one of the important unsolved problems in physics. In nonlinear physics obviously nonlinear terms such as the advective term in hydrodynamics play an important role. However, the presence of a very small but non-vanishing (linear) dissipative term plays as crucial a role for turbulence. Dissipation is recognized as the catalyst for pattern formation (and perhaps life formation in recent studies of nonlinear dynamics.[30]) Without a viscous term the hydrodynamic equation $\partial u/\partial t + u\partial u/\partial x = 0$ is exactly integrable to give an implicit solution $u = f(x - ut)$, where f is an arbitrary function. With a viscous term the fluid can exhibit a more complicated structure, including turbulence. The Reynolds number R characterizes the importance of the nonlinear term relative to the dissipative term. In general, the larger the Reynolds number is, the more complicated and turbulent the structure of the system becomes. At the same time the larger the Reynolds number is, the more nonlinear and stiff the numerical integration of the fluid equation becomes. This situation is a highly turbulent state as $R^{-1} \to 0$, while no turbulence exists when $R^{-1} = 0$. See also the discussion in Chapter 1.

A situation similar to this appears in the resistive MHD equation. The ideal MHD without the resistive term is characterized by the Lundquist number S (the magnetic Reynolds number) being infinite (or $S^{-1} = 0$), which is the ratio of the nonlinear magnetic induction term to the linear resistive dissipation. In this, however, no change in the topology of magnetic fields takes place, as no magnetic field reconnection is allowed (see Chapter 14). On the other hand, with a small but nonvanishing resistivity, active changes in magnetic topology and sometimes explosive processes of magnetic energy converting into kinetic energy occur. Again the larger S is, the more difficult the simulation becomes.

In hydrodynamic turbulence simulation Moin and Kim[31] recently carried out large simulation runs on a grid of $128 \times 64 \times 64$, where the first direction is treated by finite differences and other two directions by the spectral method. Kuwahara, Kawamura and their colleages,[32] however, developed a higher order upwind (or upstream) finite difference method and applied it to turbulence with a much larger Reynolds number. This method is based on the following observation in hydrodynamics: "Main information comes from the upstream." The upwind finite difference of the advective term is written as

$$\left(u\frac{\partial u}{\partial x}\right)_i = \begin{cases} u_i(u_i - u_{i-1})/\Delta & (\text{if } u_i \geq 0) \\ u_i(u_{i+1} - u_i)/\Delta & (\text{if } u_i < 0) \end{cases} . \tag{6.200}$$

This may be rewritten as

$$\left(u\frac{\partial u}{\partial x}\right)_i = u_i(u_{i+1} - u_{i-1})/2\Delta - \Delta \cdot |u_i|(u_{i+1} - 2u_i + u_{i-1})/\Delta^2 \ , \quad (6.201)$$

in which the second term on the RHS is approximately a diffusion term $D(|u_i|)\partial^2 u/\partial x^2$ with $D(|u_i|)$ being $\Delta|u_i|$. A higher order upwind difference is

$$\left(u\frac{\partial u}{\partial x}\right)_i = \begin{cases} u_i(3u_i - 4u_{i-1} + u_{i-2})/2\Delta & (u_i \geq 0) \ , \\ u_i(-3u_i + 4u_{i+1-U_{i+2}})/2\Delta & (u_i < 0) \ , \end{cases} \quad (6.202)$$

which may be rewritten as

$$\left(u\frac{\partial u}{\partial x}\right)_i = u_i\left[-u_{i+2} + 4(u_{i+1} - u_{i-1}) + u_{i-2}\right]/4\Delta$$

$$+|u_i|(u_{i+2} - 4u_{i+1} + 6u_i - 4u_{i-1} + u_{i-2})/4\Delta \ . \quad (6.203)$$

The first term on the RHS produces a second-order error $O(\Delta^2)$ and the second term, a third order error $O(\Delta^3)$:

$$\left(u\frac{\partial u}{\partial x}\right)_i \sim u\frac{\partial u}{\partial x} - \frac{1}{3}\Delta^2\frac{\partial^3 u}{\partial x^3} + O(\Delta^4) + \Delta^3 u\frac{\partial^4 u}{\partial x^4} + O(\Delta^5) \ . \quad (6.204)$$

The error [the second term on the RHS of Eq. (6.204)] may be eliminated by the following: replace the first term on the RHS of Eq. (6.204) by

$$u_i\left[-u_{i+2} + 8(u_{i+1} - u_{i-1}) + u_{i-2}\right]/12\Delta \ . \quad (6.205)$$

This eliminates the above-mentioned error of $O(\Delta^2)$ and the error in the scheme (6.205) becomes a fourth-order derivative of $O(\Delta^3)$. By this method they[32] obtained a simulation of turbulence with up to $R = 670000$.

6.11 Discussion of Various Methods

So far we have reviewed only some of the numerical methods that simulate the MHD (or hydrodynamical system) equations. Because of the inherent importance in predicting gross behavior of plasmas directly relevant to desired applications, the fairly long time scale and large spatial scale characteristics of the MHD, and the lack of complexities associated with particles, MHD simulations have been very popular and thus the techniques have been extensively developed. There are many other interesting techniques to tackle this problem. An interested reader may want to read the book by P. Roache[33] for a classical review of the various techniques. Some more recent attempts we have not mentioned so far include the EPIC (Ephemeral Particle-in-Cell) method[34] and the finite element method.[35]

We examine the finite element method by considering the following ordinary differential equation:

$$-p\frac{d^2u}{dx^2} + qu = f(x) \tag{6.206}$$

under the boundary condition $u(0) = u(1) = 0$ and f is a given function. Let us try to solve Eq. (6.206) by expansion by eigenfunctions. For a standard eigenvalue problem

$$-p\frac{d^2u}{dx^2} + qu = \lambda u \ , \tag{6.207}$$

$$(u(0) = u(1) = 0) \ ,$$

the eigenvalues and eigenfunctions are given by

$$\begin{aligned} \lambda_k &= p\pi^2k^2 + q \ , \\ \phi_k &= \sqrt{2}\sin k\pi x \ , \end{aligned} \qquad (k = 1, 2, \ldots) \tag{6.208}$$

where $\sqrt{2}$ is the normalization constant. For the solution of Eq. (6.206), we want to expand in terms of the eigenfunctions

$$u(x) = \sqrt{2}\sum_{j=1}^{\infty} a_j \sin j\pi x \ . \tag{6.209}$$

By inserting Eq. (6.209), it is easy to see

$$a_k = \frac{\sqrt{2}}{\lambda_k}\int_0^1 f(x)\sin k\pi x dx \ . \tag{6.210}$$

In reality (i.e., more complex problems) we may not know the eigenfunctions, so we introduce (or fix) basis functions $\{\phi_k\}(k = 1, 2, \ldots, n)$, instead. We express an approximate solution by expansion of ϕ_k as

$$\tilde{u}(x) = \sum_{j=1}^{n} a_j\phi_j(x) \ . \tag{6.211}$$

The problem is to determine the coefficients a_j. This is sometimes called the Galerkin method. Equation (6.211) is a generalization of a Fourier expansion. The spectral technique in Section 6.8 may thus be considered as a special case of the Galerkin method. In the typical Galerkin problem orthogonal polynomials, trigonometric functions etc. have been chosen as eigenfunctions. They are global functions whose modulus does not vanish away from some specific spatial point.

A contrasting choice of basis functions in the Galerkin method is local functions whose modulus vanish away from a particular local point.

How about choosing *triangle functions* as basis functions? The triangle function $\phi_k(x)$ is defined as

$$\phi_k(x) = \Lambda_k(x) = \begin{cases} 0, \ (0 \le x \le x_{k-1} \ , \ x_{k+1} \le x \le 1) \\ \frac{x - x_k}{\Delta} \ , \ (x_{k-1} \le x < x_k) \\ \frac{x_{k+1} - x}{\Delta} \ (x_k \le x < x_{k+1}) \end{cases} \tag{6.212}$$

The triangle function is a localized basis function. Let us expand on these as

$$\tilde{u}(x) = \sum_{j=1}^{n-1} a_j \phi_j(x) \ , \tag{6.213}$$

where terms with $j = 0, n$ are taken out for boundary conditions and $\tilde{u}$ is a piecewise linear function.

Consider an example of Eq. (6.206). After integrating by parts, we obtain

$$\sum_{j=1}^{n-1} a_j \int_0^1 \left(p\phi_k'\phi_j' + q\phi_k\phi_j\right) dx = \int_0^1 f\phi_k dx \ (k = 1, \ldots, n-1) \ . \tag{6.214}$$

Equation (6.214) is a matrix equation for $\{a_j\}$. By writing $\mathbf{a} = (a_j)$ and $\mathbf{f} = \left(\int_0^1 f\phi_j dx\right)$. Equation (6.214) can be written as

$$(K + M)\mathbf{a} = \mathbf{f} \ , \tag{6.215}$$

where

$$K_{ij} = \int p\phi_i'\phi_j' dx \ , \tag{6.216}$$

$$M_{ij} = \int q\phi_i\phi_j dx \ . \tag{6.217}$$

A major advantage of choosing triangle functions as basis functions is that the matrix is tridiagonal, which is quite easy to solve.

Consider Poisson's equation as an example

$$\begin{cases} \frac{\partial^2 u}{\partial x^2} + \frac{\partial^2 u}{\partial y^2} = f \ , \\ u(\Gamma) = 0 \quad (\Gamma \text{ is the boundary}) \ . \end{cases} \tag{6.218}$$

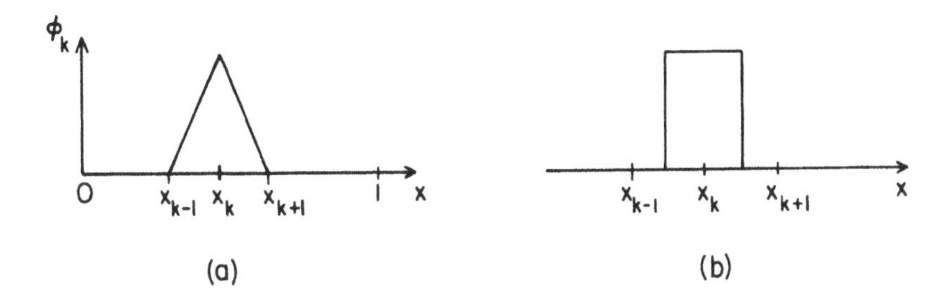

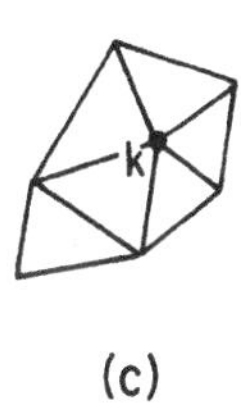

FIGURE 6.12 Local functions for basis functions. (a) the triangular function, (b) the Π function.

The basis function ϕ_k is defined as 1 at the kth node (or grid point) and otherwise is zero. And it takes a triangular shape in two dimensions. See Fig. 6.12. We solve $\{a_j\}$ by the Galerkin method. Such an approach to solving the equations by cutting the domain into finite elements and representing ϕ_k in these elements is called the finite element method.

We may be able to choose other basis functions such as

$$\phi_k(x) = \Pi(x - x_k) , \tag{6.219}$$

where $\Pi(x - x_k)$ is pictorially given in Fig. 6.12(b). This function, however, does not behave well for derivatives in Eq. (6.214). Thus Eq. (6.219) is only used for terms like M_{ij} in Eq. (6.215). The finite *element* method equation, Eq. (6.215), can be written as

$$-p\frac{u_{j-1} - 2u_j + u_{j+1}}{\Delta^2} + qu_j = f_j , \tag{6.220}$$

where $u_j = u(x_j)$. This equation is equivalent to the finite *difference* equation. The only differences are: the coefficients $(p\phi'_j, \phi'_j)$, and the RHS (f, ϕ_j) are integrals in the finite element method, while in the finite difference the coefficients $p(x_j)$ and the RHS $f(x_j)$ are functions.

We apply the finite element method to MHD. The main motivation of this for MHD is that often boundary conditions and mesh selection become easier in the finite element method in comparison with the finite difference method. On the other hand, the finite element method is less efficient or difficult in comparison with finite difference method in a regular shape boundary or in highly nonlinear problems. As an example, let us consider solving an MHD stability problem by the variational principle.[36] The stationary point of the Lagrangian displacement function is sought in variations of the plasma displacement $\boldsymbol{\xi}(\mathbf{r}, t) = \boldsymbol{\xi}(\mathbf{r})e^{-i\omega t}$, and the vector potential $\mathbf{A}(\mathbf{v}, t) = \mathbf{A}(\mathbf{r})e^{-i\omega t}$. The well-known form[36] of the Lagrangian is

$$\mathcal{L} = \omega^2 K - W \tag{6.221}$$

where

$$W = W_p + W_s + W_V ,$$

$$W_p = \frac{1}{2}\int_P d\mathbf{r}\Big\{|\mathbf{Q}|^2 - J_0 \cdot (\mathbf{Q} \times \boldsymbol{\xi}) + \gamma p_0 |\boldsymbol{\nabla} \cdot \boldsymbol{\xi}|^2 + \boldsymbol{\nabla} \cdot \boldsymbol{\xi}\boldsymbol{\xi} \cdot \boldsymbol{\nabla} p_0\Big\} , \tag{6.222}$$

$$\mathbf{Q} \equiv \boldsymbol{\nabla} \times (\boldsymbol{\xi} \times \mathbf{B}_0) , \tag{6.223}$$

$$W_s = \frac{1}{2}\oint_{\Gamma_p} ds(\mathbf{n} \cdot \boldsymbol{\xi})^2 \left\langle \boldsymbol{\nabla}\left(p_0 + \frac{1}{2}|\mathbf{B}|^2\right)\right\rangle , \tag{6.224}$$

$$W_v = \frac{1}{2}\int_V d\mathbf{r}(\boldsymbol{\nabla} \times \mathbf{A})^2 , \tag{6.225}$$

$$K = \frac{1}{2}\int_P d\rho_0 |\boldsymbol{\xi}|^2 , \tag{6.226}$$

where p designates plasma volume, V the vacuum and $\mathbf{n}$ is the normal vector on the surface of plasma Γ_p. The angular brackets $\langle X \rangle$ denote the difference of equilibrium quantity X on two surfaces before and after variation ξ. The boundary conditions are:
on the surface of plasma

$$-\gamma p_0 \boldsymbol{\nabla} \cdot \boldsymbol{\xi} + \mathbf{B}_0 \cdot (\mathbf{Q} + \boldsymbol{\xi} \cdot \boldsymbol{\nabla}\mathbf{B}_0) = \mathbf{B}_{v0} \cdot (\boldsymbol{\nabla} \times \mathbf{A}) + \boldsymbol{\xi} \cdot \boldsymbol{\nabla}\mathbf{B}_{v0} , \tag{6.227}$$

$$\mathbf{n} \times \mathbf{A} = -(\boldsymbol{\xi} \cdot \mathbf{n})\mathbf{B}_{v_0} \tag{6.228}$$

on the metallic surface

$$\mathbf{n}_v \times \mathbf{A} = 0 \ , \tag{6.229}$$

where B_{v_0} is the vacuum magnetic field in equilibrium. In the variational principle, however, Eq. (6.228) is not needed. This is in clear contrast to the initial value problem.

In order to solve $\boldsymbol{\xi}$ and ω^2 for the Lagrangian extremum under K =constant, [Eq. (6.221)] we expand $\boldsymbol{\xi}$ by the basis function

$$\boldsymbol{\xi} = \sum_j \boldsymbol{\xi}_j \phi_j(\mathbf{r}) \ . \tag{6.230}$$

By writing the set of expansion coefficients as $\boldsymbol{\zeta} \equiv (\boldsymbol{\xi}_j)$ and $\boldsymbol{\zeta}^+ \equiv (\boldsymbol{\xi}_j^+)$, Eq. (6.221) becomes $\mathcal{L} = \omega^2 \zeta^+ \widehat{K} \zeta - \zeta^+ \widehat{W} \zeta$. Thus the variational equation is

$$\widehat{W}\boldsymbol{\zeta} - \omega^2 \widehat{K} \boldsymbol{\zeta} = 0 \ , \tag{6.231}$$

which presents an eigenvalue problem for ω^2.

An example for the basis functions of the finite element method for Alfvén waves is considered. Since the shear Alfvén wave demands $\nabla \cdot \boldsymbol{\xi} = 0$, the basis functions have to contain a property that $\nabla \cdot \boldsymbol{\xi}$ is constant ($= 0$) within an element. In order to realize this, a transformation is carried out[37] (normal finite element method):

$$\boldsymbol{\xi} = (\xi_r, i\xi_\theta, i\xi_z) \rightarrow (\xi_1, \xi_2, \xi_3) \tag{6.232}$$

$$\begin{pmatrix} \xi_1 \\ \xi_2 \\ \xi_3 \end{pmatrix} = \begin{pmatrix} 1 & 0 & 0 \\ \frac{1}{r} & \frac{m}{r} & 0 \\ 0 & 0 & 1 \end{pmatrix} \begin{pmatrix} \xi_r \\ i\xi_\theta \\ i\xi_z \end{pmatrix} . \tag{6.233}$$

With this choice, $\nabla \cdot \boldsymbol{\xi}$ becomes $\nabla \cdot \xi = \frac{d\xi_1}{dr} + \xi_2 + k\xi_3$. In order to satisfy the request that $\nabla \cdot$ =constant within one element, we arrive at the conclusion that ξ_1 be linear in r; ξ_2, ξ_3 be constant in r.

The EPIC method[34] follows the particle-in-cell method (Chapters 2, 4 and 5) and transforms the governing hyperbolic equations into coordinate mappings a lá the Lagrangian displacement method. We want to find the fluid element at $\mathbf{x}$ (at $t = t$) which was at $\mathbf{x}_0$ (at $t = 0$), which may be written as $\mathbf{x} = \mathbf{x}(\mathbf{x}_0, t)$. The infinitesimal volume $d\tau_0$ at x_0 is then mapped to

$$d\tau = |J| d\tau_0 \ , \tag{6.234}$$

where

$$\overleftrightarrow{J} = \left(\frac{\partial x_i}{\partial x_{0j}} \right) = (J_{ij}) \quad \text{and} \quad J_{ij}(t=0) = \delta_{ij} \ ,$$

which is the Jacobian or the displacement gradient matrix. The transformation of a surface element $ds_0 \to ds$ is $ds = ds_0 \cdot \overleftrightarrow{J} |J|^{-1}$ and that of the density is $\rho = |J|^{-1}\rho_0$ and for the magnetic field it is $\mathbf{B} = |J|^{-1}\overleftrightarrow{J} \cdot \mathbf{B}(x_0, 0)$ (if resistivity $\eta = 0$), as $\mathbf{B} \cdot ds = \mathbf{B}_0 \cdot ds_0$.

Let us project density evolution $\rho = |J|^{-1}\rho_0$ onto a set of the basis functions $\{w_k(\mathbf{x}, t)\}$:

$$\int w_k(\tilde{\rho} - |J|^{-1}\rho_0)d\tau = -\int w_k \varepsilon d\tau = 0 \ , \tag{6.235}$$

where $\tilde{\rho}$ is the trial function in approximation of ρ, ($\rho = \tilde{\rho} + \varepsilon$) which is expanded in terms of the trial functions $\{\phi_\ell\}$ (or often taken as $\{w_k\}$ themselves).

Equation (6.235) may describe: (i) a finite element method, or (ii) an Eulerian EPIC method, or (iii) a moving mesh EPIC method. Here the Galerkin method is incorporated in the basis function technique $\{\omega_k\}$, while the particle (Lagrangian) method is in the Jacobian $\{J_{ij}\}$, so to speak.

Case (i)

When $w_k(\mathbf{x}) = w_k(\mathbf{x}_0)$, i.e., there is neither convection of the basis functions nor temporal evolution of these.

$$\sum_\ell \left(\int w_k(\mathbf{x})\phi_\ell(\mathbf{x})d\tau \right) \rho_\ell = \int w_k(\mathbf{x}_0)\rho(\mathbf{x}_0)d\tau_0 \ . \tag{6.236}$$

Case (ii)

When $\frac{\partial w_k}{\partial t} = 0$,

$$\sum_\ell \left(\int w_k(\mathbf{x})\phi_\ell(\mathbf{x})d\tau \right) \rho_\ell = \int w_k(\mathbf{x})\rho(\mathbf{x}_0)d\tau \ . \tag{6.237}$$

Equations (6.237) can be written as

$$A_{k\ell}\rho_\ell = m_k \ , \tag{6.238}$$

where

$$A_{k\ell} = \int w_k(x)\phi_\ell(x)d\tau \ , \tag{6.239}$$

$$m_k = \int w_k(x)\rho(\mathbf{x}_0)d\tau_0 \ , \tag{6.240}$$

to solve for ρ_ℓ. The new $\{\rho_\ell\}$ are determined from previous $\rho(\mathbf{x}_0)$, which was expanded by $\{w_k\}$, the basis functions.

If the basis functions are spatially localized functions such as $\Lambda_k(x)$ [Eq. (6.212)] or $\Pi(x-x_k)$ [Eq. (6.219)], then m_k may be interpreted as the mass associated with the grid at k. Then Eq. (6.238) is equivalent to Eq. (6.206). (If w_k is a spatially nonlocalized global function such as a trigonometric function, then this is more like a normal mode expansion solution.)

Many different approaches have been tried to improve numerical solutions of spatial treatment of fluid equations including the continuity equation. These include various finite difference methods (Sections 6.2-6.4), the FCT method (Section 6.5), characteristic particle methods (Section 6.6 and 6.11), spectral methods (Section 6.8) and finite element methods (Section 6.11). Spectral methods give superb results in many cases, but lack general applicability and particularly so for complicated geometry. Spectral methods can become expensive when complicated nonlinearities are present. Characteristic particle methods get over the nonlinear and diffusion-term stumbling blocks nicely, but run into massive amounts of computer time when the number of "particles" has to be made large to reduce fluctuations. Finite-element and spline methods also give excellent results where they are applicable, but their complexity and computational cost are often prohibitively high. The finite difference methods are the most traditional and easy to construct. As we have seen, they have shortcomings in accuracy and stability. Many improvements have been proposed, including the FCT and the upwind difference. In real problems we often combine these techniques, such as the finite difference in the radial direction and the spectral method in the θ and z-directions (Sections 6.8 and 6.9).

In terms of many improvements on temporal treatment of fluid equations, we discussed the explicit method, the implicit method (Section 6.9), the semi-implicit method, and the method of reducing the original equations to a set of equations that contain less time hierarchy and thus less stiff numerical properties (Sections 6.7 and 6.8). The explicit method is simple and contains all the physics delineated by the original fluid equations, while its stability condition is rigidly limited by the CFL condition, which could often set way too small a timestep for practical applications. The implicit method allows unconditional numerical stability, with certain physics with frequencies higher than the Nyquist frequency $T_N = \pi/\Delta t$ (the inverse of the timestep $\times\pi$). However, this method calls for inversion of matrix for the timestep, which is too massive and too expensive to carry out in many cases. The semi-implicit method may be a nice compromise between the two. Within a limit of certain nonlinearities and that of relevant frequencies of interest, the semi-implicit method can be "unconditionally" stable without resorting to full scale matrix inversion. The reduced set of MHD equations is an alternative method, in which the starting equations are already free of "unwanted" high frequency physics. This technique is obviously problem-dependent.

Problems

1 Discuss what kind of difficulty arises in the fluid description Eqs. (6.27) and (6.28) or Eqs. (6.14), (6.17), and (6.18) when there are two counter-streaming interpenetrating fluid components. In general, non-Maxwellian distribution in velocity space gives rise to a similar difficulty.

2 Write down the "standard" resistive MHD equations from Eq. (6.24) and Eq. (6.25) by dropping the second and fourth terms in Eq. (6.25). What are justifications to do so?

3 Show the incompressible MHD equation (6.139)-(6.143) have only the shear Alfvén branch. Note: $p_i^* = 0$.

4 Discuss the dealiasing strategy in the pseudo-spectral method for equations with cubic nonlinear terms, such as $|u|^2 u$. Generalize this argument to n^{th} order nonlinearity and show that the dealiasing cutoff is $\widehat{M} = \frac{2}{n+1} M$, where $M = N/2$.

5 Prove Eq. (6.204).

6 Burgers' equation may be looked upon as a special case of the Navier-Stokes equation: $\partial v/\partial t + v \partial v/\partial x = R^{-1} \partial^2 v/\partial x^2$. Discuss the difference in behavior of solutions when $R^{-1} = 0$ and when $0 \neq R^{-1} \ll 1$. The exact integral is known.[38]

7 By introducing Dirac's representation for basis function such as $\phi_k(x) = |k\rangle$, rewrite the procedure of the finite element method.

8 In the reduced MHD, which ratio of numbers represent the degree of nonlinearity of the system?

9 Show momentum conservation with the system of the algorithm of Eq. (6.79).

10 Study the stability of the system of Eqs. (6.161)-(6.167) and show that $\Delta t|\mathbf{k} \cdot \mathbf{B}_0| < 1$ is required.

References

1 H. Alfvén, *Cosmical Electrodynamics* (Oxford Univ. Press, London, 1950).

2 N. Krall and A. Trivelpiece, *Principle of Plasma Physics* (McGraw Hill, New York, 1973) p. 78.

3 T. Tajima, J.N. Leboeuf, and J.M. Dawson, J. Comp. Phys. **38**, 237 (1980).

4 G.A. Sod, J. Comp. Phys. **27**, 1 (1978).

5 R.D. Richtmyer and K.W. Morton, *Difference Methods for Initial-Value Problems* (Wiley and Sons, New York, 1967).

6 J.P. Boris and D.L. Book, J. Comp. Phys. **11**, 38 (1973).

7 D.E. Potter, *Computational Physics* (Wiley and Sons, London, 1973).

8 J.P. Boris and D.L. Book, J. Comp. Phys. **20**, 397 (1976).

9 M.W. Evans and F.H. Harlow, LASL Rep. LA-2139 (1957).

10 J.N. Leboeuf, T. Tajima, and J.M. Dawson, J. Comp. Physics **31**, 379 (1979).

11 J. Pasta and S. Ulam, LASL Rep. LA-1557 (1953).

12 F. Brunel, J.N. Leboeuf, T. Tajima, J.M. Dawson, J. Comp. Phys. **43**, 269 (1981).

13 A. Nishiguchi and T. Yabe, J. Comp. Phys. **47**, 297 (1982); H. Tatewaki, A. Nishiguchi, and T. Yabe, J. Comp. Phys. **61**, 26 (1985).

14 J. Brackbill, and J. Monaghan, *Particle Methods in Fluid Dynamics and Plasma Physics*, Comp. Phys. Comm. (1988).

15 J. Brackbill Comp. Phys. Comm. **47**, 1 (1987).

16 W.D. Shulz, *Methods in Computational Physics* Vol. 3 (Academic, New York, 1984).

17 A.Y. Aydemir and D.C. Barnes, J. Comp. Phys. **53**, 100 (1984).

18 A.Y. Aydemir and D.C. Barnes, J. Comp. Phys. **59**, 108 (1985).

19 H.R. Strauss, Phys. Fluids **19**, 134 (1976).

20 B.V. Waddell, M.N. Rosenbluth, D.A. Monticello, R.B. White, L.B. Carreras, *Theoretical and Computational Plasma Physics*, International Centre for Theoretical Physics, College, 22 March 1977 and Third International ("Kiev"" Conference, (IAEA, Trieste, 1978) p. 79; H.R. Hicks, B. Carreras, J.A. Holmes, D.K. Lee, and B.V. Waddell, J. Comp. Phys. **44**, 46 (1981).

21 L. Artisimovich, *Controlled Thermonuclear Reactions* (Gordon and Breach, New York, 1964).

22 S. Orszag, J. Fluid Mech. **49**, 75 (1971).

23 G.S. Patterson and S.A. Orszag, Phys. Fluids **14**, 2538 (1971).

24 A. Arakawa, J. Comp. Phys. **1**, 119 (1968).

25 C.H. Finan and J. Killeen, Comp. Phys. Comm. **24**, 441 (1981).

26 A.J. Robert, Proc. WMO/IUGG Symp. on Numerical Weather Prediction (Tokyo, 1969).

27 D.S. Harned and J.W. Kerner, J. Comp. Phys. **60**, 62 (1985).

28 D.S. Harned and D.D. Schnack, J. Comp. Phys. **65**, 57 (1986).

29 D.D. Schnack, D.C. Barnes, Z. Mikic, D. Harned, and E.J. Caramana, J. Comp. Phys. **70**, 330 (1987).

30 G. Nicolis and I. Prigogine, *Self-Organization in Non-equilibrium System* (Wiley and Sons, New York, 1977).

31 P. Moin and J. Kim, J. Fluid Mech. **118**, 341 (1982).

32 T. Kawamura and K. Kuwahara, AIAAA paper 84-0340 (1984); S. Osawa, R. Himeno, K. Kuwahara, S. Shirayama, and T. Kawamura, Proc. Euromech. No. 199 (1985).

33 P.J. Roache, *Computational Fluid Dynamics* (Hermosa, Albuquerque, 1972).

34 J.W. Eastwood and W. Arter, to be published.

35 G. Strang and G.J. Fix, *An Analysis of the Finite Element Method* (Prentice-Hall, New Jersey, 1973).

36 I.B. Bernstein, E.A. Frieman, M.D. Kruskal, and R.M. Kulsrud, Proc. Roy. Soc. **A244**, 17 (1958).

37 D. Berger, R. Gruber and F. Troyon, Comp. Phys. Comm. **11**, 313 (1976).

38 J.M. Burgers, Proc. Roy. Neth. Acad. Sci. **43**, 1 (1940).

7

GUIDING-CENTER METHOD

The time scales desired for computer simulation of plasmas cover a broad range, as discussed in Chapter 1. Various types of codes have been developed in an attempt to convey the pertinent physics, while overlooking less important phenomena. Table 7.1 lists some of these.

In general, the faster the plasma phenomenon is, the more particle-like it is; conversely, the slower the plasma variation is, the more fluid-like it is. Because of this general tendency, the appropriate numerical technique in high frequencies tends to be a particle-like approach, while that in lower frequencies tends to be fluid-like. Phenomena between these two extremes often exhibit both facets of nature. For such problems hybrid methods may be the most natural. In this chapter we particularly focus on the guiding-center method. This method exhibits a partially fluid-like behavior in the direction perpendicular to the ambient magnetic field. However, in the guiding-center method it is paramount to treat the dynamics parallel to the magnetic field as particle-like. More "fluid-like" hybrid methods will be treated in Chapter 8.

Particle Codes	Guiding-Center Codes	Quasi-neutral Codes	MHD Particle Codes	MHD/resistive Codes	Transport Codes
		Implicit particle codes			Diffusion eqs.
			Vortex-in-cell		Fokker-Planck eqs.
					Lagrangian eq. etc.
		electron-fluid ion-particle (or quasi-particle)			mapping
←Particle-like		←Hybrid→		Fluid-like→	

TABLE 7.1 Various types of plasma simulation codes corresponding to different levels of time hierarchy

7.1 $\mathbf{E} \times \mathbf{B}$ Drift

A guiding-center model may be profitably used when the motion consists of a fast oscillatory motion plus another slower motion. An example of this type of physical situation is the gyration of charged particles in a magnetic field. The motion breaks into the familiar gyro-motion with a characteristic frequency, Ω, plus a slower $\mathbf{E} \times \mathbf{B}$ drift of the guiding center. (The concept of a guiding center, a term used to denote the center of the gyro-orbit, was first introduced by H. Alfvén[1]).

$$\frac{d\mathbf{v}}{dt} = \frac{q}{m}\left(\mathbf{E} + \frac{\mathbf{v}}{c} \times \mathbf{B}\right), \tag{7.1}$$

$$\langle \mathbf{v} \rangle = \mathbf{v}_g = c\frac{\mathbf{E} \times \mathbf{B}}{B^2}. \tag{7.2}$$

Equation (7.1) describes full scales of motion, while Eq. (7.2) describes the slower drift motion obtained by taking a time average of Eq. (7.1).

A similar effect may be seen in meteorological and geophysical settings. The role of the magnetic field is played by the angular momentum vector and the role of the electric field by the pressure gradient. The Coriolis force is analogous to the $\mathbf{E} \times \mathbf{B}$ force. Assume that the rotation is constant: $\Omega = \text{const}$. We have

$$\frac{d\mathbf{v}}{dt} = -\frac{1}{\rho}\left[\boldsymbol{\nabla} p - 2\mathbf{v} \times \boldsymbol{\Omega} + \nabla\Phi + \boldsymbol{\Omega} \times (\boldsymbol{\Omega} \times \mathbf{x})\right] \tag{7.3}$$

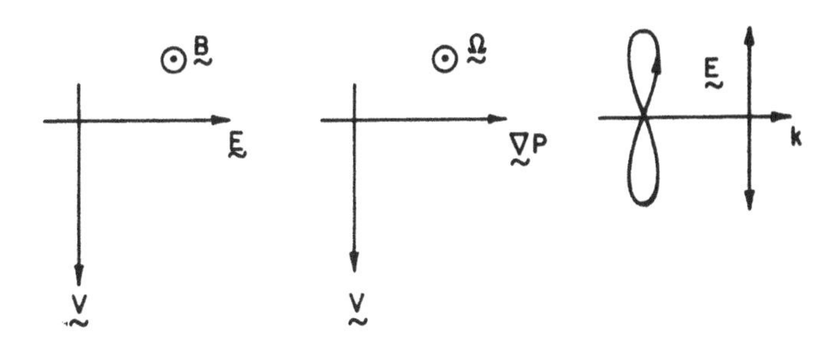

FIGURE 7.1 The $\mathbf{E} \times \mathbf{B}$ drift, the geostrophic flow, and the fast oscillating electromagnetic wave and the consequent particle orbit

where $\boldsymbol{\Omega}$ is the angular velocity of the earth's rotation. Methods like the vortex-in-cell techniques or discrete vortex technique[2] have been developed, which parallel the traditional plasma particle simulation under discussion. On the earth, the wind blows primarily perpendicular to the pressure gradient due to the second term, Coriolis effect, on the right-hand side of Eq. (7.3). The fourth term is the centrifugal force, which is usually small or can be balanced with the pressure (∇p) and the gravitational ($\nabla\Phi$) terms.

For large scale waves on the earth or other planets such as Jupiter, the ratio of the acceleration [the left-hand side of Eq. (7.3)] to the Coriolis force is ordinarily small. Letting the characteristic scale length be L and the characteristic velocity be U, this ratio is of the order of $U/2\Omega L \equiv R$, which is called the Rossby number. When the inviscid flow has a small Rossby number, such a flow is said to be geostrophic. This is equivalent to the guiding-center approximation in the case of magnetized plasmas:

$$\mathbf{v} = -\frac{1}{2\rho}\nabla(p - \rho\Phi) \times \boldsymbol{\Omega}/\Omega^2.$$

Another example occurs when a laser beam interacts with plasma. A high frequency electric field oscillates perpendicular to the direction of the wave propagation. The slower motion can be set off by the time-averaged electromagetic wave pressure called the ponderomotive force. An example of the resultant equation for the slower motion may be found in Refs. 3 and 4. See Fig. 7.1.

7.2 Guiding-Center Model

Let us concern ourselves with the slow motion produced by a system such as that of a plasma with strong external magnetic fields. The equation of motion for a particle in such a system is Eq. (7.1) . Since we are concerned with the slow time scale, the left-hand side of this equation may be approximated by zero. The expansion parameter here is ω/Ω, where ω is the typical low frequency of interest and Ω the cyclotron frequency. A more comprehensive derivation may be found in Refs. 5 and 6. The motion may then be decoupled into components perpendicular and parallel to the magnetic field.

$$\mathbf{v}_\perp = c\frac{\mathbf{E}\times\mathbf{B}}{B^2} \quad , \quad \frac{d\mathbf{x}_\perp}{dt} = \mathbf{v}_\perp \ , \tag{7.4}$$

$$\frac{dv_\parallel}{dt} = \frac{q}{m}E_\parallel \quad , \quad \frac{dx_\parallel}{dt} = v_\parallel \ . \tag{7.5}$$

For the parallel motion, the leapfrog method can be used as before. However, the time stepping is different in the perpendicular direction. If a leapfrog is attempted, one has

$$\mathbf{v}_\perp = \frac{d\mathbf{x}_\perp}{dt} \rightarrow \frac{1}{\Delta t}\left(\mathbf{x}_\perp^{n+1} - \mathbf{x}_\perp^n\right) = c\frac{\mathbf{E}(\mathbf{x}^n)\times\mathbf{B}}{B^2} \ . \tag{7.6}$$

The left-hand side is centered about time step $n+\frac{1}{2}$, while the right-hand side is at time n. This equation is, therefore, not time-centered (and proves to be unstable) and so something different must be done to push $\mathbf{x}_\perp$.

To proceed toward a possible cure, look at temporal finite differencing and its numerical stability of the equation for guiding-center motion,

$$0 = \mathbf{E} + \frac{\mathbf{v}}{c}\times\mathbf{B}_0 \ .$$

This yields the solution for the perpendicular velocity,

$$\mathbf{v}_\perp = c\frac{\mathbf{E}\times\mathbf{B}_0}{B_0^2} \ .$$

Suppose $\mathbf{B}_0$ is in the $\hat{z}$ direction. Then $\mathbf{v}_\perp$ may be written in Cartesian coordinates,

$$\frac{dx}{dt} = c\frac{E_y}{B_0} \ , \tag{7.7}$$

$$\frac{dy}{dt} = -c\frac{E_x}{B_0} \ . \tag{7.8}$$

Further, suppose

$$E_x = -\alpha x \ , \tag{7.9}$$

and

$$E_y = -\alpha y \ . \tag{7.10}$$

This is to say that we use the electrostatic approximation with $\alpha \propto \omega_p^2$. Next, finite difference in time

$$x^{n+1} - x^n = -\alpha \Delta t y^{n+1/2} , \tag{7.11}$$

$$y^{n+1} - y^n = +\alpha \Delta t x^{n+1/2} . \tag{7.12}$$

The position of the particle is not known at half time steps, so it is approximated by an implicit relation

$$x^{n+1/2} \simeq \frac{1}{2}\left[x^{n+1} + x^n\right] , \tag{7.13}$$

$$y^{n+1/2} \simeq \frac{1}{2}\left[y^{n+1} + y^n\right] . \tag{7.14}$$

Insertion of this approximation in Eqs. (7.11) and (7.12) yields

$$x^{n+1} - x^n = -\frac{\alpha \Delta t}{2}\left[y^{n+1} + y^n\right] , \tag{7.15}$$

$$y^{n+1} - y^n = +\frac{\alpha \Delta t}{2}\left[x^{n+1} + x^n\right] , \tag{7.16}$$

which is now properly time-centered. This, as we shall see below, leads to stability and neutrality.

The stability may be analyzed as before by assuming $x^n \propto g^n$, $y^n \propto g^n$ with g being the amplification factor and requiring that the determinant of the amplification matrix be zero. Rewriting the equations (7.15) and (7.16), we obtain

$$\left(x^{n+1} - x^n\right) + \frac{\alpha \Delta t}{2}\left(y^{n+1} + y^n\right) = 0 ,$$

$$-\frac{\alpha \Delta t}{2}\left(x^{n+1} + x^n\right) + \left(y^{n+1} - y^n\right) = 0 ,$$

We obtain the amplification matrix

$$\begin{pmatrix} (g-1) & \frac{\alpha \Delta t}{2}(g+1) \\ -\frac{\alpha \Delta t}{2}(g+1) & (g-1) \end{pmatrix} . \tag{7.17}$$

Equation (7.17) has a determinant equal to zero for values of g given by the quadratic equation

$$(g-1)^2 + \left(\frac{\alpha \Delta t}{2}\right)^2 (g+1)^2 = 0 , \tag{7.18}$$

with solutions given by

$$g = \frac{1 - \tilde{\alpha} \pm \left[(1 - \tilde{\alpha})^2 - (1 + \tilde{\alpha})^2\right]^{1/2}}{1 + \tilde{\alpha}} \tag{7.19}$$

where $\tilde{\alpha} = (\alpha \Delta t/2)^2$. The observation of properties of solutions, Eq. (7.19),

$$|g|^2 = g_1 g_1^* = g_2 g_2^* = g_1 g_2 = 1 , \tag{7.20}$$

shows both stability and neutrality. This corresponds to the stable and neutral algorithm.

7.3 Numerical Methods for Guiding-Center Plasmas

A. Predictor-corrector Method

We consider an equation of the form

$$\dot{f} = \Gamma(t, f) . \tag{7.21}$$

Equation (7.4) also has this form. As we discussed, when the right-hand side of Eq. (7.21) contains the quantity we wish to integrate on the left-hand side, a naive finite differenced time integration yields a numerically unstable solution. To cure this, the predictor-corrector method (see Chapters 3 and 6) is invoked.

The simplest form of the predictor-corrector method is

$$f_{\text{pred}}^{n+1} = f^n + \Delta t \Gamma^n , \tag{7.22}$$

$$f_{\text{corr}}^{n+1/2} = \frac{1}{2}\left(f_{\text{pred}}^{n+1} + f^n\right) . \tag{7.23}$$

A more satisfactory predictor-corrector is like the Lax-Wendroff scheme studied in Chapter 6 on MHD models. The version of the predictor-corrector method presented here was first done by Lee and Okuda[7] for the guiding-center equations. The equations utilized are of the form

$$f_{\text{pred}}^{n+1} = f^{n-1} + 2\Delta t \Gamma(t^n, f^n) , \tag{7.24}$$

$$f_{\text{corr}}^{n+1} = f^n + \Delta t \Gamma\left[t^{n+1/2}, \frac{1}{2}\left(f_{\text{pred}}^{n+1} + f_n\right)\right] . \tag{7.25}$$

Quantities f must be stored at two time steps.

To utilize this method for the solution of guiding-center motion, Eq. (7.4), three steps are followed.

1) Predictor step

(Notation: use $*$ for predicted values)

$$\frac{\mathbf{x}_{\perp}^{*n+1} - \mathbf{x}_{\perp}^{n-1}}{2\Delta t} = \frac{c}{B_0^2} \left[\mathbf{E}^n \left(\mathbf{x}^n \right) \times \mathbf{B}_0 \right] . \tag{7.26}$$

Note that not only does one need to know the present particle position $\mathbf{x}^n$, but also the particle position one time step in the past $\mathbf{x}_{\perp}^{n-1}$. Here, the electric field $\mathbf{E}^n$ was determined by solving Poisson's equation using known $\mathbf{x}^n$.

2) A predicted electric field is obtained by solving Poisson's equation

$$\left\{ x^{*n+1} \right\} \rightarrow \nabla \cdot E^{*n+1} = 4\pi e \sum_i \delta \left(x - x_j^{*n+1} \right) . \tag{7.27}$$

If finite sized particles are involved, $\delta \rightarrow S$, the shape factor.

3) Corrector Step

$$\frac{\mathbf{x}_{\perp}^{n+1} - \mathbf{x}_{\perp}^{n}}{\Delta t} = \frac{c}{B_0^2} \frac{1}{2} \left[\mathbf{E}^{*n+1} \left(\mathbf{x}^{*n+1} \right) + \mathbf{E}^n \left(\mathbf{x}^n \right) \right] \times \mathbf{B}_0 . \tag{7.28}$$

Note the fields are evaluated at their respective particle position. The time averaging of $\mathbf{E}$ in this step is used because we really want

$$\frac{\mathbf{x}_{\perp}^{n+1} - \mathbf{x}_{\perp}^{n}}{\Delta t} = \frac{c}{B_0^2} \mathbf{E}^{n+1/2} \left(\mathbf{x}^{n+1/2} \right) \times \mathbf{B}_0 ,$$

but this is not available. The scheme wherein $\mathbf{E}^{n+1/2}(\mathbf{x}^{n+1/2})$ is replaced by $1/2(\mathbf{E}^{*n+1} + \mathbf{E}^n)(\mathbf{x}^{n+1/2})$ in Eq. (7.28) does not work and is numerically unstable. The total position for the guiding center is

$$\mathbf{x} = \mathbf{x}_{\perp} + x_{\parallel} \hat{b}(\mathbf{x}) . \tag{7.29}$$

The method is now stable, however, slightly dissipative and dispersive. J. Denavit showed that the numerical damping for this type of algorithm is $\mathcal{O}(\Delta t^3)$.

B. Separate Treatments of Parallel and Perpendicular Directions

As we shall see in Secs. 7.4 and 7.6, the polarization drift and finite Larmor radius effects are (often) unimportant for electrons. Therefore, the guiding-center model, Eqs. (7.4) and (7.5), suffices for magnetized electrons. On the other hand, it does not suffice in ions for such problems as drift waves due to the polarization drift and finite Larmor radius effects. Thus a standard approach is to employ full dynamics for ions [Eq. (7.1)]. This combination of electron guiding centers and ions with full dynamics has been successfully and frequently employed in low frequency particle simulations[8,9] (also see Chapter 13).

Because electrons in a magnetized plasma behave quite differently in the direction parallel to the magnetic field as opposed to those perpendicular to it, we expect quite different behaviors in the parallel and perpendicular directions in a three-dimensional code. Electrons can propagate rapidly along the field line to screen the charge separation which may develop. Thus any perturbation along the magnetic field with a short wavelength may be screened rapidly except for the high frequency (ordinary) plasma oscillations, so that only relatively long wavelength modes can develop. On the other hand, perturbations across the field with short wavelengths could survive with electron motion basically perpendicular to the perturbation. Only when the wavelength becomes shorter than the ion Larmor radius, perturbations can be mixed up by the ion gyromotion.[10]

Electrostatic perturbations along (or oblique to) the field line retain the characteristics of plasma oscillatons, whose frequency is $\omega_{pe}k_{\parallel}/k$, where $k_{\parallel} = \mathbf{k}\cdot\mathbf{B}/B$ [see Fig. 7.2(a)]. Electrostatic perturbations completely normal to the field line lose the above property. Such perturbations would be of (nearly) zero frequency or of purely imaginary (damping) frequency, if the ions are given infinite mass: this is a convective cell or vortex mode.[11] If the ions are allowed to move, perturbations can take frequencies of lower hybrid oscillations, i.e., those of ion plasma oscillations ($\sim \omega_{pi}$).

If we are interested in low frequency waves such as drift waves, the high frequency plasma oscillations which develop away from the plane normal to the magnetic field are spoiling our purpose in two ways. First, they add high frequency noise that could bury the modes of interest. Second, they demand a short time step $\Delta t \lesssim \omega_{pe}^{-1}$ for numerical stability of the explicit leapfrog code, which is very restrictive for low frequency phenomena simulation. A simple cure for this may be to prohibit those modes with

$$k_{\parallel}/k < \cos\theta_c \tag{7.30}$$

(where θ_c is a number close to $\pi/2$) from being applied to particle pushing. See Fig. 7.2(b). This method is sometimes called "fanning," as we eliminate modes in a fan shape region in **k**-space. This will improve the stability condition

$$\Delta t \lesssim \omega_{pe}^{-1}k/k_{\parallel} \ , \tag{7.31}$$

while it still has to satisfy the ballistic CFL condition

$$\Delta t \lesssim \Delta_z/v_{\max} \ , \tag{7.32}$$

where Δ_z is the z-directional grid separation and $v_{\max}$ the typical maximum velocity in that direction (such as a few times thermal velocity v_e). Note that the ballistic CFL condition is not a requirement in the directions perpendicular

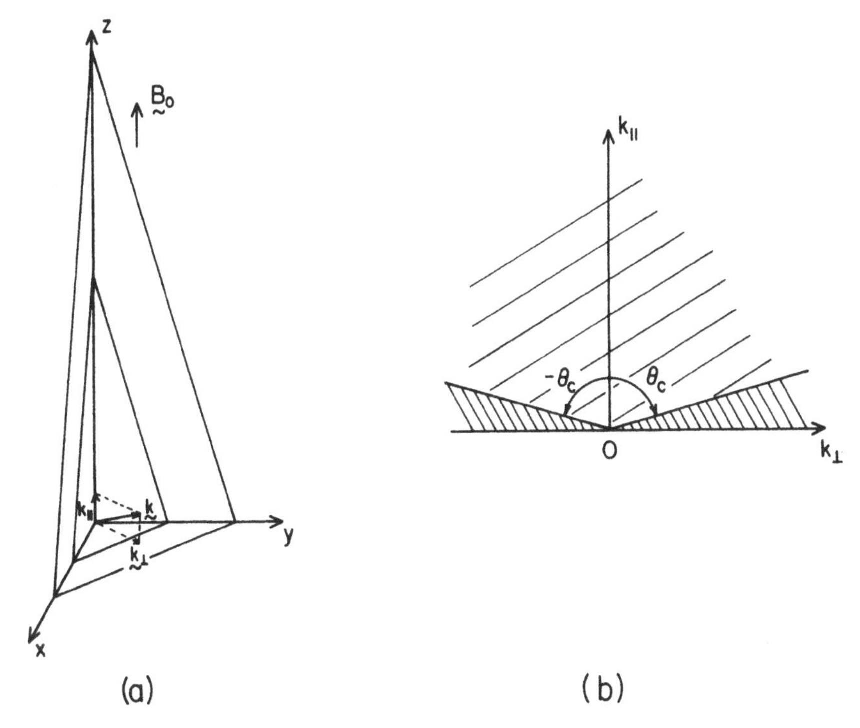

FIGURE 7.2 Typical wavenumber and elimination of mode in a fan-like region

to the magnetic field. The stability condition per waves in the perpendicular directions is

$$\Delta t \lesssim \omega_{LH}^{-1} , \tag{7.33}$$

where ω_{LH} is the lower hybrid frequency ($\sim \omega_{pi}$).

The numerical stability (the thermal instability) requires that the plasma should not be too cold because of the aliases coupling (Chapter 4) such that the Debye length should not be much less than the unit grid separation. The severity of this condition depends sensitively on the accuracy of the interpolation scheme (Chapter 4). We can write this as

$$\alpha^{-1}\lambda_{D_e} = \alpha^{-1}v_e\omega_{pe}^{-1} \gtrsim \Delta_z , \tag{7.34}$$

where α is about 0.3 for the SUDS and < 0.1 for the area weighting. [For the cubic spline α is extremely small $\sim O(10^{-6})$] (Private Comm., J.S. Wagner, see also Chapter 4). In the perpendicular directions the thermal stability condition (7.34) is not required to fulfill because of the gyromotion. Equation (7.34) means that given the thermal velocity, the cell size Δ_z cannot be chosen arbitrarily large, while Eq. (7.32) suggests that the larger Δ_z is, the better.

Equations (7.32) and (7.34) also indicate that

$$\Delta t \lesssim \alpha^{-1}\left(\frac{v_e}{v_{\max}}\right)\omega_{pe}^{-1} , \tag{7.35}$$

which is severe if α is of the order of 0.1 or larger.

Cheng and Okuda[12] devised a method to overcome this difficulty, sometimes called the mode expansion method. Since the relevant wavelengths are much larger in the z-direction than in the others, it would be profitable to take a large cell size Δ_z, consistent with Eq. (7.32). This violates Eq. (7.34), however, giving rise to the numerical thermal instability. The origin of this instability was interpolation errors and alias coupling (Chapter 4). If we use the exact particle position without resorting to the grid method and interpolation, this numerical instability is arrested. To avoid excessive collisions, particles still have a finite size in the z-direction, and that size is approximately related to the smallest wavelength of the mode we want to keep in the simulation system. Let z_j be the j-th particle's (exact) position. The electric field at the position $\mathbf{x}_j = (x_j, y_j, z_j)$ is

$$\mathbf{E}(\mathbf{x}_j) = \sum_{n=-\infty}^{\infty} \mathbf{E}(x_j, y_j, k_n)e^{ik_n z_j} , \tag{7.36}$$

where $k_n = 2\pi n/L_z$. Here the summation on n can be a partial sum over modes one wants to keep in the simulation. The charge density due to the j-th finite size particle is

$$\rho_j(\mathbf{x}) = \frac{q_j}{(2\pi)^{3/2}a_x a_y a_z}\exp\left[-\frac{(x-x_j)^2}{2a_x^2} - \frac{(y-y_j)^2}{2a_y^2} - \frac{(z-z_j)^2}{2a_z^2}\right] , \tag{7.37}$$

where a_x, a_y, and a_z are the particle size in three directions and q_j is the charge of the j-th particle, whose Fourier transform is

$$\rho_j(x, y, , n) = \frac{1}{L_z}\int_0^{L_z} \rho_j(\mathbf{x})e^{-ik_n z}dz$$

$$\cong e^{-k_n^2 a^2/2}\frac{q_j}{2\pi a_x a_y L_z}\exp\left[-\frac{(x-x_j)^2}{2a_x^2} - \frac{(y-y_j)^2}{2a_y^2}\right]e^{-ik_n z_j} . \tag{7.38}$$

The density $\rho_n(x, y)$ is defined as

$$\rho_n(x, y) = \sum_j \rho_j(x, y, n) . \tag{7.39}$$

Here we represent $\rho_n(x, y)$ only at discrete (two-dimensional) grid points $(x, y) = (x_g, y_g)$.

The density ρ and electric potential ϕ (or field $\mathbf{E}$) are related by Poisson's equation

$$\nabla^2\phi = -\boldsymbol{\nabla}\cdot\mathbf{E} = -4\pi\sum_j \rho_j(\mathbf{x}) , \tag{7.40}$$

which may be rewritten as

$$\left[\nabla_\perp^2 - \left(\frac{2\pi n}{L_z}\right)^2\right]\phi_n(x,y) = -4\pi\rho_n(x,y) , \tag{7.41}$$

where $\nabla_\perp^2 = \partial_x^2 + \partial_y^2$. Equations (7.41) are a set of Helmholtz equations in two dimensions with $-N \leqq n \leqq N$, where N is the number of kept modes, which can be solved by means of fast Fourier transforms (Chapter 4) for $\phi_n(k_x,k_y)$. With the knowledge of $\phi_n(k_x,k_y)$ ($k_x = 2\pi\ell/L_x$, $k_y = 2\pi m/L_y$) the potential on grid points can be calculated

$$\phi_n(x_g,y_g) = \sum_{m=0}^{L_y-1}\sum_{\ell=0}^{L_x-1}\phi_n(k_x,k_y)\exp\left[2\pi i(x_g\ell/L_x + y_g m/L_y)\right] . \tag{7.42}$$

In order to push particles, we need to compute the electric field in real space

$$\mathbf{E}(x_g,y_g,z_j) = \sum_{n=-N}^{N}\mathbf{E}_n(x_g,y_g)e^{+2\pi i n z_j/L_z} , \tag{7.43}$$

where $\mathbf{E}_n(x_g,y_g)$ has been calculated from $\phi_n(k_x,k_y)$ in $\mathbf{k}$ space and the inverse Fourier transform. (See Problem 4.) The force is computed using the interpolation from the grid position (x_g,y_g) to the particle position (x_j,y_j). Note that x_j and y_j have been interpolated, while z_j has been used exactly. For this reason, the algorithm is numerically stable even if L_z is taken so large with a given thermal velocity in the z-direction that the condition, Eq. (7.34), is not observed.

The disadvantage of this algorithm is the computational expense associated with the operations Eqs. (7.38) and (7.43), because the numbers of operations are proportional to the total number of particles. In the grid method (Chapter 4), on the other hand, the corresponding operations are interpolation operations which do not involve costly calculations of a transcendental function, the exponential, for each particle.

For this reason, a finer interpolation scheme has been introduced[13] to cut the costly operations involved in Eqs. (7.38) and (7.43) and, at the same time, to allow a large enough grid spacing Δ_z for the present purpose. As mentioned earlier, in the grid method Eq. (7.34) has to be maintained to avoid thermal instability. The technique developed by Okuda and Cheng[13] (see Chapter 4) allows one to decrease α in Eq. (7.34) to an extremely small value ($\sim 10^{-6}$) when we apply the cubic spline interpolation.

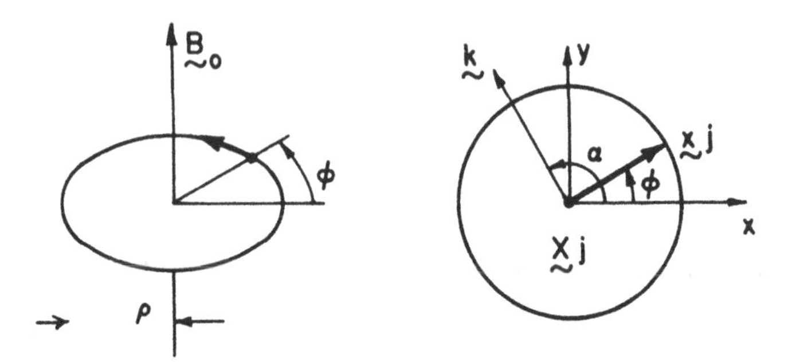

FIGURE 7.3 Gyromotion, gyrophase and guiding-center coordinate

7.4 Polarization Drift

Having outlined the basic path involved in the guiding-center method, some other points with drift approximations are considered. The starting equation, Eq. (7.1), has been looked at for strong magnetic fields to give first order results for the guiding-center motion

$$\mathbf{v}_\perp = c\frac{\mathbf{E}\times\mathbf{B}}{B^2} - \left[\frac{v_\parallel^2}{\Omega}(\hat{b}\cdot\nabla)\hat{b} + \frac{v_\perp^2}{2B\Omega}\nabla B\right]\times\hat{b} \quad , \quad \dot{v}_\parallel = \frac{q}{m}E_\parallel \, , \qquad (7.44)$$

where the terms in the angular brackets are the curvature drift and the gradient-B drift when the magnetic field is nonuniform. These terms do not cause any specific numerical difficulty, so we drop them unless they become relevant. Two major effects which are omitted in the above guiding-center description are polarization drift effects and finite Larmor radius effects. The former effects arise from the second order correction of the guiding-center approximation in terms of the ratio of the frequency of the wave to the cyclotron frequency, i.e., $O(\frac{\omega}{\Omega})^2$. The latter effect arises from the first order convection of the guiding-center approximation in terms of the ratio of the Larmor radius to the relevant wavelength, i.e., $O(\rho/\lambda)$.

The effect of a finite Larmor radius is the spreading of the charge distribution. This spreading is due to the fact that the particle with its charge is actually moving with a finite radius and thus may be influenced by variations in its surroundings which occur on this length scale. (See Fig. 7.3).

Polarization drift may be obtained by expanding the starting equation to the next higher order in $\frac{\omega}{\Omega} \ll 1 (\Omega = \frac{qB}{mc})$. To the zeroth order, the gyromotion $\mathbf{v}^{(0)} = v_\perp(\hat{x}\cos\phi + \hat{y}\sin\phi)$ is obtained. To the first order, the guiding-center description,

$$0 = \mathbf{E} + \frac{\mathbf{v}^{(1)}}{c} \times \mathbf{B} , \tag{7.45}$$

is obtained. The second order equation is

$$\frac{d\mathbf{v}^{(1)}}{dt} = 0 + \frac{q}{m}\frac{\mathbf{v}^{(2)}}{c} \times \mathbf{B} . \tag{7.46}$$

If $\mathbf{B}$ is taken to the left side of the equation, and noting that $\frac{d}{dt} \sim O(\omega)$, one may see that the equation is of order $\frac{\omega}{\Omega} v^{(1)}$, i.e., second order.

Assume

$$\mathbf{v} = \mathbf{v}^{(0)} + \mathbf{v}^{(1)} + \mathbf{v}^{(2)} + \ldots . \tag{7.47}$$

From Eq. (7.46), we obtain

$$\frac{d}{dt}\left[c\frac{\mathbf{E} \times \mathbf{B}}{B^2}\right] = \frac{q}{mc}\mathbf{v}^{(2)} \times \mathbf{B} \tag{7.48}$$

so that a solution for $\mathbf{v}_\perp^{(2)}$ is

$$\mathbf{v}^{(2)} = \frac{-c(\dot{\mathbf{E}}) \times \mathbf{B}}{\Omega B^3} , \tag{7.49}$$

and that consists of two terms (sometimes called linear and nonlinear polarization terms)

$$\dot{\mathbf{E}} = \frac{d\mathbf{E}}{dt} = \frac{\partial \mathbf{E}}{\partial t} + \mathbf{v}^{(1)} \cdot \frac{\partial}{\partial \mathbf{x}}\mathbf{E} = \frac{\partial \mathbf{E}}{\partial t} + c\frac{\mathbf{E} \times \mathbf{B}}{B^2} \cdot \frac{\partial}{\partial \mathbf{x}}\mathbf{E} . \tag{7.50}$$

The polarization drift velocity $\mathbf{v}^{(2)}$ is not contained in the simple guiding-center description, Eq. (7.4). The ion polarization drift is the mass ratio (M/m) times the electron polarization drift. Thus the latter is often negligible compared with the former.

Why is this effect called a polarization drift? Look at an electrostatic model (constant $\mathbf{B}$) $\mathbf{v}_\perp^{(1)} = c\frac{\mathbf{E} \times \mathbf{B}}{B^2}$. To the first order, the electrons and ions have the same drift velocity. Therefore, this $\mathbf{E} \times \mathbf{B}$ drift does not give rise to any charge separation. To the second order, however, $\mathbf{v}^{(2)}$ has a dependence on $\Omega = \frac{qB}{mc}$. This drift is charge dependent, so a net charge can be induced, i.e., polarization. The polarization drift term should be kept, if the primary mechanism of charge separation is this, such as in a drift wave where $k_\parallel \ll k_\perp$. A schematic representation of the drift wave is depicted in Fig. 7.4.

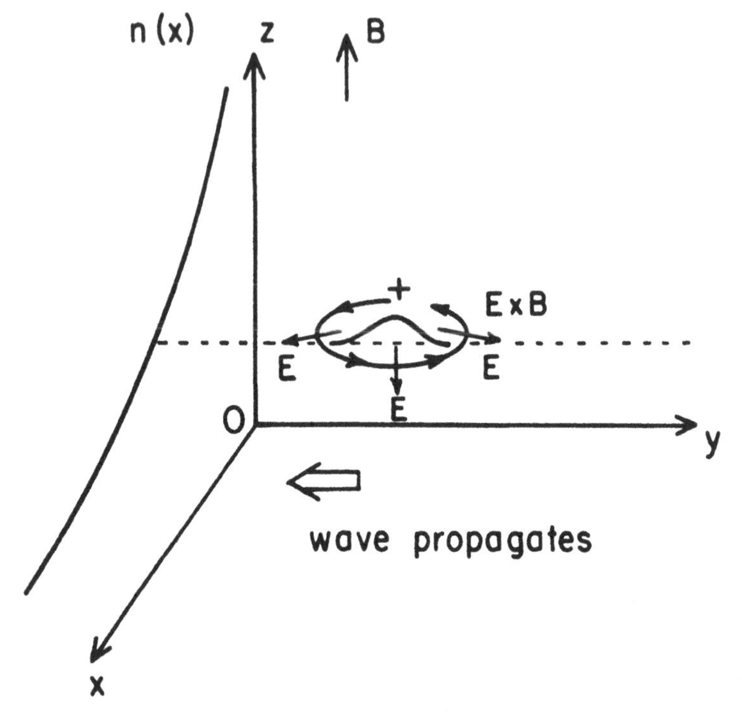

FIGURE 7.4 Schematic explanation of the drift wave excitation

7.5 Geostrophic Flows

Let us compare the polarization drift in magnetized plasmas with the geostrophic shallow water situation. In Fig. 7.5, we show the geometry of the shallow water. We assume constant, uniform density of the fluid ρ. The height of the surface is $z = h(x, y, t)$, and the angular velocity $\Omega = \Omega\hat{z}$ and $f \equiv 2\Omega$. The height of the rigid bottom is $z = h_b(x, y)$ so that the depth of water is $H = h - h_b$. We assume that when D is the characteristic depth, the shallow water condition $\delta = D/L \ll 1$ is satisfied, where L is the characteristic horizontal length such as the horizontal wavelength. In this case, the horizontal momentum equation becomes[14]

$$\frac{\partial \mathbf{v}_\perp}{\partial t} + \mathbf{v}_\perp \cdot \frac{\partial \mathbf{v}_\perp}{\partial \mathbf{x}} = 2\mathbf{v}_\perp \times \boldsymbol{\Omega} - g\nabla_\perp h \ , \tag{7.51}$$

where the subscripts $\perp$ refer to the horizontal plane perpendicular to the angular velocity of rotation $\boldsymbol{\Omega}$ and to the direction of the gravity and g is the gravitational acceleration. The "free boundary condition" $p(x, y, h) = p$ = constant yields

$$\nabla_\perp p = \rho g \nabla_\perp h \ . \tag{7.52}$$

The "rigid boundary condition" at the bottom $z = h_b$ requires the flow velocity normal to the bottom boundary to be zero, which yields $v_z = \mathbf{v}_\perp \cdot \nabla_\perp h_b$. Using these conditions, the continuity equation is cast into

$$\frac{\partial H}{\partial t} + \nabla_\perp \cdot (\mathbf{v}_\perp H) = 0 \ . \tag{7.53}$$

Let us define the vorticity by $\boldsymbol{\omega} = \nabla \times \mathbf{v}$. From Eq. (7.51), after taking the curl and eliminating h, we obtain

$$\frac{d}{dt}\left(\frac{\omega_z + 2\Omega_z}{H}\right) = 0 \ . \tag{7.54}$$

This means that the "generalized" vorticity $(\omega_z + 2\Omega_z)/H$ in the z-direction is conserved for each fluid column along its "orbit" of the fluid element. Equation (7.54) may be called potential vorticity conservation. There exists a similar relation in plasma physics,[15,16] in which the "generalized" vorticity $\omega_z + \Omega_z$ is conserved along the particle orbit, where $\Omega_z = eB_z/mc$. Equation (7.54) can be rewritten as

$$\frac{\partial}{\partial t}\left(\nabla_\perp^2 \psi - \frac{f^2}{gD}\psi\right) + \left[\psi, \nabla^2\psi - \frac{f^2}{gD}\psi + \frac{f}{D}h_b\right] = 0 \ , \tag{7.55}$$

where $f \equiv 2\Omega_z$ in Eq. (7.54) and H is replaced by the characteristic height D and the square brackets refer to the Poisson brackets, i.e., $[f, g] = \frac{\partial f}{\partial x}\frac{\partial g}{\partial y} - \frac{\partial f}{\partial y}\frac{\partial g}{\partial x}$. Here we assumed that $R \ll 1$ so that $\omega_z \ll f$.

Equation (7.55) can be written as

$$\left(\frac{\partial}{\partial t} + v_d \cdot \nabla_\perp\right)\left(\nabla_\perp^2 \eta - \left(\frac{fL}{gD}\right)^2 \eta + \eta_b\right) = 0 \ , \tag{7.56}$$

where η is the normalized h. Compare this with

$$\left(\frac{\partial}{\partial t} + v_E \cdot \nabla_\perp\right)\left(\rho_s^2 \nabla_\perp^2 \phi - \phi - \frac{T_e}{e}\ell n n_0\right) = 0 \ , \tag{7.57}$$

in the drift wave equation in a plasma.[16]

In a geostrophic fluid there exists a wave corresponding to the drift wave of a magnetized plasma. This is the Rossby wave. For small amplitude waves one can linearize equations. Let the thickness of the fluid vary as

$$H(x, y, t) = H_0(x, y) + \eta(x, y, t) \ , \tag{7.58}$$

where $H_0(x,y)$ is the fluid thickness when it is unperturbed and we assume $\eta \ll H_0$. Further, let $H_0 = D_0(1 - \frac{sx}{L})$ or $\partial H_0/\partial x = -D_0 s/L \equiv -RD_0$. We linearize Eqs. (7.51) or (7.55), which gives rise to

$$\left(\frac{\partial}{\partial t}\right)\left[\left(\frac{\partial^2}{\partial t^2} + f^2\right)\eta - \nabla \cdot (gH_0)\nabla\eta\right] - gfJ(H_0, \eta) = 0 \ , \qquad (7.59)$$

where J is a Jacobian. From Eq. (7.59) with $\frac{\omega}{f} \ll 1$, one obtains

$$\left(\frac{\partial}{\partial t}\right)\left(\nabla_\perp^2 \eta - \frac{f^2}{gh_0}\eta\right) - \frac{f}{H_0}\frac{\partial H_0}{\partial x}\frac{\partial n}{\partial y} = 0 \ . \qquad (7.60)$$

This is the equation that governs Rossby waves. The Rossby wave has the linear dispersion relation

$$\omega = -2\kappa\Omega_z k/(k^2 + n^2\pi^2/L^2 + 4\Omega_z^2/g^2 H_0^2) \ , \qquad (7.61)$$

where n is the node number in the x-direction. This relation is similar to the drift wave dispersion relation[17]

$$\omega = \frac{\omega_{*e} I_0(b) e^{-b}}{1 + \frac{T_e}{T_i}(1 - I_0(b)e^{-b})} \cong \frac{v_d k_y}{1 + k_\perp^2 \rho_s^2} \ , \qquad (7.62)$$

for small b in the last equality, where $v_{de} = kcT_e/eB$, $\omega_{*e} = k_y v_{de}$, $\rho_s^2 = (T_e/T_i)\rho_i^2$, $b = k_\perp^2 \rho_i^2$, and I_0 is the zeroth modified Bessel function of the first kind.

The similarities between the physics of the drift wave and the Rossby wave may be seen by the following argument. Let us have a charge density hump (positive) (see Fig. 7.4) for the drift wave case in a magnetized plasma. This charge hump creates an electric field directed outward from the hump in the x-y-plane, (although the quasineutrality dictates nearly cancelled charge). This radial electric field causes the $\mathbf{E} \times \mathbf{B}$ drift according to Eq. (7.4) in the azimuthal direction around the charge hump. This azimuthal flow of particles with the nonuniform $n(x)$ makes more charge convected to the negative y-direction of the hump. The positive y-direction has a density depression in turn. Thus the overall result is for the charge hump to propagate toward the negative y-direction with velocity $v_y = \frac{cT_e}{eB}\frac{1}{n}\frac{dn}{dx}$. Similarly, when there is a hump of fluids (see Fig. 7.5), the higher pressure at the point makes the $\nabla p \times \boldsymbol{\Omega}$ circulation of fluid. The circulation convects more matter from the deeper side in the x-direction to the negative y-direction than the shallower sides in the x-direction to the positive y-direction. Thus, the end result is to increase the fluid height to the negative y-direction side of the hump. The Rossby wave thus propagates in the negative y-direction with phase velocity $v_y = \frac{g}{f}\frac{1}{H_0}\frac{\partial H_0}{\partial x}$. A general discussion on geophysic fluid dynamics may be found in Ref. 14.

Equation (7.56) in fluid dynamics and Eq. (7.57) in plasma physics are often solved in the fluid method as discussed in Chapter 6. For example, the spectral method (Section 6.8) has often been used to solve these equations.

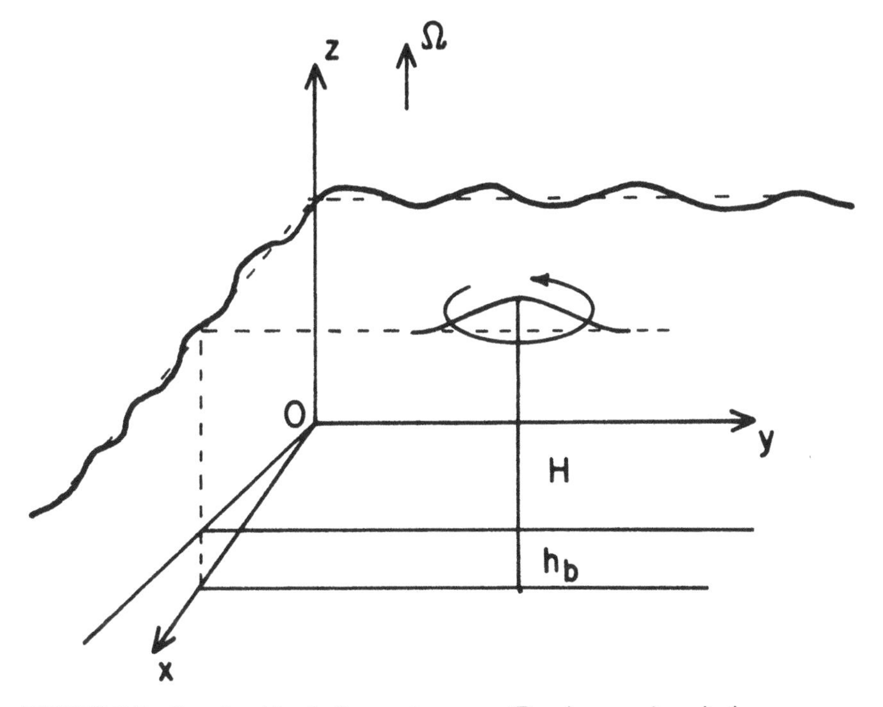

FIGURE 7.5 Geostrophic shallow water wave (Rossby wave) excitation

7.6 Finite Larmor Radius Effects

Magnetized plasmas have a property that is not seen in fluid dynamics. That is the finite gyroradius effect. Rosenbluth et al.[10] considered the finite Larmor radius effects on plasma instability in detail. We will look at the effects of finite gyroradius on orbits. The particle position and velocity may be written in terms of the guiding-center position and velocity $(\mathbf{X}_j, \mathbf{V}_j)$ and the gyroradius and gyrophase (ρ, ϕ) (Fig. 7.3):

$$\mathbf{x}_j = \mathbf{X}_j + \hat{\mathbf{x}}\rho\cos\phi + \hat{\mathbf{y}}\rho\sin\phi \tag{7.63}$$

$$\mathbf{v}_j = \mathbf{V}_j - \hat{\mathbf{x}}v_\perp\sin\phi + \hat{\mathbf{y}}v_\perp\cos\phi\ . \tag{7.64}$$

The electric field which affects the particle is

$$\mathbf{E}(\mathbf{x}_j) = \int d\mathbf{k}\mathbf{E}(\mathbf{k})e^{i\mathbf{k}\cdot\mathbf{x}_j}$$

$$= \int d\mathbf{k}\mathbf{E}(\mathbf{k}) \exp\left[i\mathbf{k}\cdot\mathbf{X}_j + i\mathbf{k}\cdot(\hat{\mathbf{x}}\rho\cos\phi + \hat{\mathbf{y}}\rho\sin\phi)\right] . \quad (7.65)$$

The wavevector is defined as

$$\mathbf{k} = k_\perp(\hat{\mathbf{x}}\cos\alpha + \hat{\mathbf{y}}\sin\alpha) , \quad (7.66)$$

where α is defined in Fig. 7.3. The electric field can then be written

$$\mathbf{E}(\mathbf{x}_j) = \int d\mathbf{k}\mathbf{E}(\mathbf{k}) \exp\left[i\mathbf{k}\cdot\mathbf{X}_j + ik_\perp\rho\cos(\phi-\alpha)\right] . \quad (7.67)$$

This may be expanded by the use of the Bessel expansion

$$e^{ik\rho\cos\phi} = \sum_{n=-\infty}^{\infty} i^n e^{in\phi} J_n(k\rho) ,$$

as

$$\mathbf{E}(\mathbf{x}_j) = \sum_{n=-\infty}^{\infty} \int d\mathbf{k}\mathbf{E}(\mathbf{k}) i^n e^{in(\phi-\alpha)} e^{i\mathbf{k}\,\mathbf{x}_j} J_n(k_\perp\rho) . \quad (7.68)$$

Since the motion of interest is on the slow time scale, we average (or coarse-grain) over the fast time scale of the gyroperiod. This is sometimes called gyrophase averaging (or gyroaverage). This average is denoted by

$$\langle\ldots\rangle \equiv \int_0^{2\pi} \frac{d\phi}{2\pi}\cdots .$$

The gyroaveraged electric field is calculated

$$\begin{aligned}\langle\mathbf{E}(\mathbf{x}_j)\rangle &= \sum_{n=-\infty}^{\infty} \delta_{n,0} \int d\mathbf{k}\mathbf{E}(\mathbf{k}) J_n(k_\perp\rho) e^{i\mathbf{k}\,\mathbf{X}_j} \\ &= \int d\mathbf{k}\mathbf{E}(\mathbf{k}) J_0(k_\perp\rho) e^{i\mathbf{k}\,\mathbf{X}_j} , \quad (7.69)\end{aligned}$$

where any harmonic n other than zero, representing gyrophase dependent terms, drops out upon gyroaveraging. A drawback of this method is that when the magnetic field varies spatially, the operation (7.69) becomes less valid. An alternative gyroaveraging is to carry the averaging in real space.

$$\langle\mathbf{E}(\mathbf{x}_j)\rangle = \left\langle \int d\mathbf{x}\mathbf{E}(\mathbf{x})\delta(\mathbf{x}-\mathbf{X}_j-\boldsymbol{\rho}_j) \right\rangle = \int d\mathbf{x}\mathbf{E}(\mathbf{x})\delta(\rho_j - |\mathbf{x}-\mathbf{X}_j|) , \quad (7.70)$$

where $\boldsymbol{\rho} = \mathbf{x}_j - \mathbf{X}_j$ [Eq. (7.63)], $\rho_j = |\boldsymbol{\rho}_j|$, and $\delta(\rho_j - |\mathbf{x}-\mathbf{X}_j|)$ is the ring distribution function. This operation is local in nature and thus readily adaptable to the situation where the magnetic field spatially varies.

The Larmor radius ρ is square-root mass ratio times larger for ions than electrons. For typical cases such as in drift waves $k_\perp \rho_i$ is of the order unity or less than unity, and $J_0(k_\perp \rho_e)$ is practically 1. If finite size particles are considered, we replace $\mathbf{E}_p$ by $\mathbf{E}_f$ as below:

$$\mathbf{E}_p(\mathbf{k}) \to \mathbf{E}_f(\mathbf{k}) = \mathbf{E}_p(\mathbf{k}) S(\mathbf{k}) \ .$$

This carries through the derivation to yield

$$\langle \mathbf{E}_f(\mathbf{x}_j) \rangle = \int d\mathbf{k} \mathbf{E}(\mathbf{k}) S(\mathbf{k}) J_0(k_\perp \rho) e^{i\mathbf{k}\cdot\mathbf{X}_j} \ . \tag{7.71}$$

The guiding-center velocity may now be written using the previously mentioned assumption

$$\mathbf{v} = \mathbf{v}^{(1)} + \mathbf{v}^{(2)} + \ldots$$

as

$$\frac{d\mathbf{X}_j}{dt} = c \int d\mathbf{k} \frac{\mathbf{E}(\mathbf{k}) \times \mathbf{B}}{B^2} J_0(k_\perp \rho) e^{i\mathbf{k}\cdot\mathbf{X}_j}$$

$$+ c \int d\mathbf{k} \frac{(\dot{\mathbf{E}}_f \times \hat{b}) \times \hat{b}}{\Omega B} J_0(k_\perp \rho) e^{i\mathbf{k}\cdot\mathbf{X}_j} \ , \tag{7.72}$$

where $\hat{b} \equiv \frac{\mathbf{B}}{B}$ and the terms on the RHS of Eq. (7.72) are the $\mathbf{E} \times \mathbf{B}$ drift term and the polarization drift term, respectively. This is what we would like to use for the perpendicular motion. For the parallel motion the leapfrog method remains adequate. The terms in Eq. (7.72) carry clear physical meanings. The application of a straightforward algorithm, however, is unusable, since $\dot{\mathbf{E}}$ creates numerical instability. This is the main reason ions are kept in full dynamics in Ref. 7. In Chapter 9 more sophistication on this will be discussed.

7.7 Gyrokinetic Model

In the previous sections we investigated the guiding-center model. In the actual codes we adopt the guiding-center approximation only for electrons and advance ions in explicit full dynamics in many cases. Such a code has coarse-grained over the electron cyclotron time. Since electrons do not have inertia in the directions perpendicular to the magnetic field in this model, the plasma oscillations have the period due to the charge separation along the magnetic field line, which goes as $2\pi/\omega = 2\pi(k/k_\parallel)\omega_{pe}^{-1}$, where $k_\parallel$ and k are the parallel and total wavenumbers. Thus the smallest time scale that limits the time step of the code, the Courant-Friedrichs-Lewy condition, is the smallest of the following four, $2\pi(k/k_\parallel)\omega_{pe}^{-1}$, $\Delta/(v_e \sin\theta)$, $k\Delta/v_i$, $\frac{2\pi}{\Omega_i}$, and $2\pi/\omega_{pi}$, where θ is the angle between the magnetic field and the z-axis (the second term is for two-dimensional case with x and y), v_e and v_i are the thermal (or typical) electron and ion speeds, and ω_{pi} is the ion plasma frequency. One would think that if we adopt ion guiding centers, the last two time scales, Δ/v_i and $2\pi/\omega_{pi}$, are substantially lengthened to $\Delta/(v_i \sin\theta)$ and $2\pi(k/k_{\|})\omega_{pi}^{-1}$, respectively. Nevertheless, the ion full dynamics has been used in Ref. 7 because when we treat both electrons and ions as guiding centers, the important charge separation effect is completely (or largely) buried. Remember that the guiding-center motion of $c\mathbf{E} \times \mathbf{B}/B^2$ is identical for both electrons and ions. This would result in the absence or severe distortion of the physics of waves related to charge separation such as drift waves. In order to overcome this difficulty, we have to look at the governing equation more carefully. The generally accepted starting point for these investigations is the gyrokinetic equation. We, therefore, consider this equation to a greater extent in the following.

A. Gyrokinetic Equation

The Vlasov equation, that describes statistical properties of collisionless plasma without the guiding-center approximation, can be transformed from particle coordinates $(\mathbf{x}, \mathbf{v}, t)$ to guiding-center coordinates $(\mathbf{X}, v_{\|}, \mathbf{v}_{\perp}, \varphi, t)$ in an exact manner. Here these coordinates are related each other by

$$\mathbf{x} = \mathbf{X} + \frac{1}{\Omega}\hat{b} \times \mathbf{v}, \tag{7.73}$$

$$\mathbf{v}_{\perp} = v_{\perp}\left(\hat{e}\cos\varphi - \hat{e}_2 \sin\varphi\right), \tag{7.74}$$

$$\hat{e}_1 \times \hat{e}_2 = \hat{b}\,, \tag{7.75}$$

and $\Omega = qB/mc$. In the electrostatic limit in slab geometry the Vlasov equation in guiding-center coordinates is

$$\left[\frac{\partial}{\partial t} + v_{\|}\hat{b} \cdot \frac{\partial}{\partial \mathbf{X}} + \Omega^2 \frac{\partial}{\partial \varphi} - \frac{c}{B}\left(\frac{\partial}{\partial \mathbf{X}}\phi \times \hat{b}\right) \cdot \frac{\partial}{\partial \mathbf{X}} \right.$$

$$- \frac{q}{m}\hat{b} \cdot \frac{\partial}{\partial \mathbf{X}}\phi \frac{\partial}{\partial v_{\|}} - \frac{q}{m}\frac{\partial}{\partial \mathbf{X}}\phi \cdot \hat{v}_{\perp}\frac{\partial}{\partial v_{\perp}}$$

$$\left. + \Omega \frac{1}{v_{\perp}}\frac{c}{B}\left(\frac{\partial}{\partial \mathbf{X}}\phi \times \hat{b}\right) \cdot \hat{v}_{\perp}\frac{\partial}{\partial \varphi}\right] f = 0\,, \tag{7.76}$$

where $f = f(\mathbf{x}, v_{\|}, \mathbf{v}_{\perp}, \varphi, t)$. Decomposing the distribution function into an ensemble averaged stationary piece plus a fluctuating piece $f = \bar{f} + \tilde{f}$, we get the lowest order equation in $\rho_s/L_n(\sim \rho_i/L_n)$

$$\Omega_i \frac{\partial}{\partial \varphi} \bar{f} = 0 ,$$

indicating that the average distribution is gyroangle independent. Decomposing the next order equation into an adiabatic piece and a nonadiabatic piece $\tilde{f} = (-\bar{f} q \tilde{\phi})/T + \tilde{h}$, we obtain

$$\Omega_i \frac{\partial}{\partial \varphi} \tilde{h} = 0 ,$$

indicating that the nonadiabatic piece is independent of the gyroangle. After integrating over the gyroangle, the next order equation along with the integrability condition results in the so-called gyrokinetic equation[18,19] for the nonadiabatic distribution

$$\frac{\partial h}{\partial t} + v_{\parallel} \hat{b} \cdot \frac{\partial}{\partial \mathbf{X}} h - \frac{c}{B} \frac{\partial}{\partial \mathbf{X}} \left\langle \tilde{\phi} \right\rangle \times \hat{b} \cdot \frac{\partial}{\partial \mathbf{X}} h = S\left(\langle f \rangle , \langle \phi \rangle\right) , \tag{7.77}$$

where

$$S\left(\langle f \rangle , \langle \phi \rangle\right) = \frac{q}{T_i} \langle f \rangle \frac{\partial \langle \phi \rangle}{\partial t} + \frac{c}{B} \frac{\partial}{\partial \mathbf{X}} \langle \phi \rangle \times \hat{b} \cdot \frac{\partial}{\partial \mathbf{X}} \langle f \rangle . \tag{7.78}$$

Here the angular bracket means the gyroaverage $\langle \cdots \rangle = \frac{1}{2\pi} \int_0^{2\pi} d\varphi \cdots$ and the source terms pertain to the convected background distribution.

B. Weighted Particle Method

Equation (7.77) is not readily amenable to the conventional particle simulation technique. One of the major reasons for this is that Eq. (7.77), an equation for quantity h, involves not only terms with h but also those with $\langle f \rangle$, which constitute source terms. Since the particle simulation in essence solves the Klimontovich-type equation, the source terms spoil the method. Equation (7.77) is an inhomogeneous differential equation with respect to h with the right-hand side being the inhomogeneous source term.

However, there have been attempts to incorporate source terms in the Vlasov or Klimontovich equation for particle simulation (see Problem 7). Let us introduce a *weighted* distribution function

$$h(\mathbf{x}, \mathbf{v}, t) = \sum_i w_i[\mathbf{X}_i(t), \mathbf{v}_i(t), t] \delta[\mathbf{x} - \mathbf{X}_i(t)] \delta[\mathbf{v} - \mathbf{v}_i(t)] . \tag{7.79}$$

If w_i is a constant of motion, Eq. (7.79) satisfies Klimontovich equation without a source term. When w_i obeys the equation,

$$\frac{dw_i}{dt} = S\left(\langle f \rangle , \langle \phi \rangle , \mathbf{X}_i\right) n^{-1} , \tag{7.80}$$

with n being the phase space density of particles, then Eq. (7.79) is a solution to Eq. (7.77) where

$$
\begin{aligned}
\dot{\mathbf{X}}_i &= \mathbf{v}_i \,, \\
\mathbf{v}_i &= v_\parallel \hat{b} - \frac{c}{B} \frac{\partial}{\partial \mathbf{X}} \langle \phi \rangle \times \hat{b} \,, \\
\dot{v}_{\parallel,} &= \frac{q}{m} E_\parallel - \frac{\mu}{m} \hat{b} \cdot \nabla B \,.
\end{aligned}
$$

This means that we can push the guiding-center particles a la Section 7.2 whose *weight* w_i is not a constant and unity anymore, but varies in time according to the formula Eq. (7.80).

In this scheme the particle pushing has to follow only the *perturbed* portion of the distribution and we can dispense with pushing the majority of uninteresting portion of the distribution. This way it is possible to realize a much better resolution with less particle noise. If n in Eq. (7.80) is fluctuating a lot, however, numerical integration of Eq. (7.80) tends to be unstable (see Problem 7). On the other hand, if n is assumed to be constant, as is reasonable for low β plasmas, one can avoid this difficulty. The time derivative on the potential on the right-hand side arises from a portion of the polarization drift. The treatment of such a term is sensitive to numerics. Discussion is deferred to Section 9.6. There is much investigation yet to be done concerning this method.

C. Gyrokinetic Equation (Continued)

Let us examine certain limits of Eq. (7.77). The simplest version of the equation for the gyroaveraged distribution is the basic $\mathbf{E} \times \mathbf{B}$ guiding-center equation

$$
\begin{aligned}
&\left[\frac{\partial}{\partial t} + v_\parallel \hat{b} \cdot \frac{\partial}{\partial \mathbf{X}} - \frac{c}{B} \left(\frac{\partial}{\partial \mathbf{X}} \left\langle \tilde{\phi} \right\rangle \times \hat{b} \right) \right. \\
&\quad \left. \times \frac{\partial}{\partial \mathbf{X}} - \frac{q}{m} \left(\hat{b} \cdot \frac{\partial}{\partial \mathbf{X}} \left\langle \tilde{\phi} \right\rangle \right) \frac{\partial}{\partial v_\parallel} \right] \langle f \rangle = 0 \,. \qquad (7.81)
\end{aligned}
$$

The second and last terms commute:

$$
\begin{aligned}
v_\parallel \hat{b} \cdot \frac{\partial}{\partial \mathbf{X}} \langle f \rangle &= \frac{\partial}{\partial \mathbf{X}} \cdot \left(\hat{b} v_\parallel \langle f \rangle \right) \,, \\
-\frac{q}{m} \left(\hat{b} \cdot \frac{\partial}{\partial \mathbf{X}} \left\langle \tilde{\phi} \right\rangle \right) \frac{\partial}{\partial v_\parallel} \langle f \rangle &= \frac{\partial}{\partial v_\parallel} \left[-\frac{q}{m} \left(\hat{b} \cdot \frac{\partial}{\partial \mathbf{X}} \left\langle \tilde{\phi} \right\rangle \right) \langle f \rangle \right] . \qquad (7.82)
\end{aligned}
$$

The third term, however, commutes only when the magnetic field in the plasma is shearless, i.e., $\frac{\partial}{\partial \mathbf{X}} \times \hat{b} = 0$. Or

$$\frac{\partial}{\partial \mathbf{X}} \cdot \left(\left\langle \tilde{\mathbf{E}} \right\rangle \times \hat{b} \right) = 0 \tag{7.83}$$

$$\frac{\partial}{\partial \mathbf{X}} \cdot \left(\left\langle \tilde{\mathbf{E}} \right\rangle \times \hat{b} \langle f \rangle \right) = \left\langle \tilde{\mathbf{E}} \right\rangle \times \hat{b} \cdot \frac{\partial}{\partial \mathbf{X}} \langle f \rangle \ . \tag{7.84}$$

Then the guiding-center Vlasov equation can be put in a continuity form

$$\frac{\partial}{\partial t} \langle f \rangle + \frac{\partial}{\partial \mathbf{X}} \cdot \left[\left(v_{\|} \hat{b} - \frac{c}{B} \frac{\partial}{\partial \mathbf{X}} \left\langle \tilde{\phi} \right\rangle \times \hat{b} \right) \langle f \rangle \right]$$
$$+ \frac{\partial}{\partial v_{\|}} \left[-\frac{q}{m} \left(\hat{b} \cdot \frac{\partial}{\partial \mathbf{X}} \left\langle \tilde{\phi} \right\rangle \right) \langle f \rangle \right] = 0 \ . \tag{7.85}$$

However, if there is a shear in magnetic field such as $\mathbf{B} = B(\hat{\mathbf{y}}x/L_s + \hat{\mathbf{z}})$, Eq. (7.83) does not hold and we now have

$$\frac{\partial}{\partial \mathbf{X}} \cdot \left(\left\langle \tilde{\mathbf{E}} \right\rangle \times \hat{b} \right) = \frac{1}{L_s} \hat{b} \cdot \frac{\partial}{\partial \mathbf{X}} \left\langle \tilde{\phi} \right\rangle \ . \tag{7.86}$$

Then the continuity equation in phase space, Eq. (7.85), should have an extra term

$$-\frac{1}{L_s} \frac{c}{B} \left(\hat{b} \cdot \frac{\partial}{\partial \mathbf{X}} \left\langle \tilde{\phi} \right\rangle \right) \langle f \rangle \ . \tag{7.87}$$

The size of the linear part of this term is of the order of $(L_n/L_s)^2$ times the dominant linear term $-\frac{c}{B}(\frac{\partial}{\partial \mathbf{X}} \left\langle \tilde{\phi} \right\rangle \times \hat{b}) \cdot \frac{\partial}{\partial \mathbf{X}} \bar{f}$, where L_n is the density gradient scale length and L_s the shear length.

This noncommutability in the gyrokinetic equation in general and the guiding-center kinetic equation with magnetic shear in particular is one of the significant differences from the Vlasov and Klimontovich equations, in which commutability is guaranteed because of the Hamiltonian nature of the basic force laws. See Chapter 1 for discussion and Eq. (1.5) in particular. Even though the equation for the gyroaveraged distribution cannot be put in the continuity form, it can still be solved by the particle simulation method. This is because the particle pushing method is equivalent to the method of characteristics. The gyrokinetic equation Eq. (7.85) can, in fact, be put in a form that conserves the phase space density along the characteristics

$$\frac{d}{dt} \langle f \rangle = \left[\frac{\partial}{\partial t} + \frac{d\mathbf{X}}{dt} \cdot \frac{\partial}{\partial \mathbf{X}} + \frac{d\mathbf{v}}{dt} \cdot \frac{\partial}{\partial \mathbf{v}} \right] \langle f \rangle = 0 \ . \tag{7.88}$$

The characteristics for Eq. (7.88) are

$$\frac{d\mathbf{X}}{dt} = v_{\parallel}\hat{b} - \frac{c}{B}\left(\frac{\partial}{\partial \mathbf{X}}\left\langle\tilde{\phi}\right\rangle \times \hat{b}\right) , \qquad (7.89)$$

$$\frac{dv_{\parallel}}{dt} = -\frac{q}{m}\hat{b}\cdot\frac{\partial}{\partial \mathbf{X}}\left\langle\tilde{\phi}\right\rangle . \qquad (7.90)$$

In contrast to this the original Vlasov equation satisfied the continuity of phase space density (six-dimensional phase space). With $f(\mathbf{x}, \mathbf{v}, t)$ the phase space density the continuity equation (or the phase space volume incompressibility) is

$$\frac{\partial}{\partial t}f + \frac{\partial}{\partial \mathbf{x}}\cdot(\mathbf{v}f) + \frac{\partial}{\partial \mathbf{v}}\cdot(\mathbf{a}f) = 0 . \qquad (7.91)$$

The structure of this equation is almost identical to Eq. (1.3), with the only difference being that f in Eq. (7.91) is a smooth function of $(\mathbf{x}, \mathbf{v})$ while $\mathcal{N}$ in Eq. (1.3) is a collection of many δ-functions. When the acceleration of the particles $\mathbf{a}$ is due to the Lorentz force $\mathbf{a} = q/m[\mathbf{E}(\mathbf{x}, t) + \frac{1}{c}\mathbf{v} \times \mathbf{B}(\mathbf{x}, t)]$, for example, the derivatives with respect to the phase space coordinates commute with the operand as follows

$$\frac{\partial}{\partial \mathbf{x}}\cdot(\mathbf{v}f) = \mathbf{v}\cdot\frac{\partial}{\partial \mathbf{x}}f + f\frac{\partial}{\partial \mathbf{x}}\cdot\mathbf{v} = \mathbf{v}\cdot\frac{\partial}{\partial \mathbf{x}}f , \qquad (7.92)$$

$$\frac{\partial}{\partial \mathbf{v}}\cdot(\mathbf{a}f) = \mathbf{a}\cdot\frac{\partial}{\partial \mathbf{v}}f + f\frac{\partial}{\partial \mathbf{v}}\cdot\mathbf{a} = \mathbf{a}\cdot\frac{\partial}{\partial \mathbf{v}}f \qquad (7.93)$$

because $\frac{\partial}{\partial \mathbf{v}}\cdot\mathbf{E} = 0$ and $\frac{\partial}{\partial \mathbf{v}}\cdot(\mathbf{v}\times\mathbf{B}) = \mathbf{B}\cdot\frac{\partial}{\partial \mathbf{v}}\times\mathbf{v} - \mathbf{v}\cdot\frac{\partial}{\partial \mathbf{v}}\times\mathbf{B} = 0$.

D. Multiple Spatial Expansion

In many problems, perturbed density fluctuations are much smaller than the density variation of the background plasma; at the same time the wavelength of the perturbed density fluctuations is much smaller than the background density scale length. For example, a steady state of drift wave turbulence of a magnetically confined plasma is typically in under these conditions where it may be convenient to expand the gyrokinetic equation in multiple spatial scales.[20] Equation (7.85) reduces to the gyrokinetic equation expanded in multiple spatial scales

$$\frac{\partial f}{\partial t} + v_{\parallel}\hat{b}\cdot\frac{\partial f}{\partial \mathbf{X}} - \frac{q}{m}\frac{1}{\Omega}\left(\frac{\partial \phi}{\partial \mathbf{X}}\times\hat{b}\right)\cdot\left(\frac{\partial}{\partial \mathbf{X}} - \kappa\right)f$$

$$-\frac{q}{m}\frac{\partial \phi}{\partial \mathbf{X}}\cdot\hat{b}\frac{\partial f}{\partial v_{\parallel}} = 0 , \qquad (7.94)$$

where $\boldsymbol{\kappa}$ is the density gradient vector. On the other hand, when the continuity equation in phase space Eq. (7.85) is expanded in multiple spatial scales, it reads

$$\begin{aligned}\frac{\partial f}{\partial t} &+ v_{\|}\hat{b}\cdot\frac{\partial f}{\partial \mathbf{X}} - \frac{q}{m}\frac{1}{\Omega}\left(\frac{\partial\phi}{\partial \mathbf{X}}\times\hat{b}\right)\cdot\frac{\partial f}{\partial \mathbf{X}} - \frac{q}{m}\frac{1}{\Omega}\frac{\partial}{\partial \mathbf{X}}\cdot\left(\phi\boldsymbol{\kappa}\times\hat{b}\right)f \\ &-\frac{q}{m}\frac{\partial\phi}{\partial \mathbf{X}}\cdot\hat{b}\frac{\partial f}{\partial v_{\|}} - \left[\frac{q}{m}\frac{1}{\Omega}\phi(\boldsymbol{\kappa}\times\hat{b})\cdot\frac{\partial f}{\partial \mathbf{X}}\right. \\ &\left.-\frac{q}{m}\frac{1}{\Omega}\frac{\partial\phi}{\partial \mathbf{X}}\cdot\left(\frac{\partial\phi}{\partial \mathbf{X}}\times\hat{b}\right)f\right] = 0\,. \end{aligned} \tag{7.95}$$

After using vector identities, we obtain from the continuity equation

$$\begin{aligned}&\left[\frac{\partial f}{\partial t} + v_{\|}\hat{b}\cdot - \frac{q}{m}\frac{1}{\Omega}\left(\frac{\partial\phi}{\partial \mathbf{X}}\times\hat{b}\right)\cdot\left(\frac{\partial}{\partial \mathbf{X}} - \boldsymbol{\kappa}\right)f - \frac{q}{m}\frac{\partial\phi}{\partial \mathbf{X}}\cdot\hat{b}\frac{\partial f}{\partial v_{\|}}\right] \\ &+\left[-\frac{q}{m}\frac{1}{\Omega}\phi(\boldsymbol{\kappa}\times\hat{b})\cdot\frac{\partial}{\partial \mathbf{X}}f + \frac{q}{m}\frac{1}{\Omega}\left(\frac{\partial}{\partial \mathbf{X}}\times\hat{b}\right)\right. \\ &\left.\cdot\left(\frac{\partial\phi}{\partial \mathbf{X}} + \boldsymbol{\kappa}\phi\right)f\right] = 0\,. \end{aligned} \tag{7.96}$$

The terms in the first square brackets are those that appear in Eq. (7.94), the multiple expanded gyrokinetic equation. The terms in the second square brackets are extra terms that have now appeared because the phase space convective derivatives do not commute the phase space flow terms in the multiple expansion equation.

Although the continuity equation does not in general reduce to the gyrokinetic equation expanded in multiple spatial scales, they are approximately equal in some limits. For a shearless case ($\frac{\partial}{\partial \mathbf{X}}\times\hat{b} = 0$) the only extra term is the convective derivative term with respect to the background density gradient $-\frac{q}{m}\frac{1}{\Omega}\phi(\mathbf{k}\times\hat{b})\cdot\frac{\partial}{\partial \mathbf{X}}f$. The typical size of this term is $\frac{q}{m}\frac{1}{\Omega}\frac{k_y}{L_n}\frac{e\phi}{T_e}\phi\bar{f}$ in contrast to the dominant $\mathbf{E}\times\mathbf{B}$ term $\frac{q}{m}\frac{1}{\Omega}\left(\frac{\partial\phi}{\partial \mathbf{X}}\times\hat{b}\right)\cdot\boldsymbol{\kappa}f$, whose size is typically $\frac{q}{m}\frac{1}{\Omega}\frac{\phi}{L_n}k_y\bar{f}$. The ratio of the former to the latter is $e\phi/T_e$, which is $\sim 10^{-2}$ in a tokamak plasma example. Including the shear terms, an extra term due to the noncommutability is

$$\frac{q}{m}\frac{1}{\Omega}\left(\frac{\partial}{\partial \mathbf{X}}\times\hat{b}\right)\cdot\left(\frac{\partial\phi}{\partial \mathbf{X}} + \boldsymbol{\kappa}\phi\right)f = \frac{q}{m}\frac{1}{\Omega}\left(\frac{\partial}{\partial \mathbf{X}}\times\hat{b}\right)\cdot\frac{\partial\phi}{\partial \mathbf{X}}f\,, \tag{7.97}$$

which is of the order of $\frac{q}{m}\frac{1}{\Omega}\frac{\phi}{L_s}k_z\bar{f}$ and whose ratio to the $\mathbf{E}\times\mathbf{B}$ term is $L_n k_z/L_s k_y \sim L_n x/L_s^2 \lesssim (L_n/L_s)^2 \ll 1$. Thus, even in the magnetically sheared slab case the extra terms are small compared to the dominant terms and the continuity equation can reduce to the multiple scaled gyrokinetic equation.

However, solving the characteristic equations

$$\frac{d\mathbf{X}}{dt} = v_\parallel\hat{b} - \frac{q}{m}\frac{1}{\Omega}\left(\frac{\partial\phi}{\partial\mathbf{X}} - \boldsymbol{\kappa}\phi\right)\times\hat{b} \tag{7.98}$$

$$\frac{dv_\parallel}{dt} = \frac{q}{m}\frac{\partial\phi}{\partial\mathbf{X}}\cdot\hat{b} \tag{7.99}$$

in Eq. (7.88) does not give rise to Eq. (7.94). This is because

$$\frac{q}{m}\frac{1}{\Omega}\left(\frac{\partial}{\partial\mathbf{X}}\times\hat{b}\right)\cdot\boldsymbol{\kappa}f \neq -\frac{q}{m}\frac{1}{\Omega}(\boldsymbol{\kappa}\phi\times\hat{b})\cdot\frac{\partial f}{\partial\mathbf{X}}\,.$$

We, therefore, get errors occurring from using the particle pushing equations Eqs. (7.98) and (7.99) to simulate the physics with magnetic shear.

E. Lee's Model

Lee[20] and Dubin et al.[21] introduced an expression for the gyrokinetic equation an alternative to Eq. (7.79). By introducing the Lie transformation to deform the original equation into a suitable form for gyroaveraging, their equations for the Vlasov-Poisson system $[f,H]=0$ and $\nabla^2\phi = 4\pi e(\int f(x,p,t)d^3p - n_e)$, where electrons are treated as drift kinetic (i.e., zero Larmor radius) and f is the ion distribution function in $(\mathbf{x},\mathbf{p})$ space, H the Hamiltonian and $[\quad]$ is the Poisson bracket.[22] By transforming $\mathbf{z}=(\mathbf{x},\mathbf{p})$ into $\mathbf{Z}=(\mathbf{X},\mu,v_\parallel,\varphi)$ and $f(\mathbf{x},\mathbf{p})$ into $g(\mathbf{X},\mu,v_\parallel,\varphi)$, we have

$$\begin{aligned}[g,\hat{H}] &= 0\\ \nabla^2\phi &= -4\pi e\left(\int g(\mathbf{Z})\delta(\mathbf{X}-\mathbf{x}+\boldsymbol{\rho})d^6\mathbf{Z} - n_e\right)\,,\end{aligned} \tag{7.100}$$

where $\hat{H}$ is the gyroaverage of H and $d^6\mathbf{Z} = J(Z/z)d^3X d\mu dv_\parallel d\varphi$ with $J(Z/z)$ being the Jacobian. After averaging over the gyroangle, they obtained for $F=\langle g\rangle$

$$\begin{aligned}\frac{\partial F}{\partial t} &+ \left(v_\parallel\hat{b} - \frac{q}{m\Omega}\frac{\partial}{\partial\mathbf{X}}\psi\times\hat{b}\right)\cdot\frac{\partial}{\partial\mathbf{X}}F\\ &-\frac{q}{m}\hat{b}\cdot\frac{\partial}{\partial\mathbf{X}}\psi\frac{\partial F}{\partial v_\parallel} = 0\,,\end{aligned} \tag{7.101}$$

$$\psi = \bar{\phi} - \frac{q\rho_i^2}{2T}\left(\frac{\partial}{\partial \mathbf{X}_\perp}\phi\right)^2 , \tag{7.102}$$

$$\nabla^2\phi + \frac{\rho_i^2}{n_0\lambda_{Di}^2}\boldsymbol{\nabla}_\perp \cdot (n_i \boldsymbol{\nabla}_\perp \phi) = -4\pi q(\hat{n}_i - n_e) , \tag{7.103}$$

where $\bar{\phi}$ is the gyroaveraged ϕ and

$$n_i = \int d^6\mathbf{Z}\delta(\mathbf{X} - \mathbf{x})F ,$$

$$\hat{n}_i = \int 2\pi\Omega d\mu dv_\parallel \frac{d^3k}{(2\pi)^3} e^{i\mathbf{k}\cdot\mathbf{X}} J_0(k_\perp \rho)F .$$

The second term on the left-hand side of Eq. (7.103) is due to the polarization drift of ions. The ion density on the right-hand side of Eq. (7.103) incorporates the finite Larmor radius effect. In Eq. (7.101) the electric potential is gyroaveraged. The second term on the right-hand side of Eq. (7.102) is introduced to make Eq. (7.101) energy-conservative.

A succinct way to advance these equations is

$$\frac{d\mathbf{X}}{dt} = v_\parallel \hat{b} - \frac{q}{m}\frac{1}{\omega}\frac{\partial \psi}{\partial \mathbf{X}} \times \hat{b} \tag{7.104}$$

$$\frac{dv_\parallel}{dt} = -\frac{q}{m}\frac{\partial \psi}{\partial \mathbf{X}} \times \hat{b} \tag{7.105}$$

in place of Eqs. (7.4) and (7.5), while we replace usual Poisson's equation by Eq. (7.103). Lee has utilized this algorithm to investigate low frequency plasma phenomena such as drift waves.

Further discussion on handling the gyrokinetic equation is presented in Chapter 9.

7.8 Guiding-Center Magnetoinductive Model

A magnetoinductive model of plasma was introduced in Chapter 5 in order to coarse-grain the fast radiation time scale for electromagnetic simulation. For a magnetized plasma we want to incorporate guiding-center electrons into the magnetoinductive algorithm.[23,24] Besides the obvious change in electron pushers from the full dynamics to a set of Eqs. (7.4) and (7.5), the field solver for the transverse electric field is modified as

$$\nabla^2\mathbf{E}_T - \frac{1}{c^2}\left[\omega_{pe}^2 E_\parallel(\mathbf{x})\hat{b} + \omega_{pi}^2\mathbf{E}(\mathbf{x})\right]_T$$

$$= 4\pi \left[\mathbf{J}_i \times \mathbf{\Omega}_i(\mathbf{x}) - \nabla \cdot \overleftrightarrow{R}(\mathbf{x}) \right]_T \qquad (7.106)$$

where

$$\overleftrightarrow{R}(\mathbf{x}) \equiv en_0 \left[\langle \mathbf{vv} \rangle_i (\mathbf{x}) - \langle \mathbf{vv} \rangle_e (\mathbf{x}) \right] , \qquad (7.107)$$

with $\langle \mathbf{vv} \rangle \equiv \int d\mathbf{v} f(\mathbf{v}) \mathbf{vv}$ and $\mathbf{\Omega}_i = e\mathbf{B}/Mc$. Note that a term like $\mathbf{J}_e \times \mathbf{\Omega}_e(\mathbf{x})$ does not appear on the RHS of Eq. (7.106) because the electron velocity obeys the guiding-center dynamics. The present model naturally filters out fast time scales of radiation and electron gyration. Under appropriate conditions it is again possible to construct the incompressible model, as we did in Chapter 5.

It is of considerable interest to see how the particle code behaves as a many-body system, in particular how the screening of some of the high frequency modes affect the properties of the system. A many-body system in (or close to) thermal equilibrium satisfies the fluctuation-dissipation theorem.[25,26] (See also Chapter 2.) It is also instructive to see which modes are excited and at which amplitudes they are excited in the thermal computational plasma, as no equivalence exists for fluid simulation in which all modes are "quiet" (zero amplitude). The following discussion parallels that of Ref. 23.

To obtain the dispersion relation, the electric and magnetic fields are assumed to have the phase dependence $e^{i(\mathbf{k}\ \mathbf{x} - \omega t)}$. The electric field is expressed in terms of the current density when $\mathbf{B}$ is eliminated by Maxwell's equations to yield

$$(n^2 + 1) \frac{\mathbf{k}(\mathbf{k} \cdot \mathbf{E})}{k^2} - n^2 \mathbf{E} = \frac{-4\pi i}{\omega} \mathbf{J} \qquad (7.108)$$

where the index of refraction n is related as $n^2 = k^2 c^2 / \omega^2$. In terms of indicial notation, we have

$$\gamma_{ij} E_j = \frac{-4\pi i}{\omega} J_i , \qquad (7.109)$$

where

$$\gamma_{ij} = (n^2 + 1) \frac{k_i k_j}{k^2} - n^2 \delta_{ij}$$

and δ_{ij} is the Kronecker delta function. Using linearized theory, the Fourier transformed plasma current of particle species α can be expressed as a linear function of the electric field through the susceptibility tensor χ_{ij}^{α};

$$J_i^{\alpha}(\omega, \mathbf{k}) = -i\omega \chi_{ij}^{\alpha}(\omega, \mathbf{k}) E_j(\omega, \mathbf{k}) . \qquad (7.110)$$

The current densities of all species can be summed and combined with Eq. (7.109) to yield

$$\Lambda_{ij}(\omega, \mathbf{k}) E_j(\omega, \mathbf{k}) = 0 , \qquad (7.111)$$

where Λ_{ij} is the dispersion tensor defined by $\Lambda_{ij} = \gamma_{ij} + 4\pi \chi_{ij}$ and $\chi_{ij} = \sum_{\alpha} \chi_{ij}^{\alpha}$. The dispersion relation is obtained by setting the determinant of the dispersion tensor to zero

$$\Lambda \equiv \det |\Lambda_{ij}| = 0 . \qquad (7.112)$$

The roots of Eq. (7.112) for a given set $(\omega, \mathbf{k})$ determine the normal modes of oscillation.

The dispersion relation of a cold, homogeneous plasma in the magnetoinductive limit with perpendicular guiding-center electron motion is given by

$$\tan^2 \phi = \frac{-\left(1+\frac{\omega_p^2}{\omega^2}\right)\left[\frac{k^2c^2}{\omega^2} - \frac{\omega_{pi}^2}{\Omega_i(\omega+\Omega_i)}\right]\left[\frac{k^2c^2}{\omega^2} + \frac{\omega_{pi}^2}{\Omega_i(\omega-\Omega_i)}\right]}{\left(\frac{\omega_p^2+k^2c^2}{\omega^2}\right)\left[\left(1-\frac{\omega_{pi}^2}{\omega^2-\Omega_i^2}\right)\frac{k^2c^2}{\omega^2} + \frac{\omega_{pi}^2}{\omega^2-\Omega_i^2}\left(1+\frac{\omega_{pi}^2}{\Omega_i^2}\right)\right]} , \qquad (7.113)$$

where ϕ is the angle between $\mathbf{B}$ and $\mathbf{k}$, and $\omega_p^2 = \omega_{pe}^2 + \omega_{pi}^2$. Gaussian-shaped finite size particle effects are again included in Eq. (7.113) by simply replacing ω_{pi}^2 and ω_{pe}^2 by $\omega_{pi}^2 e^{-k^2a^2}$ and $\omega_{pe}^2 e^{-k^2a^2}$. The dispersion relation can be expressed as a cubic polynomial in ω^2 whose roots determine the eigenfrequencies of the normal mode oscillations.

For propagation parallel to the magnetic field, as in Chapter 2, the dispersion relation for electrostatic plasma oscillations is

$$\omega^2 = \omega_p^2 e^{-k^2a^2} + 3\frac{k^2 T_e}{m} . \qquad (7.114)$$

The first order thermal correction and finite size particle effects are included in Eq. (7.114). For propagation perpendicular to the magnetic field, the electrostatic spectrum exhibits peaks in the vicinity of the lower hybrid frequency, $\omega_{LH}^2 = \omega_{pi}^2 + \Omega_i^2$, and of harmonics of the cyclotron frequency. These peaks correspond to the ion Bernstein modes, which are predicted from warm plasma theory. The analysis of the electron Bernstein modes for the full dynamics electron model has been investigated by Kamimura et al.[27] The dispersion relation for perpendicular propagation of the ion Bernstein modes is given in Ref. 28 as

$$1 - \frac{e^{-k^2a^2}}{k^2\lambda_{Di}^2} \sum_{n=-\infty}^{\infty} \Gamma_n(\beta_i) \frac{n\Omega_i}{\omega - n\Omega_i} = 0 , \qquad (7.115)$$

where $\lambda_{Di}^2 = T_i/M\omega_{pi}^2$, and $\Gamma_n(\beta_i) = I_n(\beta_i)e^{-\beta_i}$ where I_n is the modified Bessel function of the first kind. The parameters for the run are: $L_x \times L_y = 128\Delta \times 320\Delta$, $a_x = 1.5\Delta$, $a_y = 15\Delta$, $v_e = 1.0\omega_{pe}\Delta$, $M/m = 1600$, $T_i/T_e = 1.0$, $\Omega_i/\omega_{pi} = v_A/c = 1/3$ and $\theta = 3.1°$. The frequency peaks of the power spectra for the $(k_x, k_y = 0)$ modes plotted as a function of $k\Delta$ in Fig. 7.6 are in good agreement with Eq. (7.115).

For purely parallel propagation, the two electromagnetic modes predicted by the cold plasma analysis are the circularly polarized whistler waves and the shear Alfvén waves. The whistler wave rotates in the same direction as the electron cyclotron motion whereas the shear Alfvén wave rotates in the same direction as the ion cyclotron motion. The shear Alfvén wave frequency

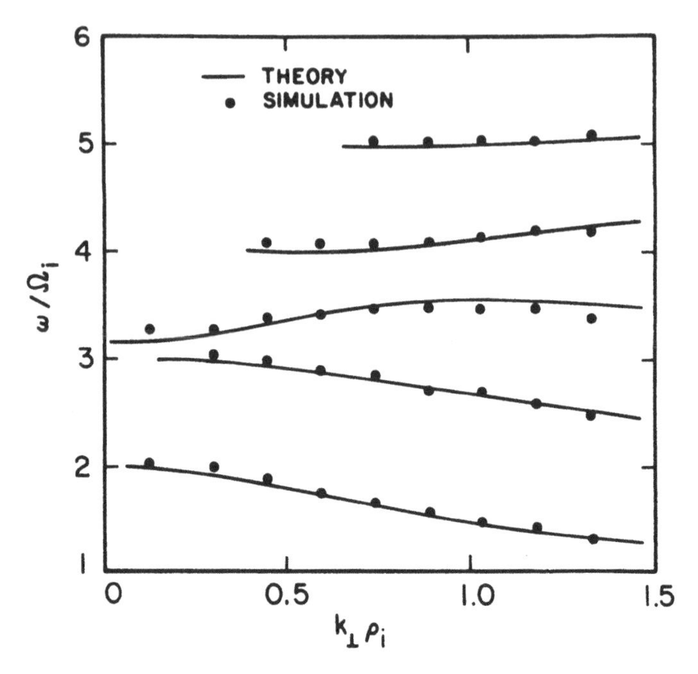

FIGURE 7.6 Dispersion relation for electrostatic ion Bernstein and lower hybrid modes obtained from the guiding-centered magnetoinductive code (after Ref. 23)

is given by

$$\omega_s = \frac{k^2 v_A^2}{2\Omega_i} e^{k^2 a^2} \left[\left(1 + \frac{4\Omega_i^2}{k^2 v_A^2} e^{-k^2 a^2} \right)^{1/2} - 1 \right] . \tag{7.116}$$

For small k, the frequency ω has an approximately linear relationship to k, $\omega = k v_A$ and, for large k, the frequency approaches the ion cyclotron frequency $\omega = \Omega_i$. The peaks of the simulation frequency spectrum for $(k_x = 0, k_y)$ modes showed reasonable correspondence between the simulation points and the theory. The measured dispersion relation of the shear Alfvén wave for oblique propagation is displayed in Fig. 7.7. The $(k_x = 0, k_y)$ modes are plotted from a simulation run[23] with parameters: $L_x \times L_y = 128\Delta \times 320\Delta$, $a_x = 1.5\Delta$, $a_y = 15\Delta$, $v_T = 1.0\omega_{pe}\Delta$, $M/m = 1600$, $T_i/T_e = 1.0$, $\Omega_i/\omega_{pi} = v_A/c = 1/3$, and $\theta = 3.1°$. The agreement between the simulation results and theory in Fig. 7.7 which shows the curve that is the prediction from the full cold plasma dispersion relation, Eq. (7.113), is good. The whistler wave is not observed in the thermal spectrum.

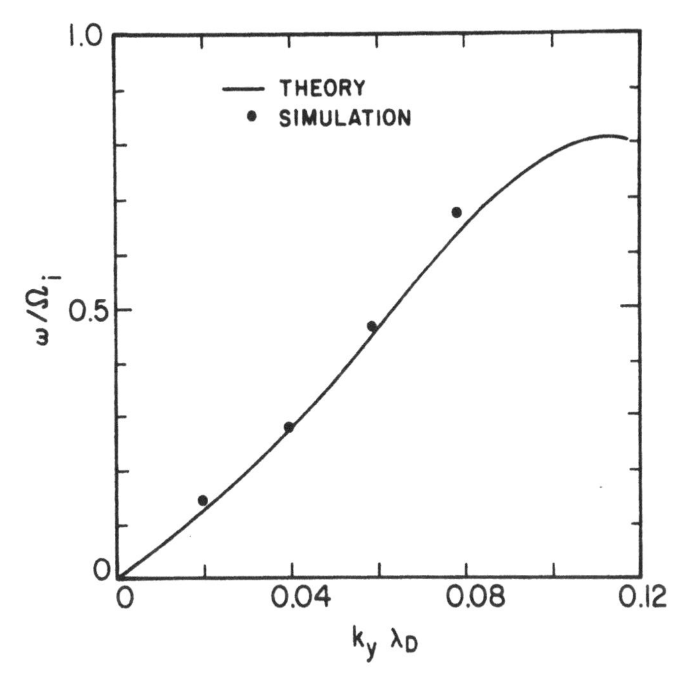

FIGURE 7.7 Dispersion relation for electromagnetic shear Alfvén waves from the same code as in Fig. 7.6 (after Ref. 23)

In the Vlasov theory the plasma dielectric function may be written in the form,

$$\epsilon(\mathbf{k},\omega) = 1 - \frac{k_{\parallel}^2 c^2}{\omega^2} + \frac{\omega_{pe}^2}{\omega^2}\frac{\omega}{\sqrt{2}k_{\parallel}v_{Te}} Z\left(\frac{\omega+\Omega_e}{\sqrt{2}k_{\parallel}v_{Te}}\right)$$

$$+\frac{\omega_{pi}^2}{\omega^2}\frac{\omega}{\sqrt{2}k_{\parallel}v_{Ti}} Z\left(\frac{\omega-\Omega_i}{\sqrt{2}k_{\parallel}v_{Ti}}\right) . \qquad (7.117)$$

For a strongly magnetized plasma, $\omega_{ce} \gg \omega$ and in the magnetoinductive approximation $k_{\parallel}^2/\omega^2 \gg 1$. This reduces the dielectric function to the form

$$\epsilon(\mathbf{k},\omega) = \frac{-k_{\parallel}^2 c^2}{\omega^2} - \frac{\omega_{pe}^2}{\omega\Omega_e} + \frac{\omega_{pi}^2}{\omega^2}\frac{\omega}{\sqrt{2}k_{\parallel}v_{Ti}} Z\left(\frac{\omega-\Omega_i}{\sqrt{2}k_{\parallel}v_{Ti}}\right) . \qquad (7.118)$$

The dispersion relation is given by the equation for the zeros of the dielectric function $\epsilon(\mathbf{k},\omega) = 0$.

The level of the magnetic field fluctuations is a function of the angle ϕ between $\mathbf{k}$ and the magnetic field. The magnetic field fluctuations are obtained in terms of the current density fluctuations. The fluctuations of the current density are related to the linear response function using the fluctuation-dissipation theorem. According to Sitenko,[26] the current density fluctuations in an infinite, nonisothermal plasma can be given in terms of the imaginary (dissipative) part of the response function by

$$\langle J_i J_j \rangle_{\mathbf{k},\omega} = i\omega \left[\gamma_{ik} \Lambda_{km}^{-1}\right]^* \times \left[\gamma_{j\ell} \Lambda_{\ell n}^{-1}\right] \sum_{\alpha} T^{\alpha} \left(\chi_{mn}^{*\alpha} - \chi_{nm}^{\alpha}\right) , \qquad (7.119)$$

where γ_{ij} and Λ_{ij} are given by Eqs. (7.109) and (7.111), where T^{α} is the temperature of species α.

Let us introduce the tensor λ_{ij} whose elements are the cofactors of the dispersion tensor Λ_{ij},

$$\Lambda_{ij} \lambda_{jk} = \Lambda \delta_{ik} \; . \qquad (7.120)$$

The current density fluctuations from Eq. (7.119) can then be expressed as

$$\langle J_i J_j \rangle_{\mathbf{k},\omega} = \frac{i\omega}{|\Lambda|^2} \left\{ T_{\|}^{e} \gamma_{ik} \lambda_{k3}^{*} \gamma_{j\ell} \lambda_{\ell 3} \left(\chi_{33}^{e*} - \chi_{33}^{e}\right) \right.$$

$$\left. + T^{i} \gamma_{ik} \lambda_{kn}^{*} \gamma_{j\ell} \lambda_{\ell n} \left(\chi_{mn}^{i*} - \chi_{nm}^{i}\right) \right\} , \qquad (7.121)$$

where the superscripts, e and i, refer to electrons and ions respectively. Inspection of the above formula indicates that the fluctuation spectrum has sharp peaks near the normal modes of plasma oscillation because of the zeros of the dispersion relation in the denominator, given by $\Lambda = 0$. For simplicity, we adopt the cold plasma theory to evaluate magnetic field fluctuations. The magnetic field fluctuation spectrum can be computed from the relation

$$\frac{\langle B^2 \rangle_{\mathbf{k},\omega}}{8\pi} = \frac{2\pi}{k^2 c^2} \left[\langle J^2 \rangle_{\mathbf{k},\omega} - \frac{\langle k_i J_i k_j J_j \rangle_{\mathbf{k},\omega}}{k^2} \right] , \qquad (7.122)$$

once the current density fluctuation spectrum is known. The integration of Eq. (7.122) over all frequencies yields the time averaged fluctuation spectrum of the magnetic field

$$\frac{\langle B^2 \rangle_{\mathbf{k}}}{8\pi} = \int \frac{\langle B^2 \rangle_{\mathbf{k},\omega}}{8\pi} \frac{d\omega}{2\pi} \; . \qquad (7.123)$$

The dispersion tensor contains the information for determining the current density fluctuations which, subsequently, can yield the magnetic field fluctuation spectrum. We shall examine the special cases where wave propagation is either completely parallel or perpendicular to the background magnetic field.

Let us first investigate fluctuation spectra for propagation parallel to the magnetic field, which is assumed to lie in the z-direction. The magnetic field fluctuation spectrum in Eq. (7.122) reduces to

$$\frac{\langle B^2\rangle_{\mathbf{k},\omega}}{8\pi} = \frac{2\pi}{k^2c^2}\left[\langle J_x^2\rangle_{\mathbf{k},\omega} + \langle J_y^2\rangle_{\mathbf{k},\omega}\right] . \tag{7.124}$$

Here we evaluate the elements of the susceptibility tensor using the particle equations of motion in a cold fluid approximation.[23] The coordinate axes are defined such that

$$\mathbf{B}_0 = B_0\hat{\mathbf{z}} \quad , \quad \mathbf{k} = k_x\hat{\mathbf{x}} + k_z\hat{\mathbf{z}} .$$

From Eqs. (7.4) and (7.5), the electron conductivity tensor elements are given by

$$4\pi\chi_{ij}^e = \begin{pmatrix} 0 & \frac{-i\omega_{pi}^2}{\Omega_i\omega} & 0 \\ \frac{i\omega_{pi}^2}{\omega\Omega_i} & 0 & 0 \\ 0 & 0 & -\omega_{pe}^2/\omega^2 \end{pmatrix} . \tag{7.125}$$

The ion conductivity tensor is given by

$$4\pi\chi_{ij}^i = \begin{pmatrix} \frac{-\omega_{pi}^2}{\omega^2-\Omega_i^2} & -i\frac{\Omega_i}{\omega}\frac{\omega_{pi}^2}{\omega^2-\Omega_i^2} & 0 \\ \frac{i\Omega_i}{\omega}\frac{\omega_{pi}^2}{\omega^2-\Omega_i^2} & \frac{-\omega_{pi}^2}{\omega^2-\Omega_i^2} & 0 \\ 0 & 0 & -\frac{\omega_{pi}^2}{\omega^2} \end{pmatrix} . \tag{7.126}$$

Using the cold-plasma equations, the expression for the magnetic field fluctuation is

$$\frac{\langle B^2\rangle_{\mathbf{k},\omega}}{8\pi} = \frac{iT^i}{2\omega}\frac{k^2c^2}{\omega^2}\left[\frac{\lambda_{11}+\lambda_{22}}{\Lambda} - c.c.\right] . \tag{7.127}$$

When $\phi = 0$, it can be seen that

$$\frac{\lambda_{11}+\lambda_{22}}{\Lambda} = \frac{\Lambda_{11}+\Lambda_{22}}{\Lambda_{11}\Lambda_{22}+\Lambda_{12}^2} .$$

We then have

$$\frac{\langle B^2\rangle_{\mathbf{k},\omega}}{8\pi} = \frac{iT^i}{2\omega}\frac{k^2c^2}{\omega^2}\left\{\frac{2\left(\frac{k^2c^2}{\omega^2} + \frac{\omega_{pi}^2}{\omega^2-\Omega_i^2}\right)}{\left[\frac{\omega_{pi}^2}{\Omega_i(\omega+\Omega_i)} - \frac{k^2c^2}{\omega^2}\right]\left[\frac{\omega_{pi}^2}{\Omega_i(\omega-\Omega_i)} + \frac{k^2c^2}{\omega^2}\right]} - c.c\right\} , \tag{7.128}$$

where c.c. denotes the complex conjugate.

The observed power spectra for the magnetic field reveal that, for parallel propagation, most of the energy is deposited in the shear Alfvén mode.

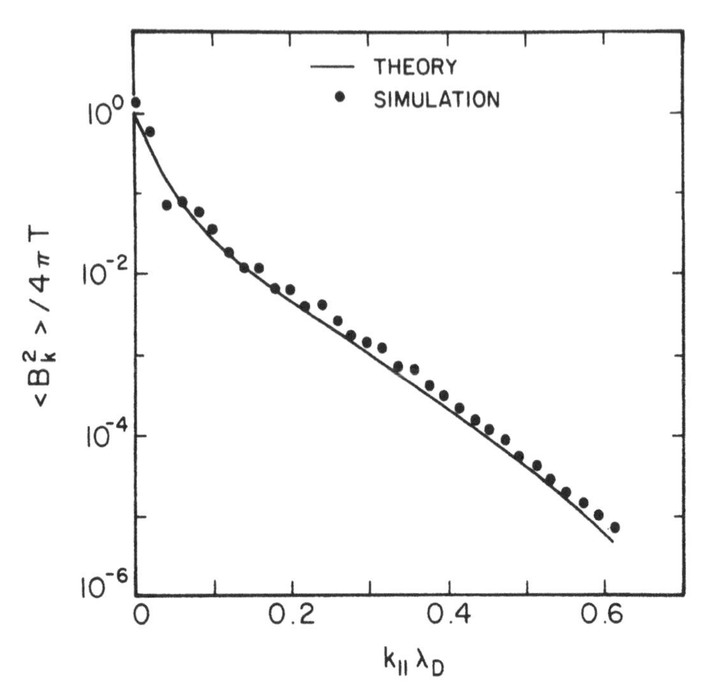

FIGURE 7.8 Magnetic fluctuations associated with the shear Alfvén waves (after Ref. 23)

Therefore only the shear Alfvén resonance of Eq. (7.128) has been picked up among the poles. The magnetic field fluctuation spectrum is found to be

$$\frac{\langle B^2\rangle_{\mathbf{k}}}{8\pi} = T^i\left(1+\frac{4\Omega_i^2}{k^2v_A^2}\right)^{-1/2}\left[\frac{\Omega_i^2}{\omega_s^2}-\left(1+\frac{\Omega_i^2}{k^2v_A^2}\right)\right] . \tag{7.129}$$

The inclusion of finite size particle effects and the periodicity of the simulation plasma modifies Eq. (7.129) to

$$V\frac{\langle B^2\rangle_{\mathbf{k}}}{4\pi T^i} = 2\left[1+\frac{4\Omega_i^2}{k^2v_A^2}e^{-k^2a^2}\right]^{-1/2}\left[\frac{\Omega_i^2}{\omega_s^2}-\left(1+\frac{\Omega_i^2e^{-k^2a^2}}{k^2v_A^2}\right)\right] . \tag{7.130}$$

where ω_s is given by Eq. (7.116). There is good agreement between the simulation results in Fig. 7.8 and the prediction from Eq. (7.130). If the contribution from the pole representing the whistler wave is evaluated along with the shear Alfvén wave, on the other hand, the fluctuation spectra is given by

$$\frac{\langle B^2\rangle_{\mathbf{k}}}{8\pi} = T^i .$$

Let us now investigate the magnetic field fluctuation spectrum for wave propagation perpendicular to the magnetic field. The magnetic field fluctuations are given by

$$\frac{\langle B^2\rangle_{\mathbf{k},\omega}}{8\pi} = \frac{2\pi}{k^2c^2}\left[\langle J_y^2\rangle_{\mathbf{k},\omega} + \langle J_z^2\rangle_{\mathbf{k},\omega}\right] \tag{7.131}$$

from Eq. (7.122).

For the test results displayed here,[23] the electron temperature is equal to the ion temperature such that $T_{\parallel}^e = T^i = T$. In this instance, the magnetic field fluctuation spectrum for perpendicular propagation is given by

$$\frac{\langle B^2\rangle_{\mathbf{k},\omega}}{8\pi} = \frac{iT}{2\omega}\frac{k^2c^2}{\omega^2}\left[\Lambda_{22}^{-1} + \Lambda_{33}^{-1} - c.c\right] . \tag{7.132}$$

The observations of the power spectra indicate that most of the energy is deposited in a zero-frequency mode when $k_{\parallel} = 0$. The zero-frequency wave is not one that is derived from a collisionless model. This mode is proposed as a mechanism responsible for electron diffusion across the magnetic field by Chu, Chu, and Ohkawa.[29] When the electron parallel momentum equation, Eq. (7.5), has dissipation included as follows[29]

$$\frac{dv_{ez}}{dt} = \frac{-e}{m}E_z + \mu\nabla^2 v_{ez} - \nu\, v_{ez} , \tag{7.133}$$

where ν is the electron-ion collision frequency and μ is the kinematic electron viscosity.

The zz-element of the electron susceptibility tensor, χ_{33}, is given by

$$4\pi\chi_{33} = \frac{-\omega_{pe}^2}{\omega(\omega + i\eta)} , \tag{7.134}$$

where $\eta = \nu + \mu k^2$.

The magnetic field fluctuation spectrum from Eq. (7.132) becomes

$$\frac{\langle B^2\rangle_{\mathbf{k},\omega}}{8\pi} = T\frac{k^2c^2}{\omega_p^2 + k^2c^2}\frac{\eta - \gamma}{\omega^2 + \gamma^2} \tag{7.135}$$

where

$$\gamma = \frac{\eta(k^2c^2 + \omega_{pi}^2)}{k^2c^2 + \omega_p^2} .$$

Integrating over frequency, the time averaged fluctuation spectrum is given by

$$\frac{\langle B^2\rangle_{\mathbf{k}}}{8\pi} = \frac{T}{2}\frac{k^2c^2\omega_{pe}^2}{(k^2c^2 + \omega_{pi}^2)}\frac{1}{(\omega_{pe}^2 + k^2c^2 + \omega_{pi}^2)} . \tag{7.136}$$

If the ion response is ignored, Eq. (7.136) reduces to the more familiar form

$$V \frac{\langle B^2 \rangle_{\mathbf{k}}}{4\pi T} = \frac{1}{1 + k^2 c^2 e^{k^2 a^2} / \omega_{pe}^2} , \tag{7.137}$$

when finite size particle effects and the periodicity of the simulation volume are included.

We have discussed the guiding-center method to describe plasma behaviors coarse-grained over the cyclotron periods. This model retains an accurate description of kinetic plasma behavior along the magnetic field line, while it follows the MHD-like $\mathbf{E} \times \mathbf{B}$ motion in the perpendicular directions. In this way this model stands at a crossroad of the particle approach and the fluid approach.

Problems

1. The Schrödinger equations have a pure imaginary coefficient, while Eqs. (7.7) and (7.8) combined have a real coefficient: $\dot{f} = -i\omega f$, where f is complex and ω is real, which can be decomposed into real and imaginary components $\dot{f}_r = \omega f_i$ and $\dot{f}_i = -\omega f_r$. Show that the Schrödinger equation can be readily leapfrogged.
2. The bounded electrostatic plasma with a metallic wall (the electric potential $\phi = 0$ on the boundary) or an alternate ($\phi' = 0$) or the bounded electromagnetic plasma are often treated by imposing appropriate symmetry (even or odd) across the boundary. (See Refs. 12 and 23.) For full dynamics particles the method developed in Ref. 30 works appropriately. What happens in the guiding-center model? (Customarily,[31] a method of specular reflection has been employed). If $\phi = 0$ on the boundary, show that the $\mathbf{E} \times \mathbf{B}$ drift into the boundary vanishes, which is convenient. However, show that the polarization drift into the boundary does not vanish and evaluate its size.
3. Derive the characteristic frequencies in a code with guiding-center electrons and ions with full dynamics.
4. In deriving Eq. (7.43) we ordinarily compute the "force" which includes the finite size particle effect as distinct from the "field." Indicate what operation is necessary to obtain $\mathbf{E}_n(x_g, y_g)$.
5. Construct an algorithm to solve for the geostrophic fluid based on the fluid approach or based on the particle technique. This is a close parallel to the guiding-center method of a magnetized plasma.
6. Construct an algorithm for the Hasegawa-Mima equation[16] for drift-waves based either on the fluid approach or on the particle approach. This is a

close parallel to Problem 5. In this model the continuity equation for ions, neglecting the parallel ion inertia, is $\partial_t n + \nabla_\perp \cdot n_0(\mathbf{v}_E + \mathbf{v}_p) = 0$, where $\mathbf{v}_E$ is the $\mathbf{E} \times \mathbf{B}$ drift and $\mathbf{v}_p$ is the polarization drift $-[\partial_t \nabla_\perp \phi + (\mathbf{v}_E \cdot \nabla)\nabla_\perp \phi]/\Omega B$. Electrons follow along the field line so that they obey the Boltzmann distribution $n/n_0 = e\phi/T_e$.

7. In many problems of subtle plasma instabilities, developments in particle phase space are significant only for resonant particles and those for adiabatic particles may be more readily described. This feature is manifested in Eq. (7.79) for the resonant distribution $\tilde{h}$. For simplicity, consider a non-magnetized Vlasov equation $\partial_t \delta f + \mathbf{v} \cdot \nabla \delta f + q\mathbf{E}/m \cdot \nabla_v \delta f = -q\mathbf{E}/m \cdot \nabla_v \langle f \rangle$, with $\partial_t \langle f \rangle = 0$. The solution to this equation[32] is represented by $\delta f = \sum_i w_i[\mathbf{x}_i(t), \mathbf{v}_i(t), t]\delta[\mathbf{x} - \mathbf{x}_i(t)]\delta[\mathbf{v} - \mathbf{v}_i(t)]n^{-1}$, where $\dot{w}_i = -q\mathbf{E}/m \cdot \nabla \langle f \rangle$, $\dot{\mathbf{x}}_i = \mathbf{v}_i$, $\dot{\mathbf{v}}_i = q\mathbf{E}/m$, and $n = \sum_i \delta[\mathbf{x} - \mathbf{x}_i(t)]\delta[\mathbf{v} - \mathbf{v}_i(t)]$. Consider a method of solution via particle simulation to solve the above Vlasov equation with a "source term" (the term on the RHS). Then generalize it to the resonant part of the gyrokinetic equation. Discuss possible difficulties associated with this method and a way to alleviate it.

8. Derive Eq. (7.106).

9. Include finite size particle effects in Eqs. (7.125) and (7.126).

10. Derive Eq. (7.70). Describe the strategy of how to handle the spatially sheared magnetic field in this method.

11. Using the method discussed in Problem 7 in the above, devise a method to advance particles which may be annihilated and/or created. Relate this problem to Problem 5 in Chapter 5.

References

1. H. Alfvén, *Cosmical Electrodynamics* (Clarendon Press, Oxford, 1950).
2. J.P. Christiansen, J. Comp. Phys. **13**, 363 (1973).
3. V.E. Zakharov, Sov. Phys. JETP **35,** 908 (1972) [Zh. Eksp. Teor. Fiz. **62**, 1745 (1972)].
4. T. Kurki-Suonio and T. Tajima, Proc. 12th Numerical Simulation Conf. (San Francisco, 1987).
5. T.G. Northrop, *The Adiabatic Motion of Charged Particles*, (Interscience, New York, 1963).
6. M.D. Kruskal, Trieste Proceedings TC/C 1/WP. 17 (1964).
7. W.W. Lee and H. Okuda, J. Comp. Phys. **26**, 139 (1978).

8. W.W. Lee and H. Okuda, Phys. Rev. Lett. **36**, 870 (1976).
9. C.Z. Cheng and H. Okuda, Phys. Rev. Lett. **38**, 308 (1977).
10. M.N. Rosenbluth, N. Krall, and N. Rostoker, Nucl. Fusion Suppl. **1**, 143 (1962).
11. H. Okuda and J.M. Dawson, Phys. Rev. Lett. **28**, 1625 (1972).
12. C.Z. Cheng and H Okuda, J. Comp. Phys. **25**, 133 (1977).
13. H. Okuda and C.Z. Cheng, Comp. Phys. Comm. **14**, 169 (1978).
14. J. Pedlosky, *Geophysical Fluids Dynamics* (Springer-Verlag, New York, 1979) p. 81.
15. K. Mima, T. Tajima and J.N. Leboeuf, Phys. Rev. Lett. **41**, 1715 (1978).
16. A. Hasegawa and K. Mima, Phys. Rev. Lett. **39**, 205 (1977); Phys. Fluids **21**, 87 (1978).
17. W. Horton, in *Handbook of Plasma Physics* (North-Holland, Amsterdam, 1984), vol. 2, p. 383.
18. J.B. Taylor and R.J. Hastie, Plasma Physics **10**, 4769 (1968).
19. P.J. Catto, Plasma Physics **20**, 719 (1978); P.H. Rutherford and E.A. Frieman, Phys. Fluids **11**, 569 (1968).
20. W.W. Lee, Phys. Fluids **26**, 556 (1983).
21. D.H.E. Dubin, J.A. Krommes, C. Oberman, and W.W. Lee, Phys. Fluids **26**, 3524 (1983).
22. H. Goldstone, *Classical Mechanics* (Addison Wesley, Reading, Mass., 1950).
23. J.L. Geary, T. Tajima, J.N. Leboeuf, E.G. Zaidman, and J.H. Han, Comp. Phys. Comm. **42**, 313 (1986).
24. S.Y. Kim and H. Okuda, J. Comp. Phys. **65**, 215 (1986).
25. R. Kubo, J. Phys. Soc. Jpn. **12**, 570 (1957).
26. A.G. Sitenko, *Electromagnetic Fluctuations in Plasma* (W.A. Benjamin, Reading, Mass., 1973).
27. T. Kamimura, T. Wagner, and J.M. Dawson, Phys. Fluids **21**, 1151 (1978).
28. S. Ichimaru, *Basic Principles of Plasma Physics* (W.A. Benjamin, Reading, Mass., 1973).
29. C. Chu, M.S. Chu and T. Ohkawa, Phys. Rev. Lett. **41**, 653 (1978).
30. H. Naito, S. Tokuda and T. Kamimura, J. Comp. Phys. **33**, 86 (1979).
31. R.D. Sydora, J.N. Leboeuf and T. Tajima, Phys. Fluids **28**, 528 (1985).
32. T. Tajima and F.W. Perkins, Proc. 1983 Sherwood Theory Meeting (University of Maryland, Arlington, VA, 1983) 2P9.

8 HYBRID MODELS OF PLASMAS

As the time scale gets larger, the overall plasma behavior becomes more fluid-like. At intermediate time scales, a mixing of the particle-like and the fluid-like behaviors of a plasma often emerges. Such phenomena may be best described by a hybrid of the particle method and the fluid approach. A natural example of a hybrid picture of a plasma is that of fluid-like electrons and particle-like ions. This may happen because of the wide range of frequencies separated by the mass difference between electrons and ions. (See Table 1.2.)

8.1 Quasineutral Electrostatic Model

For low frequency waves in an electrostatic problem, electrons respond to the field almost instantaneously (compared with the wave period); the ions are thereby shielded and charge separation effects become much reduced. When the phenomenon under study is in low frequency, the high frequency oscillations associated with the electron inertia (such as the Langmuir plasma oscillations) can be neglected. The charge neutrality is, therefore, approximately satisfied. Electrons are adjusting quickly to the charge separation along the field line. If, on the other hand, the wave develops perpendicular or nearly perpendicular to the magnetic field direction, the quasineutrality condition may not hold even for low frequency waves (See Problem 1). Examples of the quasineutral phenomena are ion sound waves, electrostatic ion cyclotron (Bernstein) waves, drift waves, trapped particle modes in a toroidal system, and magnetohydrodynamic oscillations.

Electrons under the present condition adjust to the electric field along the field line so quickly that they behave adiabatically. This leads to the well-known Boltzmann equilibrium distribution for the electron density n_e

$$n_e = n_0 \exp(e\phi/T_e) \ , \tag{8.1}$$

where n_0 is the equilibrium density, ϕ the electric potential, and T_e the electron temperature. Poisson's equation is

$$\nabla^2\phi = -4\pi e(n_i - n_e) \ , \tag{8.2}$$

the ion density $n_i = n_0 + \delta n_i$. If the nonlinearity is small, the exponential function in Eq. (8.1) may be expanded to give $1 + e\phi/T_e + O(e\phi/T_e)^2$. Substitution of Eq. (8.1) into Eq. (8.2) yields[1]

$$\nabla^2\phi = -4\pi e\delta n_i + k_{De}^2\phi \ , \tag{8.3}$$

where $k_{De}^2 = 4\pi n_0 e^2/T_e$. [The leading terms cancel each other on the right-hand side of Eq. (8.3).] Equation (8.3) has the same form as the equation for electric field in the magnetoinductive model Eq. (5.47). This may be readily solved by the now familiar fast Fourier transform technique, as we have done in Chapter 2:

$$(k^2 + k_{De}^2)\phi(k) = 4\pi e\delta n_i(k) \ . \tag{8.4}$$

Equations for ions are fully dynamical:

$$\begin{aligned} \frac{d\mathbf{v}_i}{dt} &= \frac{e}{M}\mathbf{E} \ , \\ \frac{d\mathbf{x}_i}{dt} &= \mathbf{v}_i \ . \end{aligned} \tag{8.5}$$

Combining Eqs. (8.4) and (8.5) can yield the well-known ion-acoustic waves, whose dispersion relation is

$$(k^2 + k_{De}^2) = k^2\omega_{pi}^2/\omega^2 \ , \tag{8.6}$$

which reduces to

$$\omega^2 = k^2 c_s^2 \tag{8.7}$$

for small k, where $c_s^2 = T_e/M$. This plasma model eliminates the high frequency plasma oscillations; its highest oscillations are now those of ion-acoustic waves. This is a remarkable feature, because in the standard explicit code it is so difficult to obtain a clear ion-acoustic wave that one has to use an extremely large number of particles to achieve a nice result.[2] On the other hand, the present quasineutral code[1] neglects the resonant electron effect and retains only the adiabatic electron effect.

A few remarks are now due. A generalization to the nonlinear case ($e\phi/T_e$ is not small) is straightforward. Instead of the linearized electron density, we can use the original nonlinear expression Eq. (8.1) to obtain

$$\nabla^2\phi = 4\pi e n_i - 4\pi e n_0 \exp(e\phi/T_e) \; . \tag{8.8}$$

A possible method of solution is to iterate; the $\ell+1$-st value of ϕ may be determined in terms of the ℓ-th value of ϕ

$$\nabla^2\phi^{\ell+1} = 4\pi e n_i - 4\pi e n_0 \exp(e\phi^{\ell}/T_e) \; . \tag{8.9}$$

A faster convergence may be achieved by adopting Newton's method.[3]

So far we tacitly assumed an unmagnetized plasma. It is, however, possible to employ the present method even for a magnetized plasma.[1] As long as we ignore the nonadiabatic (resonant) effect of electrons, the main electron effect is considered to be the screening of charge; once this screening is done the electron obeys the Boltzmann law, Eq. (8.1). The condition for this is to satisfy

$$\omega \ll k_{\parallel} v_e \; , \tag{8.10}$$

where $k_{\parallel} = \mathbf{k}\cdot\hat{b}$.

8.2 Quasineutral Electromagnetic Model

In low frequency electromagnetic problems, electrons exhibit fluid-like behavior, just as we discussed in the electrostatic problems above.

A. Electromagnetic Particle Code

In a problem such as the ion-cyclotron resonance heating, it is important to retain the full dynamics of the ions and electromagnetic interaction, while we can reduce the dynamics of electrons. In a model developed by Riyopoulos and Tajima,[4] electrons are advanced using the drift equation Eqs. (7.4) and (7.5). This is equivalent to having the following cold electron contribution to the dielectric tensor instead of the full kinetic plasma dielectric tensor:

$$\epsilon_{xy} = -\epsilon_{yx} = -i\sum_{\sigma}\frac{\Omega_\sigma}{\omega}\frac{\omega_\sigma^2}{\omega^2-\Omega_\sigma^2} \; , \tag{8.11}$$

while it neglects the electron polarization current that would contribute to a part of the diagonal components of the tensor

$$\epsilon_{xx} = \epsilon_{yy} = 1 - \sum_{\sigma} \frac{\omega_{\sigma}^2}{\omega^2 - \Omega_{\sigma}^2} . \tag{8.12}$$

The electron drift contribution is of the same order as the ion drift and should be kept, while from Eq. (8.12) the electron polarization drift is m/M times the ion polarization drift and can be neglected.

The guiding-center model, Eqs. (7.4) and (7.5), does not allow electron Landau damping. This is of no great importance for $k_{\parallel} \cong 0$. For finite $k_{\parallel}$, it is appropriate to split the electron motion into two parts: the perpendicular to the magnetic field by the $\mathbf{E} \times \mathbf{B}$ drift, and the parallel to the magnetic field treated with the exact equation $dv_{\parallel}/dt = (e/m)E_{\parallel}$.

However, for small $k_{\parallel}$ and neglecting the electron resonance effect (Landau damping) neglected we can still avoid following the exact parallel electron dynamics and retain the one-dimensional modeling. The electron-shielding effect due to the parallel motion to the magnetic field can be incorporated into Poisson's equation, using the electron dielectric response, giving

$$i\mathbf{k} \cdot \mathbf{E}(\mathbf{k}) = -4\pi \left[\rho_i(\mathbf{k}) + \rho_{de}(\mathbf{k})\right] / \epsilon_e(\mathbf{k}) , \tag{8.13}$$

generalizing the method we discussed in Section 8.1. Here, ρ_{de} is the electron charge perturbation due to the electron drift motion perpendicular to the transverse electric field and the magnetic field. The electron response parallel to the electrostatic field is, by definition, built into the dielectric response function

$$\epsilon_e(k) = \frac{1}{k^2} \mathbf{k} \cdot \overleftrightarrow{\epsilon}_e \cdot \mathbf{k} .$$

For oblique wave propagation with $\omega/k_{\parallel} \ll v_e$, $\epsilon_e \to 1 + k^2/k_{\mathrm{De}}^2$ and the electrostatic response comes mainly from the electron motion parallel to the magnetic field. In the case of nearly perpendicular propagation, $\omega/k_{\parallel} \gg v_e$, the electrostatic response diminishes to $\epsilon_e \to 1 + \omega_e^2/\Omega_e^2$, coming from the polarization drift across the magnetic field. Now the main electron charge perturbation comes from the $\mathbf{E} \times \mathbf{B}$ drift in response to the transverse electric field.

B. Quasineutral Magnetoinductive Model

Many algorithms and codes have been developed utilizing this hybrid of particle and fluid descriptions of plasmas.[5-8] In these codes the electrons are treated as a fluid and ions are retained as particles. As an example, let us follow Harned's model.[8] Ampere's law may be decomposed into the longitudinal and transverse components (Chapter 5)

$$\nabla \times \mathbf{B} = \frac{4\pi}{c}\mathbf{J}_T , \tag{8.14}$$

$$0 = \frac{4\pi}{c}\mathbf{J}_L + \frac{1}{c}\frac{\partial \mathbf{E}_L}{\partial t} , \tag{8.15}$$

where subscripts L and T are defined in Section 5.1. Here we dropped the displacement current according to Darwin's model (Section 5.4). The quasineutrality condition $n_i = n_e$ implies $n = 0$ and thus

$$\nabla \cdot \mathbf{J} = 0 . \tag{8.16}$$

If $\mathbf{J}_L$ vanishes on the boundaries, or the system is periodic, $\mathbf{J}_L = 0$ throughout the system. The right-hand side of Eq. (8.14) is thus given by the total current $\mathbf{J} = \mathbf{J}_i + \mathbf{J}_e$, the sum of the ion and the electron current densities.

Electrons are treated as a massless fluid, just as we have done in the magnetohydrodynamics (see Chapter 6)

$$0 = n_e m_e \frac{d\mathbf{v}_e}{dt} = -en_e\left(\mathbf{E} + \frac{\mathbf{v}_e}{c} \times \mathbf{B}\right) - \nabla P_e , \tag{8.17}$$

where n_e, m_e, $\mathbf{v}_e$ and P_e are the electron density, velocity, and scalar pressure. The electron current is expressed as

$$\mathbf{J}_e = -n_e e \mathbf{v}_e . \tag{8.18}$$

Using Eq. (8.18) and the quasineutrality $n_e = n_i \equiv n$, we can rearrange Eq. (8.17) combined with Eq. (8.14) to give

$$\mathbf{E} = \frac{1}{nec}\mathbf{J}_i \times \mathbf{B} + \frac{1}{4\pi ne}(\nabla \times \mathbf{B}) \times \mathbf{B} - \frac{1}{ne}\nabla(nT_e) , \tag{8.19}$$

where T_e is the electron temperature. This equation is essentially equivalent to Eq. (6.21). We have neglected collisional effects here for simplicity. Equation (8.19) determines the electric field in the plasma within the quasineutrality approximation as a function of the magnetic field, electron temperature, ion current, and plasma density. The magnetic field is advanced by Faraday's law $\partial\mathbf{B}/\partial t = -c\nabla \times \mathbf{E}$ as

$$\frac{\partial \mathbf{B}}{\partial t} = \nabla \times (\mathbf{v}_i \times \mathbf{B}) - \nabla \times \left(\frac{1}{ne}\mathbf{J} \times \mathbf{B}\right) + \nabla \times \left[\frac{1}{ne}\nabla(nT_e)\right] , \tag{8.20}$$

where $\mathbf{J} = 4\pi\nabla \times \mathbf{B}/c$ from Eq. (8.14) and the subsequent arguments. The first term on the right-hand side of Eq. (8.20) is the typical MHD induction term, the second term is often called the Hall term, and the third term vanishes if the density and temperature gradients are parallel. Equation (8.20) is once again very similar to Eq. (6.25). Except for the lack of collisional effects the only difference will be the treatment of the ion velocity in the first term on the right-hand side of Eq. (8.20). In the magnetohydrodynamical approach, $\mathbf{v}_i$ is the fluid velocity, determined by the fluid ion equation. In the hybrid code of Harned's[8] $\mathbf{v}_i$ is determined by the characteristic equations for ion particles

$$\frac{d\mathbf{v}_i}{dt} = \frac{q}{M}\left(\mathbf{E} + \frac{\mathbf{v}_i}{c} \times \mathbf{B}\right) , \tag{8.21}$$

$$\frac{d\mathbf{x}_i}{dt} = \mathbf{v}_i \; . \tag{8.22}$$

According to Eq. (8.21), ions execute gyromotion due to the magnetic field and the electric acceleration. In Refs. 6 and 7 the authors also push particles according to Eqs. (8.21) and (8.23). An appropriate weighting such as linear weighting is used to calculate $\mathbf{J}_i$ and $n = n_i$ in Eq. (8.19) from the particle quantities.

On the other hand, Ref. 9 uses the ion fluid equation for "particle ion elements" as

$$\frac{d\mathbf{v}_i}{dt} = \frac{1}{4\pi\rho}(\nabla \times \mathbf{B}) \times \mathbf{B} - \frac{1}{\rho}\nabla \cdot \overleftrightarrow{P} , \tag{8.23}$$

where ρ is the ion mass density and $\overleftrightarrow{P}$ is the sum of electron and ion pressures. The difference between Eqs. (8.21) and (8.23) arises from the degree of coarse-graining of the ions in the ion pushing equation. Equation (8.21) resolves the gyromotion. On the other hand, for Eq. (8.23), Eq. (8.19) is coarse-grained in order to compute $\mathbf{E}$. Thus, the code in Ref. 9 does not carry ion gyromotion anymore and corresponds to even lower frequencies than the codes in Refs. 6-8. By defining different levels of coarse-graining in Eq. (8.19) one can, in principle, bring in a gradual transition between the model by Harned et al.[6–8] and that by Tajima et al.[9] Actual timesteppings, etc. can be found in these original papers.

One of the important effects introduced by this model is the Hall term. It has been shown[9] that the Hall term brings in the bifurcation of the Alfvén wave into two branches, i.e., the Alfvén-ion cyclotron wave and the whistler wave. This split of frequencies $\Delta\omega$ is negligible for small wavenumbers where the ideal MHD should be valid

$$\Delta\omega = k_{\parallel} v_A (k_{\parallel} v_A / \Omega_i) , \tag{8.24}$$

but becomes pronounced for wavenumbers of the order

$$k_{\parallel} \gtrsim \Omega_i / v_A . \tag{8.25}$$

The Hall term is often neglected in the MHD study and may give rise to an important effect such as the possible stabilization of the tilt mode[10,11] of a field reverse configuration (FRC).[12] In an FRC, magnetic fields can be weak locally and nonideal MHD effects should be important. In the model of Refs. 6-8 we also have finite Larmor radius effects.

8.3 Particle Electron-Fluid Ion Model

The model we discussed in Section 8.1 eliminates high frequency waves associated with the electron gyration while keeping the electron motion along the magnetic field exact. The model was particularly useful for low frequency waves propagating nearly perpendicular to an external field. Since the full ion dynamics was used, lower hybrid waves ($\omega \approx \omega_{pi}$) and ion cyclotron waves ($\omega \approx \Omega_i$) could be studied using the model.

When considering very low frequency waves whose frequency is below the ion gyrofrequency $\omega \ll \Omega_i$, it is possible to use the guiding-center fluid approximation for the motion of ions to retain the lowest order finite Larmor radius effects, while using the guiding-center drift approximation for the electrons as before.[13] Since the fluid approximation is used for the ions, resonant wave-ion interactions are neglected in this model, while the electron-wave interactions along the magnetic field are retained. It is well known that the resonant wave-particle interactions are very important for the stability of low frequency microinstabilities such as drift waves and shear Alfvén waves.[14] Since we are considering the modes with frequencies below Ω_i, we are restricting $k_\parallel/k$ to be even smaller than $(m_e/M)^{1/2}$ for this model more useful than the previously described model[1] where full dynamics was used for the ion particles.

Let us consider, as an example, an electrostatic drift wave destabilized by the Landau damping of the electrons (see Chapters 6 and 13). The frequency of the wave is given[14] by

$$\omega \sim \omega^* = k_\perp \rho_s (\rho_s / L_n) \Omega_i \ , \tag{8.26}$$

where ρ_s is the ion gyroradius at the electron temperature defined by $\rho_s = (T_e/M)^{1/2}/\Omega_i$ and L_n is the density scale length of the plasma perpendicular to an external magnetic field. For $k_\parallel \rho_s \lesssim 1$, which are the most important wavenumbers for drift instabilities, it is clear $\omega \ll \Omega_i$ since $\rho_s/L_n \ll 1$ is usually satisfied. The drift waves can become unstable in the presence of electron Landau damping, and the resonance condition is given by

$$\omega \sim \omega^* \approx k_\parallel v_e \ . \tag{8.27}$$

Then it is found that

$$\frac{k_\parallel}{k_\perp} = \left(\frac{\rho_s}{L_n}\right)\left(\frac{c_s}{v_e}\right) = \left(\frac{m}{M}\right)^{1/2}\left(\frac{\rho_s}{L_n}\right) \ll \left(\frac{m}{M}\right)^{1/2} \ , \tag{8.28}$$

as expected. While the motion of electrons along the magnetic field is followed exactly in this model, one can use a very large time step of integration valid for $\omega \ll \Omega_i$ and $k_{\parallel}/k_{\perp} \ll (m/M)^{1/2}$.

Let us consider an electrostatic drift wave in a uniform magnetic field $\mathbf{B}_0$. Since the quasineutrality is satisfied for a low frequency wave, $\nabla \cdot \mathbf{J} = 0$ follows. From this we find

$$\nabla_{\parallel} \mathbf{J}_{\parallel} = -\nabla_{\perp} \cdot \mathbf{J}_{\perp} = -\nabla_{\perp} \cdot \left(\mathbf{J}_p + \mathbf{J}_{\mathbf{E}\times\mathbf{B}} \right) , \tag{8.29}$$

where the $\mathbf{E}\times\mathbf{B}$ drift current $\mathbf{J}_{\mathbf{E}\times\mathbf{B}}$ and the polarization current are given[13,15] by

$$\mathbf{J}_p = \frac{Mn}{B_0^2}\frac{dE_{\perp}}{dt} = \frac{Mn_0}{B_0^2}\left[\frac{\partial}{\partial t} + \left(\frac{c\mathbf{E}\times\mathbf{B}_0}{B_0^2}\right)\cdot\nabla_{\perp}\right]\mathbf{E}_{\perp} , \tag{8.30}$$

$$\begin{aligned} \mathbf{J}_{\mathbf{E}\times\mathbf{B}} &= en\left(1 + \rho_i^2\nabla^2\right)\frac{c\mathbf{E}\times\mathbf{B}_0}{B_0^2} - en\frac{c\mathbf{E}\times\mathbf{B}_0}{B_0^2} \\ &= en_0\rho_i^2\nabla^2\left(\frac{c\mathbf{E}\times\mathbf{B}_0}{B_0^2}\right) . \end{aligned} \tag{8.31}$$

Since we are considering a plasma in a nearly uniform magnetic field, currents due to the gradient and curvature of the magnetic field can be neglected, as long as we consider the electrostatic waves.

Substituting $\mathbf{J}_p$ and $\mathbf{J}_{\mathbf{E}\times\mathbf{B}}$ to $\nabla \cdot \mathbf{J} = 0$, we find

$$\begin{aligned} &\frac{\partial}{\partial t}\nabla_{\perp}^2\phi + \nabla_{\perp}\cdot\left(en_0\rho_i^2\nabla_{\perp}^2\frac{e\mathbf{E}\times\mathbf{B}_0}{B_0^2}\right) + \nabla_{\perp}\cdot\left[\left(\frac{c\mathbf{E}\times\mathbf{B}_0}{B_0^2}\cdot\nabla_{\perp}\right)\nabla_{\perp}\phi\right] \\ &= \frac{B_0^2}{Mn_0}\nabla_{\parallel}\mathbf{J}_{\parallel}^e , \end{aligned} \tag{8.32}$$

where $\mathbf{J}_{\parallel} = \mathbf{J}_{\parallel}^e$, is assumed, neglecting the field aligned current associated with the ions. As mentioned earlier, $\mathbf{J}_{\parallel}^e$ is determined from the electrons simulated by the guiding-center drift approximations. It is clear from Eq. (8.32) that the ions are now represented by a fluid[13] moving under the influence of an electric field perpendicular to magnetic field with the finite gyroradius effects given by the second term of the left-hand side of Eq. (8.32). In addition to Eq. (8.32), which determines the potential ϕ, Eqs. (7.4) and (7.5) are used for the electron dynamics. Equation (8.32) corresponds to Poisson's equation, where the electrostatic potential is determined from the quasineutrality condition $\nabla \cdot \mathbf{J} = 0$.

In order to study the properties of Eq. (8.32), we consider a dispersion relation for cold electrons to determine $\mathbf{J}_{\parallel}^e$ using a set of electron guiding-center equations. It is found[13] that

$$\omega^2 = \frac{k_\parallel^2}{k_\perp^2}\Omega_e\Omega_i = \frac{k_\parallel^2}{k_\perp^2}\Omega_e^2\frac{m}{M}, \tag{8.33}$$

giving the highest frequency of the system. In order for the drift approximations of the ions to be valid, $\omega^2 \ll \Omega_i^2$ must be satisfied, so that

$$\frac{k_\parallel}{k_\perp} \ll \left(\frac{m}{M}\right)^{1/2} \tag{8.34}$$

which is consistent with Eq. (8.28) valid for drift waves. For the case of warm electrons, it is well known that Eq. (8.32) gives rise to a dispersion relation for drift waves

$$\omega = \frac{\omega^*}{1 + k_\perp^2\rho_s^2}, \tag{8.35}$$

and are destabilized in the presence of wave-electron resonant interactions, included in the present model.

Note, however, since the parallel currents associated with the ions are neglected, ion acoustic waves ($\omega^2 = k_\parallel^2 c_s^2$) are not retained in the model. Furthermore, in the presence of ∇B and curvature drifts, it is necessary to determine the currents perpendicular to magnetic field associated with these drifts. Since those drifts include the particle energy parallel and perpendicular to magnetic fields, equations which determine the ion pressure $P_\perp^i$ and $P_\parallel^i$, must be supplemented,[15] while the pressure of the electrons, $P_\perp^e$ and $P_\perp^e$, may be calculated from the moments of the particle distribution function determined from the electron particles. A set of Eqs. (7.4), (7.5) and (8.32) can be used to study drift wave instabilities in the presence of a density gradient in a uniform magnetic field.

The quasineutral hybrid models we presented in this chapter are conceptually fairly simple and powerful. Thus they have been employed in varieties of application with great success in recent years.

Problems

1. For low frequency waves the quasineutral condition and the adiabatic Boltzman law of the electron density are adequate if the angle between the wavenumber vector of the electrostatic perturbation and the plane normal to the magnetic field is larger than the square root of the mass ratio of electrons to ions. Show why. What happens in the alternative case?
2. Derive Eq. (8.6).

3. Prove the statement associated with Eq. (8.10).
4. Derive the coarse-grained Eq. (8.19) by gyroaveraging. Examine various "hybrids" between Ref. 8 and Ref. 9 by changing the coarse-grainedness.
5. Devise an algorithm for a hybrid code in which the background plasma is treated as an MHD fluid, while there exists a hot component of plasma treated by the particle method. Write down the modified MHD equation for the background and an appropriate kinetic hot particle equation. See an example in Ref. 16.
6. In Problem 5 consider applying the method of subcycling.[17]
7. Discuss the modification of Eq. (8.35) due to the finite Larmor radius effect. Compare the algorithm in Chapter 7.

References

1. H. Okuda, J.M. Dawson, A.T. Lin, and C.C. Lin, Phys. Fluids **21**, 476 (1978).
2. R.H. Berman, D.J. Tetreault, T.H. Dupree, and T. Boutros-Ghali, Phys. Rev. Lett. **48**, 1249 (1982); R.H. Berman, D.J. Tetreault, and T.H. Dupree, Phys. Fluids **26**, 2437 (1983).
3. R.W. Hockney and J.W. Eastwood, *Computer Simulation Using Particles* (McGraw-Hill, New York, 1981), p. 171.
4. S. Riyopoulos and T. Tajima, Phys. Fluids **29**, 4161 (1986).
5. A. Sgro and C.W. Nielson, Phys. Fluids **19**, 126 (1976).
6. J.A. Byers, B.I. Cohen, W.C. Condit, and J.D. Hanson, J. Comp. Phys. **27**, 363 (1978).
7. D.W. Hewett and C.W. Nielson **29**, 219 (1978); D.W. Hewett, J. Comp. Phys. **38**, 378 (1980).
8. D.S. Harned, J. Comp. Phys. **47**, 452 (1982).
9. T. Tajima, J.N. Leboeuf, and J.M. Dawson, J. Comp. Phys. **38**, 237 (1980).
10. M.N. Rosenbluth and M.N. Bussac, Nucl. Fusion **19**, 489 (1979).
11. F. Brunel and T. Tajima, Phys. Fluids **26**, 535 (1983).
12. for review, J.M. Finn and R.N. Sudan, Nucl. Fusion **22**, 1443 (1982).
13. H. Okuda, Space Sci. Rev. **42**, 41 (1985).
14. B.B. Kadomtsev, *Plasma Turbulence* (Academic, New York, 1965) p. 78.
15. A. Hasegawa and M. Wakatani, Phys. Fluids **26**, 2770 (1983).

16. F. Brunel, J.N. Leboeuf, D.P. Stotler, H.L. Berk, and S.M. Mahajan, in *Supercomputers*, eds. F.A. Matsen and T. Tajima (Univ. of Texas Press, Austin, 1986) p. 59.

17. B.I. Cohen, A.B. Langdon, and A. Friedman, J. Comp. Phys. **45**, 15 (1982); J.C. Adam, A. Gourdin Serveniere, and A.B. Langdon, J. Comp. Phys. **47**, 229 (1982).

9

IMPLICIT PARTICLE CODES

There are many problems in plasma physics, such as drift waves and collisionless tearing modes, in which the behavior of a thermal plasma is sensitive to the details of single particle motion, and yet the plasma mode evolves very slowly compared to the characteristic time scales (ω_p^{-1}, Ω^{-1}) of particle motion. The particle simulation methods studied thus far are not efficient enough in the sense that the time step has to be shorter than $\sim \Delta/v_e$, where Δ is the grid spacing and v_e the electron thermal velocity. Explicit time integration methods such as the leap-frog scheme are numerically stable only if the time step Δt is small enough to resolve the fastest physical time scales of waves such as ω_p^{-1} and Ω^{-1}, in addition to the fastest particle motion time scale Δ/v_e. We have considered many algorithms that reduce the physical effects modelled with the objective of eliminating fast time scales. The electrostatic guiding-center model[1] averages over the gyromotion of particles removing the necessity of resolving the gyrofrequency Ω (see Chapter 7). The magnetoinductive model (Chapter 5) neglects radiative effects present in electromagnetic codes by omitting the time derivative of the transverse electric field in Ampere's law. Another model takes the statistical average and derives a fluid description of the plasma (Chapter 6).

Many of the particle simulation techniques we have discussed employ basically explicit time integration of the particle characteristics. In Chapter 6, however, implicit and semi-implicit techniques for fluid simulation were considered. Implicit time integration schemes generally damp the high frequency modes selectively and can be numerically stable for large time steps. It has, however, been acknowledged[2] that such implicit schemes for a particle code would require the unusually costly operation of inverting the matrix associated with particles, in contrast to the usual implicit operation of inverting the matrix associated with grid qualities for a fluid code. This is the main reason we have postponed the discussion of the implicit particle simulation until this chapter although we have already discussed the implicit fluid simulation. However, recent progress[3,4] in implicit particle codes has changed this situation. This progress stems from the technique of extrapolating the particle positions into the future and solving implicitly for the future electric fields instead of the future particle positions themselves.

The various implementations of implicit techniques may generally be put into two broad categories, the so-called moment and direct methods. The moment methods[4,5] use the continuity and momentum moment equations to solve for the electromagnetic fields. The particles are then accelerated by the predicted electromagnetic fields. The source terms in the moment equations are updated from the new particle positions and velocities. The direct methods[6,7] involve the expansion of the different variables at future times in terms of current and previously known values. The equations for the time advanced fields can be solved recursively similarly to the methods previously encountered in the magnetoinductive model. There is an excellent review in this field found in the book *Multiple Time Scales*,[8] including the article by Langdon and Barnes.[9] Reference 9 in particular discusses the direct implicit method in detail.

The remainder of the chapter will investigate more fully the direct methods of implicit particle simulations. Section 9.1 will discuss schemes of first-order accuracy and apply them in detail to a one-dimensional, homogeneous, electrostatic plasma without any magnetic fields. Section 9.2 will discuss second order accurate schemes whose main advantage is improved time filtering and the following sections will discuss the gyrokinetic corrections resulting from the addition of a background magnetic field.[7,9] The final section 9.7 will discuss the important implication of the implicit particle simulation, i.e., the stretch of spatial scales becomes possible, along with the stretch of temporal scales, which has been the original purpose of the implicit simulation. This additional benefit is as important, because the explicit particle simulation has been hampered by the small spatial scales basically tied to the Debye length.

9.1 First Order Accurate Methods

In Chapter 3 we examined a simple example of the implicit time stepping and showed that it is unconditionally stable [see Eq. (3.36)]. Next we consider a one-dimensional, unmagnetized, homogeneous plasma of finite size particles allowing only electrostatic interactions. One form that the equations of motion can take for an implicit scheme is

$$x_j^{n+1} = x_j^n + v_j^{n+1/2}\Delta t \tag{9.1}$$

$$v_j^{n+1/2} = v_j^{n-1/2} + \Delta t \frac{q_j}{m_j} \int dx' S\left(x' - x_j^{n+1}\right) E^*(x') , \tag{9.2}$$

where E^* is some implicit electric field. Poisson's law is used to solve for the electric field

$$\frac{\partial}{\partial x} E^{n+1}(x) = 4\pi \sum_j q_j S\left(x - x_j^{n+1}\right) .$$

The implicit electric field E^* is a function of E^{n+1} and for direct methods is a function of x^{n+1}. We then have the dilemma of trying to solve for x^{n+1} using an electric field that depends on x^{n+1}. The direct methods try to predict E^{n+1} by expansion in terms of previously known values. This expansion places a limit on the size of Δt for accuracy which would otherwise be unconditional.

For example Eq. (9.2) for the fully implicit case becomes

$$v_j^{n+1/2} = v_j^{n-1/2} + \Delta t \frac{q_j}{m^i} \int dx' S\left(x' - x_j^{n+1}\right) E^{n+1}(x') , \tag{9.3}$$

and Poisson's equation becomes

$$\frac{\partial}{\partial x} E^{n+1}(x) = 4\pi \sum_j q_j S\left[x - x_j^n - v_j^{n-1/2}\Delta t \right.$$

$$\left. - \Delta t^2 \frac{q_j}{m_j} \int dx' S\left(x' - x_j^{n+1}\right) E^{n+1}(x')\right] . \tag{9.4}$$

We can expand $S(x^{n+1})$ in Eq. (9.4) by a Taylor series about its free-streaming position $\tilde{x}_j$ where

$$\tilde{x}_j = x_j^n + v_j^{n-1/2}\Delta t . \tag{9.5}$$

Better than that, one can expand Eq. (9.4) around $x_j' = x_j^n + v_j^{n-1/2}\Delta t + \frac{1}{2}E^n \Delta t^2$, including the known acceleration extrapolated into the future. Keeping only the first order terms in the expansion, we obtain

$$\frac{\partial E^{n+1}(x)}{\partial x} + \sum_j 4\pi \frac{q_j^2}{m_j} (\Delta t)^2 \frac{\partial}{\partial x} S(x - \tilde{x}_j) \int dx' S\left(x' - x_j^{n+1}\right) E^{n+1}(x')$$

$$= 4\pi \sum_j q_j S(x - \tilde{x}_j) \equiv 4\pi\tilde{\rho}(x) . \tag{9.6}$$

The truncation of the Taylor series to exclude second and higher order terms is valid only if

$$k\frac{q_j}{m_j}\Delta t^2 \int dx' S\left(x' - x_j^{n+1}\right) E^{n+1}(x')dx' \ll 1 , \tag{9.7}$$

where k is a typical wavenumber. This implies the following restriction on the size of the time step.

$$\frac{\Delta x}{(\Delta t)^2} > \left\{\frac{q_j}{m_j} \int dx' S\left(x' - x_j^{n+1}\right) E^{n+1}(x')\right\}_{\max} , \tag{9.8}$$

or

$$\frac{\Delta x}{\Delta t^2} > (a_j)_{\max} , \tag{9.9}$$

where a_j is the acceleration on the j-th particle. This expansion technique might produce erroneous results if the square of the time step is larger than the grid spacing divided by the maximum acceleration.

The linear stability of these schemes is similar to what we earlier obtained in Chapter 3.

9.2 Implicit Time Filtering

The primary goal of the implicit methods is to model low frequency phenomena accurately without laboring on description of high frequency effects. We want to improve the implicit schemes of Section 9.1 such that damping of unwanted high frequency modes will increase and damping of the desired low frequency modes will decrease. One way to improve the behavior of the frequency is to replace the fully implicit acceleration in the equations of motion with a time averaged acceleration which approximately filters out high frequency responses. One of the schemes[7] is

$$x^{n+1} = x^n + \Delta t v^{n+1/2} \tag{9.10}$$

$$v_j^{n+1/2} = v_j^{n-1/2} + \Delta t \frac{q_j}{m_j} \int dx' S\left(x' - x_j^{n+1/2}\right) \overline{E^n(x')} , \tag{9.11}$$

where the upper bar denotes the recursive averaging process

$$\overline{E^n} = \frac{1}{2}\left(E^{n+1} + \overline{E^{n-1}}\right) . \tag{9.12}$$

It is important to note that Eqs. (9.10) and (9.11) are virtually time centered, thus preserving the second order accuracy present in the leapfrog scheme. These equations of motion along with Poisson's law, Eq. (9.4), form a complete set for advancing the particles for electrostatic problems. The same basic techniques are used again to solve recursively for the time advanced electric field. This method requires more memory and more computation, but has improved frequency filtering over the first-order accurate methods. We can see this by performing another linear stability analysis

$$v_j^{n+1/2} = v_j^{n-1/2} + \frac{q\Delta t}{m}\frac{(E^{n+1} + \overline{E^{n-1}})}{2} ,$$

$$v_j^{n+1/2} \simeq v_j^{n-1/2} - \omega_p^2 \Delta t(x_j^{n+1} + \overline{x_j^{n-1}}) , \tag{9.13}$$

and

$$x_j^{n+1} = x_j^n + v_j^{n+1/2}\Delta t .$$

Again letting x_j^{n+1}, $v_j^{n+1/2} \propto g^n$, then

$$(g-1)x^n - g\Delta t v^{n-1/2} = 0 , \tag{9.14}$$

and

$$(g-1)v^{n-1/2} = -\frac{\omega_p^2\Delta t}{2}\left[g + \frac{1}{2}\left(1 + \frac{1}{2g} + \frac{1}{(2g)^2} + \cdots\right)\right]x^n ,$$

or

$$(g-1)v^{n-1/2} = -\frac{\omega_p^2}{2}\left[g + \frac{1}{2}\sum_{j=0}^{m}\left(\frac{1}{2g}\right)^j\right]x^n . \tag{9.15}$$

If we assume that $|g| > 1/2$ and take the limit where m is very large , then the sum in the above expression reduces to

$$(g-1)v^{n-1/2} + \omega_p^2\Delta t\frac{g^2}{2g-1}x^n = 0 . \tag{9.16}$$

The determinant of the amplification matrix formed by Eqs. (9.14) and (9.16) is set to zero to yield the solution for g

$$(2+\alpha^2)g^3 - 5g^2 + 4g - 1 = 0 , \tag{9.17}$$

where $\alpha \equiv \omega\Delta t$ and ω is the characteristic frequency of the system. The full solution of this stability relation for g is complicated and will not be shown here. A plot comparing the solutions of g of both the fully implicit and the time averaged schemes presented here is shown in Fig. 9.1. Both solutions coincide with the unit circle at $\omega\Delta t = 0$ and therefore neither scheme shows numerical

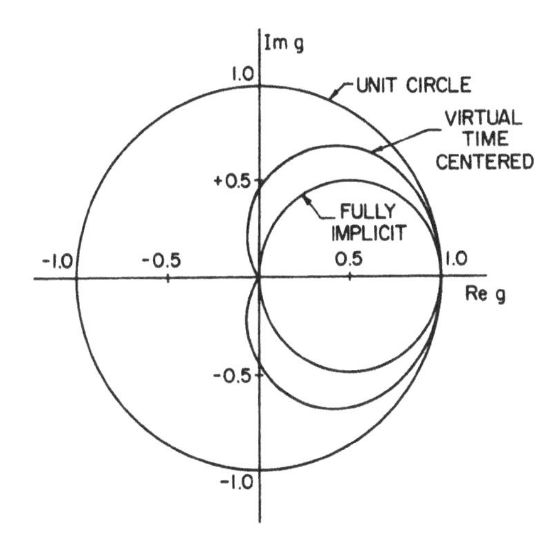

FIGURE 9.1 The complex amplification factor g for the implicit scheme (Ref. 7)

damping in the limit $\omega\Delta t$ goes to zero. Everywhere else, both schemes show damping since $|g|^2 < 1$. As $\omega\Delta t$ increases, the curves move to the left and, as can be seen, damping increases. The virtually time centered second order scheme discussed here has less damping at the lower frequencies.

The above implicit time filtering can be generalized.[7] Consider the following model equations:

$$x^{n+1} = x^n + v^{n+1/2}\Delta t \ , \tag{9.18}$$

$$v^{n+1/2} = v^{n-1/2} - \omega_m^2 \overline{x^n} \Delta t \ , \tag{9.19}$$

where ω_m is the (true) mode frequency and the super bar again represents a combination of various time levels of x^n. This gives rise to the filtering of a certain frequency domain.[10]

For low-frequency simulation, the filtering $\overline{x^n}$ should be chosen so that $\overline{x^n} \sim x^n$ for small frequencies but so that the usual stringent stability condition $|\omega_m| \leq 2\Delta t^{-1}$ is avoided. In this way, desirable properties of the usual explicit leapfrog scheme ($\overline{x^n} = x^n$) are retained for the modes of interest. On the other hand, modes for which $\omega_m \Delta t \to \infty$ (which are not representable on the time grid chosen) should be strongly damped and removed from the simulation. Thus the optimum filtering of x^n to $\overline{x^n}$ would be such that $\overline{x^n} = x^n$ for $|\omega_m| < \omega_{\max}$ with the system response zero for $|\omega_m| > \omega_{\max}$.

This may be approximated by an implicit scheme in which $\overline{x^n}$ depends on time advanced information. Thus, $\overline{x^n}$ should depend on x^{n+1}. (It is shown below that this is necessary for stability as $\omega_m \Delta x \to \infty$.) The desired filtering of x^n is accomplished by a linear filter with a finite number of poles.[11] Thus,

$$\overline{x^n} + \alpha_x \overline{x^{n-1}} + \alpha_x \overline{x^{n-2}} + \cdots + \alpha_{I-1} \overline{x^{n-I+1}}$$

$$= c x^{n+1} + \beta_0 x^n + \beta_1 x^{n-1} + \cdots + \beta_{J-1} x^{n-J+1} . \qquad (9.20)$$

Equations (9.18)-(9.20) are analyzed by the Z transform. It is assumed that $(x^n, v^{n+1/2}) = Z^n(x, v)$, where $Z = \exp(-i\omega\Delta t)$ and ω is the normal mode frequency of the difference system. The following dispersion relation results:

$$(Z-1)^2 + \omega_m^2 Z f(Z) = 0 . \qquad (9.21)$$

In Eq. (9.21) $f(Z)$ is the transfer function from x to $\bar{x}$. Suppose without loss of generality that $I = J$. Equation (9.21) becomes the polynomial equation

$$\left(Z^{I-1} + \alpha_1 Z^{I-2} + \cdots + \alpha_{I-1}\right)(Z-1)^2$$

$$+ \omega_m^2 Z \left(c Z^I + \beta_0 Z^{I-1} + \cdots + \beta_{I-1}\right) = 0 . \qquad (9.22)$$

Equation (9.22) gives the requirements for the undetermined coefficients. For $\omega_m = 0$, there are $I+1$ roots and the α's must be chosen so that the zeros of the first factor in Eq. (9.22) lie inside the unit circle. As $\omega_m \Delta t \to \infty$, there are $I+1$ roots at finite Z provided $c \neq 0$. If c vanishes, at least one of the root loci extends to ∞ and produces an unstable root for $\omega_m \Delta t$ sufficiently large. Thus, $c \neq 0$ is necessary for stability at large $\omega_m \Delta t$. For $\omega_m \Delta t$ large, the roots are $Z = 0$ and the zeros of the last factor in Eq. (9.22). Choosing all β's=0 causes all root loci to approach zero as $\omega_m \Delta t \to \infty$ and produces the maximum desired damping of high-frequency modes.

Finally, for second-order accuracy at $\omega_m \Delta t \sim 0$, $f \sim 1 + O(Z-1)^2$ as $Z \to 1$. These conditions imply that f may be written as

$$f(Z) = \frac{Z^I}{Z^I - (Z-1)^I + \alpha'_1 (Z-1)^{I-1} + \cdots + \cdots \alpha'_{I-2} (Z-1)^2} \qquad (9.23)$$

for $I \geq 2$, where the α'_1 are such that all zeros of the denominator lie inside the unit circle. The transfer function Eq. (9.23) satisfies all of the above requirements and corresponds to a linear filter of the type given by Eq. (9.20).

The choice of $I = 2$ gives the simplest zero parameter scheme[7] in which the bar operation on a quantity Q is given by

$$\bar{Q}^n = \frac{1}{2}\left(Q^{n+1} + \bar{Q}^{n-1}\right) , \qquad (9.24)$$

as we discussed in Eq. (9.12).

For $I = 3$ there is one free parameter α'_1. The solution of the dispersion relation for finite $\omega_m \Delta t$ shows that $\alpha'_1 \leq -0.5$ is required for stability.

The marginal case that minimizes damping of low-frequency modes gives the scheme

$$\bar{Q}^n = \frac{2}{5}\left(Q^{n+1} + 2\bar{Q}^{n-1} - \frac{1}{2}\bar{Q}^{n-2}\right) . \tag{9.25}$$

9.3 Decentered Lorentz Pusher

In a magnetized plasma Eqs. (9.1) and (9.2) are written in the following form[7]

$$\mathbf{x}_j^{n+1} = \mathbf{x}_j^n + \mathbf{v}_j^{n+1/2}\Delta t \tag{9.26}$$

$$\mathbf{v}_j^{n+1/2} = \mathbf{v}_j^{n-1/2} + \overline{\mathbf{A}_\alpha^n(\mathbf{x}_j^n)}\Delta t^2 + \mathbf{v}_j^{n+\gamma} \times \boldsymbol{\Omega}_\alpha \Delta t \ , \tag{9.27}$$

where $\mathbf{A}_\alpha = e_\alpha \mathbf{E}/m_\alpha$ and $\boldsymbol{\Omega}_\alpha = e_\alpha \mathbf{B}/m_\alpha c$ with α referring to the species of particles.

Consider first the last term in Eq. (9.27). Let $\mathbf{v}^{n+\gamma}$ be defined by linear interpolation[7] (a weighted average)

$$\mathbf{v}^{n+\gamma} = \left(\frac{1}{2} + \gamma\right)\mathbf{v}^{n+1/2} + \left(\frac{1}{2} - \gamma\right)\mathbf{v}^{n-1/2} . \tag{9.28}$$

This difference scheme is called decentering and is absolutely stable and second-order accurate for $\Omega_\alpha \Delta t \lesssim 1$. For $\gamma = 0$, Eq. (9.28) reduces to the usual time centered[12,13] Lorentz force equation considered in Chapter 3.

For $\Omega_\alpha \Delta t \gg 1$, Eqs. (9.26) and (9.27) no longer correctly represent gyromotion. Rather, the cyclotron frequency aliases to the Nyquist frequency, $\omega_0 = \pi/\Delta t$, and the orbit degenerates to two positions alternately assumed on alternate time steps. In this case $\gamma > 0$ can be chosen to damp this aliased motion by decentering the difference equations. All effects of gyration are then eliminated from the model and Eqs. (9.26) and (9.27) describe the motion of the zero gyroradius guiding center.[7] Since the perpendicular temperature $T_\perp$ decays (numerical plasma cooling), the energy loss happens. To avoid this, it is customary to assign zero thermal energy in the directions perpendicular to the magnetic field. If the effects of $T_\perp$ are necessary, one has to assign quantities such as magnetic moment μ to particles.

For low-frequency phenomena ($\mathbf{A}_\alpha(\cdot, t)$ slowly varying), Eqs. (9.26) and (9.27) become a difference approximation to the Lorentz force equation. Northrup[14] has shown that this differential equation describes the motion of a guiding center when appropriate gyroaveraged forces are included on the right. The drift motions retained when these additional forces are omitted are those which persist at zero gyroradius. These are the electric drift and the inertial drifts of which the two most significant are the centrifugal and polarization drifts.

The effect of introducing damping for $\gamma > 0$ is to select that solution of the differential equation which corresponds to the initial conditions appropriate for the drift motion solution.[14] Low-frequency motion is little affected by this damping and, while the resulting difference approximation is formally only first-order accurate for $\gamma > 0$, γ may be chosen sufficiently small so that second-order accuracy is effectively attained.

Indeed, upon expanding $\mathbf{v}$ in a Taylor series in t, Eqs. (9.26)-(9.28) give rise to a differential equation whose solution for $\Omega_\alpha \Delta t \gg 1$ is

$$\mathbf{v}_\perp = \frac{\mathbf{E}\times\mathbf{B}}{B^2} + \frac{m\dot{\mathbf{E}}_\perp}{qB^2} - \gamma\Delta t \frac{d}{dt}\left[\frac{\mathbf{E}\times\mathbf{B}}{B^2} + 2\frac{m\dot{\mathbf{E}}_\perp}{qB^2}\right] + O(\Delta t^2) \quad (9.29)$$

$$\dot{v}_\parallel = \frac{q}{m}E_\parallel + O(\Delta t^2) \; . \quad (9.30)$$

Thus, second order accuracy is attained for modes such that

$$\gamma \lesssim |\omega|\Delta t < 1 \; , \quad (9.31)$$

and γ may be chosen very small (typically $\gamma \sim 10^{-1}$) so that the condition of Eq. (9.31) is easily satisfied.

An accomplishment by the decentered Lorentz pusher is to be able to incorporate the polarization drift in $\dot{\mathbf{E}}_\perp/qB^2$ correctly, in addition to the $\mathbf{E}\times\mathbf{B}$ drift. This is significant because a naive addition of the polarization drift to the predictor-corrector scheme in Section 7.6 [Eq. (7.72)] is numerically unstable. An alternate method was developed by W.W. Lee to modify Poisson's equation (7.103), but this latter method becomes complicated when the magnetic field varies spatially. The decentered Lorentz pusher method is general and can handle general geometry without excessive complications.

9.4 Techniques for Direct Implicit Advancing

When Eq. (9.27) is applied to the plasma-field simulation system, a very large system of coupled difference equations results. It is impractical to solve such a large implicit system. Fortunately, it is only necessary to solve a much smaller system for the time filtered potential $\overline{\phi^{n+1/2}}$ without a large matrix inversion. The time-advanced particle information can be obtained in a predictor-corrector iteration.

Field corrector equations may be obtained either by the implicit moment equation method[3,4] or by the direct method,[6] as we discussed in Section 9.2 to a certain extent.

In the predictor-corrector method the particle equations are first solved with an estimated $\overline{\phi^n}$. The first estimate of the potential is provided by $\overline{\phi^n} \sim \phi^{n-1/2}$, then the error in the field equation $\nabla^2\phi = -4\pi e(n_i - n_e)$ is corrected by adjusting $\overline{\phi^n}$, taking account of the change in the right-hand side induced by a change in $\overline{\phi^n}$.

The predictor-corrector iteration level is denoted by a superscript ℓ. Then ${}^\ell n_\alpha^n$ is the raw density resulting from pushing the particles with $\overline{{}^\ell\phi^n}$. That is,

$$ {}^\ell n_\alpha^{n+1} = \sum_{j^\alpha} S\left(\mathbf{x}_g - {}^\ell\mathbf{x}_j^{n+1}\right) , \tag{9.32} $$

where S is the interpolation function and/or shape factor with $\overline{{}^\ell n_\alpha^n}$ given from Eqs. (9.24) or (9.25) with n_α^{n+1} replaced by the above estimate.

Associated with the ℓ-th iteration, there is an error in Poisson's equation

$$ {}^\ell\epsilon = \nabla^2 \overline{{}^\ell\phi^n} - 4\pi e\left(\overline{{}^\ell n_e^n} - \overline{{}^\ell n_i^n}\right) . \tag{9.33} $$

A correction to $\overline{{}^\ell\phi^n}$ is sought such that ${}^{\ell+1}\epsilon \sim 0$.

Let δ denote the first order changes associated with $\ell \to \ell+1$, so that ${}^{\ell+1}\phi^n \sim^\ell \phi^n + \delta\phi$, etc. The change $\delta\epsilon$ is computed to give a linear corrector equation from which $\delta\phi$ is determined. For this δn_α is estimated from Eq. (9.32) using Eqs. (9.26)-(9.28). This technique was discussed in Section 5.4.

For $\gamma \geq 0$, the solution of the difference Eqs. (9.27)-(9.28) may be written as

$$ \mathbf{v}_j^{n+1/2} = \overleftrightarrow{R}_\alpha \mathbf{v}_j^{n-1/2} + \overleftrightarrow{S}_\alpha \overline{\mathbf{A}_\alpha(\mathbf{x}_j^n)}\Delta t . \tag{9.34} $$

The matrices $\overleftrightarrow{R}$ and $\overleftrightarrow{S}$ are given (in diadic notation) by

$$ \overleftrightarrow{R} = d_\alpha\left\{ \left[\overleftrightarrow{I} - (1/4-\gamma^2)\,\Omega_\alpha^2\Delta t^2\right]\overleftrightarrow{I} - \mathbf{\Omega}_\alpha \times \overleftrightarrow{I}\Delta t \right. $$

$$ \left. +(1/2+\gamma)\mathbf{\Omega}_\alpha\mathbf{\Omega}_\alpha\Delta t^2 \right\} , \tag{9.35} $$

$$ \overleftrightarrow{S} = \left\{\overleftrightarrow{I} - (1/2+\gamma)\mathbf{\Omega}_\alpha\Delta t \times \overleftrightarrow{I} + (1/2+\gamma)^2\mathbf{\Omega}_\alpha\mathbf{\Omega}_\alpha\Delta t^2\right\} , \tag{9.36} $$

where

$$ d_\alpha = \left\{1 + (1/2+\gamma)^2\Omega_\alpha^2\Delta t^2\right\}^{-1} . \tag{9.37} $$

From Eqs. (9.26) and (9.34) the estimate is obtained as

$$ \delta^\ell \mathbf{x}_j^{n+1/2} = \delta^\ell \mathbf{v}_\alpha^{n+1/2}\Delta t = \overleftrightarrow{S}_\alpha \frac{(q/m)_\alpha}{(q/m)_e}\nabla\delta\phi\left(\mathbf{x}_j^n\right)(\Delta t)^2 . \tag{9.38} $$

Thus from Eq. (9.32) we obtain

$$\delta\bar{n}_\alpha = -\frac{(q/m)_\alpha}{(q/m)_e} f_n \nabla \cdot \overleftrightarrow{S}_\alpha \sum_{j^\alpha} S\left(\mathbf{x}_g - {}^\ell \mathbf{x}_j^{n+1/2}\right) \nabla\delta\phi(\mathbf{x}_j^{n+1/2}) . \tag{9.39}$$

The field corrector is obtained from ${}^{\ell+1}\epsilon \sim {}^\ell\epsilon + \delta\epsilon = 0$. Thus the equation $\delta\phi$ should satisfy is given by

$$\nabla^2\delta\phi + \nabla \cdot S^2 * \left[\overleftrightarrow{S} \nabla\delta\phi\right] = -{}^\ell\epsilon , \tag{9.40}$$

where $*$ indicates the convolution in real space. In the above $f_n = 1/2$ for (9.24) case and 2/5 for (9.25) case.

An effective method of solving Eq. (9.40) is obtained by writing Eq. (9.40) in **k**-space and "renormalizing" as discussed in Chapter 5. Its Fourier transform is

$$k^2\delta\phi(\mathbf{k}) + |S(\mathbf{k})|^2 \sum_{\mathbf{k}'} \mathbf{k} \cdot \overleftrightarrow{S}(\mathbf{k} - \mathbf{k}') \cdot \mathbf{k}'\delta\phi(\mathbf{k}') ={}^\ell \epsilon(\mathbf{k}) . \tag{9.41}$$

An iterative $m+1$-th approximation of Eq. (9.41) may be written as

$${}^{m+1}\delta\phi(\mathbf{k}) = \frac{{}^\ell\epsilon - |S(\mathbf{k})|^2 \sum_{\mathbf{k}'\neq\mathbf{k}} \mathbf{k} \cdot S(\mathbf{k} - \mathbf{k}') \cdot \mathbf{k}'{}^m\delta\phi(\mathbf{k}')}{k^2 + |S(\mathbf{k})|^2\mathbf{k} \cdot \overleftrightarrow{S}(0) \cdot \mathbf{k}} . \tag{9.42}$$

In implementing the iteration of (9.42), a combination of real and **k** space operations are used to evaluate the convolution sums conveniently. Let ${}^m\xi ={}^{m+1} \delta\phi -{}^m \delta\phi$. Equation (9.48) may be rewritten as

$${}^m\xi(\mathbf{k}) = \frac{{}^m s(\mathbf{k})}{k^2 + |S(\mathbf{k})|^2\mathbf{k} \cdot S(0) \cdot \mathbf{k}} , \tag{9.43}$$

where the residual ${}^m s$ is the difference of the right-hand and left-hand sides of Eq. (9.42) with ${}^m\delta\phi$. See also Section 5.4.

9.5 Direct Implicit Electromagnetic Algorithm

The implicit simulation of the electromagnetic dynamics of plasmas is now considered. The set of field equations are those of Maxwell's equations (5.1)-(5.3) and the equation of motion for a plasma particle is Eq. (5.15). The treatment of Eq. (5.15) is the same as in the case of the electrostatic implicit simulation discussed so far. We decenter the Lorentz force term as discussed in Section 9.3, but the only difference is that we now allow a temporal change in Ω_α (the magnetic fields) in Eq. (9.27). The particle pushing algorithm, however, is identical to Eq. (9.28).

Next, consider the differencing of the field equations. Thus, the difference approximation of Eqs. (5.1) and (5.2) may be written as[15]

$$\mathbf{B}^{n+1/2} = \mathbf{B}^{n-1/2} - \boldsymbol{\nabla} \times \mathbf{E}^n \Delta t \ , \tag{9.44}$$

$$\mathbf{E}^{n+1} = \mathbf{E}^n + c\nabla \times \overline{\mathbf{B}^{n+1/2}} \Delta t - S^2 {*} \mathbf{J}^{n+1/2} \Delta t \ , \tag{9.45}$$

where

$$\mathbf{J}^{n+1/2}(\mathbf{x}) = \frac{1}{2} \sum_j \pm \left[S(\mathbf{x} - \mathbf{x}_j^n) + S(\mathbf{x} - \mathbf{x}_j^{n+1}) \right] \mathbf{v}_j^{n+1/2} \ , \tag{9.46}$$

and

$$\overline{\mathbf{B}^{n+1/2}} = \frac{1}{2} \left(\mathbf{B}^{n+3/2} + \overline{\mathbf{B}^{n-1/2}} \right) \tag{9.47}$$

as in $\bar{E}$.

In Eq. (9.46), the upper (lower) sign is chosen for electrons (ions).

The conservation of charge will not be exact[16] due to truncation errors. To prevent the accumulation of these errors, the electric field $\mathbf{E}^{n+1}$ appearing in Eq. (9.45) is replaced by $\mathbf{E}'^{n+1} = \mathbf{E}^{n+1} + \nabla \psi$ where

$$\nabla^2 \psi = \nabla \cdot \mathbf{E}'^{n+1} - S^2 {*} \rho^{n+1} \ , \tag{9.48}$$

with the normalized density

$$\rho^{n+1}(\mathbf{x}) = \sum_j \pm S(\mathbf{x} - \mathbf{x}_j^{n+1}) \ . \tag{9.49}$$

In this way, the finite difference form of Gauss' law is satisfied exactly; that is,

$$\boldsymbol{\nabla} \cdot \mathbf{E}^{n+1} = S^2 {*} \rho^{n+1} \ . \tag{9.50}$$

The difference scheme consists of Eqs. (9.26) and (9.27) for advancing the particles, Eqs. (9.44) and (9.45) for advancing the fields, and Eqs. (9.46)-(9.49) for determining the source, ${}^{n+1/2}\mathbf{J}$, for the field advance. Particle pushing requires the quantities $\mathbf{A}_\alpha^n$ and $\boldsymbol{\Omega}_\alpha^n$ these are given by

$$\overline{\mathbf{A}_e^n} = \overline{\mathbf{E}^n} \ , \tag{9.51}$$

$$\overline{\mathbf{A}_i^n} = -\frac{m_e}{m_i} \overline{\mathbf{E}^n} \ , \tag{9.52}$$

$$\boldsymbol{\Omega}_e^n = \frac{1}{2} \left(\mathbf{B}^{n-1/2} + \mathbf{B}^{n+1/2} \right) \ , \tag{9.53}$$

$$\boldsymbol{\Omega}_i^n = -\frac{m_e}{m_i} \boldsymbol{\Omega}_e^n \ . \tag{9.54}$$

The electric field used to accelerate the particles is a time-filtered electric field which contains time-advanced information. This removes the constraint associated with the electron plasma oscillations. The Courant condition associated with the propagation of light waves is also removed by introducing time-filtered information into the field advance equations. This time-filtering is introduced into Ampere's law rather than into Faraday's law.[15] In this way, the magnetic field advance remains completly explicit, and the **B**-field at time n needed to advance the particles does not need to be estimated from an implicit prediction. This has the important effect that there are no terms arising in the plasma response from the variation of $\mathbf{B}^n$ with the time advanced electric field.

These implicit equations cannot be solved directly. Either additional information about higher moments needs to be introduced[3,4] or a predictor-corrector method can be developed which directly expresses the plasma response to the time advanced electric field (see Section 9.4). The latter is known as the direct method[15] in analogy to the method given for electrostatic fields only (Section 9.4). For this the particles are pushed at least twice. First, a prediction is made by pushing the particles using some guess for the unknown time-advanced field. The error in satisfying the implicit field equation is computed, and a field adjustment is computed in which the plasma response is estimated from the particle equations of motion. Then a correction to the particle data is made by again pushing the particles with the corrected electric field. While this iteration could be repeated with further field corrections, it has not been found necessary to use more than one corrector pass in ordinary applications.[15]

This method leads to the following algorithm. First, approximate $\mathbf{E}^{n+1}$ by $\tilde{\mathbf{E}} \equiv \bar{\mathbf{E}}^{n-1}$, then all of the particle and field dynamical equations except Eq. (9.45) may be stepped forward in time. Denote the resulting quantities by writing them with a tilde. There will be some errors ϵ^n in the remaining equation which will be defined by

$$2\epsilon^n = \mathbf{E}^n - \tilde{\mathbf{E}} + c\nabla \times \tilde{\mathbf{B}}\Delta t - S^2{*}\tilde{\mathbf{J}}\Delta t = \mathbf{E}^n - \bar{\mathbf{E}}^{n-1} + \frac{c}{2}\nabla$$

$$\times \left(\bar{\mathbf{B}}^{n-1/2} + \mathbf{B}^{n+1/2} - \Delta t \nabla \times \bar{\mathbf{E}}^{n-1}\right) \Delta t - f_n S^2{*}\tilde{\mathbf{J}}\Delta t \,. \qquad (9.55)$$

A correction to $\mathbf{E}'^{n+1}$ is sought so that repushing the particles with $\mathbf{E}'^{n+1} = \tilde{\mathbf{E}} + 2\delta\mathbf{E}^n$ will reduce ϵ^n to zero (note that this definition gives $\mathbf{E}'^n = \bar{\mathbf{E}}^{n-1} + \delta\mathbf{E}^n$) . The retention of only the first order (linear) terms in $\delta\mathbf{E}^n$ leads to the linear field corrector equation

$$\mathcal{L}(\delta\mathbf{E}^n) = \delta\mathbf{E}^n + \frac{\Delta t^2 c^2}{2}\nabla \times \nabla \times \delta\mathbf{E}^n + \frac{f_n}{2}S^2{*}\frac{\delta\mathbf{J}\Delta t}{\delta\mathbf{E}} \cdot \delta\mathbf{E}^n = \epsilon^n \,, \qquad (9.56)$$

where the last term on the left symbolically represents the plasma susceptibility. Equation (9.56) is solved for the adjustment $\delta\mathbf{E}^n$ and all of the dynamical equations, except (9.45), are advanced to the next time step.

The plasma susceptibility is expressed in terms of the perturbation of the particle orbits caused by a change in the time-filtered electric field, $\delta\mathbf{E}^n$. Apparently from Eqs. (9.26) and (9.27), the change in the time-advanced particle data is given by

$$\delta\mathbf{x}_j = \delta\mathbf{v}_j\Delta t \; ,$$

$$\left[\overleftrightarrow{I} + \left(\frac{1}{2}+\gamma_\alpha\right)\Omega_\alpha^n(\mathbf{x}_j^n)\Delta t \times \overleftrightarrow{I}\right]\cdot\delta\mathbf{v}_j = \frac{q_\alpha}{q_e}\frac{m_e}{m_\alpha}\delta\mathbf{E}^n(\mathbf{x}_j^n)\Delta t \; ,$$

or

$$\delta\mathbf{x}_j = \delta\mathbf{v}_j\Delta t = \pm\frac{m_e}{m_\alpha}\overleftrightarrow{S}_\alpha\cdot\delta\mathbf{E}^n\Delta t^2 \; , \tag{9.57}$$

where the upper (lower) sign, as before, is for electrons (ions) and the single particle susceptibility tensor $\overleftrightarrow{S}_\alpha$ was defined in Eq. (9.36).

Combining Eqs. (9.57) and (9.46) gives the relation between the current, whose ratio determines the electric field and the susceptibility $\overleftrightarrow{\chi}$ through

$$\delta\mathbf{J}(\mathbf{x}) = \frac{1}{2}\sum_j\left[S(\mathbf{x}-\mathbf{x}_j^n)+S(\mathbf{x}-\tilde{\mathbf{x}}_j)\right]\overleftrightarrow{S}_\alpha\cdot\delta\mathbf{E}^n(\mathbf{x}_j^n)\Delta t + \mathcal{O}(kv\Delta t) \; . \tag{9.58}$$

The terms omitted from Eq. (9.58) are smaller than those written by the order of $kv\Delta t$, where k is the wavevector.

As in Section 9.4, the plasma response term in Eq. (9.57) may be approximated by simplifying the expression given in Eq. (9.58). The resulting field corrector equation[15] is

$$\mathcal{L}(\delta\mathbf{E}^n) = \delta\mathbf{E}^n + \frac{\Delta t^2c^2}{2}\nabla\times\nabla\times\delta\mathbf{E}^n + \frac{f_n}{2}S^2{*}\overleftrightarrow{\chi}\cdot\delta\mathbf{E}^n = \epsilon^n \; , \tag{9.59}$$

where the susceptibility $\overleftrightarrow{\chi}$ is given by

$$\overleftrightarrow{\chi} = \sum_\alpha\left(\frac{m_e}{m_\alpha}\right)\bar{n}_\alpha\overleftrightarrow{S}_\alpha \; , \tag{9.60}$$

and $\bar{n}_\alpha$ is the point particle number density of species α,

$$\bar{n}_\alpha(\mathbf{x}) = \frac{1}{2}\sum_{j\epsilon\alpha}\left[S(\mathbf{x}-\mathbf{x}_j^n)+S(\mathbf{x}-\tilde{\mathbf{x}}_j)\right] \; . \tag{9.61}$$

Equation (9.59) is equivalent to Eq. (9.40) except for the second term on the left-hand side of Eq. (9.59), which is the vacuum electromagnetic susceptibility (see Section 5.4.)

The field corrector given by Eq. (9.59) is a variable-coefficient, elliptic equation for $\delta\mathbf{E}^n$. The vector corrector, Eq. (9.59) may be solved by a scheme similar to the scalar corrector Eq. (9.40).

The operator of Eq. (9.59) is very anisotropic in both $\mathbf{x}$ and $\mathbf{k}$ space. This anisotropy arises from the variation of the plasma response in the direction parallel and perpendicular to the magnetic field, $\mathbf{B}$, and from the separation of the transverse and longitudinal responses of the vacuum described by Maxwell's equations. This separation may be used advantageously in developing an iterative method for inverting Eq. (9.59). First, note that the operator of Eq. (9.59) may be written as the sum of two operators, $\mathcal{L} = \mathcal{L}_1 + \mathcal{L}_2$, where

$$\mathcal{L}_1 = 1 + \frac{\Delta t c^2}{2} \nabla \times \nabla \ , \tag{9.62}$$

and

$$\mathcal{L}_2 = \frac{f_n}{2} S^2 * \overleftrightarrow{\chi} \ . \tag{9.63}$$

Because of the anisotropy in the operator Eq. (9.59), one of the operators $\mathcal{L}_i$ will be strongly dominant on part of the solution. An approximate adjustment is given by inverting each of the $\mathcal{L}_i$'s on that part of the error where $\mathcal{L}_i$ dominates and selecting the significant portion of the approximate solution obtained. If the magnitude of each operator is estimated parallel and perpendicular to $\mathbf{B}$ and longitudinal and transverse to $\mathbf{k}$, this selection is easily accomplished. There are crossover regions where both operators are comparable. It is easy to see that in these regions, one half of the average of the two inverses will provide a good approximation to the solution.

The estimates for the magnitudes of $\mathcal{L}_i$ on the various parts of the solution can be calculated. All vectors are divided into parts parallel and perpendicular to $\mathbf{B}$ and longitudinal and transverse to $\mathbf{k}$.[15] Writing $\mathcal{F}$ for Fourier analysis (transformation from $\mathbf{x}$ to $\mathbf{k}$ space), the error $\boldsymbol{\epsilon}^n$ is decomposed as

$$\varepsilon^n_{\parallel L} \equiv \mathcal{P}_{\parallel L}\varepsilon^n = \mathcal{F}^{-1}\frac{\mathbf{kk}}{k^2} \cdot \mathcal{F}\frac{\mathbf{BB}}{B^2} \cdot \boldsymbol{\epsilon}^n \ , \tag{9.64}$$

$$\varepsilon^n_{\parallel T} \equiv \mathcal{P}_{\parallel T}\varepsilon^n = \mathcal{F}^{-1}\left[\overleftrightarrow{I} - \frac{\mathbf{kk}}{k^2}\right] \cdot \mathcal{F}\frac{\mathbf{BB}}{B^2} \cdot \boldsymbol{\epsilon}^n \ , \tag{9.65}$$

$$\varepsilon^n_{\perp L} \equiv \mathcal{P}_{\perp L}\varepsilon^n = \mathcal{F}^{-1}\frac{\mathbf{kk}}{k^2} \cdot \mathcal{F}\left[\overleftrightarrow{I} - \frac{\mathbf{BB}}{B^2}\right] \cdot \boldsymbol{\epsilon}^n \ , \tag{9.66}$$

and

$$\varepsilon^n_{\perp T} \equiv \mathcal{P}_{\perp T}\varepsilon^n = \mathcal{F}^{-1}\left[\overleftrightarrow{I} - \frac{\mathbf{kk}}{k^2}\right] \cdot \mathcal{F}\left[\overleftrightarrow{I} - \frac{\mathbf{BB}}{B^2}\right] \cdot \boldsymbol{\epsilon}^n \ , \tag{9.67}$$

where $\mathcal{P}$ denotes a projection operator. The solution $\delta\mathbf{E}^n$ is similarly decomposed.

Only the diagonal parts of $\mathcal{L}_i$ are retained in the iteration. That is, after inverting $\mathcal{L}_i$ on, say, $\epsilon^n_{\|L}$, only the $\|, L$ portion of the solution is retained. Thus, an estimate is required for the eight projections $\mathcal{L}_{i\|L}, \mathcal{L}_{i\|T}, \mathcal{L}_{i\perp L}, \mathcal{L}_{i\perp T}$. These are estimated as

$$\mathcal{L}_{1\|L} \sim 1 , \tag{9.68}$$

$$\mathcal{L}_{1\|T} \sim 1 + \frac{k^2c^2}{2} , \tag{9.69}$$

$$\mathcal{L}_{1\perp L} \sim 1 , \tag{9.70}$$

$$\mathcal{L}_{1\perp L} \sim 1 + \frac{k^2c^2}{2} , \tag{9.71}$$

$$\mathcal{L}_{2\|L} \sim \frac{f_n S^2 \bar{\rho}}{2} , \tag{9.72}$$

$$\mathcal{L}_{2\|T} \sim \frac{f_n S^2 \bar{\rho}}{2} , \tag{9.73}$$

$$\mathcal{L}_{2\perp L} \sim \frac{f_n S^2 \bar{\rho}}{2\left[1 + \left(\frac{1}{2} + \gamma\right) \bar{\Omega}_e\right]} , \tag{9.74}$$

and

$$\mathcal{L}_{2\perp T} \sim \frac{f_n S^2 \bar{\rho}}{2\left[1 + \left(\frac{1}{2} + \gamma\right) \bar{\Omega}_e\right]} , \tag{9.75}$$

where $\bar{\rho}$, $\bar{\Omega}_e$ are mean values for ρ and Ω_e. Using these estimates for the eight projections, an approximate inverse to $\mathcal{L}$ may be written as

$$\mathcal{L}^{-1} \sim \mathcal{M} = \mathcal{P}_{\|L} \left[\left(\frac{\mathcal{L}_{1\|L}}{\mathcal{L}_{1\|L} + \mathcal{L}_{2\|L}} \right)^2 \mathcal{L}_1^{-1} + \left(\frac{\mathcal{L}_{2\|L}}{\mathcal{L}_{1\|L} + \mathcal{L}_{2\|L}} \right)^2 \mathcal{L}_2^{-1} \right] \mathcal{P}_{\|L}$$

$$+ \{\|T\} + \{\perp L\} + \{\perp T\} \tag{9.76}$$

where the terms shown schematically are the same as the first with the appropriate subscripts substituted. This approximate inverse $\mathcal{M}$ is then used to construct an iterative scheme for the solution of Eq. (9.59),

$${}^{l+1}\delta\mathbf{E}^n - {}^{l}\,\delta\mathbf{E} = \mathcal{M}\left\{\varepsilon^n - \mathcal{L}^l \delta\mathbf{E}^n\right\} , \tag{9.77}$$

where l labels the iteration level.

9.6 Gyrokinetic Model (Revisited)

The flexible method of the decentered Lorentz pusher discussed in Section 9.3 has one shortcoming. The decentering algorithm damps out the gyromotion and thus there remains no finite gyroradius effect. Although the algorithm is capable of capturing the polarization drift effect, it fails to retain the finite Larmor radius effect, which leads to some inaccuracies in a certain class of problems including the universal drift waves [see Eq. (7.62)]. In this section we try to retain the finite Larmor radius effect in the algorithm that either dispenses the gyromotion such as the guiding-center approximation (Chapter 7) or damps out the gyromotion such as the decentered Lorentz pusher (Section 9.3).

We want to develop an algorithm as general and flexible as possible that can handle, for example, spatially varying magnetic fields and noncartesian geometry (Chapter 10). An electromagnetic code such as that in Section 9.5 by definition has spatially varying magnetic fields. For this reason, it is often the case that the spatial gyroaveraging [Eq. (7.70)] instead of the Fourier space gyroaveraging [Eq. (7.69)] can be more flexible and general.

It may be instructive to note the following. Let the particle position $\mathbf{x}_j = \mathbf{X}_j + \mathbf{r}_j$, where $\mathbf{X}_j$ is the guiding-center position and $\mathbf{r}_j$ the gyrating position vector for the jth particle around the guiding center (see Fig. 9.2). The charge density contribution due to the jth particle is

$$\rho_j(\mathbf{x}) = \delta(\mathbf{x} - \mathbf{x}_j) = \delta(\mathbf{x} - \mathbf{X}_j - \rho_j) \ . \qquad (9.78)$$

The Fourier space representation of the charge density (9.78) is

$$\rho_j(\mathbf{k}) = e^{-i\mathbf{k}\cdot\mathbf{X}_j} e^{-i\mathbf{k}\cdot\mathbf{r}_j} = e^{-i\mathbf{k}\,\mathbf{X}_j} \sum_{n=-\infty}^{\infty} J_n(k_\perp \cdot r_j) \int e^{-in(\Omega t + \theta_0)} \ , \qquad (9.79)$$

where $\theta(t) = \Omega t + \theta_0$. A time average over the period of the cyclotron motion yields

$$\langle \rho_j(\mathbf{k}) \rangle = e^{-i\mathbf{k}\,\mathbf{X}_j} J_0(k_\perp r_j) \ , \qquad (9.80)$$

where r_j is the Larmor radius of the jth particle. This is the familiar result that the density is represented at the guiding center with the Fourier space form factor $J_0(k_\perp r_j)$. The inverse Fourier transform of Eq. (9.80) is

$$\begin{aligned} \int \langle \rho_j(\mathbf{k}) \rangle e^{i\mathbf{k}\,\mathbf{x}} d\mathbf{k} &= \int d\mathbf{k} J_0(k_\perp r_j) e^{i\mathbf{k}(\mathbf{x}-\mathbf{X}_j)} \\ &= \int dkk\, d\theta J_0(k_\perp r_j) e^{ikR_j \cos(\theta - \varphi)} \ , \end{aligned} \qquad (9.81)$$

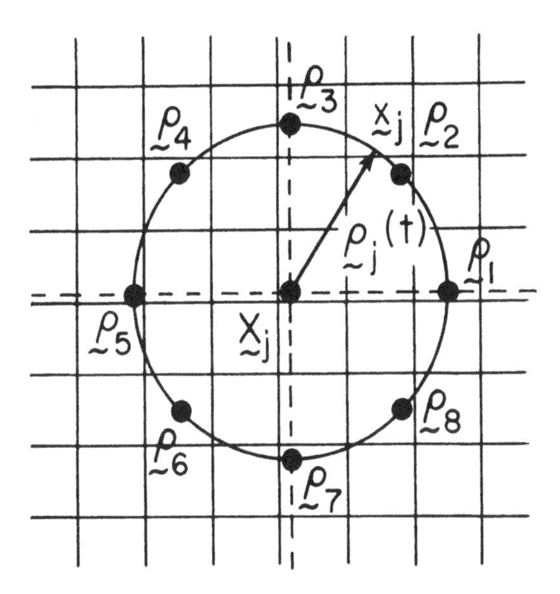

FIGURE 9.2 The ring distribution and its discrete point approximation for an ion with a finite Larmor radius (Ref. 15)

where R_j is the modulus of $\mathbf{R}_j = \mathbf{x} - \mathbf{X}_j$ and $\theta - \varphi$ is the angle between $\mathbf{k}$ and $\mathbf{R}_j$ (see Fig. 9.2). Equation (9.81) is integrated as

$$\int \langle \rho_j(k) \rangle e^{i\mathbf{k}\cdot\mathbf{x}} d\mathbf{k} = \int dk_\parallel dk_\perp dk_\perp d\theta J_0(k_\perp r_j) \sum_n J_n(k_\perp R_j) e^{in(\theta-\varphi)} e^{ik_\parallel x_\parallel}$$

$$= 2\pi \int dk_\parallel dk_\perp k_\perp J_0(k_\perp r_j) J_0(k_\perp R_j) e^{ik_\parallel x_\parallel} = \delta(r_j - R_j)\, \delta\left(x_\parallel - x_{\parallel j}\right) . \tag{9.82}$$

This means that the real space accumulation of the time-averaged charge is expressed as the ring distribution

$$R_j \equiv |\mathbf{x} - \mathbf{X}_j| = r_j . \tag{9.83}$$

This calculation suggests that the gyroaveraged charge density may be best expressed by Eq. (9.83) in real space rather than Eq. (9.80) in Fourier space. The ring charge assignment has to be reciprocated by the ring force approximated by the point distribution on the ring such as 4, 8, or 16 points distributions (see Fig. 9.2). This method has several advantages. The first is that when the magnetic field is spatially dependent such as sheared field, the charge assignment Eq. (9.83) is straightforward and simple, while Eq. (9.80) has the difficulty of the mixture of $\mathbf{k}$ and $\mathbf{x}$. The second is that this method is independent of any small parameter expansion such as the finite Larmor

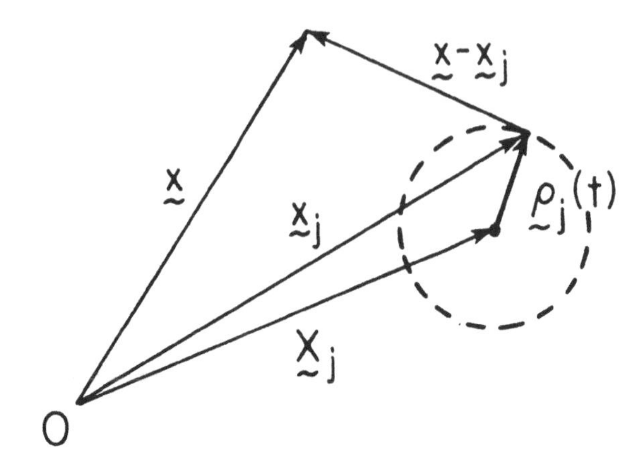

FIGURE 9.3 The field position $\mathbf{x}$, the particle position $\mathbf{x}_j$, and the guiding-center position $\mathbf{X}_j$ (Ref. 15)

radius, amplitude of waves, the plasma β, etc., except that the frequencies of the waves have to be much less than the cyclotron frequency.

Next, let us consider Poisson's equation

$$\nabla^2\phi = 4\pi e \sum_j \delta(\mathbf{x} - \mathbf{X}_j - \boldsymbol{\rho}_j) \ , \tag{9.84}$$

where we temporarily neglect the motion parallel to the magnetic field for simplicity. Following the comment just after Eq. (9.83), we may be able to gyroaverage Eq. (9.84)

$$\begin{aligned} \nabla^2\phi &= -4\pi e \sum_j \langle \delta(\mathbf{x} - \mathbf{X}_j) \rangle \\ &\cong -4\pi e \sum_j \left[\sum_\alpha \delta(\mathbf{x} - \mathbf{X}_j - \boldsymbol{\rho}_j) \right] \ , \end{aligned} \tag{9.85}$$

where $\boldsymbol{\rho}_{j\alpha}$ stands for αth point on the ring $|\mathbf{x} - \mathbf{X}_j| = r_j$ (see Fig. 9.3). Here the perpendicular position of the guiding center $\mathbf{X}_j$ may be split into the $\mathbf{E} \times \mathbf{B}$ drift component and the polarization drift component (if there is no other drift such as the ∇B drift or the curvature drift)

$$\mathbf{x}_j = \mathbf{X}_j^E + \mathbf{X}_j^p \ ,$$

where

$$\dot{\mathbf{X}}_j^E = c\mathbf{E}\times\mathbf{b}/B \quad \text{and} \quad \dot{\mathbf{X}}_j^p = c\dot{\mathbf{E}}_\perp/\Omega B \,. \tag{9.86}$$

If we use the decentered Lorentz pusher, the equation of motion for the guiding center includes the polarization drift

$$\dot{\mathbf{X}}_j = c\frac{\mathbf{E}\times\mathbf{b}}{B} + \frac{c\dot{\mathbf{E}}_\perp}{\Omega B} \,.$$

We then need to write Poisson's equation as

$$\nabla^2\phi = -4\pi e\left(\langle n_i\rangle - n_e\right) \,, \tag{9.87}$$

where $\langle n_i\rangle = \sum_j\sum_\alpha \delta(\mathbf{x}-\mathbf{X}_j-\boldsymbol{\rho}_{j\alpha})$. All we have to do here is to accumulate the ion charge according to the ring distribution or Eq. (9.85). The equation of motion is the decentered Lorentz pusher. Compare this with Poisson's equation in Section 7.7, Eq. (7.103). In Eq. (9.87) we do not need the second term on the left-hand side of Eq. (7.103).

On the other hand, if the equation of motion for the guiding center does not include the polarization drift,

$$\dot{\mathbf{X}}_j = c\frac{\mathbf{E}\times\mathbf{b}}{B} \,,$$

by the algorithm in Section 7.2, the effect of the polarization drift has to be accounted for not in the charge density but in the modified Poisson field. This can be seen as follows. From Eq. (9.84) we have

$$\nabla^2\phi = -4\pi e\sum_j \delta\left(\mathbf{x}-\mathbf{X}_j^E-\rho_j-\mathbf{X}_j^p\right)$$

$$= -4\pi e\sum_j\left[\delta\left(\mathbf{x}-\mathbf{X}_j^E-\rho_j\right) - \frac{\partial}{\partial\mathbf{x}}\mathbf{X}_j^p\cdot\delta\left(\mathbf{x}-\mathbf{X}_j^E-\rho_j\right)\right]$$

$$= -4\pi e\sum_j\sum_\alpha\left[\delta\left(\mathbf{x}-\mathbf{X}_j^E-\rho_{j\alpha}\right) - \frac{\partial}{\partial\mathbf{x}}c\frac{\mathbf{E}(\mathbf{x}_j)}{\Omega B}\delta\left(\mathbf{x}-\mathbf{X}_j^E-\rho_{j\alpha}\right)\right] \,. \tag{9.88}$$

Equation (9.88) can be cast into

$$\nabla^2\phi + \frac{\omega_{pi}^2}{\Omega_i^2}\boldsymbol{\nabla}_\perp\cdot\left[\langle\boldsymbol{\nabla}_\perp\phi(\mathbf{X}_j)/n_0\rangle\sum_j\sum_\alpha\delta\left(\mathbf{x}-\mathbf{X}_j^E-\rho_{j\alpha}\right)\right]$$

$$= -4\pi e\left[\sum_j\sum_\alpha\delta\left(\mathbf{x}-\mathbf{X}_j^E-\rho_{j\alpha}\right) - n_e\right] \,, \tag{9.89}$$

where $\langle\phi(\mathbf{x}_j)\rangle = \int \sum_\alpha \phi(\mathbf{x}')\delta(\mathbf{x}' - \mathbf{X}_j^E - \boldsymbol{\rho}_{j\alpha})d\mathbf{x}'$. Equation (9.89) can be put into the form

$$\nabla^2\phi + \frac{\omega_{pi}^2}{\Omega_i^2}\nabla_\perp \cdot \int d\mathbf{x}' M(\mathbf{x}', \mathbf{x})\nabla_\perp\phi(\mathbf{x}') = -4\pi e\left[\sum_j S\left(\mathbf{x} - \mathbf{X}_j^E\right) - n_e\right], \tag{9.90}$$

where

$$\begin{aligned} M(\mathbf{x}', \mathbf{x}) &= \sum_j n(\mathbf{x}')S\left(\mathbf{x}' - \mathbf{X}_j^E\right) S\left(\mathbf{x} - \mathbf{X}_j^E\right) \\ &\cong \sum_G n(\mathbf{x}_G)S(\mathbf{x}' - \mathbf{x}_G)S(\mathbf{x} - \mathbf{x}_G), \end{aligned} \tag{9.91}$$

where $\mathbf{x}_G$ is the nearest (or weighted averaged) grid position to the $\mathbf{E} \times \mathbf{B}$ guiding-center position $\mathbf{X}_j^E$ and the ring form factor S is now defined as

$$S\left(\mathbf{x} - \mathbf{X}_j^E\right) \equiv \left\langle\delta\left(\mathbf{x} - \mathbf{X}_j^E\right)\right\rangle = \sum_\alpha \delta\left(\mathbf{x} - \mathbf{X}_j^E - \boldsymbol{\rho}_{j\alpha}\right). \tag{9.92}$$

Equation (9.90) takes a form similar to Eq. (7.103).

When the implicit algorithm is taken in the direction parallel to the magnetic field, Poisson's equation takes the form

$$\left(\nabla^2 + \nabla \cdot \int d\mathbf{x}'(\overleftrightarrow{M} + \overleftrightarrow{N}) \cdot \nabla\right)\phi = -4\pi e\left(\langle n_i\rangle - n_e\right), \tag{9.93}$$

where $\overleftrightarrow{N} \propto \frac{\Delta t^2}{2}\delta\mathbf{E}_\parallel n$.

This algorithm now allows the finite gyroradius effect in the charge accumulation (and the force assignment) and depending upon the perpendicular pusher it can accommodate the polarization effect in the pusher or in Poisson's equation. The algorithm is in the same spirit as the implicit algorithm (see Section 9.1), and the resultant algorithm can be totally implicit, including the gyrokinetic effects.

9.7 Large Time Scale – Large Spatial Scale Simulation

Temporal and spatial hierarchies exist in plasma physics, as we mentioned in Chapter 1 and on other chapters. We have perhaps come to the point where we can see an example of this type of structure through the technique of the implicit particle simulation. In the implicit simulation, the time-step determines how much temporal coarse-graining the code carries out. Since the code is, in principle, unconditionally stable regardless of the timestep (as long as appropriate computational treatments have been incorporated), the stability consideration does not determine the choice of timestep, as opposed to the case of explicit simulation. If we choose a large timestep Δt, we damp out oscillations of phenomena faster than the timescale of Δt, while slower oscillations or phenomena are kept intact.

In most physics including plasma physics, solid state physics, or high energy physics, generally the larger the wavenumber of oscillation is, the higher its frequency. (This is, however, only a general tendency, as one may find many examples contrary to this statement in such complex systems as magnetized plasmas.) When we examine slower timescales, therefore, the characteristic spatial scales tend to be larger. The computer simulation code that handles large time scales is required to handle large spatial scales as well. As it turns out, the implicit particle simulation algorithm embodies this desirable property in itself.

In the explicit particle code (see Chapters 3 and 4) the fundamental spatial scales in the simulation system are of the same order of magnitude: $\lambda_{De} \simeq \Delta \simeq a$, where λ_{De} is the electron Debye length, Δ the grid spacing, and a the particle size. In the implicit particle code the size of particles has to be slightly larger than the grid spacing for numerical stability (typically a is 1.5 to 4 times the grid size Δ). An important difference, however, is the possibility of choosing λ_{De} much smaller than the grid spacing:

$$\lambda_{De} \ll \Delta < a \; . \tag{9.94}$$

The precise reason for this is not well understood at present. This newly acquired ability to choose λ_{De} much smaller than Δ allows us to simulate a system physically much larger than otherwise. The total length of the system is given by $M\Delta$; increasing the system length by increasing the number of grid points M is computationally expensive. However, if the size of the unit grid spacing Δ is physically large compared with the Debye length λ_{De}, the simulation system length $M\Delta$ is physically quite large. Thus we are able to simulate large spatial scale systems; however, the number of degrees of freedom remains the same. This resembles the action of a microscope or telescope. As we zoom in towards smaller scales or farther structures, our peripheral vision becomes narrower, while we can discern finer structures. The implicit code can freely zoom in or out by decreasing or increasing the time step, respectively.

There are two ways to make λ_{De} much smaller than Δ. The first method is, effectively, to "cool" the plasma by making the electron thermal velocity v_e small and thus λ_{De} small, while keeping the wavenumber k fixed (or Δ kept fixed). The other method is, effectively, to "stretch" the grid. Fixing the velocity v_e, in this technique we increase the grid spacing Δ so that the wavenumber $k = 2\pi m/L = 2\pi m/M\Delta$ decreases. In a one-dimensional unmagnetized plasma these two methods are essentially identical albeit there are some differences in terms of actual coding. In a higher dimensional magnetized plasma they can be different, as there are other fundamental scale lengths present such as the ion Larmor radius. The dynamics is also quite different in the directions parallel and perpendicular to the ambient magnetic field. Even in the explicit code in this circumstance the Debye length can be substantially small compared with the grid spacing in the perpendicular direction.

In order to examine the timestep filtering effects and the ability of stretching the spatial scales in conjunction with the temporal scale stretch, the plasma dispersion relations obtained from the autocorrelation of time series of fields and the fluctuations of the thermal plasma are examined[15]. A similar study was conducted in Chapter 5. Figures 9.4, 9.6, and 9.7 show the dispersion relations, while Fig. 9.5 displays the spectrum of fluctuations.

The test parameters[15] are (system length) $N_x = 256\Delta$, (number of particles) $N_0 = 10240$, (particle size) $a = 1.5\Delta$, where Δ is the grid spacing with Gaussian particle shape represented by $S = \exp(-k^2a^2/2)$, (ion-to-electron mass ratio) $m_i/m_e = 100$, (electron-to-ion temperature ratio) $T_e/T_i = 20$. The constant magnetic field $\mathbf{B}_0$ lies in the x-z plane at an angle ϕ from the x-axis which is the electric field wavevector direction. Its strength is such that $\Omega_e/\omega_{pe} = 1.0$. The speed of light c is set at $c/\omega_{pe}\Delta = 1.0$ and the Alfvén velocity scales as $v_A/c = \Omega_i/\omega_{pi} = 0.1$. Electrons and ions are initially loaded uniformly on the one-dimensional spatial grid with zero perpendicular thermal velocity. The time decentering parameters are such that $\gamma_e = \gamma_i = 0.1$ and both electrons and ions are treated implicitly.

The parallel electron and ion distribution functions are taken to be Maxwellian. The thermal velocity of the electrons is chosen such that $v_e/\omega_{pe}\Delta = 5 \times 10^{-2}$, so that the electron Debye length is $\lambda_{De}/\Delta = 5 \times 10^{-2}$. The timestep can then be chosen as $\omega_{pe}\Delta t = 10, 10^2, 10^3$ with grid spacing $\Delta = \delta, 10\delta, 100\delta$, where δ is the original grid spacing. This represents a factor of 10^2 to 10^4 increase over that allowed for an explicit code in which ω_{pe} has to be resolved. The calculation comprises 8,192 time steps so that many ion-acoustic and Alfvén wave periods are resolved.

For purely parallel propagation ($\phi = 0$), the electrostatic and electromagnetic low frequency modes predicted by a theoretical analysis are the ion-acoustic waves, the whistler waves and the shear Alfvén waves. Higher frequency modes are also written in Figs. 9.4, 9.6 and 9.7. However, no mode energy in high frequency branches are observed. For one set of parameters, the simulation plasma can carry waves only within a range of wavelengths delimited by the system length and spatial grid size or particle size. Since we use two independent values of Δ or two sets of values of k, but with the same ratio $\lambda_{De}/\Delta = 0.05$, all of the results from these two runs are combined into one figure over a range of three decades of $k\lambda_{De}$. The measured frequencies thus obtained are shown by triangles ($\Delta/\delta = 10$) and circles ($\Delta/\delta = 1$) in Fig. 9.4. The frequency, ω/ω_{pe}, varies over a four decade range. Such a wide

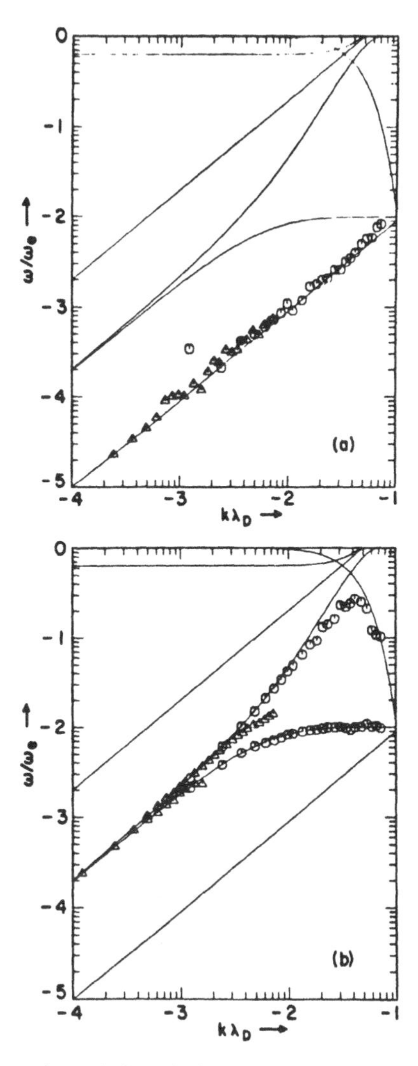

FIGURE 9.4 The dispersion relations in the parallel propagation (a) the electrostatic mode, (b) the electromagnetic modes (Ref. 15)

range of simulation in wavenumber and frequency has become possible for the first time in particle simulation by the method of the implicit particle simulation.

The electromagnetic modes can be extracted from the cold plasma dis-

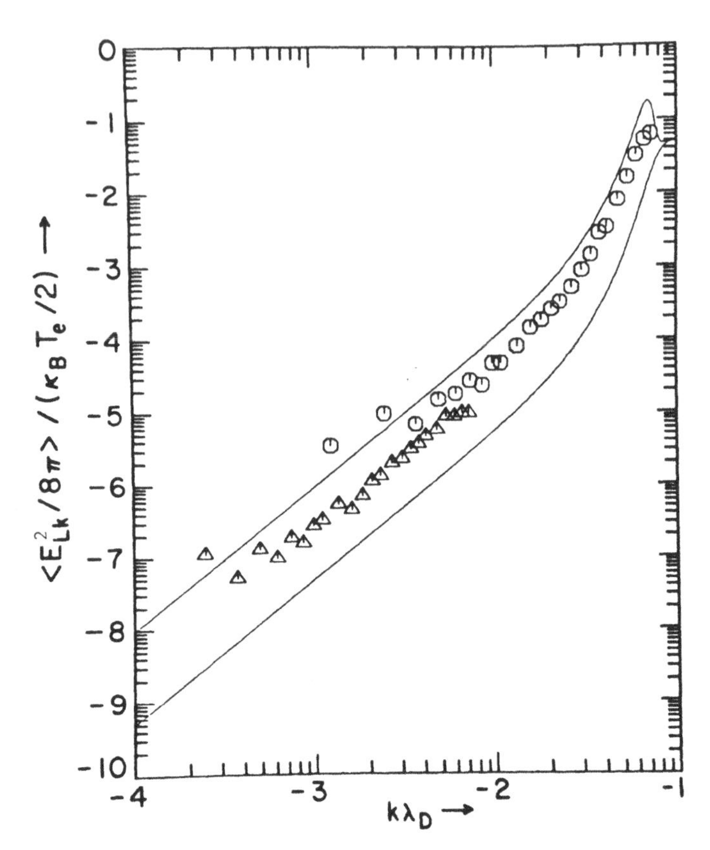

FIGURE 9.5 Electrostatic field fluctuation spectrum as a function of wavenumber. The upper curve is a theory curve with guiding-center electrons, while the lower one with quasineutral approximation (Ref. 15)

persion relation

$$\tan^2\phi = -\frac{\left(\frac{1}{n^2}-\frac{1}{\varepsilon_R}\right)\left(\frac{1}{n^2}-\frac{1}{\varepsilon_L}\right)}{\left(\frac{1}{n^2}-\frac{1}{\varepsilon}\right)\left(\frac{1}{n^2}-\frac{1}{2}\left(\frac{1}{\varepsilon_R}+\frac{1}{\varepsilon_L}\right)\right)}, \tag{9.95}$$

$$n^2 = k^2c^2/\omega^2 \tag{9.96}$$

$$\varepsilon_{R,L} = 1-\frac{S^2\omega_e^2}{\omega(\omega\mp\Omega_e)}-\frac{S^2\omega_i^2}{\omega(\omega\pm\Omega_i)}, \tag{9.97}$$

$$\varepsilon = 1 - (\omega_{pe}^2 + \omega_{pi}^2)S^2/\omega^2 . \tag{9.98}$$

The ion-acoustic dispersion relation for $\omega \ll \Omega_i$ is obtained as

$$\varepsilon = 1 + \frac{S^2}{k^2\lambda_{De}^2}\left(1 + \frac{\omega}{\sqrt{2}k_{\|}v_e} Z\left(\frac{\omega}{\sqrt{2}k_{\|}v_e}\right)\right)$$
$$+ \frac{S^2}{k^2\lambda_{Di}^2}\left(1 + \frac{\omega}{\sqrt{2}k_{\|}v_i} Z\left(\frac{\omega}{\sqrt{2}k_{\|}v_i}\right)\right) , \tag{9.99}$$

with $k_{\|} = k\cos\phi$. An approximate solution to Eq. (9.99) gives $\omega/\omega_{pe} \simeq (m_e/m_i)^{1/2} k_{\|}\lambda_{De}[(1 + 3T_i/T_e)/(1 + k^2\lambda_{De}^2 S^{-2})]^{1/2}$. The dispersion relations ω/ω_{pe} versus $k\lambda_{De}$ obtained from numerical solutions of these equations with $\phi = 0$ are shown as solid curves in Fig. 9.4. Excellent agreement is evident with the low frequency branches, the first curve from the bottom being the ion-acoustic one, the second the shear Alfvén or Alfvén ion cyclotron, and the third the whistler branch.

The time averaged electrostatic energy per wavenumber $\langle E_{Lk}^2/8\pi\rangle$ normalized to the thermal energy per degree of freedom $\kappa_B T_e/2$ (with κ_B Boltzmann's constant), or fluctuation spectrum, is shown in Fig. 9.5 for the same two runs with $\phi = 0$. For $T_e \gg T_i$ and $\omega \ll \omega_{pe}, \Omega_e$, the fluctuation spectrum of a two temperature Maxwellian plasma can be written as[7]

$$\frac{\langle E_{Lk}^2/8\pi\rangle}{\kappa_B T_e/2} \sim \frac{k^2\lambda_{De}^2 S^2}{1 + k^2\lambda_{De}^2 S^2} . \tag{9.100}$$

The fluctuation spectrum predicted by Eq. (9.100) is plotted as the upper curve in Fig. 9.5. The fluctuation spectrum measured in the simulations indicated by triangles ($\Delta/\delta = 10$) and circles ($\Delta/\delta = 1$) closely follows the prediction of Eq. (9.100). The lower curve is the spectrum one would obtain with adiabatic Boltzman electrons, i.e., in the absence of electron Landau damping.[7] This curve is indeed reproduced by the quasineutral electrostatic hybrid code discussed in Section 8.1.

As the angle ϕ between the wavenumber (or x-axis) and the magnetic field is increased from 0° to 90°, the whistler wave merges with the compressional Alfvén wave. The shear Alfvén wave frequency goes to zero in the limit of purely perpendicular propagation. For $0 < \phi < 90°$, the three branches coexist. Results for propagation with $\phi = 45°$ are displayed in Fig. 9.6. The simulation frequencies, measured in the three components of the electric field E_x, E_y and E_z are plotted as circles. Only one value for the grid spacing, Δ, $\Delta/\delta = 1$, is used so that the wavevector varies over two decades and the frequency over four. Excellent agreement is obtained between simulation and the theoretical dispersion relations of Eqs. (9.95) and (9.99) represented by curves in Fig. 9.6, the first from the bottom being the ion-acoustic branch, the

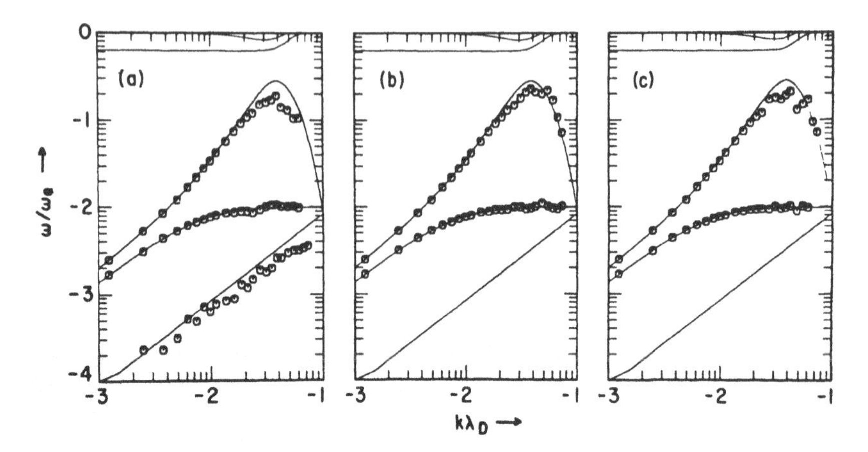

FIGURE 9.6 Dispersion relations for modes with the angle between **k** and **B** as 45° (a) $E_L(= E_x)$, (b) E_y, and (c) $E_{\|}(= E_z)$ (Ref. 15)

second the Alfvén (or Alfvén ion cyclotron) branch and the third the whistler or magnetosonic branch.

For purely perpendicular propagation, only the compressional Alfvén or magnetosonic wave is predicted as a low frequency mode. The dispersion relation for perpendicular propagation ($\phi = 90°$) is presented in Fig. 9.7. Three values of the stretch factor, $\Delta/\delta = 1$, 10 and 100, are used in the simulations[15] and the results from these three runs again combined into one diagram. The wavevector varies over four decades, the frequencies represented by plusses ($\Delta/\delta = 100$), triangles ($\Delta/\delta = 10$) and circles ($\Delta/\delta = 1$).

The following observations are suggested by the results shown in Figs. 9.4-9.7. As the time step Δt is raised (as is possible here when the unit grid spacing Δ is increased or the electron thermal velocity v_e is lowered), the observable frequency range, $\omega_{\min} < \omega < \omega_{\max}$, shifts toward smaller ω in accordance with $\omega_{\max}\Delta t < 1$ and $\omega_{\min} = \omega_{\max}/N_t$, with N_t the total number of time steps in a run. This is the natural frequency filtering intrinsic to the implicit algorithm. Raising the time step Δt also has the consequence that the resolvable wavevector range $k_{\min} < k < k_{\max}$ shifts toward smaller k. The maximum wavevector is set by $k_{\max} v \Delta t < 1$ and for waves such that $\omega = k v_{ph}$, (as is the case here with v_{ph} the phase velocity of the wave), also by $k_{\max} v_{ph} \Delta t < 1$ since $\omega_{\max}$ $(= k_{\max} v_{ph})$ $\Delta t < 1$. The minimum resolvable wavenumber $k_{\min}$ is set either by the limit of the spatial resolution $k_{\min} = 2\pi/N_x\Delta$ or by the limit of the temporal resolution $k_{\min} = \omega_{\min}/v_{ph}$ for waves such that $\omega = k v_{ph}$, whichever is larger. Therefore, increasing Δt within the constraints of stability and accuracy provides a natural zoom towards

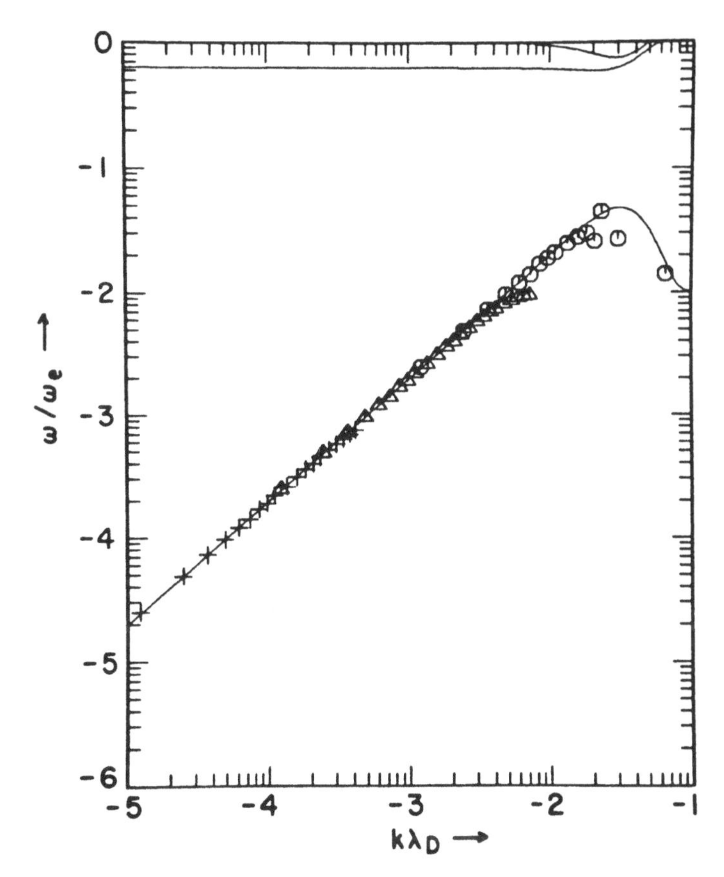

FIGURE 9.7 Dispersion relation of modes propagate perpendicular to **B**. The theory curve (magnetosonic mode) agrees with four decades of simulation points. The implicit method opened up such a wide range (Ref. 15)

both the longer time scales and the larger space scales, as we have discussed earlier. We note that raising Δt with lowering v_e has the consequence that the wavevector k of the scope of the simulation becomes small with respect to $k_{De}(=\lambda_{De}^{-1}) = \omega_{pe}/v_e$ since $k_{De} \to \infty$ as $v_e \to 0$. Raising Δt with increasing Δ makes $k = 2\pi m/N_x\Delta$ small with respect to k_{De} simply because Δ is larger.

The method of implicit simulation is showing at this time great promise to attain two of the goals of simulation outlined in Chapter 1 [Eqs. (1.53) and (1.54)]. Future efforts will help this powerful new technique to mature. Meanwhile, the discussion of the third goal, Eq. (1.55), will be picked up in Chapter 11.

Problems

1. Solve a harmonic oscillator $\ddot{x} = -kx$ by an implicit time stepping and study the numerical stability.
2. Equation (9.6) is similar to Eq. (5.47). Devise an iterative solution (renormalization) similar to one employed in the magnetoinductive code.
3. The derivation from Eq. (9.33) to Eq. (9.40) involves an approximation related to the term with the form $\sum_j S(x - x_j)S(x' - x_j)f(x')$. Clarify the approximation and justify it or express when it is justifiable.
4. Include finite gyroradius effects in Eqs. (9.26) and (9.27) by adding additional operation to Eq. (9.27) and simultaneously modifying the charge-current source equations.
5. Instead of Eq. (9.45) we could time-filter Eq. (9.44). Discuss some possible ramifications in this case.
6. Cite some counterexamples to the statement made in the beginning of Section 9.7. That is, show cases of the plasma dispersion relation where the group velocity of the particular oscillation is negative.
7. Derive the lower theoretical curve in Fig. 9.5.
8. Demonstrate the assertion that the decentered Lorentz force algorithm yields the correct $\mathbf{E} \times \mathbf{B}$ drift and polarization drift in a large Δt.
9. Since the decentering scheme (Section 9.3) eliminates the gyromotion, it becomes necessary to put in the (electron) magnetization current $\mathbf{J}_m = e\nabla \times \sum_j \mathbf{b}\mu_j S(\mathbf{x} - \mathbf{x}_j)$ in high β plasmas, where μ_j is the j-th particle's magnetic moment in the magnetic field. Discuss a possible implementation of this current in the implicit electromagnetic code.
10. List the advantages of the real space gyroaveraging over the Fourier space gyroaveraging.
11. Discuss the form of the operator $\overleftrightarrow{N}$ in Eq. (9.93).
12. Execute a part of integral corresponding to the energy integral in Eq. (9.91), assuming the Maxwellian distribution.

References

1. W.W. Lee and H. Okuda, J. Comp. Phys. **26**, 139 (1978).
2. A.B. Langdon, J. Comp. Phys. **30**, 202 (1979).
3. R.J. Mason, J. Comp. Phys. **41**, 233 (1981).

4. J. Denavit, J. Comp. Phys. **42**, 337 (1981).

5. J.U. Brackbill and D.W. Forslund, J. Comp. Phys. **46**, 271 (1982).

6. A. Friedman, A.B. Langdon and B.I. Cohen, Comments on Plasma Phys. and Contr. Fusion **6**, 225 (1981).

7. D.C. Barnes, T. Kamimura, J.N. Leboeuf, and T. Tajima, J. Comp. Phys. **52**, 480 (1983).

8. J.U. Brackbill and B.I. Cohen, eds. *Multiple Time Scales* (Academic, New York, 1985).

9. A.B. Langdon and D.C. Barnes, in *ibid.* p. 335.

10. B. Gold and C.M. Rader, *Digital Processing of Signals* (McGraw-Hill, New York, 1969).

11. B.I. Cohen, A.B. Langdon, and A. Friedman, J. Comp. Phys. **46**, 15 (1982).

12. O. Buneman, J. Comp. Phys. **1**, 517 (1967).

13. J.P. Boris, in *Proc. Fourth Conf. on Numerical Simulation of Plasma*, Boris and Shanny eds. (U.S. Gov. Print., Washington, D.C., 1970) p. 4.

14. T.G. Northrup, Ann. Phys. **15**, 19 (1961).

15. T. Kamimura, E. Montalvo, D.C. Barnes, J.N. Leboeuf, and T. Tajima, to be published in J. Comp. Phys.

16. A.B. Langdon and B. Lasinski, in *Methods in Computational Physics*, B. Alder et al., eds., vol. 16 (Acadedmic, New York, 1976) p. 327.

10

GEOMETRY

In the previous chapters, we discussed various basic algorithms and techniques. These are not complicated by the otherwise necessary additional attention to noncartesian geometrical effects. For example, in cartesian geometry, the finite differencing of magnetic fields to make them divergence free is trivial (see Chapter 6); the Fourier transform technique is another evident example of simplicity in cartesian geometry. In this chapter, we shall concentrate on methods to treat noncartesian geometrical effects in curvilinear coordinates. In plasma physics, geometry plays an essential role partly because the Coulombic interaction is long-ranged; the influence of boundaries and the shape of the plasma and device is directly related to the properties of the state of the plasma. In this sense, the behavior of plasmas can be said to be (often) essentially global. In contrast, in many examples of solid-state matter in which the interaction is short-ranged, i.e., between adjacent lattice points, the behavior of the material can be said to be essentially local. A very important exception to this in a force field that is short-ranged is the case of phase transiton. In this case, even when the interaction is short-ranged, the correlation length becomes long-ranged;[1] the boundary and also the geometrical effects could become important. A similar emphasis on geometry could come from astrophysics, where the interaction of gravitational force is again Coulombic and long-ranged. In fact, it can be argued that the theory of gravitation, in general, and the general theory of relativity, in particular, are essentially geometrical.[2,3,4]

We will discuss applications of generalized coordinates and tensor analysis to the three dimensional particle and magnetohydrodynamic codes. Here we discuss the toroidal case in particular, then, we shall touch on flux coordinates. The emphasis on toroidal geometry is partly due to the experimental importance of toroidal plasma devices such as tokamaks,[5] stellarators,[6,7] betatrons,[8] toroidal z-pinches,[9] and ELMO bumpy tori.[10] It is partly due to the belief that this geometry is intrinsically important next to the uniform system (no such terrestrial device exists), the spherically symmetric system (again no terrestrial plasma applies here because of the so-called bad curvature trouble everywhere in this geometry, besides the impossibility of such a field structure), and the cylindrically symmetric system (which may suffer from the open ends).

Although no one knows at this moment what the best geometric configuration is, it may be argued that the more symmetric the configuration is, the easier the plasma confinement. Because of the inherent finiteness and stability of plasmas, the most symmetrical systems (the uniform system and the spherically symmetric system) are rejected. With the confining magnetic field along the axis, the (infinitely long) cylindrical system loses only one directional symmetry, i.e., the axial directon (one dimension) is separate from the others (two dimensions). The otherwise symmetric particle dynamics now bifurcates into the axial motion and the perpendicular gyrational motion. When we twist the straight cylinder into a toroid, we further destroy the symmetry which was preserved in the cylindrical system, namely the split of the radial direction and the vertical direction. The only remaining symmetry is in the toroidal direction in this case. An example is an axisymmetric tokamak. One can further break down the symmetry by introducing "irregularity" along the toroidal direction. Examples are a non-axisymmetric tokamak and a stellarator. The above-mentioned steps are the process of symmetry breaking. The first step introduces a straight orbit along the axial direction and gyromotion in two other directions, a breakdown of three dimensional uniform motion. The second step creates toroidally untrapped particles and trapped particles. The third step destroys these (basically) closed orbits. Toroidal devices may be classified into three broad categories based on the dominant magnetic field component. The first class is the one with the toroidal magnetic field as the dominant component: this includes tokamaks, stellarators, and ELMO bumpy tori. The second class is the one with the poloidal magnetic field dominating: toroidal z-pinches belong to this. The third class is the one with the vertical magnetic field as the main field: betatrons, large aspect ratio field reversed configurations (FRC), and mirrors are in this category.

In the following sections we develop techniques to handle such geometry in MHD, kinetic, and transport levels.

10.1 MHD Particle Code

Useful reviews of tensor analysis necessary for general coordinates may be found in Refs. 11-13. Using the results of these, we shall discuss the method of general coordinates and particularly that of torus in the following for plasma simulation. In this section we discuss MHD simulation. We discuss the MHD equations in curvilinear coordinates[14] in general. We refer the reader to Chapter 6 for discussions of basic MHD. As an example, we shall focus on the MHD particle code,[15] one of the methods developed to simulate the magnetohydrodynamic equations. The fluid elements are treated like finite size particles that behave in a Lagrangian fashion and the momentum equation of MHD is used as the force equation to push the fluid quasiparticles. This section discusses the techniques of formulating and numerically solving the MHD equations in a MHD particle code. In the first part we will employ the formalism of tensor analysis. In the next part we will apply this analysis to the specific case of cylindrical coordinates and in the final part we will consider the role of boundary conditions in finite difference methods. Although because of the author's familiarity we specifically discuss the MHD particle code, this discussion applies in other cases equally well.

The equation for pushing the fluid-like MHD particles combines the momentum equation with Ampere's Law to obtain

$$\rho \frac{d\mathbf{v}}{dt} = -\nabla p + \frac{1}{4\pi}(\nabla \times \mathbf{B}) \times \mathbf{B} \ . \tag{10.1}$$

Since $\mathbf{B}$ is divergence-free, Eq. (10.1) can be rewritten

$$\rho \frac{d\mathbf{v}}{dt} = -\nabla \left(p + \frac{B^2}{8\pi} \right) + \nabla \cdot \left(\frac{\mathbf{BB}}{4\pi} \right) \ . \tag{10.2}$$

The equation for calculating the magnetic field combines Faraday's Law with Ohm's Law (including resistivity) to yield

$$\frac{\partial \mathbf{B}}{\partial t} = \nabla \times (\mathbf{v} \times \mathbf{B}) + \frac{c^2}{4\pi} \nabla \times (\eta \nabla \times \mathbf{B}) \ , \tag{10.3}$$

where η is the resistivity. An equation of state for the pressure p, such as the ideal gas law

$$p = cn^{\gamma} \ , \tag{10.4}$$

where c is a constant, n is the density, and γ is the specific heat, closes these equations.

These expressions will be evaluated in a contravariant basis. The full time derivative on the right-hand side of Eq. (10.2) is identified with the intrinsic derivative (equivalent to the Lagrangian or convective derivative)

$$\left(\frac{d\mathbf{v}}{dt}\right)^i = \frac{\delta v^i}{\delta t} = v^{i,k}\frac{dx^k}{dt} = \frac{dv^i}{dt} + \left\{ \begin{matrix} i \\ rk \end{matrix} \right\} v^r v^k \ . \tag{10.5}$$

Here the Christoffel symbols have been introduced:

$$\text{First kind} \quad [ij,k] \equiv \frac{1}{2}\left\{ \frac{\partial g_{ik}}{\partial x^j} + \frac{\partial g_{jk}}{\partial x^i} - \frac{\partial g_{ij}}{\partial x^k} \right\} \tag{10.6}$$

$$\text{Second kind} \quad \left\{ \begin{matrix} \ell \\ ij \end{matrix} \right\} \equiv g^{\ell k}[ij,k] \ . \tag{10.7}$$

The contravariant fundamental tensor (or metric tensor) g^{ik} is conjugate to g_{ij}, which is defined by the line element (distance) ds

$$ds^2 = g_{ij}dx^i dx^j \ , \tag{10.8}$$

and

$$g_{ij}g^{ik} = \delta_j^k \ . \tag{10.9}$$

The first term on the right-hand side of Eq. (10.2) is written in a covariant basis as

$$\left[\nabla\left(p + \frac{B^2}{8\pi}\right)\right]_k = \frac{\partial}{\partial x^k}\left(p + \frac{B^2}{8\pi}\right) \ . \tag{10.10}$$

The second term on the right-hand side is written in a contravariant basis as

$$\begin{aligned} [\nabla\cdot(\mathbf{BB})]^\ell &= (B^k B^\ell)_{,k} = \frac{\partial}{\partial x^k}(B^k B^\ell) + \left\{ \begin{matrix} k \\ rk \end{matrix} \right\} B^r B^\ell + \left\{ \begin{matrix} \ell \\ rk \end{matrix} \right\} B^k B^r \\ &= \frac{\partial}{\partial x^k}(B^k B^\ell) + \frac{\partial}{\partial x^r}(\ell n\sqrt{g})\, B^r B^\ell + \left\{ \begin{matrix} \ell \\ rk \end{matrix} \right\} B^k B^r \\ &= \frac{1}{\sqrt{g}}\frac{\partial}{\partial x^k}(\sqrt{g}B^k B^\ell) + \left\{ \begin{matrix} \ell \\ rk \end{matrix} \right\} B^k B^r \end{aligned} \tag{10.11}$$

where the comma followed by a subscript indicates a differentiation and

$$g = \det(g_{ij}) = g_{i\alpha}G^{i\alpha} \ ,$$

where the sum is only in α and $G^{i\alpha}$ is the cofactor of $g_{i\alpha}$. Using Eq. (10.5) and (10.11) and a contravariant form for Eq. (10.10), Eq. (10.2) becomes

$$\begin{aligned} \rho\frac{dv^i}{dt} &+ \left\{ \begin{matrix} i \\ rk \end{matrix} \right\} v^r v^k = -g^{ik}\frac{\partial}{\partial x^k}\left(p + \frac{B^2}{8\pi}\right) \\ &+ \frac{1}{4\pi\sqrt{g}}\frac{\partial}{\partial x^k}(\sqrt{g}B^k B^i) + \frac{1}{4\pi}\left\{ \begin{matrix} i \\ rk \end{matrix} \right\} B^r B^k \ . \end{aligned} \tag{10.12}$$

For many physical problems, it is more convenient to use an orthogonal coordinate system. The off-diagonal elements of the metric tensor are zero for orthogonal systems, i.e.,

$$
\begin{aligned}
g_{ii} &= h_i^2 \ , \\
g_{ij} &= 0 \quad , \quad (i \neq j)
\end{aligned}
\tag{10.13}
$$

where h_i is the differential scale factor. It can also be shown for orthogonal systems that

$$
\begin{aligned}
\left\{ \begin{array}{c} i \\ jj \end{array} \right\} &= -\frac{h_j}{h_i^2}\frac{\partial h_j}{\partial x^i} \ , \\
\left\{ \begin{array}{c} i \\ ij \end{array} \right\} &= \frac{1}{h_i}\frac{\partial h_i}{\partial x^j} \ , \\
\left\{ \begin{array}{c} i \\ jk \end{array} \right\} &= 0 \quad , \quad \sqrt{g} = h_1 h_2 h_3 \equiv h \ ,
\end{aligned}
\tag{10.14}
$$

where i, j and k are not equal and where the summation convention does not apply.

At this point it is convenient to define physical components of a tensor $\bar{A}_j$, such that

$$A^2 = \bar{A}_j \bar{A}_j = h_j^2 A^j A^j = \frac{1}{h_j^2} A_j A_j \ ,$$

and such that

$$\bar{A}_j = h_j A^j \equiv \frac{1}{h_j} A_j \ . \tag{10.15}$$

The physical components are a useful basis for physically interpreting a tensor quantity. Their units are the same as the corresponding physical quantity.

In an orthogonal coordinate system, Eq. (10.12) becomes

$$
\begin{aligned}
&\rho \left[\frac{dv^i}{dt} - \frac{h_j}{h_i^2}\frac{\partial h_j}{\partial x^i} v^{j2} + \frac{2}{h_i}\frac{\partial h_i}{\partial x^j} v^i v^j \right] \\
&= \frac{1}{4\pi h} \left\{ \frac{\partial}{\partial x^k}\left(\frac{h}{h_k h_i} \bar{B}_k \bar{B}_i \right) - \frac{h}{h_i^2 h_j}\left(\frac{\partial h_j}{\partial x^i} \right) \bar{B}_j^2 \right. \\
&\left. + \frac{2h}{h_i^2 h_j}\left(\frac{\partial h_i}{\partial x^j} \right) \bar{B}_i \bar{B}_j \right\} - \frac{1}{h_i^2}\frac{\partial}{\partial x_i}\left(p + \frac{B^2}{8\pi} \right) \ .
\end{aligned}
\tag{10.16}
$$

An alternate way to derive a tensorial expression for the MHD terms is discussed. Let

$$\mathbf{F}_M = (\boldsymbol{\nabla} \times \mathbf{B}) \times \mathbf{B} \ . \tag{10.17}$$

This is for the magnetic force term. This force term can be expressed as a contravariant vector. First note that

$$\nabla \times \mathbf{B} = \varepsilon^{\ell mn} \frac{\partial}{\partial x^m} B_n \ , \tag{10.18}$$

where $\varepsilon^{\ell mn}$ is $e^{\ell mn}/\sqrt{g}$ with $e^{\ell mn}$ being the permutation. Now by taking the cross product of this contravariant vector with $\mathbf{B}$ one obtains

$$(\nabla \times \mathbf{B}) \times \mathbf{B} = \left(\varepsilon_{\ell jk} \varepsilon^{\ell mn} \frac{\partial}{\partial x^m} B_n B^k \right) \ . \tag{10.19}$$

Since we want the force to transform as a contravariant vector, we multiply by $g^{i\ell}$ and obtain

$$F_M^i = g^{i\ell} \varepsilon_{\ell jk} \varepsilon^{\ell mn} \frac{\partial}{\partial x^m} B_n B^k \ . \tag{10.20}$$

Using the identity

$$\varepsilon_{\ell jk} \varepsilon^{\ell mn} = \delta_k^m \delta_\ell^n - \delta_k^n , \delta_\ell^m \ ,$$

we get the following equation

$$F_m^i = g^{i\ell} \left(\delta_k^m \delta_\ell^n - \delta_k^n \delta_\ell^m \right) \left(\frac{\partial}{\partial x^m} B_n \right) B^k = g^{i\ell} \frac{\partial B_\ell}{\partial x^k} B^k - g^{i\ell} \frac{\partial B_k}{\partial x^\ell} B^k \ . \tag{10.21}$$

As was done earlier, we shall go to an orthogonal coordinate system. That is

$$g_{ii} = h_i^2 \quad , \quad g_{ij} = 0 (i \neq j) \ ,$$
$$g^{ii} = \tfrac{1}{h_i^2}, h = \sqrt{g} = h_1 h_2 h_3 \ .$$

Using this, the force equation becomes

$$F_m^i = \frac{1}{h_i^2} \frac{\partial B^i}{\partial x^k} B^k - \frac{1}{h_i^2} \frac{\partial B_k}{\partial x^i} B^k = F_{M_1}^i - F_{M_2}^i \ . \tag{10.22}$$

Each force term can be dealt with separately

$$F_{M_1}^i = \frac{1}{h_i^2} \frac{\partial B_i}{\partial x^k} B^k$$

$$= \left\{ \frac{1}{h} \frac{\partial}{\partial x^k} (h B^i B^k) - \frac{1}{h} B^i \frac{\partial}{\partial x^k} (h B^k) \right\} + \left(\frac{2}{h_i} \frac{\partial h_i}{\partial x^k} \right) B^i B^k \ . \tag{10.23}$$

Notice that the divergence is given as

$$\text{div} A^j = \frac{1}{\sqrt{g}} \frac{\partial}{\partial x^j} \sqrt{g} A^j \ . \tag{10.24}$$

Therefore, in fact

$$\nabla \cdot \mathbf{B} = \frac{1}{h} \frac{\partial}{\partial x^k} h B^k = 0 \ . \tag{10.25}$$

So the first force term reduces to

$$F^i_{M_1} = \frac{1}{h}\frac{\partial}{\partial x^k}(hB^iB^k) + \frac{2}{h_i}\frac{\partial h_i}{\partial x^k}B^iB^k \ , \tag{10.26}$$

$$\begin{aligned} F^i_{M_2} &= \frac{1}{h_i^2}\left(\frac{\partial}{\partial x^i}B_k\right)B^k \\ &= \frac{1}{h_i^2}\frac{\partial}{\partial x^i}\left(h_k^2B^kB^k\right) - \frac{1}{h_i^2}\frac{\partial B^k}{\partial x^i}h_k^2B^k \ . \end{aligned} \tag{10.27}$$

Note that

$$B^2 = h_k^2B^kB^k \tag{10.28}$$

$$F^i_{M_2} = \frac{1}{h_j^2}\frac{\partial}{\partial x^i}B^2 - F^i_{M_2} + \left(\frac{2h_k}{h_i^2}\frac{\partial h_k}{\partial x^i}\right)B^kB^k \ , \tag{10.29}$$

or

$$F^i_{M_2} = \frac{1}{h_i^2}\frac{\partial}{\partial x^i}\frac{B^2}{2} + \frac{h_k}{h_i^2}\frac{\partial h_k}{\partial x^i}B^kB^k \ . \tag{10.30}$$

Adding Eqs. (10.26) and (10.30), we get:

$$F^i_M = -\frac{1}{h_i^2}\frac{\partial}{\partial x^i}\left(\frac{B^2}{2}\right) - \frac{h_k}{h_i^2}\frac{\partial h_k}{\partial x^i}B^kB^k + \frac{1}{h}\frac{\partial}{\partial x^k}(hB^iB^k) + \left(\frac{2}{h_i}\frac{\partial h_i}{\partial x^k}\right)B^iB^k \ . \tag{10.31}$$

Using the transformation

$$A^j = \bar{A}_j/h_j \ , \tag{10.32}$$

we obtain:

$$\begin{aligned} F^i_M = &-\frac{1}{h_i^2}\frac{\partial}{\partial x^i}\frac{B^2}{2} - \frac{h_k}{h_i^2}\frac{\partial h_k}{\partial x^i}\frac{\bar{B}_k\bar{B}_k}{h_k^2} + \frac{1}{h}\frac{\partial}{\partial x^k}\left(\frac{h}{h_ih_k}\bar{B}_i\bar{B}_k\right) \\ &+2\left(\frac{1}{h_i}\frac{\partial h_i}{\partial x^k}\right)\frac{\bar{B}_i\bar{B}_k}{h_ih_k} \ . \end{aligned} \tag{10.33}$$

This equation is equivalent to Eq. (10.16).

A similar analysis will now be applied to the equation for the magnetic field Eq. (10.3). The contravariant expressions for the cross products in parentheses on the right-hand side of Eq. (10.3) are

$$\begin{aligned} (\mathbf{v}\times\mathbf{B})^i &= \varepsilon^{ijk}v_jB_k \ , \\ (\boldsymbol{\nabla}\times\mathbf{B})^i &= \varepsilon^{ijk}B_{k,j} \ , \end{aligned}$$

which for orthogonal systems reduces to

$$(\nabla \times \mathbf{B})^i = \varepsilon^{ijk} \frac{\partial B_k}{\partial x^j} . \tag{10.34}$$

The full contravariant expression for Eq. (10.3) becomes

$$\frac{\partial B^i}{\partial t} = \varepsilon^{ijk} \left\{ g_{kl} \varepsilon^{\ell mn} v_m B_n - \frac{c^2}{4\pi} \eta g_{k\ell} \varepsilon^{\ell mn} B_{n,m} \right\}_j . \tag{10.35}$$

This, again, for orthogonal systems reduces to

$$\frac{\partial B^i}{\partial t} = e^{ijk} e^{kmn} \sqrt{g} \frac{\partial}{\partial x^i} \left(h_k^2 \sqrt{g} v_m B_n - \frac{\eta h_k^2}{\sqrt{g}} \frac{\partial B_n}{\partial x^m} \right) . \tag{10.36}$$

This expression can be simplified by noting the formula for the product of the permutation symbols yields

$$\frac{\partial B^i}{\partial t} = \frac{1}{h} \frac{\partial}{\partial x^j} \frac{h_k^2}{h} (v_i B_j - v_j B_i) - \frac{1}{h} \frac{\partial}{\partial x^j} \left\{ \eta \frac{h_k^2}{h} \left(\frac{\partial}{\partial x i} B_j - \frac{\partial B_i}{\partial x^j} \right) \right\} , \tag{10.37}$$

where k is not equal to i and j. Here note relations

$$\frac{\partial g}{\partial x^i} = g g^{\ell m} \left(g_{mk} \left\{ \begin{matrix} k \\ \ell i \end{matrix} \right\} + g_{\ell r} \left\{ \begin{matrix} r \\ mi \end{matrix} \right\} \right) = g \left\{ \begin{matrix} \ell \\ \ell i \end{matrix} \right\} + g \left\{ \begin{matrix} m \\ mi \end{matrix} \right\} ,$$

$$\frac{\partial g}{\partial x^i} = 2g \left\{ \begin{matrix} j \\ ji \end{matrix} \right\} ,$$

$$\frac{1}{2g} \frac{\partial g}{\partial x^i} = \left\{ \begin{matrix} j \\ ji \end{matrix} \right\} ,$$

$$\frac{\partial}{\partial x^i} (\ell n \sqrt{g}) = \left\{ \begin{matrix} j \\ ji \end{matrix} \right\} . \tag{10.38}$$

In physical components, Eq. (10.37) becomes

$$\frac{\partial \bar{B}_i}{\partial t} = \frac{1}{h_j h_k} \frac{\partial}{\partial x^j} \left[h_k \left(\bar{v}_i \bar{B}_j - \bar{v}_j \bar{B}_i \right) \right]$$

$$+ \frac{1}{h_j h_k} \left\{ \eta \frac{h_k}{h_i h_j} \left[\frac{\partial}{\partial x^j} (h_i \bar{B}_i) - \frac{\partial}{\partial x^i} (h_j \bar{B}_j) \right] \right\} . \tag{10.39}$$

The general analysis given thus far will be applied to a cylindrical coordinate system (r, z). This coordinate system is suited for investigating some experimental devices, such as mirror devices and theta pinches, and it can also be a useful coordinate system for studying toroidal devices, where the z-direction of the cylindrical coordinate system is identified with an idealized toroidal direction. The metric for the cylindrical coordinate system chosen here is

$$ds^2 = dr^2 + \frac{r^2}{r_0^2} d\zeta^2 + dz^2 \ , \tag{10.40}$$

where ζ replaces the usual polar coordinate θ such that $\zeta = r_0\theta$. The differential scale factors h_1, h_2, h_3, and h are given by

$$h_1 = 1 \ ; \ h_2 = \frac{r}{r_0} \ ; \ h_3 = 1 \ ; \ h = h_2 = \frac{r}{r_0} \ . \tag{10.41}$$

Since the cylindrical coordinate system is an orthogonal system, the expressions in Eq. (10.39) are applicable here. These expressions when explicitly evaluated become

$$\frac{1}{h}\frac{\partial h_j}{\partial x_i} \quad = \quad \frac{1}{h_2}\frac{\partial h_2}{\partial x^1} = \frac{r_0}{r}\frac{\partial}{\partial r}(r/r_0) = \frac{1}{r} \ , \tag{10.42}$$

$$\frac{h_j}{h_i^2}\frac{\partial h_j}{\partial x^i} \quad = \quad \frac{h_2}{h_1}\frac{\partial h_2}{\partial x^1} = \frac{r}{r_0}\frac{\partial}{r_0}\frac{\partial}{\partial r}(r/r_0) = \frac{r}{r_0^2} \ . \tag{10.43}$$

The explicit forms for the components of the force equation and the magnetic field equation, Eqs. (10.16) and (10.39), in this cylindrical coordinate system can be written down. Using Eq. (10.41) for h_ℓ, we obtain the following form for the force equations

$$\rho\left(\frac{d^2r}{dt^2} - \frac{r}{r_0^2}\frac{d\zeta^2}{dt}\right) = -\frac{\partial}{\partial r}\left(p + \frac{B^2}{8\pi}\right)$$

$$+\frac{1}{4\pi h}\left[\frac{\partial}{\partial r}(h\bar{B}_r^2) + \frac{\partial}{\partial\zeta}(\bar{B}_r\bar{B}_\zeta) + \frac{\partial}{\partial z}(h\bar{B}_r\bar{B}_z)\right] - \frac{r}{r_0^2}\frac{1}{4\pi}\frac{\bar{B}_\zeta^2}{h^2} \ ,$$

$$\rho\left(\frac{d^2\zeta}{dt^2} + \frac{2}{r}\frac{d\zeta}{dt}\frac{dr}{dt}\right) = -\frac{1}{h^2}\frac{\partial}{\partial\zeta}\left(p + \frac{B^2}{8\pi}\right)$$

$$+\frac{1}{4\pi h}\left[\frac{\partial}{\partial r}(\bar{B}_r\bar{B}_\zeta) + \frac{\partial}{\partial\zeta}\left(\frac{\bar{B}_\zeta^2}{h}\right) + \frac{\partial}{\partial z}(\bar{B}_\zeta\bar{B}_z) + \frac{2}{r_0}\frac{\bar{B}_r\bar{B}_\zeta}{h}\right] \ ,$$

$$\rho\frac{d^2z}{dt^2} = -\frac{\partial}{\partial z}\left(p + \frac{B^2}{8\pi}\right)$$

$$+\frac{1}{4\pi h}\left\{\frac{\partial}{\partial r}(h\bar{B}_r\bar{B}_z) + \frac{\partial}{\partial\zeta}(\bar{B}_\zeta\bar{B}_z) + \frac{\partial}{\partial z}(h\bar{B}_z^2)\right\} \ . \tag{10.44}$$

We also obtain the following form for the magnetic equations

$$\frac{\partial \bar{B}_r}{\partial t} = \frac{1}{h}\left[\frac{\partial}{\partial \zeta}(\bar{v}\bar{B}_\zeta - \bar{v}_\zeta \bar{B}_r) + \frac{\partial}{\partial z} h(\bar{v}_r \bar{B}_z - \bar{v}\bar{B}_r)\right]$$

$$+\frac{c^2}{4\pi}\left\{\frac{1}{h}\left[\frac{\partial}{\partial \zeta}\eta\frac{1}{h}\left(\frac{\partial}{\partial \zeta}\bar{B}_r - \frac{\partial}{\partial r}h\bar{B}_\zeta\right)\right.\right.$$

$$\left.\left.+\frac{\partial}{\partial z}h\eta\left(\frac{\partial}{\partial z}\bar{B}_r - \frac{\partial}{\partial r}\bar{B}_z\right)\right]\right\} \tag{10.45}$$

$$\frac{\partial \bar{B}_\zeta}{\partial t} = \frac{\partial}{\partial r}(\bar{v}_\zeta \bar{B}_r - \bar{v}_r \bar{B}_\zeta) + \frac{\partial}{\partial z}(\bar{v}_\zeta \bar{B}_z - v_\zeta \bar{B}_z)$$

$$+\frac{c^2}{4\pi}\left[\frac{\partial}{\partial r}\frac{\eta}{h}\left(\frac{\partial}{\partial r}h\bar{B}_\zeta - \frac{\partial}{\partial \zeta}\bar{B}_r\right)\right.$$

$$\left.+\frac{\partial}{\partial z}\frac{\eta}{h}\left(\frac{\partial h}{\partial z}\bar{B}_\zeta - \frac{\partial}{\partial \zeta}\bar{B}_z\right)\right], \tag{10.46}$$

$$\frac{\partial \bar{B}_z}{\partial t} = \frac{1}{h}\left[\frac{\partial}{\partial r}h(\bar{v}_z \bar{B}_r - \bar{v}_r \bar{B}_z) + \frac{\partial}{\partial \zeta}(\bar{v}_z \bar{B}_\zeta - \bar{v}_\zeta \bar{B}_z)\right]$$

$$+\frac{c^2}{4\pi}\frac{1}{h}\left[\frac{\partial}{\partial r}\eta h\left(\frac{\partial}{\partial r}\bar{B}_z - \frac{\partial}{\partial z}\bar{B}_r\right)\right.$$

$$\left.+\frac{\partial}{\partial \zeta}\frac{\eta}{h}\left(\frac{\partial}{\partial \zeta}\bar{B}_z - \frac{\partial}{\partial z}h\bar{B}_\zeta\right)\right]. \tag{10.47}$$

To obtain an expression similar to Schnack and Killeen[14] for the force equation, the following relation is used:

$$\frac{1}{h_i^\alpha}\frac{1}{h}\frac{\partial}{\partial x^k}(hh_i^\alpha B^i B^k) = \frac{1}{h}\frac{\partial}{\partial x^k}(hB^i B^k) + \alpha\left(\frac{1}{h_i}\frac{\partial h_i}{\partial x^k}\right)B^i B^k . \tag{10.48}$$

For $\alpha = 1$ Eq. (10.33) becomes

$$F_M^i = -\frac{1}{h_i^2}\frac{\partial}{\partial x^i}\left(\frac{B^2}{2}\right) - \frac{h_k}{h_i^2}\frac{\partial h_k}{\partial x^i}B^k B^k + \frac{1}{hh_i}\frac{\partial}{\partial x^k}(hh_i B^i B^k)$$

$$+\frac{1}{h_i}\frac{\partial h_i}{\partial x^k}B^i B^k . \tag{10.49}$$

The expressions for the r- and ζ-components of the forces, Eqs. (10.45) and (10.46), are not readily solvable numerically in this form *per se*, because they contain quadratic velocity terms due to the centrifugal force effect. Defining the right-hand side of Eq. (10.45) as F_r and the right-hand side of Eq. (10.46) as $\frac{F_\zeta}{h^2}$, these equations simplify to

$$\frac{d^2r}{dt^2} - \frac{r}{r_0^2}\left(\frac{d\zeta}{dt}\right)^2 = F_r, \tag{10.50}$$

$$\frac{d^2\zeta}{dt^2} + \frac{2}{r}\frac{d\zeta}{dt}\frac{dr}{dt} = \frac{F_\zeta}{h^2}. \tag{10.51}$$

This difficulty may be tackled by defining a new variable, the angular momentum, ℓ,

$$\ell = h^2\frac{d\zeta}{dt} = \frac{r^2}{r_0^2}\frac{d\zeta}{dt} \tag{10.52}$$

such that

$$\frac{d\ell}{dt} = \frac{r^2}{r_0^2}\left(\frac{d^2\zeta}{dt^2} + \frac{2}{r}\frac{dr}{dt}\frac{d\zeta}{dt}\right). \tag{10.53}$$

The force equations, Eqs. (10.50) and (10.51), assume the more tractable form of

$$\frac{d^2r}{dt^2} - \frac{r_0^2}{r^3}\ell^2 = F_r, \tag{10.54}$$

$$\frac{d\ell}{dt} = F_\zeta. \tag{10.55}$$

The force equations and the positions may be numerically integrated by a time-centered algorithm. The variables r, ζ, F_r, and F_ζ are defined at times t^n and the variables ℓ and $\frac{dr}{dt}$ are defined at times $t^{n+1/2}$. The force equations are advanced in time by the scheme

$$\dot{r}^* = \dot{r}^{n-1/2} + \left\{\frac{r_0}{r^{(n)3}}\left[\ell^{(n-1/2)}\right]^2 + F_r^n\right\}\frac{\Delta t}{2}, \tag{10.56}$$

$$\ell^{n+1/2} = \ell^{n-1/2} + F_\zeta^n\Delta t, \tag{10.57}$$

$$\dot{r}^{n+1/2} = \dot{r}^* + \left\{\frac{r_0^2}{r^{(n)3}}\left[\ell^{(n+1/2)}\right]^2 + F_r^n\right\}\frac{\Delta t}{2}. \tag{10.58}$$

The positions are then advanced in time by the scheme

$$\zeta^* = \zeta^n + \ell^{n+1/2}\frac{r_0^2}{r^{(n)2}}\frac{\Delta t}{2}, \tag{10.59}$$

$$r^{n+1} = r^n + \dot{r}^{n+1/2}\frac{\Delta t}{2} , \tag{10.60}$$

$$\zeta^{n+1} = \zeta^* + \ell^{n+1/2}\frac{r_0^2}{r^{(n+1)^2}}\frac{\Delta t}{2} . \tag{10.61}$$

This algorithm has proven to be accurate and stable as long as $v\Delta t \ll \Delta r$ where v is the particle velocity, Δt the time step, and Δr the radial mesh size. This means that near $r = 0$, the finite differencing in time becomes inaccurate and introduces fictitious collisions. We, then, should avoid this spot in the computational region.

Advancing the force equation and the magnetic field equation requires the evaluation of spatial derivatives as well as time derivatives. Most finite difference schemes require the formulation of boundary conditions to completely specify the values of quantities on the mesh. A physically desirable property of a finite difference method is that it have the same conservative properties as the original differential equations. These conditions are usually easily satisfied by periodic boundary conditions. However, if the boundary conditions are different from the periodic boundary conditions, the way to preserve the conservative nature of equations becomes non-trivial. In the remainder of this section we will discuss an example of how these conditions are imposed at a perfectly conducting wall. A more thorough discussion is given by Schnack and Killeen.[14]

The boundary conditions for the fluid elements at a hard nonporous wall is

$$\hat{\mathbf{n}} \cdot \mathbf{v} = 0 , \tag{10.62}$$

where $\hat{\mathbf{n}}$ is a unit vector normal to the wall. The boundary conditions on the electromagnetic fields at a perfectly conducting wall are

$$\hat{\mathbf{n}} \cdot \mathbf{B} = 0 , \tag{10.63}$$

$$\hat{\mathbf{n}} \times \mathbf{E} = 0 , \tag{10.64}$$

i.e., the normal component of the magnetic field and the tangential component of the electric field at the wall are zero. The condition on the tangential electric field can be used to derive a condition on the tangential magnetic field at the wall. Combining Ohm's Law and Ampere's Law yields

$$\frac{\eta c}{4\pi}\mathbf{v} \times \mathbf{B} = \mathbf{E} + \frac{\mathbf{v}}{c} \times \mathbf{B} . \tag{10.65}$$

Taking $\hat{\mathbf{n}}\times$, this equation gives

$$\frac{\eta c}{4\pi}\hat{\mathbf{n}} \times (\boldsymbol{\nabla} \times \mathbf{B}) = \mathbf{n} \times \mathbf{E} + (\hat{\mathbf{n}} \cdot \mathbf{B})\mathbf{v} - (\hat{\mathbf{n}} \cdot \mathbf{v})\mathbf{B} , \tag{10.66}$$

which means that at the conducting wall,

$$\hat{\mathbf{n}} \times (\nabla \times \mathbf{B}) = 0 \tag{10.67}$$

is satisfied.

These conditions imply that magnetic flux does not cross wall boundaries such that the total magnetic flux is preserved, i.e.,

$$\frac{\partial \Phi}{\partial t} = \int ds \cdot \frac{\partial \mathbf{B}}{\partial t} = \int ds \cdot (\nabla \times \mathbf{E}) = \oint d\boldsymbol{\ell} \cdot \mathbf{E} = 0 \ . \tag{10.68}$$

To satisfy Eq. (10.68) in a curvilinear coordinate system, Eq. (10.39) is rewritten as

$$\frac{\partial \bar{B}_i}{\partial t} = \frac{1}{h_j h_k} \frac{\partial}{\partial x^j}(h_k G_{ij}) \ , \tag{10.69}$$

where, again, k is not equal to i and j, and where

$$\bar{G}_{ij} = \bar{v}_i \bar{B}_j - \bar{v}_j \bar{B}_i + \frac{\eta}{h_i h_j} \left[\frac{\partial}{\partial x^j}(h_i \bar{B}_i) - \frac{\partial}{\partial x^i}(h_j \bar{B}_j) \right] \ . \tag{10.70}$$

Note that $\bar{G}_{ij}$ is an antisymmetric tensor.

Consider the case where the quantity to be conserved is the magnetic flux through a surface of constant coordinate x_i. The flux through a small element, e.g., one grid point, is given by

$$\Phi_{i,j,k} = h_j h_k \Delta x^j \Delta x^k \bar{B}_i \ . \tag{10.71}$$

The total flux through the i plane is obtained by summing over all grid points perpendicular to i such that

$$\frac{\partial \Phi_i}{\partial t} = \sum_{\mathrm{grid} \perp i} h_j h_k \Delta x^j \Delta x^k \frac{\partial \bar{B}_k}{\partial t} = \sum_{\mathrm{grid} \perp i} \Delta x^j \Delta x^k \frac{\partial (h_k G_{ij})}{\partial x^j} \ . \tag{10.72}$$

Let $i = 1$ and assume a constant mesh spacing in the perpendicular directions, then

$$\frac{\partial \Phi_1}{\partial t} = \sum_{\substack{j=1,J \\ k=1,K}} \Delta x^2 \Delta x^3 \left\{ \frac{h_3 G_{12}(j+1,k) - h_3 G_{12}(j-1,k)}{2\Delta x^2} \right.$$

$$\left. + \frac{h_2 G_{13}(j,k+1) - h_2 G_{13}(j,k-1)}{2\Delta x^3} \right\} , \tag{10.73}$$

where J, K are the total number of grid points in the j, k directions. Summation over one of the variables yields

$$\frac{\partial \Phi_1}{\partial t}$$

$$= \sum_{k=1,K} \frac{\Delta x^3}{2} \left\{ h_3 G_{12}(J+1,k) + h_3 G_{12}(J,k) - h_3 G_{12}(0,k) - h_3 G_{12}(1,k) \right\}$$

$$+ \sum_{j=1,J} \frac{\Delta x^2}{2} \left\{ h_2 G_{13}(j,K+1) + h_2 G_{13}(j,K) - h_2 G_{13}(j,0) \right.$$

$$\left. - h_2 G_{13}(j,1) \right\} , \tag{10.74}$$

where all the terms in the sums from the interior cells cancel exactly. In order for the total flux to be conserved, $\frac{\partial \Phi_1}{\partial t} = 0$, a common requirement for the finite difference scheme to satisfy is

$$\begin{aligned} h_3(J+1)G_{12}(J+1,k) &= -h_3(J)G_{12}(J,k) , \\ h_3(0)G_{12}(0,k) &= -h_3(1)G_{12}(1,k) , \\ h_2(K+1)G_{12}(j,K+1) &= -h_2(K)G_{13}(j,K) , \\ h_2(0)G_{13}(j,0) &= -h_2(1)G_{13}(j,1) . \end{aligned} \tag{10.75}$$

10.2 Toroidal Corrections

Many devices including tokamaks, however, are best described in toroidal geometry. Let us consider corrections that arise from the transition from cylindrical geometry to toroidal geometry. We shall see the so-called toroidal corrections. We concentrate here on the magnetic induction equation. The toroidal-polar coordinates, (r, ζ, θ), are pictured in the accompanying Fig. 10.1. Relationships among the coordinates follow

$$R = R_0 + r\cos\theta , \tag{10.76}$$

$$\mathbf{r} = R\cos\zeta\hat{x} + R\sin\zeta\hat{y} + r\sin\theta\hat{z} . \tag{10.77}$$

Here the major radius R is a function of the minor radius coordinate r and the poloidal angle θ, while ζ is the poloidal angle. Note that in Section 10.1 ζ

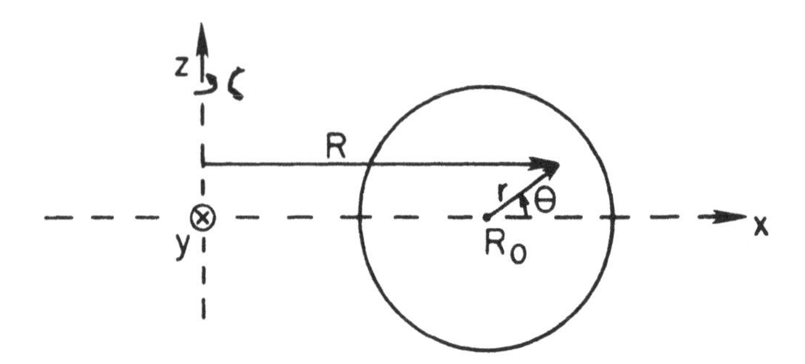

FIGURE 10.1 Toroidal coordinates (r, θ, ζ)

was used for the cylindrical angle variable. The set $(\frac{\partial \mathbf{r}}{\partial r}, \frac{\partial \mathbf{r}}{\partial \theta}, \frac{\partial \mathbf{r}}{\partial \zeta})$ defines a set of contravariant basis vectors. The metric elements in this case are

$$g_{rr} = \frac{\partial \mathbf{r}}{\partial r} \cdot \frac{\partial \mathbf{r}}{\partial r} = 1 \;, \tag{10.78}$$

$$g_{\theta\theta} = \frac{\partial \mathbf{r}}{\partial \theta} \cdot \frac{\partial \mathbf{r}}{\partial \theta} = r^2 \;, \tag{10.79}$$

$$g_{\zeta\zeta} = \frac{\partial \mathbf{r}}{\partial \zeta} \frac{\partial \mathbf{r}}{\partial \zeta} = R^2 \;. \tag{10.80}$$

For the Jacobian, we have

$$J = \frac{\partial \mathbf{r}}{\partial r} \cdot \left(\frac{\partial \mathbf{r}}{\partial \zeta} \times \frac{\partial \mathbf{r}}{\partial \theta} \right) = rR = [\nabla r \cdot \nabla \zeta \times \nabla \theta]^{-1} \;, \tag{10.81}$$

where

$$\nabla r = \frac{1}{J} \frac{\partial \mathbf{r}}{\partial \zeta} \times \frac{\partial \mathbf{r}}{\partial \theta} \;, \tag{10.82}$$

$$\nabla \theta = \frac{1}{J} \frac{\partial \mathbf{r}}{\partial r} \times \frac{\partial \mathbf{r}}{\partial \zeta} \;, \tag{10.83}$$

$$\nabla \zeta = \frac{1}{J} \frac{\partial \mathbf{r}}{\partial \theta} \times \frac{\partial \mathbf{r}}{\partial r} \;. \tag{10.84}$$

These define a reciprocal basis or are covariant basis vectors.

The magnetic field can be expressed in this coordinate system with the components, from $\mathbf{B} = \boldsymbol{\nabla} \times \mathbf{A}$, as

$$B_r = \frac{1}{rR} \frac{\partial}{\partial \theta} \psi - \frac{1}{R} \frac{\partial}{\partial \zeta} A_\theta \;, \tag{10.85}$$

$$B_\theta = \frac{1}{R}\frac{\partial}{\partial\zeta}A_r - \frac{1}{R}\frac{\partial}{\partial r}\psi \ , \tag{10.86}$$

$$B_\zeta = \frac{1}{r}\frac{\partial}{\partial r}rA_\theta - \frac{1}{r}\frac{\partial}{\partial\theta}A_r \ , \tag{10.87}$$

where the flux function $\psi = RA_\zeta$.

Since $R = R(r,\theta)$, all of the poloidal harmonics are coupled. This is the so-called poloidal mode coupling phenomenon. Example:

$$iB_r(r;mn) = \frac{1}{2\pi}\int_0^{2\pi} d\theta \left\{ \frac{1}{rR(r,\theta)} \sum_{-M\le m'\le M} im'\psi(r;m',n)e^{im'\theta} - \frac{in}{R}\sum_{-M\le m'\le M} A_\theta(r;m',n)e^{im'\theta} \right\} e^{-im\theta} \ , \tag{10.88}$$

or

$$B_r(r;mn) = \frac{1}{2\pi}\int_0^{2\pi} d\theta \left[\frac{m'\psi(m',n)}{rR_0} - \frac{n}{R_0}A_\theta(m',n) \right] \frac{e^{i(m'-m)\theta}}{1+\varepsilon r\cos\theta} \ , \tag{10.89}$$

where $\epsilon \equiv \frac{a}{R_0}$ the inverse aspect ratio. And, finally, we have the form

$$B_r(r;mn) = \sum_{-M(n)\le m'\le M(n)} \left[\frac{m'}{rR_0}\psi(m',n) - \frac{n}{R_0}A_\theta(r;m',n) \right] H(r,m'-m) \ , \tag{10.90}$$

where

$$H(r,m'-m) \equiv \frac{1}{2\pi}\int_0^{2\pi} \frac{e^{i(m'-m)\theta}}{1+\epsilon r\cos\theta} d\theta \tag{10.91}$$

is a coupling coefficient between different poloidal mode numbers. The inverse aspect ratio ϵ represents the strength of poloidal mode coupling in Eq. (10.91). In the limit of $\epsilon = 0$, the coupling vanishes.

10.3 Electrostatic Particle Code

With the advent of supercomputers, such as the Cray-XMP and others, it has become increasingly feasible to perform three-dimensional plasma particle simulations. This allows us to simulate with increasing realism the conditions that occur in many modern plasma confinement devices. The systems that are to be examined here are cylindrical (R, ϕ, z) and toroidal (r, θ, ϕ). Note the change of labels. (See Figs. 10.1 and 10.2.) Here we concentrate on electrostatic interactions only. In a system with a strong magnetic field in a given direction (e.g., tokamak) often only a small number of modes parallel to the magnetic field are important. We may use a Fourier expansion in this direction, keeping only a small number of modes and expect fairly realistic results. In this section we follow the development by LeBrun and Tajima.[16]

A. Coordinates and the Grid System

In addition to the angle-like variables (θ, ϕ) we introduce the length-like variables (χ, ζ), of which, ζ, has been already used in Section 10.2:

$$\chi = r_0\theta \quad , \quad \zeta = -R_0\phi \ , \tag{10.92}$$

where R_0 is the major radius at the magnetic axis and r_0 is the mean minor radius and note a minus sign in Eq. (10.92). The differential scale factors are

$$h_r = 1 \ , \ h_\chi = r/r_0 \ , \quad \text{and} \quad h_\zeta = R/R_0 \ , \tag{10.93}$$

where R is given by Eq. (10.78). The "inverse aspect ratios" and curvatures are introduced as

$$\epsilon_\zeta = r_0/R_0 \quad , \quad \epsilon_\chi \ = \ L_r/2r_0 \ , \tag{10.94}$$

and

$$\kappa_\zeta = R_0^{-1} \quad , \quad \kappa_\chi \ = \ r_0^{-1} \ . \tag{10.95}$$

Since χ and ζ are periodic variables, the field variables such as the electric potential can be expressed and calculated in the Fourier representation

$$\Phi(r, k_\chi, k_\zeta) = \sum_{m,n} \Phi(r, \chi, \zeta) \exp(ik_\chi \chi + ik_\zeta \zeta) \ , \tag{10.96}$$

where the wavevector components are given by

$$k_\chi \ = \ \frac{2\pi m}{L_\chi} \quad , \quad L_\chi = 2\pi r_0$$

$$k_\zeta \ = \ \frac{2\pi n}{L_\zeta} \quad , \quad L_\zeta = 2\pi R_0 \ ,$$

or equivalently

$$\Phi(r, m, n) = \sum_{m,n} \Phi(r, \theta, \phi) \exp(im\theta - in\phi) \ . \tag{10.97}$$

In the radial direction, however, no periodicity is expected and finite differencing in r is adopted. In many problems it is desirable to resolve a certain portion of the plasma in detail and others in less detail. For example, tearing modes and drift waves are localized around the rational surfaces and interesting and rapid variations take place in their neighborhood. This phenomenon belongs to the boundary layer problem. The boundary-fitted coordinates have been introduced to describe the generated-grid system for studying, for example, flows around an obstacle.[17] In the plasma simulation this boundary is manifested by the magnetic field. Generated grids that conform to the shape of the magnetic field are often employed in some of the modern MHD and transport simulations. These most often employ orthogonal or non-orthogonal (e.g., Hamada coordinates) flux coordinates; adaptive system is also used.[18] This method of grid generation is a systematic approach to nonuniform grids.

Let us consider a one-dimensional nonuniform grid in the x-direction. It is helpful to adopt the notation

$$\begin{aligned} \bar{\Delta}_i &\equiv (\Delta_{1+1/2} + \Delta_{i-1/2})/2 \,, \\ \delta_i &\equiv (\Delta_{1+1/2} - \Delta_{i-1/2}) \,, \\ \gamma_i &\equiv \delta_i(2\bar{\Delta}_i) \,, \end{aligned} \tag{10.98}$$

where $\bar{\Delta}_i$ is the mean cell size and δ_i is the change in grid spacing at the ith grid point. The quantity γ_i is approximately 1/2 the rate of change of the grid spacing, a useful measure of the nonuniformity of the grid. With these definitions, the finite difference approximation for the first derivative is

$$\frac{f_{i+1} - f_{i-1}}{2\bar{\Delta}_i} = f_i' + \gamma_i \bar{\Delta}_i f_i'' + (1 + 3\gamma_i^2)\frac{\bar{\Delta}_i^2}{6} f_i''' + \cdots \tag{10.99}$$

and for the second derivative is

$$\frac{(1-\gamma_i)f_{i+1} - 2f_i + (1+\gamma_i)f_{i-1}}{\bar{\Delta}_i^2(1-\gamma_i^2)} = f_i'' + \gamma_i \frac{2\bar{\Delta}_i}{3} f_i''' + (1 + 3\gamma_i^2)\frac{\bar{\Delta}_i^2}{12} f_i^{\mathrm{IV}} + \cdots \tag{10.100}$$

which, aside from notation, match the expressions given by De Rivas.[19] These approximations are no longer formally second-order accurate in the grid spacing. By changing the grid spacing sufficiently slowly, γ_i can be neglected and formal second order accuracy will be retained. This condition is given by Blottner and Roache[20]

$$\gamma_i \sim \bar{\Delta}_i \,. \tag{10.101}$$

The nonuniform grid spacing may be conveniently defined by the stretching function that relates the real coordinates x to transformed ones $\xi(x)$

$$\Delta(x) = \frac{\Delta x}{\Delta \xi}(x) \,, \tag{10.102}$$

or

$$\xi(x) = \int \frac{dx}{\Delta(x)} .$$

Various choices for $\xi(x)$ are discussed in Ref. 16.

B. Field Solver

For the electrostatic model, the only field equation that must be solved is the Poisson equation

$$\nabla^2 \Phi = -4\pi\rho , \tag{10.103}$$

$$\mathbf{E} = -\boldsymbol{\nabla}\Phi . \tag{10.104}$$

Given the differential scale factors (metric coefficients) for the toroidal system [Eq. (10.93)], the Laplacian is given by

$$\nabla^2 = \nabla_c^2 + \mathbf{C} \tag{10.105}$$

with ∇_c^2 and $\mathbf{C}$ given by

$$\nabla_c^2 = \frac{\partial^2}{\partial r^2} + \frac{1}{r}\frac{\partial}{\partial r} + \frac{r_0^2}{r^2}\frac{\partial^2}{\partial \chi^2} + \frac{\partial^2}{\partial \zeta^2} , \tag{10.106}$$

$$\mathbf{C} = \frac{\cos\theta}{R}\frac{\partial}{\partial r} - \frac{\sin\theta}{R}\frac{r_0}{r}\frac{\partial}{\partial \chi} + \frac{R_0^2 - R^2}{R^2}\frac{\partial^2}{\partial \zeta^2} . \tag{10.107}$$

Here ∇_c^2 is the cylindrical Laplacian, and $\mathbf{C}$ represents the toroidal corrections. Cheng and Okuda's[21] torus was in cylindrical geometry (Fig. 10.2), while the torus under consideration is shown in Fig. 10.3. The cylindrical Laplacian can be inverted by a standard technique — finite differencing r then applying the recurrence solution for a tridiagonal matrix. However, the toroidal Laplacian as a whole is not directly invertible. This leads us to introduce an approximate method, based on the observation that

$$\nabla^2 = \nabla_c^2(1 + \mathcal{O}(\epsilon_\zeta)) , \tag{10.108}$$

that is, the correction terms are formally of order ϵ_ζ compared to the cylindrical Laplacian. Since ϵ_ζ tends to be small, we expand the potential in powers of ϵ_ζ,

$$\Phi = \sum_n \epsilon_\zeta^n \phi^n \tag{10.109}$$

and substitute into the Poisson equation, obtaining

$$\sum_n \epsilon_\zeta^n \phi^n = \sum_n (\nabla_c^2)^{-1} \left[-4\pi\rho - \mathbf{C}(\epsilon_\zeta^n \psi^n)\right] . \tag{10.110}$$

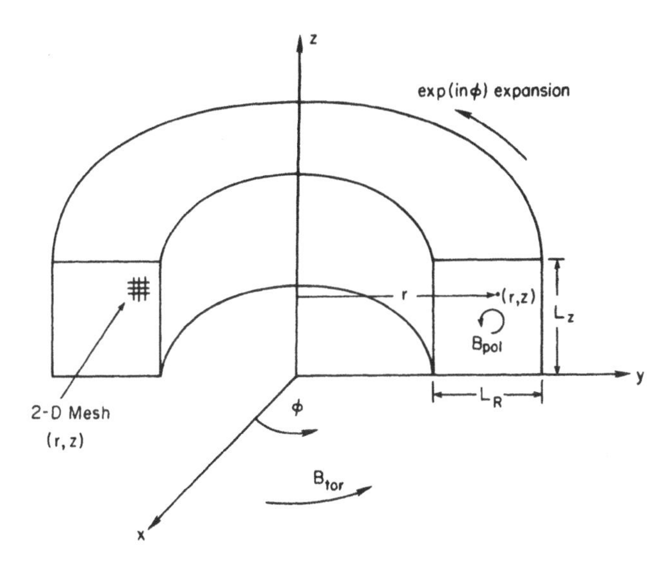

FIGURE 10.2 Cylindrical geometry for a torus (Ref. 21)

Equating terms of the same order in ϵ_ς yields the following chain of equations:

$$\begin{aligned}
\psi^0 &= (\nabla_c^2)^{-1}[-4\pi\rho] \, , \\
\psi^1 &= -\frac{1}{\epsilon_\varsigma}\mathbf{C}[\psi^0] \, , \\
&\vdots \\
\psi^n &= -\frac{1}{\epsilon_\varsigma}\mathbf{C}[\psi^{n-1}] \, .
\end{aligned}$$

The sum is truncated upon reaching the desired order in ϵ_ς for convergence. This method is closely related to the fixed point iteration scheme

$$\Phi^{p+1} = (1-\alpha)\Phi^p + \alpha(\nabla_c^2)^{-1}[-4\pi\rho - \mathbf{C}(\Phi^p)] \, , \tag{10.111}$$

where α is a relaxation parameter, typically on the order of one, and p represents the iteration count. When no relaxation is used ($\alpha = 1$), the two methods are identical. In practice, the code uses Eq. (10.111) in the field solver. The convergence criterion for the potential is

$$\frac{||\Phi^{p+1} - \Phi^p||}{||\Phi||} \leq \epsilon \, .$$

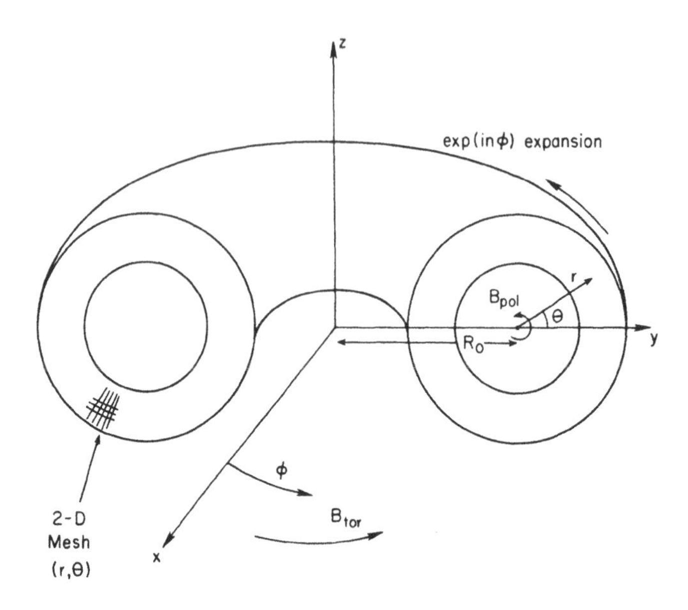

FIGURE 10.3 Toroidal geometry for a torus

[The inversion of the cylindrical Laplacian is done in (ξ, k_χ, k_ζ) space, where ξ is the transformed coordinate for the nonuniform grid.[16]] The electric fields are calculated from the potential as

$$E_r = -\frac{\partial \Phi}{\partial r}\,,\quad E_\chi = -\frac{r_0}{r}\frac{\partial \Phi}{\partial \chi}\,,\quad E_\zeta = -\frac{R_0}{R}\frac{\partial \Phi}{\partial \zeta}\,. \tag{10.112}$$

The radial component is specifically given by

$$(E_r)_i = \left(\frac{\partial \xi}{\partial r}\right)_i \frac{\Phi_{i+1} - \Phi_{i-1}}{2}\,, \tag{10.113}$$

where the finite difference of Φ is in ξ-space.

C. Charge Accumulation and Interpolation

The charge density ρ on the nonuniform grid in curved metric is related to that (ρ') of the transformed grid ξ by

$$\rho = \frac{\partial V'}{\partial V}\rho' = \frac{1}{J}\rho' = \frac{\partial \xi/\partial r}{h_\chi h_\zeta}\rho'\,. \tag{10.114}$$

where the Jacobian J is given by $J = h_r h_\chi h_\zeta$.

The charge accumulation from particles to the grid involves both interpolation and smoothing (filtering or the finite size particle shaping). On a uniform grid a commonly-used filter in particle simulation is the method of finite size particles (Chapter 2). However, for a (highly) nonuniform grid an alternative is the digital filtering.[22] The digital filter employed here is written as $\hat{S}_D^N$, denoting the N-fold application of the simple filter

$$\hat{S}_D(\rho_i) = \frac{\rho_{i-1} + 2\rho_i + \rho_{i+1}}{4} \tag{10.115}$$

where i stands for the ξ-coordinate. This is known as a binomial filter, and is equivalent to a gaussian shape factor in the limit of $N \to \infty$. A detailed discussion on interpolation and self-charge question may be found in Ref. 16.

D. Particle Pusher

The equations of motion for an arbitrary metric is given by

$$h_k \ddot{q}_k + \sum_j h_j \dot{q}_j \left(q \dot{q}_k \frac{\partial h_k}{h_j \partial q_j} - \dot{q}_j \frac{\partial h_j}{h_k \partial q_k} \right) = \frac{q}{m} \left\{ E_k + \sum_{ij} h_i \dot{q}_i B_j \epsilon_{ijk} \right\} , \tag{10.116}$$

which can be directly from the Lagrange equations, using the transformed Lorentz force law. A simpler representation is obtained in terms of the angular momenta, giving $q_k = l_k / h_k^2$ for the coordinates,

$$\frac{\dot{l}_k}{h_k} - \sum_j \left(\frac{l_j}{h_j} \right) \left(\frac{l_j}{h_j^2} \frac{\partial h_j}{h_k \partial q_k} \right) = \frac{q}{m} \left\{ E_k + \sum_{ij} \frac{l_i}{h_i} B_j \epsilon_{ijk} \right\} . \tag{10.117}$$

If the toroidal metric is used, the orbit equations become

$$\ddot{u} - \epsilon_\chi \frac{l_\chi^2}{(1 + \kappa_\chi u)^3} - \epsilon_\zeta \frac{l_\zeta^2 \cos\theta}{(1 + \epsilon_\zeta (1 + \kappa_\chi u)\cos\theta)^3} = F_u , \tag{10.118}$$

$$\frac{\dot{l}_\chi}{(1 + \kappa_\chi u)} + \epsilon_\zeta \frac{l_\zeta^2 \sin\theta}{(1 + \epsilon_\zeta (1 + \kappa_\chi u)\cos\theta)^3} = F_\chi , \tag{10.119}$$

$$\frac{\dot{l}_\zeta}{(1 + \epsilon_\zeta (1 + \kappa_\chi u)\cos\theta)} = F_\zeta , \tag{10.120}$$

in which the right-hand terms are determined from Eq. (10.117) and $u = r - r_0$.

A straightforward implementation of the time-centered method of finite differencing Eq. (10.118) in Section 3.3 will not work in a general coordinate system, due to the nonlinear terms in the equations of motion (10.118). In this case, there is a choice of three quantities that may be considered given at the half-integer timestep and correspondingly finite-differenced. These are the time derivative of the coordinates $\dot{q}_k$, the velocities $h_k \dot{q}_k$, or the angular momenta $h_k^2 \dot{q}_k$. Regardless of which variable is differenced, the nonlinearities remain. This may be solved by adopting a predictor-corrector approach based on the smallness of the nonlinear terms in Eq. (10.118). This has also been done in Section 10.1. In the full toroidal system, however, this type of pusher becomes ill-defined near $r = 0$. Tests show[16] that even when the particle is well kept away from the origin, significant loss of energy occurs. The problem with this method is that the second term in Eq. (10.118) is not small and is a poor expansion parameter in the typical toroidal case. The resulting decenteredness causes the cyclotron motion to slowly decay. Although the damping of cyclotron motion is often an intended effect in an implicit code (Chapter 9), it is undesirable for it to be an unavoidable by-product of the metric. Since *any* predictor-corrector method will be at least partially decentered in time, the integration of the cyclotron orbit will be subject to damping, and may cause problems depending on the time scale of the simulation. A solution[16] to this problem is to push in the cartesian coordinate system and transform quantities back to the toroidal geometry. In the direction of ζ, the usage of the canonical angular momentum push may help.

For magnetized particles such as electrons, the guiding-center method[23] can be adopted. Here the equations of motion are

$$\mathbf{v}_d = c\mathbf{E} \times \mathbf{b}/B + \frac{\mathbf{b}}{\Omega} \times \left[\frac{\mu}{m}\nabla B + v_\parallel^2(\mathbf{b}\cdot\nabla\mathbf{b})\right] , \qquad (10.121)$$

$$\frac{dv_\parallel}{dt} = \frac{q}{m}E_\parallel - \frac{\mu}{m}\mathbf{b}\cdot\nabla B , \qquad (10.122)$$

$$\frac{d\mathbf{x}}{dt} = \mathbf{v}_d + v_\parallel \mathbf{b} . \qquad (10.123)$$

The polarization drift may be added in Eq. (10.121) by adopting the decentered Lorentz pusher (Chapter 9).

Applications of the code in the present section may be found in Chapter 13 and Ref. 24.

10.4 Method of Flux Coordinates

This section presents a method for solving MHD and transport equations in an axisymmetric geometry in flux coordinates.[18] We introduce the flux coordinates which are that of constant flux surfaces. This assumes a good flux

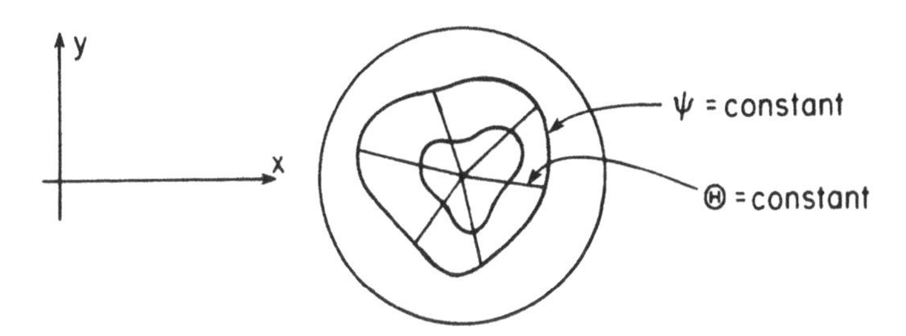

FIGURE 10.4 Cross-section of a torus and the flux coordinate ψ (Ref. 18)

system where flux surfaces are nested and are not interconnected. We follow the development of Ref. 18 here.

In these coordinates, the ideal MHD motion is "frozen," since the ideal MHD equations suggest that the fluid is tied to the field line. Even if the fluid is not ideal, only a slight deviation from this will be registered. Thus, this description of the MHD fluid can be efficient. A similar idea was used in the particle-moving mesh-finite element method MHD code[25] in Chapter 6 as well as in Brackbill et al.[26] The coordinate system for the cylindrical case is expressed in terms of (ψ, θ, ϕ). See Fig. 10.4. Here the flux coordinate ψ is proportional to the area inside the magnetic surface ($\psi = \pi$ at the outermost surface). Instead of taking the radial coordinate r we have taken in the previous sections (Sections 10.1-10.3), we are taking the magnetic flux as a spatial coordinate. This, of course, assumes that flux is well-defined and nested. No reconnection or X-point is allowed. θ is a periodic coordinate, $0 < \theta < 2\pi$. The rays where $\theta =$ constant divide the flux surfaces into equal areas. ϕ is the ignorable (toroidal) coordinate, normalized such that $|\nabla\phi|^2 = k^2$, $k = 2\pi/L$, i.e., $k =$ (major radius)$^{-1}$.

Advantages in using this coordinate system include:

1. Less numerical diffusion occurs if the coordinates are in the "same direction" as the field lines.
2. It is easier to separate the slow and fast time scales if the direction of **B** is known accurately.
3. Since the location of the plasma-vacuum boundary is known, free boundary problems can be calculated accurately.

The coordinate system is, however, non-orthogonal, $\nabla\psi\cdot\nabla\theta \neq 0$. However, $\nabla\psi \cdot \nabla\phi = 0$ and $\nabla\theta \cdot \nabla\phi = 0$. There is a constraint on the Jacobian of the form

$$J^{-1} = \nabla\phi \cdot \nabla\psi \times \nabla\theta = 2\pi k T(t) \; , \tag{10.124}$$

so $J = J(t)$. If the computational zones have equal areas at $t = 0$, then this remains the case for all t. This continual rezoning through $J(t)$ eliminates the need for an explicit rezoning step, i.e., the mesh is always "nice" (no reconnection).

The constraint on J is satisfied by treating the grid positions as dynamical variables with (x, y) being the dependent variables and (ψ, θ, ϕ) being the independent variables. The Jacobian J can then be calculated,

$$J = \left(\frac{\partial \mathbf{r}}{\partial \psi} \times \frac{\partial \mathbf{r}}{\partial \theta}\right) \cdot \frac{\partial \mathbf{r}}{\partial \phi} . \tag{10.125}$$

Note $\frac{\partial \mathbf{r}}{\partial \phi} = \frac{1}{k}\hat{z}$, because $|\nabla\phi|^2 = k^2$. Then

$$J = \frac{1}{k}\hat{z} \cdot \left\{\left(\frac{\partial x}{\partial \psi}\hat{x} + \frac{\partial y}{\partial \psi}\hat{y}\right) \times \left(\frac{\partial x}{\partial \theta}\hat{x} + \frac{\partial y}{\partial \theta}\hat{y}\right)\right\} , \tag{10.126}$$

or

$$J = \frac{1}{k}\left\{\frac{\partial x}{\partial \psi}\frac{\partial y}{\partial \theta} - \frac{\partial x}{\partial \theta}\frac{\partial y}{\partial \psi}\right\} = \frac{1}{k}[x, y]_{\psi,\theta} , \tag{10.127}$$

where [,] is the Poisson brackets. The coefficients of the metric tensor can also be calculated

$$g^{\psi\psi} = \frac{1}{k^2 J^2}\left[\frac{\partial x}{\partial \psi}\frac{\partial y}{\partial \theta} - \frac{\partial x}{\partial \theta}\frac{\partial y}{\partial \psi}\right] , \tag{10.128}$$

and similarly,

$$g^{\theta\theta} = \frac{1}{k^2 J^2}\left\{\left(\frac{\partial x}{\partial \psi}\right)^2 + \left(\frac{\partial y}{\partial \psi}\right)^2\right\} , \tag{10.129}$$

$$g^{\phi\phi} = \nabla\phi\nabla\phi = k^2 . \tag{10.130}$$

There are also two off-diagonal elements (since non-orthogonal coordinate system),

$$g^{\psi\theta} = g^{\theta\psi} = \nabla\theta \cdot \nabla\psi = \frac{-1}{k^2 J^2}\left[\frac{\partial x}{\partial \theta}\frac{\partial x}{\partial \psi} + \frac{\partial y}{\partial \theta}\frac{\partial y}{\partial \psi}\right] . \tag{10.131}$$

The area of the plasma column can then be expressed as

$$\begin{aligned} A &= \int_0^{\pi}\int_0^{2\pi}\left(\frac{\partial \mathbf{r}}{\partial \psi} \times \frac{\partial \mathbf{r}}{\partial \theta}\right) \cdot \frac{\nabla\phi}{|\nabla\phi|} d\psi d\theta \\ &= \int\int J|\nabla\phi| d\psi d\theta = \frac{(k)(\pi)(2\pi)}{2\pi k T(t)} . \end{aligned} \tag{10.132}$$

Then

$$A(t) = \frac{\pi}{T(t)} \ . \tag{10.133}$$

The magnetic field in this coordinate system (expressed in contravariant form) is

$$\mathbf{B} = B^{\psi}\frac{\partial \mathbf{r}}{\partial \psi} + B^{\theta}\frac{\partial \mathbf{r}}{\partial \theta} + B^{\phi}\frac{\partial \mathbf{r}}{\partial \phi} \tag{10.134}$$

(in covariant form, $\mathbf{B} = B_{\psi}\nabla\psi + B_{\theta}\nabla\theta + B_{\phi}\nabla\phi$). Since $\mathbf{B}\cdot\nabla\psi = 0$, due to the way the coordinate system is defined, $B^{\psi} = 0$,

$$B^{\theta} = \mathbf{B}\cdot\nabla\theta \ , \tag{10.135}$$

$$B^{\phi} = \mathbf{B}\cdot\nabla\phi \ , \tag{10.136}$$

$$\frac{\partial \mathbf{r}}{\partial \theta} = J\nabla\phi \times \nabla\psi \ , \tag{10.137}$$

$$\frac{\partial \mathbf{r}}{\partial \phi} = J\nabla\psi \times \nabla\theta \ . \tag{10.138}$$

Use of $\nabla\psi \times \nabla\theta \cdot \nabla\phi_{\|} = \frac{1}{J}$, and $\nabla\phi\nabla\psi \times \nabla\theta$ (also $|\nabla\phi| = k$) yields

$$\nabla\psi \times \nabla\theta = \frac{\nabla\phi}{k^2 J} \ . \tag{10.139}$$

Substituting Eq. (10.139) into Eq. (10.134),

$$\mathbf{B} = JB^{\theta}\nabla\phi \times \nabla\psi + \frac{B^{\phi}}{k^2}\nabla\phi \ . \tag{10.140}$$

Let us define two scalars f, g:

$$\mathbf{B} = \mathbf{B}_0\left[f(\psi,t)\nabla\psi \times \nabla\phi + g(\psi,\theta,t)\frac{\nabla\phi}{k}\right] \ , \tag{10.141}$$

where $\mathbf{B}_0 =$ constant. That f is not a function of θ can be seen from $\nabla\cdot\mathbf{B} = 0$ as

$$\frac{\partial}{\partial\psi}(JB^{\psi}) + \frac{\partial}{\partial\theta}(JB^{\theta}) + \frac{\partial}{\partial\phi}(JB^{\phi}) = 0 \ . \tag{10.142}$$

We then have $B^{\psi} = 0$, $\frac{\partial}{\partial\phi} = 0$ and therefore

$$\frac{\partial B^{\theta}}{\partial\theta} = 0 \ . \tag{10.143}$$

Therefore, f is not a function of θ, but g is a function of ψ, θ, and t. The magnetic field $\mathbf{B}$ as written in this form is the most general axisymmetric solution under the conditions

$$\mathbf{B}\cdot\nabla\psi = 0 \quad \text{and} \quad \nabla\cdot\mathbf{B} = 0 \ . \tag{10.144}$$

Since in the present formulation the provision is made that J is a constant in space, the grid does not necessarily move with the same velocity as the fluid so that there is some slippage between the fluid frame and the grid frame. The fluid velocity is written as

$$\mathbf{v} = \mathbf{v}^L + \mathbf{v}^E \ , \tag{10.145}$$

where

$\mathbf{v}^L$ = velocity of the grid

$\mathbf{v}^E$ = Eulerian velocity of the fluid with respect to the moving grid .

The main objective here is to take up a part of fluid motion into the grid motion so that the rest of the fluid motion relative to the grid is slow and describable in accurate detail. The velocity of the grid coordinates is written

$$\mathbf{v}^L = \frac{\partial x}{\partial t}\hat{x} + \frac{\partial y}{\partial t}\hat{y} \ , \tag{10.146}$$

where $x = x(\psi, \theta, t)$, $y = y(\psi, \theta, t)$. The Eulerian fluid velocity is

$$\mathbf{v}^E = u^\psi \frac{\partial \mathbf{r}}{\partial \psi} + u^\theta \frac{\partial \mathbf{r}}{\partial \theta} + u^\phi \frac{\partial \mathbf{r}}{\partial \phi} \ . \tag{10.147}$$

If isotropic expansion of the plasma column $\left(\frac{\partial T}{\partial t} \neq 0\right)$ is allowed, $x(\psi, \theta, t)$ and $y(\psi, \theta, t)$ still evolve on a slow time scale. The motion of the grid is decoupled from the fast wave motion (to the lowest order in k^2). The velocities $\frac{\partial x}{\partial t}$, $\frac{\partial y}{\partial t}$ are given by the constraint of the Jacobian

$$J^{-1} = 2\pi k T(t) \ . \tag{10.148}$$

Substituting Eq. (10.148) into Eq. (10.127), we obtain

$$\frac{\partial x}{\partial \psi}\frac{\partial y}{\partial \theta} - \frac{\partial x}{\partial \theta}\frac{\partial y}{\partial \psi} = \frac{1}{2\pi T(t)} \ . \tag{10.149}$$

If x and y satisfy this equation at $t = 0$ and if we let

$$\frac{\partial x}{\partial t} = \frac{\partial t}{\partial \psi}\Lambda + \frac{\partial x}{\partial \theta}\Omega \ , \tag{10.150}$$

$$\frac{\partial y}{\partial t} = \frac{\partial y}{\partial \psi}\Lambda + \frac{\partial y}{\partial \theta}\Omega \ , \tag{10.151}$$

where

$$\Lambda \equiv T\frac{\partial \zeta}{\partial \theta} - \pi\frac{\partial T}{\partial t}\left\{[x - x(0)]\frac{\partial y}{\partial \theta} - [y - y(0)]\frac{\partial x}{\partial \theta}\right\} \ , \tag{10.152}$$

$$\Omega \equiv -T\frac{\partial \zeta}{\partial \psi} + \pi\frac{\partial T}{\partial t}\left\{[x - x(0)]\frac{\partial y}{\partial \psi} - [y - y(0)]\frac{\partial x}{\partial \psi}\right\} \tag{10.153}$$

and $x(0)$, $y(0)$ denote the magnetic axis coordinated and ζ is some stream function to be determined, then the constraint $J^{-1} = 2\pi k T(t)$ is satisfied at all times.[18]

The variables which are to be advanced in time in the MHD problem[18] are $x, y, T(t), \zeta, \rho, f, g, u^{\psi}, u^{\theta}, u^{\phi}$. The plasma column area determines $T(t)$. The stream function, ζ, is utilized in determining x and y, ρ is the mass density, f and g are to determine $\mathbf{B}$. To separate the fast time scale from the solution, solve for Δ and P implicitly, where the compressibility and the total pressure

$$\Delta = \boldsymbol{\nabla} \cdot \mathbf{u} \quad \text{and} \quad P = p + 1/2B^2$$

are the driving terms for the compressional waves. The equations

$$\frac{\partial \Delta}{\partial t} + \nabla \cdot \frac{1}{\rho} \nabla P + Q = 0 \,, \tag{10.154}$$

$$\frac{\partial}{\partial t} P + (\gamma p + g^2 B_0^2)\Delta + N = 0 \,, \tag{10.155}$$

are, therefore, solved implicitly,[18] holding the metric terms constant (fixed grid) since they evolve on a slower time scale (the Lagrangian velocity of the grid is incompressible except for an isotropic expansion). Q and N are defined in Ref. 18.

This method is based on the nested good magnetic surface, since J has to be well-behaved. In this way, when magnetic fields reconnect, the method breaks down or becomes very cumbersome. A good application of the present method is to couple this with transport phenomena[27] (Chapter 15). However, a very active MHD plasma can be ill-suited for this method.

Generalized geometric treatments of plasmas as discussed in this chapter are an indispensable component for building a numerical laboratory of plasma research.

Problems

1. Discuss the benefits and difficulties of the ballooning representation[28] as the simulation coordinate.
2. Express toroidal differential scale factors in terms of the inverse aspect ratios and curvatures, Eqs. (10.94) and (10.95). How are the slab and cylindrical limits taken from the toroidal particle code?
3. Compare the cartesian push and curvilinear coordinate push. Cite the reasons for more accuracy for the cartesian push for the toroidal particle code.

4. Discuss what kind of difficulties are expected for $r = 0$ in the toroidal particle codes.
5. Discuss the difficulties associated with non-orthogonal coordinates.
6. Explain the advantages of the flux coordinates for a transport code and some of the disadvantages of the flux coordinates for a MHD code.
7. Show that the digital filter $(S_D)^N$ Eq. (10.115) becomes equivalent to a gaussian shape factor as $N \rightarrow \infty$.

References

1. S.K. Ma, *Modern Theory of Critical Phenomena* (Benjamin, Reading, 1976).
2. L.D. Landau and E.M. Lifshitz, *The Classical Theory of Fields* (Pergamon Press, Oxford, 1975).
3. J.H. Sloan and L.L. Smarr, *Numerical Astrophysics*, eds. J.M. Centrella, J.M. LeBlanc, and R.L. Bowers (Jones and Bartlett, Boston, 1983) p. 52.
4. C.W. Misner, K.S. Thorne, and J.A. Wheeler, *Gravitation* (Freeman, San Francisco, 1973).
5. L.A. Artsimovich, *Controlled Thermonuclear Reactions* (Gordon and Breach, New York, 1964).
6. L. Spitzer, Phys. Fluids **1**, 253 (1958).
7. A.I. Morosov and L.S. Soloviev, *Review of Plasma Physics* vol. 2 ed. by M.A. Leontovich (Consultant Bureau, New York, 1966) p. 1.
8. D. Kerst, *Handbuch der Physik* ed. S. Flugge, vol. 44 (Springer Verlag, Berlin, 1959) p. 13.
9. D.C. Robinson, Plasma Phys. **13**, 439 (1971).
10. R.A. Dandl and G.E. Guest, *Fusion* ed. E.A. Teller (Academic, New York, 1981) vol. 1, Part 13, p. 79.
11. B. Spain, *Tensor Calculus* (Interscience Publishing, New York, 1960).
12. I.S. Sokolnikov, *Tensor Analysis, Theory and Applications* (J. Wiley and Sons, New York, 1951).
13. J.L. Synge and A. Schild, *Tensor Calculus* (Inver Publishing, New York, 1978).
14. D. Schnack and J. Killeen, J. Comp. Phys. **35**, 110 (1980).

15. F. Brunel, J.N. Leboeuf, T. Tajima, J.M. Dawson, M. Makino, and T. Kamimura, J. Comp. Phys. **43**, 968 (1981); also, see F. Brunel, J.N. Leboeuf, D.P. Stotler, H.L. Berk, and S.M. Mahajan, in *Supercomputers* eds. F.W. Matsen and T. Tajima (Univ. of Texas Press, Austin, 1986) p. 59.

16. M.J. LeBrun and T. Tajima, to be published: also M.J. LeBrun, Ph.D. dissertation (Univ. of Texas, Austin, 1988).

17. J.F. Thompson and Z.U.A. Warsi, J. Comp. Phys. **47**, 1 (1982).

18. S.C. Jardin, J.L. Johnson, J.M. Greene, and R.C. Grimm, J. Comp. Phys. **29**, 101 (1978).

19. E.K. DeRivas, J. Comp. Phys. **10**, 202 (1972).

20. F.G. Blottner and P.J. Roache, J. Comp. Phys. **8**, 498 (1971).

21. C.Z. Cheng and H. Okuda, J. Comp. Phys. **25**, 133 (1977).

22. C.K. Birdsall and A.B. Langdon, *Plasma Physics via Computer Simulation* (McGraw-Hill, New York, 1985).

23. W.W. Lee and H. Okuda, J. Comp. Phys. **26**, 139 (1978).

24. H.L. Berk, H. Momota, and T. Tajima, Phys. Fluids **30**, 3548 (1987).

25. A. Nishiguchi and T. Yabe, J. Comp. Phys. **47**, 247 (1982).

26. J. Brackbill, *Methods in Computational Physics* vol. 16 (Academic, New York, 1976) p. 1.

27. S.P. Hirshman and S.C. Jardin, Phys. Fluids **22**, 731 (1979).

28. A. Glasser, *Proc. The Finite Beta Theory Workshop*, Varenna, eds. B. Coppi and W. Sadowski (U.S. Department of Energy, Washington, D.C., 1977) p. 55; Y.C. Lee and J. Van Dam, *ibid.* p. 93; J.W. Connor, R.J. Hastie, and J.B. Taylor, Phys. Rev. Lett. **40**, 396 (1978).

11

INFORMATION AND COMPUTATION

Having surveyed various computational plasma physics methods and techniques in the preceding ten chapters, we come to the point where any sophisticated software now has to rely on the particular computing hardware — the computer — to carry out the calculation. In this chapter we will discuss the present and future (anticipated) computers and computer architecture for scientific computation and, furthermore, their prospect in a wider perspective of the general theory of information. We will also discuss the technique of information or data processing and its applications in handling simulation data ("postprocessing").

11.1 The Future of Computers and Scientific Computation

As Fig. 1.1 indicated, the trend of increasing supercomputer speed is close to exponential growth (or at least some kind of geometrical growth). Similarly, the trend of increasing supercomputer memory size is also close to exponential growth [see Fig. 11.1].[1] Equally important is the storage capac-

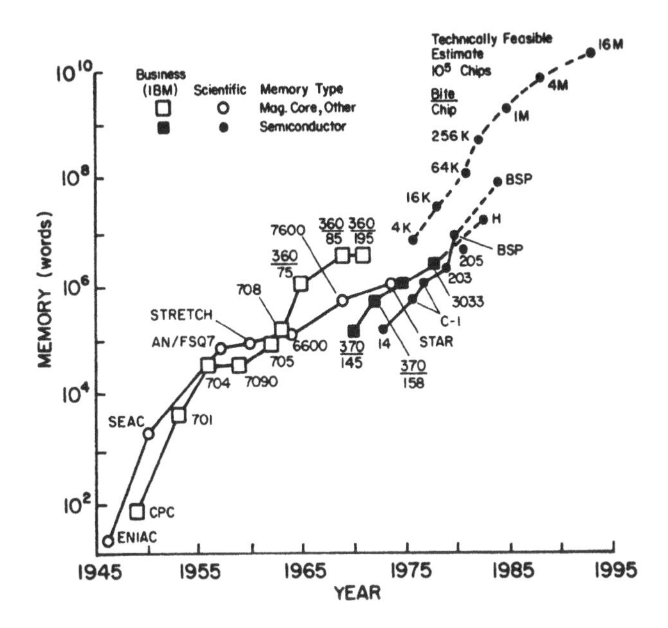

FIGURE 11.1 Progress of computer memory over the years (Ref. 1)

ity per memory circuit. Since the advent of an integrated circuit by a Texas Instruments engineer, the IC technology has become as impressive as supercomputer technology. In fact, the IC technology provides the basis for the growth of supercomputer technology. However, the IC (and VLSI, very large scale integrated systems) provides an opportunity to build an inexpensive computer much smaller than previously envisaged before the IC technology, but equally (or more) capable of computational execution. This provides the basis for the era of personal computers and systems with many processing units of considerable capability.

Today's supercomputer is built on processors of information, called semiconductor chips whose memory size is of the order 100 M words and the fast solid state device (SSD) memory of the order of 200 M words. A typical clock cycle is a few nanoseconds (or a few hundred M flops). The propagation speed of information is that of an electromagnetic pulse—close to the speed of light $c = 3 \times 10^{10}$ cm/sec, and the mode of operation is digital, i.e., all or nothing by switching the electricity on or off. The typical amount of contacts with other elements (chips) is several per unit. Its strength is in its speed and accuracy. In comparison with today's supercomputer, a human brain operates on chemical processors of information. The number of unit processors (neurons) numbering, at birth, on the order of 10 billion (10^{10}) and synaptic contacts per neuron may range from several hundreds to 20 thousands. This gives rise to a very large number of configurations of connections. The recovery frequency of synaptic membrane impulses is less than several hundred per second. The impulse travels in the neural cord with the speed $\lesssim 100\ m/\sec$. The synaptic response is digital, i.e., all or nothing. Its strength is in its tremendous capacity and the complexity of connection. It is thus clear that the operation of a human brain is quite different from that of a present day supercomputer.

In the respect of information processing, the present day supercomputers (and ordinary computers) are basically the von Neumann machine.[2,3] Although recent computers of the von Neumann type tend to have more processors, they basically have only one central processing unit (CPU). The CPU is simply an electronic *soroban* (or abacus). Both the data and the instruction are coded in binary words in the same way and they are in sequence. The data or instruction is fetched to the CPU one at a time and it is identified by time and location. The data processing is carried out in an appointment style. This is the traditional view of computer design as promulgated by such people as Charles Babbage, Alan Turing and John von Neumann and as built into computers ranging from ENIAC and UNIVAC (see Fig. 1.1) and today's pocket calculators.

Because the speed of the CPU is a crucial determining factor of the computational speed in the von Neumann computer architecture, the overall speed of the computer of this type is essentially determined by the CPU speed. At present the silicon semiconductor chip technology is capable of achieving the speed of 10^3 Mflops, while the emerging GaAs technology and the superconducting Josephson junction technology may be able to deliver the speed of one and two orders of magnitude faster, respectively. (Surely, the emerging new high temperature superconductors[4] will change all these stated here drastically in due time.) Nevertheless, the speed of the CPU chip determines the computer performance and the upper limit of the chip computational speed is believed not to increase as rapidly as in the recent past. This suggests that in order to capture further increases in computational speed like those experienced in recent years, we should now seriously begin to contemplate utilizing multiple processors instead of relying solely on improving the speed of one CPU.

People frequently do more than one thing at a time, like driving a car while listening to the radio, etc. Our brains control separately but in a coordinated fashion breathing, heartbeat, and several different motor activities. Separate units of activity are carried out by separate processors that work simultaneously (at the same time, but not in lockstep) and interact to produce the final effect. In contrast to the von Neumann sequential computer, one can think of a series of concurrent processors in this situation.

The vector architectures that dominate today's very high-performance computers incorporate processors that operate on vectors by performing the required operation simultaneously on several elements, or components, of the vector. Furthermore, because one knows in advance the sequence in which operands are required by the program, one can arrange to fetch them in advance and present them to the CPU in order. Such operations can be performed while the CPU is working on other data — a special kind of parallelism called pipelining. In many cases, some units of activity of a task must be performed in a specified order: You can't steam rice until the rice is soaked in water. The difference is in what is called the architecture of the computer, the way it is designed to perform tasks: Does it perform several units of activity at once, using interacting processors, or does it use a single processor to perform one unit of activity at a time?

By the way the computer handles data we distinguish computer architectures (within the von Neumann type or its extension) in the following four categories.[5] (i) SISD: In the original and most primitive type of the von Neumann architecture the computer applies a *single* stream of *instructions* to a *single* stream of *data*; (ii) SIMD: the computer applies a *single* stream of *instructions* to *multiple* streams of *data*; (iii) MISD: the computer applies *multiple* streams of *instruction* to a *single* stream of *data*; (iv) MIMD: the computer applies *multiple* streams of *instruction* to *multiple* streams of *data*. The processor that handles type (i) (the most traditional von Neumann type) is called the scalar processor. The other types of processors in (ii)-(iv) are called the array processors. Sometimes people call the MISD type as the pipeline processor and the SIMD and MIMD types as the parallel ones. In the traditional architecture we have a clock and all operations are synchronized to the clock. This way we can guarantee that the computations will be carried out in the proper order and at the proper time. With explicit synchronization the instruction streams are regulated either by an external intervention or by a mutual protocol. With implicit synchronization instruction streams are regulated by the flow of data: the processor either waits to perform its computation until all the necessary inputs are ready, or it waits to call for inputs until the results of its computation are called elsewhere.

In spite of the tremendously rapid progress in digital computers lately which enabled us to look for some meaningful solutions to nonlinear physics problems that otherwise could not have been resolved, answers to tough nonlinear physics questions are far from satisfactory. At the same time, as the capability of the computer improves, the treatment of more ambitious problems and new types of problems become feasible. This imposes even more demands on the computer. Because of the hardware limitation of the ultimate speed (the speed of light) and the software bottleneck due to the von Neumann architecture, a need for the development of computers with massively parallel processors or perhaps with architecture quite different from the von Neumann's has become evident. Several such attempts may be found in Ref. 6. At the same time there are attempts to construct algorithms that are suited to the computer architecture. As we have seen in the previous chapters, the main strategy of computational physics is to simulate the natural system as closely as possible to the original nature one wants to emulate. It often involves finite differencing in the place of differentiation. It often involves macroparticles in the place of microparticles. These attempts are then accommodated and carried out within the constraints of the digital computer. These algorithms often "force" or favor a certain computer architecture. The example of the cellular automaton simulation of fluid,[7,8] however, starts from the computer constraints and "forces" the computing algorithm to adjust to the architecture. We will discuss the latter in some detail later.

Attempts to cut the embryonic tie of the computer architectural thinking with von Neumann's precept have been investigated in recent years. The data flow type[9] of computers is an attempt to drop the "appointment" concept of the von Neumann type, in which the data and the operator are put to the central processor (the given place) at the specified time (the given time) in a given order. The data flow concept is to assign a "token" to each data so that the data carries certain information about how it should be processed. In this way the strict "appointment" like protocol becomes unnecessary and more parallel and flexible computation may become possible. Experimentation in this direction has begun and is in an early stage. One can extend this concept further. In the von Neumann computer the data and the instruction take the same form of a word: they cannot be distinguishable by themselves and only by the order or protocol in principle. In the data flow type of computer, the token (or instruction) is attached to the data as a header: a word now consists of two parts, one the content (or data) and the other the label (or instruction etc.). This is similar to the information processing of DNA molecules. A DNA (deoxyribonucleic acid) molecule consists of a strand made up of four bases A (adenine), T (thymine), G (guanine), C (cytosine), while an RNA (ribonucleic acid) molecule has A, T, G, and U (uracil). The DNA replication of one strand for the other strand is done by the pairing affinity of adenine and thymine and that of cytosine and guanine.[10]: Adenine is a token to call for thymine and guanine is a token for cytosine. In the extreme of the data flow type of computer a DNA like computer may be constructed. (This type of computer is also called a Brownian computer, not from its algorithmic point of view, but from the energetic point of view[11]). Here a word becomes lengthy and contains all the necessary information including the order of operation. Thus in this computer the synchronization which is generally assumed in a data flow type of computer becomes relaxed or unnecessary, as in the information processing in DNA molecules.

An attempt to emulate the brain is an alternative deviation from the von Neumann type. This approach may be called the artificial intelligence approach, the connectionist's approach etc. The history of the study of neural network is old for its intrinsic importance and its possible application (or spinoff) to a new computational scheme.[12–14] The strategy of this type of computer is to have a very large number of processors ("neurons") and an even larger number of interconnections among these processors whose structure of connection dictates efficient processing and flow of data. Such data handling may be synchronous or asynchronous. A possible advantage of this type is the massive concurrent computation. A further possibility is that this computation may be "smartly" adjusted to the experience of the past computation. Active investigations in this field have also begun.[15–18]

The other important recent development that should not escape our attention is the rapid decline of cost and equally rapid increase of speed and memory of personal computers (PC's) that are due mainly to the tremendous leap in the technology of microchips. The PC is now so inexpensive that a student can afford to buy one for himself, just as a student a generation ago bought a typewriter for himself. The PC is, at the same time, as powerful as a medium frame computer a generation ago. Along with this hardware development the interface between the human and the machine has greatly improved due to the software development. Not only do we no longer see card decks, but also the communication between man and machine is increasingly interactive, as evidenced in such an invention as a "mouse" which interactively controls and instructs through the items on the cathode tube screen. These factors make it possible for computational physics to become an easily accessible subject and to be a part of main line physics, particularly that of nonlinear physics. One can see many efforts to enhance and enrich this new possibility and experience for wider communities.[19]

11.2 Computation on a Cellular Automaton

The dynamical simulation of multidimensional field theories, such as the high Reynolds-number Navier-Stokes equation, demands an extreme capacity of computers. As we discussed in Section 11.1, massively parallel architectures and algorithms are likely to be necessary to carry out this task to a satisfactory degree, because the ultimate computational limits set by the speed of light and various solid-state constraints are believed to be encroaching upon us. A cellular automaton algorithm[7] has been recently applied[8] to simulate the Navier-Stokes equation to meet this challenge for algorithms. The cellular automaton algorithm for the fluid simulation[8] aims at both natural parallelism of computation and treatment of all bits on an equal footing. The cellular automaton[7] by definition has only local interaction. For example, a common computational rule of thumb for the conventional algorithm (Chapter 6) based on floating point representations requires at least 32 bits per word and preferably 64 bits just to compute a modest accuracy of 5 bits for a drag by a turbulent flow past an obstacle. Floating point representations hierarchically favor bits in the most significant places,[20] which often leads stability. Schemes which give bits equal weight would be preferred. Because of round-off noise on floating point calculation can run away to unphysical regimes in an attempt to treat each bit equally.[8]

The cellular automaton for a lattice gas has been shown to produce asymptotically the incompressible two and three-dimensional Navier-Stokes systems.[8] The cellular automaton of a lattice gas model may start from that of Hardy and Pomeau,[21] in which they constructed a regular, square, two-dimensional lattice with unit link length. At each cell vertex there are up to four "molecules" of equal mass with unit speed whose velocity points to one of the four link directions. Time is discrete. The simultaneous occupation of a vertex by the identical state (position and velocity) of molecules is forbidden (a Fermi exclusion principle for this Boolean gas). Thus when we push molecules by one link to the nearest vertex to the direction of the velocity vector, a new configuration is realized except when more than one particle occupies the same vertex with opposite directions of velocities. When two particles coming to the same vertex with opposite velocities, which corresponds to a head-on collision, their velocities are replaced by a velocity at a right angle respectively to the original (collision rule). This model has been demonstrated to possess several important properties. Relaxation to thermodynamical equilibrium has been numerically shown[8] with the average density and momentum. When density and momentum are varied slowly in space and time, macroscopic dynamical equations emerge. They are, however, different from the Navier-Stokes equations in three respects: (i) Lack of Galilean invariance, (ii) lack of isotropy, and (iii) a cross-over dimension problem. Galilean invariance is broken by the lattice. This is manifested in the nonlinear momentum flux tensor in the momentum equation, which not only has quadratic terms in the hydrodynamic velocity $\mathbf{u}$, as it should in the Navier-Stokes equation, but also has nonlinear corrections to arbitrarily high orders in the velocity, which are negligible at low Mach numbers (a condition that is a requirement for incompressibility). The square automaton is invariant under $\pi/2$ rotations. Such lattice symmetry, however, is not sufficient to insure isotropy of the fourth degree tensor relating momentum flux to quadratic terms in velocity. The cross-over dimension occurs in two-dimensional hydrodynamics; the viscosity develops a logarithmic scale dependence,[22] common in phase transitions and field theory. In three-dimensions this difficulty does not occur.

The momentum flux tensor in Ref. 21

$$P_{\alpha\beta} = p\delta_{\alpha\beta} + T_{\alpha\beta\gamma\epsilon}v_\gamma v_\epsilon + O(u^4) \,, \tag{11.1}$$

can be cast into the following form[8]

$$P_{\alpha\beta} = (p + \mu v^2)\delta_{\alpha\beta} + \lambda v_\alpha v_\beta + O(u^4) \tag{11.2}$$

with λ and μ being scalars in a two-dimensional, hexagonally rotationally symmetric lattice cellular automaton (hexagonal lattice gas). Here the tensor

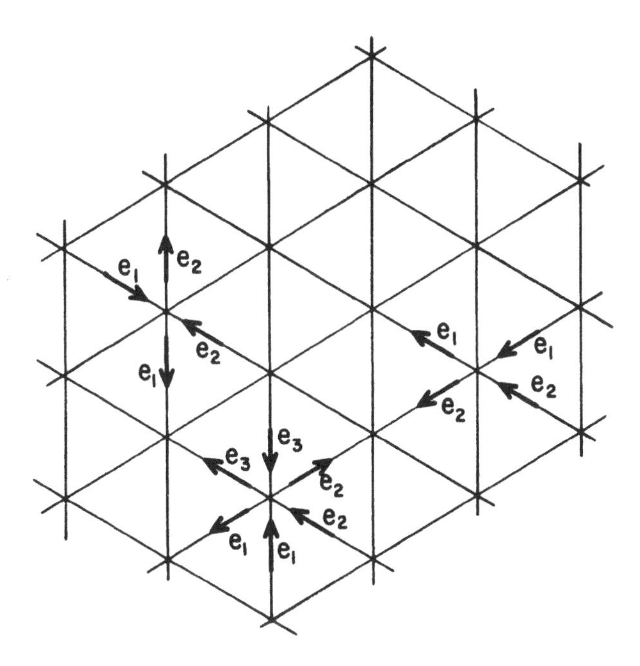

FIGURE 11.2 Collision rules for the cellular automaton algorithm

T is isotropic. At low Mach numbers, Eq. (11.2) is the correct form for the Navier-Stokes equation. The automaton rules are identical to those of the square lattice gas except for the collision rule. Name the six links out of any vertex counterclockwise with an index i. These can be both two and three-body collisions. For two-body collisions we have $(i, i+3)$ goes to (a) $(i+1, i-2)$ or (b) $(i-1, i+2)$. Types a and b have equal *a priori* weights. For three-body collisions we have $(i, i+2, i-2)$ goes to $(i+3, i+1, i-1)$. These rules conserve particle number and momentum at each vertex. Without three-body collisions there would be four scalar conservations (mass and three-directional momenta); with three-body collisions there are three scalars conserved (mass and two momenta). See Fig. 11.2.

The hexagonal lattice gas system can be shown to lead to a system of the two-dimensional Navier-Stokes equation[8] in an asymptotic limit. Let N_i be the population at a vertex with velocity in the direction i, averaged over a macroscopic space and time scale. Let us define a slowly varying density ρ and momentum $\rho\mathbf{v}$ by

$$\rho = \sum_i N_i \quad , \quad \rho\mathbf{v} = \sum_i N_i \mathbf{e}_i \tag{11.3}$$

where $\mathbf{e}_j$ is a unit vector in the direction j, $\mathbf{e}_j = (\cos 2\pi j/6, \sin 2\pi j/6)$. From the definition Eq. (11.3) and the continuity equations at thermal equilibrium one obtains a Fermi-Dirac distribution[8]

$$N_i = \{1 + \exp\left[\alpha(\rho, \mathbf{v}) + \beta(\rho, \mathbf{v})\mathbf{e}_i \cdot \mathbf{v}\right]\}^{-1} , \tag{11.4}$$

which is a manifestation of the microscopic Fermi exclusion rule of the cellular automaton. For $\mathbf{v} = 0$ α and β are expanded around $\mathbf{v} = 0$ and $N_i = \rho/6$. Second order terms in gradients (viscous terms) are obtained by the Kubo relation[23] or by a Chapman Enskog expansion.[24] By so doing the following set of hydrodynamic equations is obtained

$$\frac{\partial \rho}{\partial t} + \nabla \cdot (\rho \mathbf{v}) = 0 , \tag{11.5}$$

$$\frac{\partial}{\partial t}(\rho v_\alpha) + \frac{\partial}{\partial x_p}\left[g(\rho)\rho v_\alpha v_\beta + O(v^4)\right]$$

$$= -\frac{\partial}{\partial x_\alpha} p + \eta_1(\rho)\nabla^2 v_\alpha + \eta_2(\rho)\frac{\partial}{\partial x_\alpha}(\nabla \cdot \mathbf{v}) , \tag{11.6}$$

where $g(\rho) = (\rho - 3)/(\rho - 6)$ and $p = \rho/2$, $\eta_1(\rho)$ and $\eta_2(\rho)$ are the shear and bulk viscosities. These equations approach the incompressible Navier-Stokes equation in a limit that the Mach number $M = \sqrt{2}\ v$ and L goes to the thermodynamic limit ∞ with ML kept constant.

It is noted that the implementation of boundary conditions for cellular automaton is often straightforward by construction. Some practical limitations on lattice gas cellular automaton can be mentioned. The Reynolds number associated with a d-dimensional lattice with N sites in each direction is $O(N)$. The kinematic viscosity of the hexagonal lattice gas is $O(1)$. From standard turbulence theory[25] the dissipation scale is $O(N^{1/2})$ in two dimensions. This insures the required scale separation at large Reynolds numbers if N is sufficiently large. Thus, this is the requirement for the hydrodynamic description, as it requires a scale separation between the hydrodynamic scale and the lattice link length.

A generalization of the above mentioned cellular automaton model of the fluid in terms of a hexagonal lattice gas has been formulated by Montgomery and Doolen[25] to include magnetohydrodynamics (MHD) in two dimensions. They imposed all the same rules and, in addition, they put the "spin magnetization" $\sigma(1, 0$ or $-1)$ on the vertex for z-component of magnetic vector potential A_z. The number of particles with a given σ-index is preserved in each scattering, except when two-body scattering events occur between particles with σ-indices $(1, -1)$, $(-1, 1)$, and $(0, 0)$. In these three cases one of the three possibilities $(1, -1)$, $(-1, 1)$, and $(0, 0)$ is chosen randomly as the outcome with all three possibilities equally probable. This method can produce the Fermi-Dirac distribution for the zeroth-order local thermal equilibrium and appropriate conservation laws of mass and flux. The momentum conservation law, however, does not contain the magnetic acceleration term $\mathbf{j} \times \mathbf{B}$ and this has to be added *a posteriori*; this arises because $\mathbf{j}$ and $\mathbf{B}$ involves spatial derivatives of A_z, a nonlocal operation.

Chen and Matthaeus overcome this difficulty by instituting tensorial scattering rules for bidirectional links.[26] In the MHD cellular automaton, two state vectors $\mathbf{e}_i$ and $\mathbf{e}_j$ are assigned to each particle at a vertex, where $\mathbf{e}_i$ and $\mathbf{e}_j$ $(i, j = 1 - 6)$ are unit vectors on the hexagonal vertex introduced earlier. Roughly speaking, when $\mathbf{e}_i$ corresponds to the microscopic velocity, $\mathbf{e}_j$ to the microscopic magnetic induction. Again, no more than one particle in each cell is allowed to occupy a state with a given pair of i and j, so that a maximum of 36 particles may reside simultaneously in a cell. Letting $N_i^j (= 0 \text{ or } 1)$ denote the occupation number, we define $f_i^j = \langle N_i^j \rangle$ to be the ensemble (or spatio-temporal) average particle distribution. According to Chen and Matthaeus[26] the "magnetic acceleration" effect is implemented by the following: the particle at the vertex with state (i, j) moves to the adjacent vertex in the direction $\mathbf{e}_i$ with probability $1 - |P_{ij}|$; alternatively with probability $|P_{ij}|$ the particle moves to the adjacent vertex in the direction $\mathbf{e}_j P_{ij} / |P_{ij}|$. This leads to a kinetic equation for the tensor particle distribution f_i^j,

$$\frac{\partial f_i^j(\mathbf{x}, t)}{\partial t} = -\left[(1 - |P_{ij}|)\, \mathbf{e}_i + P_{ij} \mathbf{e}_j\right] \cdot \boldsymbol{\nabla} f_i^j(\mathbf{x}, t) + \Omega_{ij} , \tag{11.7}$$

where Ω_{ij} represents the effect of all collisions that modify f_i^j.

The macroscopic number density n, fluid velocity $\mathbf{v}$, magnetic field $\mathbf{B}$, are related to the microstate by

$$n = \sum_{i,j} f_i^j , \tag{11.8}$$

$$n\mathbf{v} = \sum_{i,j} \left[(- |P_{ij}|)\, \mathbf{e}_i + P_{ij} \mathbf{e}_j\right] f_i^j , \tag{11.9}$$

$$n\mathbf{B} = \sum_{i,j} \left(Q_{ij} \mathbf{e}_j + R_{ij} \mathbf{e}_i\right) f_i^j , \tag{11.10}$$

where the 6×6 matrices $\mathbf{P}$, $\mathbf{Q}$, $\mathbf{R}$ are given below. We introduce the auxiliary fields $n\mathbf{v}^* \equiv \sum_{i,j} \mathbf{e}_i f_i^j$ and $n\mathbf{B}^* \equiv \sum_{i,j} \mathbf{e}_j f_i^j$. Bidirectional particle streaming allows the average trajectory of a given particle to vary relative to $\mathbf{e}_i$ and $\mathbf{e}_j$, which corresponds roughly to microscopic momentum and magnetic quanta, respectively. Thus we expect $\mathbf{v}^* \sim \mathbf{v}$ and $\mathbf{B}^* \sim \mathbf{B}$. Considering the most probable $\mathbf{e}_j$ to be a good estimate of the local macroscopic $\mathbf{B}$, specific symmetry constraints are imposed on the coefficients P_{ij} from the "velocity" deflections that particles with all $\mathbf{e}_i$'s will experience. If the particles that constitute a fluid element flow across a simple sheared **B**, the effect of the Lorentz force can be accounted for by deflecting the particle trajectories towards the most probable $\mathbf{e}_j$ when the angle $\theta = \cos^{-1}(\mathbf{e}_j \cdot \mathbf{e}_j)$ is acute or towards $-\mathbf{e}_j$ when the angle is obtuse. The effect of magnetic pressure would be to accelerate particles towards a weaker magnetic field region where the most probable $\mathbf{e}_j$ will be encountered less frequently. This consideration[26] yields the relations $P_{ii+1} = P_{ii+5} = -P_{ii+2} = -P_{ii+4}$ so that **P** are completely determined by P_{ii} and P_{ii+1}. Similar considerations for the field stretching term $\mathbf{B} \cdot \nabla \mathbf{v}$ leads to a choice for **Q** and **R**.[26] Collision rules that randomize the microscopic state while preserving the macroscopic quantities are introduced.[26] Under these collisions the system takes the Fermi-Dirac equilibrium distribution $f_i^j(\text{eq.}) = [1 + \exp(\alpha + \boldsymbol{\beta} \cdot \mathbf{e}_i + \boldsymbol{\gamma} \cdot \mathbf{e}_j)]^{-1}$. Based on this, Chen and Matthaeus have discussed the macroscopic equations that this system exhibits,[26] which is approximately that of two-dimensional incompressible magnetohydrodynamics. This is accomplished without introducing any new nonlocal force. Essential to this approach are inequalities between collisional and macroscopic time and length scales that allow the local microstate to be treated as near equilibrium.

It seems that the present CA method resembles that of the particle-in-cell treatment of a fluid, developed by Harlow[27] and Leboeuf et al.[28] as discussed in Chapter 6. In the latter, particles carry certain state vectors and obey a certain collision rule within a cell. The present CA model has explicitly focused on the local interaction and its computational advantage to massively parallel computation. And thus it systematically caters to this purpose for efficient computation in large and complex systems. At the same time it has been pointed out that the CA approach may encounter severer constraints than the conventional approaches, including particle-in-cell fluid dynamics.[29]

11.3 Information Processing in Biological Systems and Artificial Intelligence

As we saw in Section 11.1, the main superiority of the brain to the present-day computer is its large number of processors. Thus it is very important to understand how and why human intelligence exhibits so many amazing capabilities such as association, language, influence, creativity, and other mental functions. An important field of the neural network research is artificial intelligence research. In the past, however, most artificial intelligence research has relied heavily on high level symbolic manipulation that favors serial computation. As we noted earlier, on the other hand, parallel computing can solve bottlenecks associated with many kinds of pattern recognition problems. Nevertheless, when it comes to knowledge processing, parallelism is considered to be a poor substitute for sequential heuristic search. A quite different approach to machine intelligence has been inspired by the study of neural networks. Many neural network researchers believe that massive parallelism, distributed information storage, and associative interconnections, as evidenced in brains, should be the base upon which to construct intelligent computers.

A most important problem should be to understand the functional organization of groups of neurons. Let us take an example of processing of sensory input in the cortex.[12,30,31] In early work Hebb[12] hypothesized that the modification of synaptic strength is dependent on correlated pre-post synaptic neuronal firing. Little and Shaw[30] mapped the neural problem onto a generalization of the Ising spin model. The analogy goes as follows. Consider a network (or cortical column) of $N(\sim 10^3)$ highly interconnected neurons with the firing $(S = +1)$ or not firing $(S = -1)$ at a given time and look at the evolution of such patterns in discrete time steps τ of the order of a few milliseconds, which is of the order of the refractory period and also the decay time of a postsynaptic potential. The spin 1/2 particle is associated with the N neurons and assign spin "up" ("down") where the neuron has fixed (not fixed) in the time step. A configuration of all spins gives the firing pattern of the network at a given time. A signal from a fired neuron propagates along the axon to synaptic junctions on other neurons, influencing the probability of these neuron firings. This leads to the next stage in which a new firing pattern will evolve which can again be represented by a set of N spins. There is an effective spin-spin interaction between sets of spins (neurons) in adjacent time steps. A function analogous to the partition of the spin system describes the time evolution of the firing patterns of the neural network.

An important element of the model is that it is *not* deterministic. It is known that there are fluctuations in the post-synaptic potentials due to the statistical nature of the release of the chemical transmitter.[32] This indeterminacy gives rise to noise in the network which is directly analogous to thermal noise in the spin system and thus an effective temperature can be defined for the network.[30] In spite of the noise, well-defined states of the system can occur which are analogous to phases like the ferromagnetic or anti-ferromagnetic domains.

The synapse is the most probable place where plasticity related to learning or memory or processing capability takes place. Little et al.[30] investigated the storage capacity of the network based on the Hebb hypothesis[12] in the spin model. For a highly interconnected network of N neurons there are about N^2 pieces of information that can be encoded in the N^2 modifiable synaptic strengths V. The 2^N firing pattern α of the network exist in terms of each neuron firing or not firing. In principle we have to calculate a $2^N * 2^N$ probability matrix P in terms of the N^2 V's that a given firing pattern α goes to α' one time step later. There are, however, enormous redundancies in the $2^N \alpha$'s. The large processing and memory storage capabilities of neural network require fluctuations. The analogy with the Ising spin model by Fisher et al.[33] is illuminating here. They considered an Ising model for a magnetic material which has three types of couplings between the spins depending upon direction and distance of separation. They found infinitely many stable phases; further, they found that (i) one of the three types of interactions must be repulsive and the other two attractive, and (ii) many states exist only at finite temperature. These results may translate to the neural properties of (i) the inhibitory as well as excitatory synaptic potentials V and (ii) the presence of the neural network's fluctuations.

In 1982 Hopfield wrote an epoch-making paper,[15] in which he mathematically modeled the neural networks with the emphasis on emergent collective computational capabilities. Since then the research in this area and its applied field, artificial intelligence research, has "taken off." Hopfield argues that computational properties of a neural network (or equivalent integrated circuit) can emerge as collective properties of the system because it has a large number of simple equivalent components (neurons). He describes the physical meaning of content-addressable memory by an appropriate phase space flow of the state of the system. His model produces through collective interaction a content-addressable memory that correctly yields an entire memory from any subpart of sufficient size. As opposed to Little et al.,[16] Hopfield's model is based on *asynchronous* parallel processing. Other emergent collective properties of his model include some capacity for generalization, familiarity recognition, categorization, error correction, and time sequence retention.

We shall discuss his model here. Consider two states V_i of neuron i with $V_i = 0$ and 1 corresponding to not firing and firing, as in the above.[30] When neuron i is connected to neuron j, the strength of connection is defined as T_{ij}. The instantaneous state of the system is specified by listing the N values of V_i, so that it is represented by a binary word of N bits. The state changes in time according to the following rule. For each neuron i there is a fixed threshold U_i. Each neuron i adjusts its state randomly in time with a mean attempt rate W, setting

$$\left.\begin{array}{l} V_i \to 1 \\ \\ V_i \to 0 \end{array}\right\} \text{ if } \sum_{i \neq j} T_{ij} V_j \left\{\begin{array}{ll} > & U_i \\ \\ < & U_i \,. \end{array}\right. \tag{11.11}$$

Each neuron randomly and asynchronously evaluates whether it fires or does not fire. U_i is taken to be zero, unless otherwise stated. This model resembles the Perceptron,[14] which has only forward coupling, while the present model[15] holds a strong back-coupling, among a few important differences. Interesting results arise from the strong back-coupling.

The important capacity of the neural network or any dynamically connected system is that it changes the response capacity according to its past experience. This gives rise to information storage or learning. Suppose we wish to store the set of states $V^s (s = 1, \ldots, n)$. Assign the synaptic interconnection tensor T_{ij} by the following prescription:[34]

$$T_{ij} = \sum_s (2V_i^s - 1)(2V_j^s - 1) , \tag{11.12}$$

with $T_{ii} = 0$. From this definition

$$H_j^s \equiv \sum_j T_{ij} V_j^s = \sum_{s'} \left(2V_i^{s'} - 1\right) \left[\sum_j V_j^s \left(2V_j^{s'} - 1\right)\right] , \tag{11.13}$$

yielding a so-called pseudo-orthonormality condition

$$\left\langle \sum_j T_{ij} V_j^s \right\rangle = \langle H_i^s \rangle \simeq (2V_i^s - 1)\frac{N}{2} , \tag{11.14}$$

because the mean value of the bracketed term vanishes unless $s = s'$, where the angular brackets indicate the average.

Such tensors T_{ij} have been used in theories of linear association nets.[34] The present model, however, uses its strong nonlinearity[30,35] (see Fig. 11.3) to make decisions, produce categories, and regenerate information and discriminate information.[15] The modification of the synaptic interconnection corresponds to learning, in the spirit of Hebb.[12] The modification ΔT_{ij} is computed by the correlations of the past neural excitations such as

$$\Delta T_{ij} = \langle V_i(t) V_j(t) \rangle , \tag{11.15}$$

where the average is a certain appropriate weighting over past history.

This model has stable limit points. Consider a special case $T_{ij} = T_{ji}$ and define a Lyapunov (energy) function[18]

$$E = -\frac{1}{2} \sum_i \sum_j^{(i \neq j)} T_{ij} V_i V_j . \tag{11.16}$$

The change in E, ΔE, due to change ΔV_i is given by

$$\Delta E = -\Delta V_i \sum_{j \neq i} T_{ij} V_j . \tag{11.17}$$

Thus the algorithm for altering V_i causes E to be a monotonically decreasing function. State changes continue until at least "energy" E is reached. See Fig. 11.3. This is isomorphic with an Ising model. T_{ij} provides the role of the exchange coupling and there is also an external local field at each site. When T_{ij} is symmetric but has a random character (the spin glass) there are known to be many locally stable states. Monte Carlo calculations were carried out

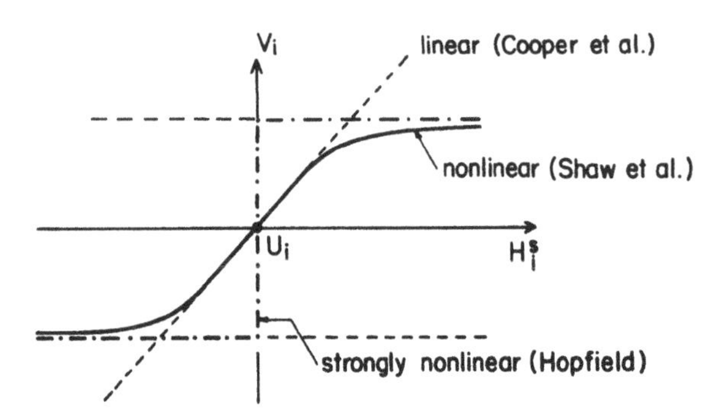

FIGURE 11.3 Linear and nonlinear synaptic response V_i (Ref. 15)

for a $N = 30$ system.[15] For $N = 30$ the system does not display an ergodic wandering through state space. Typically, within a time $4W^{-1}$, it settles into limiting behaviors. One is a stable state. The other is a simple cycle around states such as $A \rightarrow B \rightarrow A \rightarrow B \cdots$. The third is chaotic wandering in a small region of state space. The Hamming distance between two binary states A and B is defined as the number of places in which the digits are different. The chaotic wandering occurs within a short Hamming distance of one particular phase state. Statistics are collected on the probability p_i of the occurrence of a state in a time of wandering around the minimum "energy" and a measure of entropy S of the available states M is taken as

$$S = \ell n\, M = -\sum p_i \ell n\, p_i \ . \tag{11.18}$$

Hopfield found[15] $M = 25$ for a system of $N = 30$. Let us consider a pictorial example. The energy landscape of the Hopfield associative memory is shown in Fig. 11.4. Suppose V_i' approximately equals μ_i^1 but with the last 5 bits of a 32 bit word differing. Starting this system with $V_i = V_i'$ this model will move the system down to $V_i = \mu_i^1$, which is the most adjacent minimum. The flow in phase space produced by this model has the properties necessary for a physical content-addressable memory. According to Hopfield[15] about $0.15N$ states can be simultaneously remembered before the error in recall becomes severe.

Lee et al.[18] generalized and improved Hopfield's model. Their model improves the memory capacity and enhances the pattern discriminating capability of the neural system by the introduction of *higher* order tensorial memory

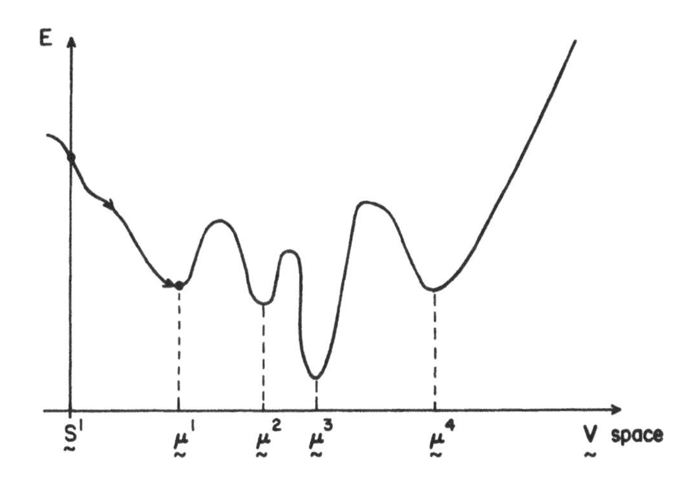

FIGURE 11.4 The energy (or Lyapunov function) of the neural network and various attractors

functions (correlations). As we shall discuss later in Section 11.5, binary correlations can describe nonlinearities and system behaviors only to a certain degree. Since binary correlation models can learn only binary correlations, they have about as little predictive ability as that of the binary correlation functions of fully developed fluid turbulence (see Chapter 13) in which higher order correlations such as skewness, kurtosis, etc. are as important as the binary correlations to describe the system. The higher the order of the interconnections, the more complex the correlations the network can learn. The energy function, Eq. (11.16) of Hopfield is generalized to have a higher order rank interconnection tensor $T_{\nu_1 \nu_2 \cdots \nu_k}$

$$E = \sum_{\nu_1=1}^{N} \sum_{\nu_2=1}^{N} \cdots\cdots \sum_{\nu_k}^{N} T_{\nu_1 \nu_2 \cdots \nu_k} V_{\nu_1} V_{\nu_2} \cdots V_{\nu_k} \, , \tag{11.19}$$

where $\mathbf{V} = (V_1, V_2, \cdots, V_N)$ is a state vector whose component $V_j (j = 1, \ldots, N)$ can only assume the values of 1 or -1, and k specifies the order of the interconnections. The neural response function, Eq. (11.11), is generalized to be

$$V_i^{(n+1)} = W \left[\sum_{\nu_2, \nu_3, \ldots, \nu_k} T_{i,\nu_2 \cdots \nu_k}^{(n)} V_{\nu_2}^{(n)} V_{\nu_3}^{(n)} \cdots V_{\nu_k}^{(n)} \right] \, , \tag{11.20}$$

where the superscripts refer to the timestep for the case of synchronous modelling (although asynchronous representation is possible), $W(x)$ is a step function whose value is 1 for $x > 0$ and -1 otherwise, and $V_i^{(n)}$ is the i-th neuron firing state at time step n. The behavioral modification of neurons is described by the change of the synaptic interconnection tensor $T^{(n)}_{\nu_1,\nu_2\cdots\nu_k}$, as before, which is also generalized from Eq. (11.15) to be of higher rank

$$\Delta T^{(n)}_{\nu_1\cdots\nu_k} = T^{(n+1)}_{\nu_1\cdots\nu_k} - T^{(n)}_{\nu_1\cdots\nu_k} = -\omega T^{(n)}_{\nu_1\cdots\nu_k}$$

$$+\omega \sum_{\mu_1\cdots\mu_k} \Delta^{\mu_1\cdots\mu_k}_{\nu_1\cdots\nu_k} V^{(n)}_{\mu_1} V^{(n)}_{\mu_2} \cdots V^{(n)}_{\mu_k} , \qquad (11.21)$$

where ω^{-1} is the characteristic time scale of long term memory and $\Delta^{\mu_1\cdots\mu_k}_{\nu_1\cdots\nu_k}$ is a positive definite matrix. The tensor T evolves slowly compared with $\mathbf{V}$, as an experience and knowledge accumulate in time scale ω^{-1}. Lee et al.[18] showed that this generalized model can simulate auto-associative, hetero-associative, as well as multi-associative memory, where auto, hetero, and multi-associative memories correspond to a single layer of primary neurons, two (input and output) layers, and many layers of neurons. Respectively, Lee et al.[18]'s simulation showed a drastic increase in the memory capacity and speed over that of the Hopfield model.[15]

11.4 Information and Entropy

The concept of entropy was introduced by Boltzmann. He realized that to a macroscopic state corresponds a number of microscopic states (*Komplexionszahl*—complexity number W) and that the system most likely evolves towards the largest of this number (or its logorithm, entropy, $S = k\ell nW$ with k being the Boltzmann constant) under a constant energy. The system relaxes from more organized to less organized states. The most disorganized chaotic state corresponds to the maximum of W and thus entropy S. The degree of the measure of organization, therefore, may be defined by the negative of entropy, called negentropy N,

$$N = -S = -k = \ell nW . \qquad (11.22)$$

Intuitively speaking, more organized states or phenomena should have larger negentropy, such as living organisms, brains, and highly developed information processors.

In relation to organization and chaos there exists the concept of information. Also related to information is algorithm complexity. Interested readers may see Refs. 36 and 11, for an example. In order to define the quantity of information, consider the following example.[37] An announcer in a television quiz game program asks the participants to guess what word or subject he has in mind. Without asking questions the participants, of course, cannot tell what it is. Their information sheet is blank. Then the announcer accepts a certain number of questions from the participants, say up to 20, to which he answers only by "yes" or "no". Through these interrogations, the participants try to guess the word or subject. To each question there exist two possible answers, "yes" or "no." For 20 questions there exist 2^{20} possible categories of answers. If a participant makes 20 questions, he can choose a specific answer out of 2^{20} possibilities. On the other hand, if he makes 10 questions, he chooses an answer out of 2^{10} possibilities. Of course, 20 questions should carry more information than 10 questions. Based on this observation, it might be reasonable to define the measure of information obtained by 20 interrogations by 2^{20}, while that by 10 interrogations by 2^{10}. However, if the participant asks 10 questions and later asks another 10 questions, the amount of information obtained from the former 10 questions and that from the other 10 should be the same and the total should be twice that obtained simply by asking 10 questions. If we express the amount of information by the number of possibilities, the above required property is not fulfilled because $2^{20} \neq 2 \times 2^{10}$. We find that if we define that by the logarithm of the number of possibilities (except for a multiplicative constant) this is fulfilled because $\ell n 2^{20} = 2\ell n 2^{10}$.

Take another example.[37] We would like to visit a friend at his hotel room. The receptionist tells us that his room is #501. Until we are told the room number, we do not know at all in which one of the available, say 240, rooms he is staying. To know the room number corresponds to obtaining information that enables us to pick one out of 240 possibilities. To gain information is to collapse a large number of possibilities into a small number of possibilities. This resembles the situation of measurement in quantum mechanics, where a "collapse of states" takes place upon measurement. On the other hand, if the receptionist tells us that his room is next to the elevator on the fifth floor, this information should amount to the same as the previous one. This information is composed of two pieces of information, one that specifies the fifth floor out of 8 possible floors and the other that specifies the location of the room out of 30 possible rooms on the floor. Indeed, the logarithms of the numbers of possibilities are compared and they agree: $\ell n 240 = \ell n 8 + \ell n 30$.

The measure of information gained by knowing that a particular thing happens out of W possible allowable cases of all is given by

$$I = K \ell n W \ , \tag{11.23}$$

where K is a constant and customarily chosen to be $\log_2 e$ so that the unit of I is bits:

$$I = \log_2 W \text{ (bits)} \ . \tag{11.24}$$

Implicitly assumed here is an equal probability for all W cases, i.e.

$$P = W^{-1} \ , \tag{11.25}$$

which leads to the expression

$$I = -K \ell n P \ . \tag{11.26}$$

If there are N events with i-th probability P_i, we have the total

$$I(\{P_i\}) = -K \sum_{i=1}^{N} \ell n P_i \ . \tag{11.27}$$

This definition, Eqs. (11.26) and (11.27), generalizes Eq. (11.23) or Eq. (11.24): to obtain the knowledge that a certain thing happens (or happened) that is supposed to happen only with probability P constitutes information whose amount is expressed by Eq. (11.26). This definition is in agreement with a common feeling that news carries a large amount of (important) information if it reports that an incident happened that was supposed to happen only rarely. The average amount of information, or the expectation value of information amount, is similarly defined after Shannon[38] as

$$H(\{P_i\}) = -K \sum_{i=1}^{N} P_i \ell n P_i \ . \tag{11.28}$$

Shannon introduced this information measure in constructing the mathematical basis for communication.[38]

Let us consider an imperfect knowledge or partial information. Suppose we know a seven-letter English word "PHYSICS". Then the amount of our information is $I = 7K\ell n 26$, since there are 26 alphabet characters and here we assume that they are equally probable. (If not, the expression is more complicated). If we know only the first character P, then our knowledge is $I = K\ell n 26$. Next, suppose we are certain only to the extent that the first character is either P, Q, R, S, or T, then $I = -K\ell n\frac{5}{26} = K(\ell n 26 - \ell n 5)$.

We can generalize the above consideration. Let W_i be the number of all possible configurations for the i-th event and w_i the number of configurations among which we cannot narrow down further. ($W_i \geqq w_i \geqq 1$). The amount of information in this case is

$$I = K \sum_{i=1}^{N} (\ell n W_i - \ell n w_i) \ . \tag{11.29}$$

A generalization of this similar to Eq. (11.27) is

$$I = -K \sum_{i=1}^{N} (\ell n P_i - \ell n p_i) \tag{11.30}$$

where P_i is the probability for one particular configuration to happen for the i-th event out of all possible configurations and p_i is the probability beyond which certainty we cannot predict the i-th outcome ($0 < P_i \leq p_i \leqq 1$).

Sometimes it is instructive to express Eq. (11.30) in terms of the phase space volumes (with the ergodicity ansatz)

$$I = K\sum_i(\ell n V_i - \ell n v_i) \tag{11.31}$$

$$= \int d\boldsymbol{\lambda}\, \ell n(V_{\boldsymbol{\lambda}}/v_{\boldsymbol{\lambda}}) \, , \tag{11.32}$$

where V_i is the volume of phase space for the i-th event to take and v_i the volume of phase space within which we have uncertainty. Equation (11.32) is the case with continuous parameters $\boldsymbol{\lambda}$ instead of discrete parameters. In a physical world, for example, v_i can be $\Delta x_i \Delta p_i$ which is $\gtrsim \hbar$ with $\hbar$ being the Planck constant $/2\pi$ according to Heisenberg's uncertainty principle. The operation $V_{\boldsymbol{\lambda}}/v_{\boldsymbol{\lambda}}$ corresponds to a collapse of the phase space volume from $V_{\boldsymbol{\lambda}}$ to $v_{\boldsymbol{\lambda}}$ by gaining knowledge and narrowing down possibilities with the true solution now being confined with the smaller volume $v_{\boldsymbol{\lambda}}$. The operation may be looked upon as a projection of the phase space volume $V_{\boldsymbol{\lambda}}$ onto narrower subspace $v_{\boldsymbol{\lambda}}$. It is imagined that in a human brain this projection onto narrower subspace by watching, hearing, touching, sniffing, etc., seems to happen very rapidly and efficiently with $\boldsymbol{\lambda}$ being a very large vector, including various signatures by sight, sound, size, smell, etc.

The above definition of the measure of information and partial information seems to make sense. It proves, however, very incomplete upon further scrutiny. Consider the following three cases: one blank magnetic tape, the second tape with utter noise for one minute and the third with a one-minute message. For the reader of the tape the first two are the same, because both carry no useful information. A paradox develops. According to the definition, Eq. (11.27), the amount of information of the first tape is zero, while that of the second tape is $I \sim \log_2 W$ bits > 0. The third tape contains approximately the same amount as well. This is clearly against our intuition. The key to resolution resides in "useful" information. Two comments are due here. First, the noise may or may not be quite random. With an appropriate deciphering device the noise may be interpreted to be a useful message. It may have been simply a noise for an untrained ear in this case. On the other hand, the noise may be completely random, as is the series of digits of the number π. In this case, it is hopeless to learn any information. (Wait! Except for what π is, of course). This is in spite of the fact that the second and the third have the same I's. This is because the definition of I was for the *gross* amount of information. Sometimes this can be said in a different way — that the quantity I only stands for a potential amount of information in this sense. The usefulness is not defined in this definition. This gross definition I of the measure of information also fails to take note of or distinguish the lack of information and misinformation (or disinformation). The essential distinction between useful and useless information one arises from the presence of correlation or a total lack of it (*randomness*).

Correlations are measured only by higher order probability functions or multi-body correlation functions. The gross quantities are computed from the one-body probability function $f(X_1)$, where X_1 stands for all variables for an event. This is similar to the one-body distribution probability function $f(X_1)$ of a plasma particle, where $X_1 = (\mathbf{x}_1, \mathbf{v}_1)$ and $f(X_1)$ means the probability of finding a particle of position $\mathbf{x}_1$ and velocity $\mathbf{v}_1$. (See Chapter 1.) In many-body physics the one-body distribution function is coupled with the two-body distribution and the two-body distribution is coupled with the three-body distribution and so on through the BBGKY-hierarchy.[39] This structure has been studied in depth in plasma physics by Rostoker and Rosenbluth.[40] They have derived the hierarchy of equations for the one-body distribution function, two-body, three-body, etc. from the Liouville equation that describes the flow of many particles in multidimensional phase space. Although the dynamics and interaction mechanisms are different from this physical description,[39,40,41] one should also be able to envision the dynamics and interaction mechanism of information processing in terms of the phase space flow, as suggested by Hopfield.[15] It is then natural to introduce the "many-body" distribution probability functions a la the BBGKY hierarchy. The s-body distribution function $f_s(X_1, \ldots, X_s; t)$ may be defined[41] by the Liouville distribution function $D(X_1, \ldots, X_N; t)$ through

$$f_s(X_1, \ldots, X_s; t) = \int d^{N-s} X_i D(X_1, \ldots, X_N; t) \ , \tag{11.33}$$

where we postulate that D obeys as yet an undefined Liouville equation of the phase space flow for information processing

$$\left(\frac{\partial}{\partial t} + \mathcal{L} \right) D(X_1, \ldots, X_n; t) = 0 \ . \tag{11.34}$$

The legitimacy of Eq. (11.34) is subtle and murky. For example, such an equation may be defined in the theory of quantum mechanical statistical physics, although in the quantum world the phase space volume can collapse upon measurement. The measure of information may be subject to a similar situation. By the attainment of knowledge, the unknown (or available) phase space volume collapses. However, just as in the quantum statistical mechanics, the *potentially* available phase space volume may be argued to be constant regardless of the gain of knowledge.

The distribution D or Eq. (11.33) generates the following chain of equations:

$$\begin{aligned} f(X_1) &\equiv f_1(X_1) \ , \\ f_2(X_1, X_2) &= f_1(X_1) f_1(X_2) + g(X_1, X_2) \ , \end{aligned} \tag{11.35}$$

$$\begin{aligned} f_3(X_1, X_2, X_3) &= f(X_1)f(X_2)f(X_3) + \sum_{\text{perm}} f_1(X_1)g(X_2, X_3) \\ &\quad + h(X_1, X_2, X_3) \;, \qquad (11.36) \\ &\;\;\vdots \end{aligned}$$

where we suppressed the time arguments and "perm" indicates summation over all possible permutations. Here $g(X_1, X_2)$ is a two-body correlation function (sometimes called an intrinsic correlation) that determines the probability to find event 1 at X_1 when event 2 is at X_2, and sometimes can be written in terms of the Bayesian joint probability function $P(X_1|X_2)$ as

$$g(X_1, X_2) = f(X_2)P(X_1|X_2) \; . \qquad (11.37)$$

Consider the following example: If we find the word "THE" often in an English message, the two-body distribution function $f_2(X_1, X_2) \equiv f_2(X_1 - X_2; \frac{1}{2}(X_1 + X_2)) \equiv f_2(X_1 - X_2; \xi)$ should be large for

$$\begin{pmatrix} X_1 \\ X_2 \end{pmatrix} = \begin{pmatrix} \cdots & T & & \cdots \\ \cdots & & H & \cdots \end{pmatrix}$$

and thus $g_2(X_1 - X_2; \xi)$ also peaks at the same "distance." Similarly $f_3(X_1, X_2, X_3)$ should have a peak at

$$\begin{pmatrix} X_1 \\ X_2 \\ X_3 \end{pmatrix} = \begin{pmatrix} \cdots & T & \cdot & \cdot & \cdots \\ \cdots & \cdot & H & \cdot & \cdots \\ \cdots & \cdot & \cdot & E & \cdots \end{pmatrix}$$

and so does h. The function h is sometimes called a bicorrelation. If one finds a large correlation at a certain "distance", one learns that there must be a certain piece of information with a meaning. Thus we define the measure of higher order correlated information by

$$\begin{aligned} I_2 &= -K\ell n f_2(12) = -K\ell n[f_1(1)f_1(2) + g(12)] \;, \qquad (11.38) \\ I_3 &= -K\ell n f_3(123) \\ &= -K\ell n \left[f_1(1)f_1(2)f_1(3) + \sum_{\text{perm}} f_1(1)g(23) + h(123) \right] , \qquad (11.39) \\ &\;\;\vdots \end{aligned}$$

where 1, 2, 3, etc., stand for X_1, X_2, X_3, etc. These definitions are a natural extension of Eq. (11.26). If the system exhibits finite but weak intrinsic correlations, one can expand Eqs. (11.38), (11.39), etc.

$$I_2 = -K\left\{\ell n f_1(1)f_1(2) + \ell n\left[1+\frac{g(12)}{f_1(1)f_1(2)}\right]\right\}$$

$$\cong -K\left[\ell n f_1(1) + \ell n f_1(2) + \frac{g(12)}{f_1(1)f_1(2)}\right] \qquad (11.40)$$

$$I_3 = -K\left\{\ell n f_1(1)f_1(2)f_1(3) + \ell n\left[1+\sum_{\text{perm}}\frac{g(12)}{f_1(1)f_1(2)}\right.\right.$$

$$\left.\left. + \frac{h(123)}{f_1(1)f_1(2)f_1(3)}\right]\right\}$$

$$\simeq -K\left[\ell n f_1(1) + \ell n f_1(2) + \ell n f_1(3)\right.$$

$$\left. + \sum_{\text{perm}}\frac{g(12)}{f_1(1)f_1(2)} + \frac{h(123)}{f_1(1)f_1(2)f_1(3)}\right] . \qquad (11.41)$$

Some approximations for $h(123)$ such as Kirkwood's superposition and the hypernetted chain approximation[42] may turn out to be useful for analyzing higher order correlation information. Such quantities may prove to be useful for the analysis of various information processing "devices" such as information in artificial intelligence machines (computers) and in brains, and in sequences of amino acids in DNA molecules, etc. Heeding one caution in interpretation is necessary. If there is a strong attractive correlation in 1 and 2, $g(12)$ takes a positive value, which contributes in reducing I_2 in Eq. (11.38), for example. This is because 1 and 2 were assumed or known to be correlated in advance (or *a priori*), but when we acquire actual knowledge about the state of 1 and 2, the collapse of the correlation of 1 and 2 is not as great as otherwise.

The generalized measures of information $I_2, I_3, \cdots$ may be utilized to obtain the most probable state etc. In thermodynamics the most probable distribution function is obtained by maximizing the entropy $S = -k\int f_1 \ell n f_1 d1$ with the energy $E = \int\left[\frac{1}{2}mv_1^2 + V(x_1)\right]d1$ kept constant. This can be done by the variational principle employing the Lagrange multiplier λ:

$$\delta S + \lambda\delta E = 0 .$$

Our generalized measures can be applied back to the usual thermodynamics as well, in which the new variational principle is to maximize the generalized

(n-th) entropy $S_n = -k \int f_n \ell n f_n d^n 1 \cdots n$ with the energy kept constant. We may require s generalized entropies maximized simultaneously, yielding s independent equations

$$\delta S_n + \lambda_n \delta E = 0 \quad , \quad (n = 1, \ldots, s) \ .$$

In the theory of information the variational principle may be introduced in many facets. One example is to describe the dynamics of the neural dynamics (Section 11.3). The "average" energy function may be defined as

$$\bar{E} = -\frac{1}{2} \sum_i \sum_j T_{ij} V_i V_j f_2(i,j) \ ,$$

while the second (generalized) "average" measure of information is

$$\bar{I}_2 = -K \sum_i \sum_j f_2(i,j) \ell n f_2(i,j) \ .$$

The most probable state may be described by extremizing $\bar{I}_2$ while $\bar{E}$ is kept constant. Another example of application of the variational principle of the measure of information is discussed in Section 11.5.

It may be worth mentioning the relation between information and negentropy.[43] The negentropy is defined by Eq. (11.22). Suppose that by applying an operation (or measurement) on a system we acquire information. With this operation we reduce the number of possible states the system can take from W_0 to W_1. The amount of information acquired by this is

$$I_1 = k(\ell n W_0 - \ell n W_1) \ , \tag{11.42}$$

where K was replaced by the Boltzmann constant k and an argument similar to derive Eq. (11.29) was invoked. Boltzmann's entropy expression is inserted on the right-hand side of Eq. (11.42) to yield

$$I_1 = S_0 - S_1 \ , \tag{11.43}$$

where $S_i = k \ell n W_i$. Equation (11.43) is rewritten as

$$\Delta N = N_1 - N_0 = I_1 \ . \tag{11.44}$$

We find that negentropy in the system increases by the amount of information acquired, i.e., the amount of information acquired is equal to the creation of negentropy. It has been found that:

(i) Information can be converted into negentropy and its reverse is possible as well. Unless it is reversible, there will be a loss in the process of conversion.

(ii) When a system acquires information, the entropy of the environment outside of the system (or the overall system) increases $\Delta S - \Delta I = \Delta(S - I) \leqq$

0, or $\Delta(N+I) \leqq 0$. The sum of negentropy and the measure of information always decreases (or remains constant).

(iii) In order to obtain a certain amount of information, there exists a minimum amount of negentropy. In order to obtain a bit of information, this amount of negentropy is approximately $k\ell n2 = 0.69k$.

A question may arise. If a biological system "eats" negentropy to survive, as Schrödinger[44] defined life, how can the biological environment guarantee that life can continue the process which is characterized by the "deteriorating direction" $\Delta(N+I) \geqq 0$? This question is important. It may be partially answered by the following observation. A bowl of ashes will be blown apart when the wind blows and thus it goes into an obviously higher entropy state. However, ash particles will not hang in the air forever, but they (or most of them) eventually come down on the surface of the earth. This process is most eloquently proved by geology and the rocks showing such history. It means that on the earth the ash particles apparently do not stay in the highest entropy state! The only possible hand that interfered with the law of maximum entropy is the earth's gravitation that eventually forced the ash particles to come down. Another observation: When you sit in a boat on the ocean, you can still sweat and the sweat still evaporates. According to the thermodynamical equilibrium the water vapor pressure should be in equilibrium with the ocean water, which means that the air would be saturated with water vapor and the sweat would not evaporate. This paradox may be resolved by the fact that the air can emit radiation into space, thereby put itself out of thermodynamical equilibrium. It is not possible to discern all aspects of entropy in the universe here (see Problem 11). Here we only mention the one peculiar aspect of the gravitational potential. That is that the attractive potential can have arbitrarily large (negative) energy states, as the particle approaches the gravity center. Lynden-Bell[45] introduced entropy in self-gravitating systems. In dissipative systems (such as radiation) the attractive potential system allows creation of negentropy (theoretically) indefinitely through gravitation. Thus ("theoretically") indefinite evolution becomes a possibility (of course, barring various cataclysmic disasters including such "man-made" disasters as nuclear wars).

11.5 Correlation Analysis and Maximum Entropy

To diagnose or analyze a complex many-body system (e.g., a plasma), we must be content with representing it by macroscopic variables (sometimes called intensive thermodynamical variables) such as density, fluid velocity, temperature etc. For a plasma we may want to add (macroscopic) electric and magnetic fields, and charge and current densities to these. In contrast, sometimes we would like to know more microscopic quantities such as the energy (or velocity) distribution of plasma particles. Of course, such microscopic data are much more difficult to obtain and even if they are obtained, they are too numerous to process. Data that are macroscopic but more detailed than intensive thermodynamical variables are correlation data. Correlation analysis is very important in cases where there are many modes or noise present. It is for this reason that in actual experiments and particle simulations this technique has been extensively utilized but rarely employed in fluid simulations so far.[46]

The (cross-)correlation function of certain quantities $A(t)$ and $B(t)$ is defined as

$$C_{AB}(\tau) = \lim_{T\to\infty} \frac{1}{T} \int_{-T/2}^{T/2} A(t)^* B(t-\tau) dt \ , \tag{11.45}$$

where the asterisk refers to a complex conjugate and τ is called the lag time. When $B = A$, the correlation is called the auto-correlation function. According to the Wierner-Khinchin theorem, the spectral intensity or density (or energy spectrum) $S(k, w)$ is expressed as

$$S(k,\omega) = \int_{-\infty}^{\infty} d\tau \ C_{AA}(\tau) e^{i\omega\tau} \ , \tag{11.46}$$

where $A = A_k(t)$. Using Eq. (11.46), the mode energy (or spectrum) is given by

$$S(k) = \int_{-\infty}^{\infty} \frac{d\omega}{2\pi} S(k,\omega) = C_{AA}(0) = \int_{-\infty}^{\infty} \frac{dt}{2\pi} S(k,t) \ , \tag{11.47}$$

where the last equality comes from Parseval's theorem (Prob. 6). When the quantity A is the electric field E, for example, $S(k, w)$ is the spectral density of the electric field. The fluctuating longitudinal electric field energy in a stable plasma in thermal equilibrium can be expressed in terms of the imaginary part of the dielectric function through the fluctuation-dissipation theorem[23,41,47] as

$$\frac{\langle E^2(k) \rangle}{8\pi} = \frac{S(k)}{8\pi} = \frac{T_e}{2} \frac{1}{1 + k^2/k_{De}^2} \ , \tag{11.48}$$

where T_e and k_{De} are the electron temperature and electron Debye wavenumber and we assumed only electron contributions here. Generalizations of this to electrons and ions and to electromagnetic cases may be found in Refs. 47 and 48. It has been quite a useful method[49] in testing a particle simulation code and analyzing its thermal equilibrium (and non-equilibrium) properties via measurements of autocorrelation functions and energy spectra. The interferometric diagnosis[49] is also useful for plasma mode analysis. For example, the electric field $(E_{k_y}(x,t))$ interferometry with the sinusoidal function is

$$C(k_y, x, \tau) = \lim_{T\to\infty} \frac{1}{T} \int_0^T E_{k_y}(x, t+\tau) \sin(\omega_0 t) dt \ , \tag{11.49}$$

where τ is the lag time, k_y the wavenumber in the y-direction, and the x-coordinate has been kept in real space when the plasma is non-uniform in that direction. An example of this application is found in Chapter 13. Leaving the coordinate x, and analyzing at a particular frequency ω_0 will give us the phase of the correlation function (as well as the amplitude) as a function of x. This method can be useful, if the plasma varies significantly in x. A particularly strong correlation should be obtained when ω_0 is chosen to be the relevant frequency of the system. This is in fact the case when the plasma contains an eigenfunction whose eigenvalue frequency is ω_0.

Most actual experimental data and all simulational data are digital. The above correlation functions are defined on continuous points, while digital data are only on discrete points. We introduce the sampling function[50,51]

$$\Delta(t) = \sum_{n=-\infty}^{\infty} \delta(t-n) . \tag{11.50}$$

Suppose there is a continuous time series of electric field [Fig. 11.5(a)], whose Fourier transform is also a continuous function over $(-\infty, \infty)$. However, our data are obtained only at discrete points separated by the sampling time t_s [Fig. 11.5(b)], where $f_N = \frac{1}{2t_s}$, the Nyquist frequency. Therefore, the actual sampled time series looks like Fig. 11.5(c), whose Fourier transform creates aliases. See Chapter 4 and Chapter 7 for more on aliases. Furthermore, the actual data is a segment of finite length [Fig. 11.5(c)], which generates secondary, tertiary,$\cdots$ peaks at $f = 1/T_0, 2/T_0, \cdots$. The Blackman-Tukey window function[52] tries to minimize these ripples and leakage. In the frequency domain, the data is also discrete. This leads to a real space data which is now periodic over the length T_0, while the original data was not. Compare Fig. 11.5(f) and Fig. 11.5(a). Since the simulation run is necessarily finite and sampled over digital steps, we must have enough steps to include all relevant periods of interest which in turn should be longer than the sample time t_s.

When many modes are excited such as in a turbulent plasma, it can be useful to measure higher order correlation functions. For example, for a time series $X(t)$, $Y(t)$, $Z(t)$,... we define the autocorrelation function of n-th order,[53–55]

$$R(t, \tau_1, \tau_2, \ldots \tau_n) = \lim_{T\to\infty} \frac{1}{T^{n-1}} \int_{-T/2}^{T/2} X(t)X^*(t-\tau_1)\cdots X^*(t-\tau_n) , \tag{11.51}$$

and a cross-correlation function. Associated cross-spectral density functions are, for example,

$$S_{YXX}(\omega_1, \omega_2) = \int_{-\infty}^{\infty} d\tau_1 \int_{-\infty}^{\infty} d\tau_2 R_{YXX}(\tau_1, \tau_2) e^{i(\omega_1\tau_1 + \omega_2\tau_2)} , \tag{11.52}$$

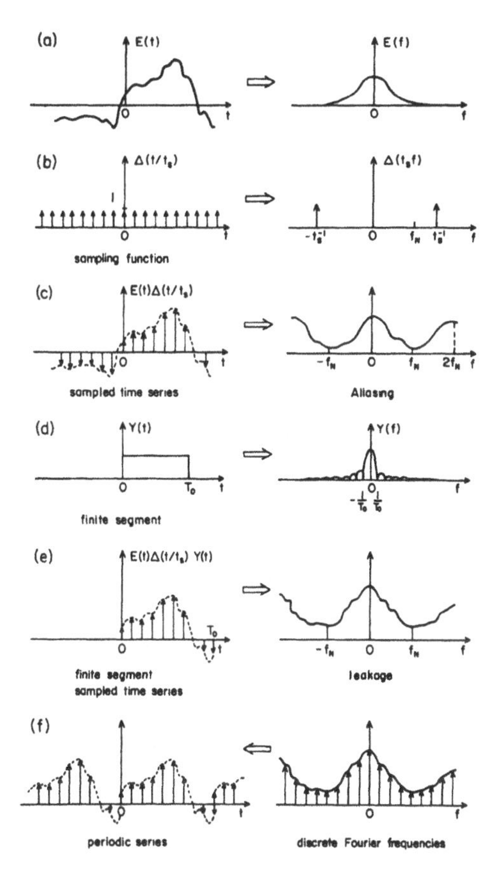

FIGURE 11.5 Time series, their Fourier transforms, and various processing, including the finite sampling points (Ref. 50)

where $\omega = \omega_1 + \omega_2$. The normalized spectral and bispectral functions are called coherency and bicoherency.[55,56] For example, the bicoherency is defined by

$$b^2(\omega_1, \omega_2) = \frac{|S_{YXX}(\omega_1, \omega_2)|}{[S_{XX}(\omega_1) S_{XX}(\omega_2) S_{XX}(\omega = \omega_1 + \omega_2)]^{1/2}} \,. \tag{11.53}$$

The bicoherency ranges from 0 to 1. The bispectrum has the following symmetry relations:

$$\begin{aligned} B(\omega_1, \omega_2) &= B(\omega_2, \omega_1) = B^*(-\omega_1, -\omega_2) \\ &= B(-\omega_1 - \omega_2, \omega_2) = B(\omega_1, -\omega_1 - \omega_2) \,, \end{aligned} \tag{11.54}$$

which allow us to look at a certain portion of quadrants.[56] The bicoherency helps to identify coupled waves. If wave a and wave b are coupled to produce wave c and $\omega_a + \omega_b = \omega_c$, the bicoherency $b^2(\omega_1, \omega_2)$ will peak at the point $(\omega_1 = \omega_a, \omega_2 = \omega_b)$ in the plane of ω_1, ω_2.

We now come back to the (ordinary second order) spectral functions. As we have seen, the digitized finite length data introduces a number of distortions on the original time series. Consider a stationary time series $X(t)(|t| \leqq N\Delta t$, where N is the maximum observed point and at the time step). Since we do not have data beyond N, we have to assume $X(t) = 0$ for $|t| > N\Delta t$ [see Fig. 11.5(d)]. In order to avoid ripples seen in the Fourier space of Fig. 11.5(d), a window function $w(t)$ is introduced to smooth the sudden cut-off of the data at $|t| = N\Delta t$.[52] The window function $w(t)$ is chosen in such a way as to optimize the two requirements: (i) to maximize the resolution of spectrum and (ii) to reduce the contaminated spectra, "side lobes", seen in Fig. 11.5(d), and, at the same time, the spectral function should stay nonnegative. Without changing the existing data, one of the choices is to choose the spectrum that is the most random time series whose autocorrelation function agrees with the known values ($|t| \leqq N\Delta t$).The other way to state this is as follows. According to the information theory[38] we discussed in Section 11.4, to know the time series means that by measurement a number of possible states have been collapsed into the particular one that was observed. During this process we gain information and increase the amount of information I. By interpolating, extrapolating, or guessing the complete time series or power spectrum from the known series, we obtain the most information if the guessed time series is most random so that it is most difficult to extract information from it. The resulting guessed information on the time series means that we have gained the maximum possible amount of information. This strategy can be mathematically stated as follows: Let $P(f)$ be the power spectrum of the time series of interest where f is the frequency. We try to find $P(f)$ that maximizes the value of

$$\int_{-W}^{W} \ell n P(f) df \ , \tag{11.55}$$

under the constraint that

$$R(n) = \int_{-W}^{W} P(f) e^{-i2\pi f_n \Delta t} df \quad , \quad (-N \leqq n \leqq N) \ , \tag{11.56}$$

where $R(n)$ is a given function and $W = f_N = \frac{1}{2\Delta t}$. With λ_n being Lagrange multipliers, the variational principle to obtain the maximizing $P(f)$ is written as

$$\delta \int_{-W}^{W} \left[\ell n P(f) - \sum_{n=-N}^{N} \lambda_n P(f) e^{-2\pi f_n \Delta t} \right] dt = 0 \ . \tag{11.57}$$

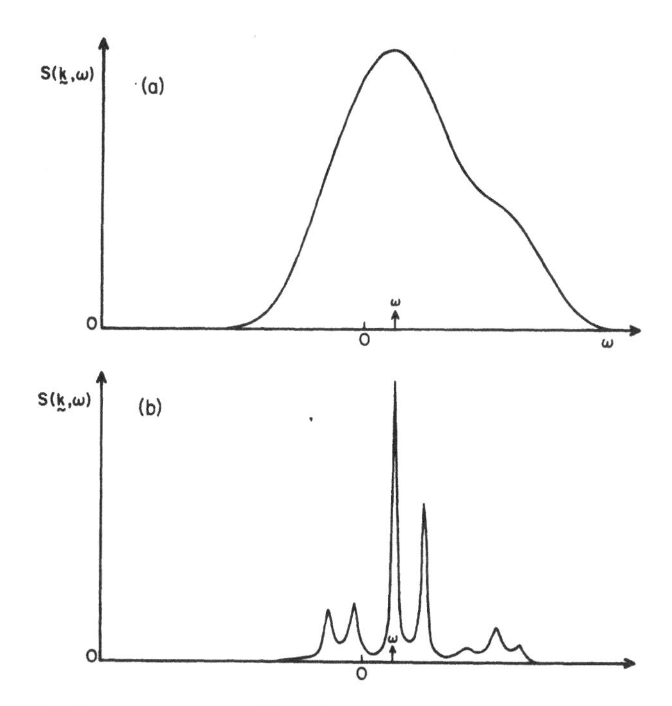

FIGURE 11.6 Power spectrum $S(\mathbf{k}, \omega)$ for the drift wave fluctuation obtained in a particle simulation, processed by the Fourier transform (Blackman-Tukey) method (a) and by the maximum entropy (Burg) method (Ref. 58)

Equation (11.57) yields

$$P(f) = \frac{1}{\sum_{n=-N}^{N} \lambda_n G_n(f)}, \tag{11.58}$$

where $G_n(f) \equiv e^{-2\pi f_n \Delta t}$.

The determination of λ_n is carried out as follows. Substitution of Eq. (11.58) into Eq. (11.56) yields $2N + 1$ equations

$$\int_{-W}^{W} \frac{e^{-2\pi f_n \Delta t}}{\sum_{s=-N}^{N} \lambda_s e^{+2\pi f s \Delta t}} df = R(n), \qquad (-N \leqq n \leqq N) . \tag{11.59}$$

Setting $z = e^{-2\pi f n \Delta t}$, we rewrite Eq. (11.59) into

$$\int_{-W}^{W} \frac{z^n df}{\sum_{s=-N}^{N} \lambda_s z^s} = \frac{1}{2\pi i \Delta t} \oint \frac{dz}{z^{n+1} \sum_{s=-N}^{N} \lambda_s z^s} = R(n) , \tag{11.60}$$

where the contour integral is on the unit circle. Since the power spectrum $P(f)$ should be real and positive on $|z| = 1$, Eq. (11.58) suggests[57] the form

$$\sum_{s=-N}^{N} \lambda_s z^s = \frac{1}{P_N \Delta t} \left(\sum_{s=0}^{N} a_s z^s \right) \left(\sum_{s=0}^{N} a_s^* z^{-s} \right) , \tag{11.61}$$

where P_N is a positive constant, and $a_0 = 1$. All of the roots of the first polynomial in z can be chosen to lie outside the unit circle. Substitution of Eq. (11.61) into Eq. (11.60) and contour integral yield after some algebraic manipulation[57]

$$\sum_{n=0}^{N} R(r-n) a_n = P_N \quad , \quad (r = 0) ,$$

$$\sum_{n=0}^{N} R(r-n) a_n = 0 \quad , \quad (r \geqq 1) . \tag{11.62}$$

These equations can be solved by inverting the matrix.[57]

Figure 11.6 compares the spectral function $S(\mathbf{k}, \omega)$ of electric potential for the drift wave (see Chapter 13) obtained by the maximum entropy method[57,58] [Fig. 1.6(b)] to the one by the regular Fourier transform method [Fig. 11.6(a)].[49] Both methods obtain the real frequency approximately correctly. The Fourier transform method has a much wider and obtuse peak, while the maximum entropy method has a sharp large peak with other smaller peaks at other fictitious frequencies and the width of the main peak does not directly relate to the damping rate of the mode. The width observed on the Fourier method is directly related to the damping rate of the mode and the width of the lag time compared with the wave period. An instructive detailed comparison of various analyzing techniques of time series may be found in Ref. 59.

Problems

1. If P, H, Y, S, I, and C occur in English with frequency (probability to find a character) p_P, p_H, p_Y, p_S, p_I, and p_C, what is the amount of knowledge to know the 7 letter word is in fact PHYSICS?
2. Define the amount of information when an information string contains misinformation.
3. Discuss the flow of negentropy and information for biological organisms.

4. Devise a cellular automaton model of a (nearly) collisionless electrostatic plasma model. Assign to a particular on a vertex $\mathbf{x}$ of a hexagonal grid, a velocity vector $\mathbf{v} = \mathbf{e}_a = (\cos 2\pi a/6, \sin 2\pi a/6)$ and a charge vector $\mathbf{e}_b = 1$ or -1 corresponding to an ion or electron. In order to define the local Coulombic interaction, define the collision rule: (i) two particles with different signs are allowed to occupy the same cell regardless of the velocity; (ii) two particles with the same sign are scattered with appropriate angles, including the head-on collision; (iii) three-body collisions are also appropriately defined, so that the total momentum is conserved. See Fig. 11.2 for example. Derive an equivalent "Vlasov" equation for this system.
5. Consider ion dynamics in the above model Prob. 4. Ion pushing is executed every so often compared to electrons. Consider how the plasma oscillations arise and what determines the plasma oscillation frequency. How does the Landau damping come into this model?
6. Prove Parseval's theorem

$$\int_{-\infty}^{\infty} |A(t)|^2 dt = \int_{-\infty}^{\infty} |A(\omega)|^2 d\omega \ ,$$

where $A(\omega) = \int_{-\infty}^{\infty} |A(t)| e^{i\omega t} dt/\sqrt{2\pi}$.
7. Obtain the power spectrum when we maximize $\int_{-W}^{W} P(f) \ell n P(f) df$ with the constraint Eq. (11.56).
8. In a plasma in thermal equilibrium show that the variational principle "maximizing the generalized two-body entropy $S_2 = -k \int f_2(12) \ell n f_2(12) d1 d2$ under the constant energy constraint" yields the two-body correlation function is proportional to the exponential of the inter-particle potential energy.
9. Obtain the electric field spectral density $S(k, \omega)$ Eq. (11.46) when the electric field goes like $E(x,t) = E_0 \cos(k_0 x - \omega_0 t) e^{-\gamma t}$.
10. Speculate the influence of the new high-temperature superconductors (Ref. 4) on the future of computers and computation.
11. Discuss entropy and generalized entropies and their evolution in the universe.
12. What are the benefits and trade-offs of the virtual memory in a computer? What is the optimal operation of the virtual memory machine? (Ref. 60).

References

1. S. Fernbach, in Appendix B of *The Influence of Computational Fluid Dy-*

namics on Experimental Aerospace Facilities (Committee on Computational Aerodynamics Simulation Technology, National Research Council, Washington, D.C. 1983), chaired by R. Smelt.

2. J. von Neumann, The Computer and the Brain (Yale University Press, New Haven, 1958).
3. H.H. Goldstone, *The Computer from Pascal to von Neumann* (Princeton Univ. Press, Princeton, 1972); J.V.V. Atanasoff, in *The Origins of Digital Computers, Selected Papers*, ed. by B. Randell (Springer Verlag, New York, 1973); also A.R. Mackintosh, Phys. Today, March, 1987, p. 25.
4. J.G. Bednorz and K.A. Müller, Z. Phys. **B64**, 189 (1986); M.K. Wu,. J.R. Ashburn, C.J. Torng, P.H. Hor, R. Meng, L. Gao, Z.J. Huang, Y.Q. Wang, and C.W. Chu, Phys. Rev. Lett. **58**, 908 (1987).
5. for example, J.C. Browne, Phys. Today, May, 1984, p. 28.
6. F.A. Matsen and T. Tajima, eds. *Supercomputers: Algorithm, Architecture, and Scientific Computation* (University of Texas Press, Austin, 1985).
7. S. Wolfram, Rev. Mod. Phys. **55**, 601 (1983).
8. U. Frisch, B. Hasslacher, and Y. Pomeau, Phys. Rev. Lett. **56**, 1505 (1986).
9. T. Shimada, S. Sekiguchi, K. Hiraki, and K. Nishida, in *Supercomputers: Algorithm, Architecture, and Scientific Computation*, F.A. Matsen and T. Tajima, eds. (University of Texas Press, Austin, 1985) p. 327 (1985).
10. J.D. Watson, *Molecular Biology of the Gene* (Benjamin, Menlo Park, 1965).
11. C.H. Bennett, Inter. J. Theor. Phys. **21**, 905 (1982).
12. D.O. Hebb, *The Organization of Behavior*, (Wiley, New York, 1949).
13. E.R. Caianiello, ed. *Neural Networks, Proceedings of the School on Neutral Networks* (Springer, Verlag, Berlin, 1967).
14. F. Rosenblatt, *Principles of Neurodynamics, Perceptions and the Theory of Brain Mechanisms* (Spartan Books, Washington, D.C., 1961).
15. J.J. Hopfield, Proc. Natl. Acad. Sci. USA **79**, 2554 (1982); *ibid. 81*, 3088 (1984).
16. W.A. Little and G.L. Shaw, Math. Biosci. **3**, 281 (1978).
17. W.D. Hillis, *Connection Machine* (MIT Press, Cambridge, 1985).
18. Y.C. Lee, G. Doolen, H.H. Chen G.Z. Sun, T. Maxwell, H.Y. Lee, and C.L. Giles, Physica **22D**, 276 (1986).
19. For example, S. Eubank, W. Miner, T. Tajima, and J. Wiley, Am. J. Phys. **57**, 457 (1989).

20. D.E. Knuth, *The Art of Computer Programming: Semi-Numerical Algorithms* (Addison-Wesley, Reading, 1981), Vol. 2, p. 238.

21. J. Hardy and Y. Pomeau, J. Math. Phys. **13**, 1042 (1972).

22. D. Forster, D.R. Nelson, and M.J. Stephen, Phys. Rev. **A16**, 732 (1977).

23. R. Kubo, J. Phys. Soc. Jpn. **12**, 570 (1957).

24. J. Rivet, U. Frisch, and C.R. Seanes, Acad. Sci. Ser. 2 **302**, 267 (1986).

25. D. Montgomery and G. Doolen, Phys. Lett. A, **120**, 229 (1987).

26. H. Chen and W.H. Matthaeus, Phys. Rev. Lett. **58**, 1845 (1987).

27. F.H. Harlow, in *Methods in Computational Physics* (B. Adler et al., eds.) Vol. 3 (Academic, New York, 1964) p. 31.

28. J.N. Leboeuf, T. Tajima, and J.M. Dawson, J. Comp. Phys. **31**, 379 (1979).

29. S.A. Orszag and V. Yakhot, Phys. Rev. Lett. **56**, 1691 (1986).

30. W.A. Little and G.L. Shaw, Behav. Biol. **14**, 115 (1975).

31. G.L. Shaw, E. Harth, and A.B. Scheibel, Exp. Neural. **77**, 324 (1982).

32. B. Katz, *Nerve, Muscle and Synapse* (McGraw-Hill, New York, 1966).

33. M.E. Fisher and W. Selke, Phys. Rev. Lett. **44**, 1502 (1980).

34. L.N. Cooper, in *Proceedings of the Nobel Symposium on Collective Properties of Physical Systems*, eds. B. Lundquist and S. Lundquist (Academic, New York, 1973), p. 252.

35. G.L. Shaw and R. Vasudevan, Math. Biosci. **21**, 207 (1974).

36. J.P. Crutchfield and N.H. Packard, Int. J. Theor. Phys. **21**, 433 (1982).

37. The example follows: E. Teramoto, *Physics of Life*, Fundamentals of Contemporary Physics Series Vol. 9, ed. by H. Yukawa (Iwanami, Tokyo, 1972).

38. C.E. Shannon and W. Weaver, *The Mathematical Theory of Communication* (Univ. of Illinois Press, Champagne, 1949).

39. B. Bogoliubov, in *Studies in Statistical Mechanics*, ed. by J. deBaer and G.E. Uhlenback (North-Holland, Amsterdam, 1962), Vol. I, p. 1.

40. N. Rostoker and M.N. Rosenbluth, Phys. Fluids **3**, 1 (1960).

41. N. Rostoker, Nucl. Fusion **1**, 101 (1961); N. Rostoker, Phys. Fluids **7**, 491 (1964).

42. S. Ichimaru, Rev. Mod. Phys. **54**, 1017 (1982).

43. L. Brillouin, *Science and Information Theory* (Academic, New York, 1956).

44. E. Schrödinger, *What is Life — The Physical Aspect of the Living Cell* (Cambridge Univ. Press, Cambridge, 1944).

45. D. Lynden-Bell, Mon. Not. R. Astr. Soc. **136**, 101 (1967).

46. An example of a notable exception: R.E. Waltz, Phys. Fluids **26**, 169 (1983).

47. A.G. Sitenko, *Electromagnetic Fluctuations in Plasmas* (Academic, New York, 1967).

48. J.L. Geary, T. Tajima, J.N. Leboeuf, E.G. Zaidman and J.H. Han, Comput. Phy. Comm. **42**, 313 (1986).

49. J.M. Dawson, Rev. Mod. Phys. **55**, 403 (1983).

50. M.B. Priestley, *Spectral Analysis and Time Series* (Academic Press, New York, 1981).

51. E.O. Brigham, *The Fast Fourier Transform*, (Prentice Hall, Englewood Cliffs, NJ (1974).

52. R.B. Blackman and J.W. Tukey, *The Measurement of Power Spectra* (Dover, New York, 1958).

53. K. Hasselmann, W.H. Munk, and G.J.F. MacDonald, in *Time Series Analysis*, ed. by M. Rosenblatt (Wiley, New York, 1963), p. 125.

54. D.R. Brillinger and M. Rosenblatt, in *Spectral Analysis of Time Series*, ed. by B. Harris (Wiley, New York, 1967), p. 153.

55. Y.C. Kim and E.J. Powers, IEEE Trans. Plas. Sci. **PS-7**, 120 (1979).

56. C.P. Ritz and E.J. Powers, Physica **20D**, 320 (1986).

57. J.P. Burg, Ph.D. Thesis (Stanford University, Palo Alto, (1975); J.P. Burg, Geophysics **37**, 375 (1972).

58. R.D. Sydora, Ph.D. Thesis (The University of Texas, Austin, Texas, 1985).

59. S.M. Kay and S.L. Marple, Proc. IEEE **69**, 1380 (1981).

60. P.C. Gray, J.S. Wagner, T. Tajima, and R. Million, Comp. Phys. Comm. **30**, 109 (1983).

12

INTERACTION BETWEEN RADIATION AND A PLASMA

In the following chapters we will discuss applications to actual physics problems for the various numerical algorithms and techniques we have presented in earlier chapters (Chapters 2-11). In ascending order, from Chapter 12 to Chapter 15, we shall present microscopic physics to macroscopic physics, smaller length scales to larger scales, and shorter time scales to longer scales. These descriptions of applications of various algorithms which may have been too dry to some readers hopefully make more vivid the implications of the algorithms. These will also introduce us to subtler interplays between algorithms and actual physical realities.

In this chapter we will discuss the interaction between electromagnetic radiation and plasma, which is currently a lively area of investigation. The time scale involved can be much shorter than the plasma oscillation period and, thus, the physics of radiation generally involves the most microscopic phenomena in the hierarchy of scales (see Chapter 1). In the following we first discuss radiation emanated from a plasma or charged particle beam, treating the radiation from beams and runaway electrons and the free electron laser. Second, we will consider laser plasma acceleration processes, in which electromagnetic radiation such as lasers accelerates particles to high energies. A third example is the problem of plasma heating by electromagnetic radiation. In a magnetized plasma an electromagnetic wave can propagate with a frequency even below the plasma frequency (the Alfvén waves).[1] The ion cyclotron resonance heating takes place when the compressional Alfvén wave resonates with the ion cyclotron frequency. This set of problems can be studied ideally by the simulation codes which resolve short time scales discussed in Chapter 5.

12.1 Radiation from Particle Beams

A. Radiation from Runaway Electrons

An electron beam propagating through a plasma is known to have the two-stream instability of both an electrostatic and an electromagnetic nature.[2,3] The electromagnetic interaction of a beam with a magnetized plasma can give rise to radiation emanating from the plasma. When a dc electric field is embedded along the field line, the beam particles are supplied energy and they can sustain, prolong, or intensify such interaction. In a magnetically confined plasma such as a tokamak plasma which has an induced electric field along the toroidal direction, be it in a steady operation or in a disruptive stage (see Chapter 14), some portion of plasma particles form a energetic tail because of the field. This class of particles is called runaway electrons, because the more energetic electrons become, the less collisional they become, and thus the more they are accelerated. Such runaway beams are detectable when they become so energetic that they become unconfined by the magnetic fields or when they emit X-rays and microwaves as results of electromagnetic coupling to the plasma. From astrophysical plasmas, on the other hand, kilometric radiation from the auroral zone in the magnetospheric plasma and decametric radiation and other types of bursts of radiation (solar bursts) from the solar coronal plasma have been detected. Many of the reasons for these types of radiation stem from a mechanism similar to the runaway radiation mentioned above.

An unmagnetized plasma supports only the electrostatic two-stream instability, while a magnetized plasma can have the electromagnetic two-stream instability which can emit radiation. When the two-stream instability develops, the beam kinetic energy is expended in supplying energy of the electrostatic and electromagnetic waves, thus it slows down. Eventually the instability saturates or disappears either due to trapping of the beam particles or to the "crashing" of the beam into the bulk distribution.[4] If one can avoid this trapping or crashing into the bulk, one would expect that the instability does not come to halt and the wave amplitude can keep increasing in time. When one applies a dc electric field parallel to the beam particle propagation direction, the field tends to counter-balance the electric field generated by the instability that would slow down the beam. By a judicious choice of strength of the dc electric field, we can expect that the growth of the wave field amplitude exceeds far beyond that without the dc field.

When the beam propagation direction and the guide magnetic field are parallel to the wavevector, the beam-plasma interaction is purely electrostatic.[4] To investigate a general situation in which the wavevector points the direction in an angle θ with the beam and the guide field direction, the tilting angle is now non-zero. This introduces electromagnetic coupling between beam and waves.

The simulation model is employed in Ref. 5 and is a one-and-two halves dimensional electromagnetic particle code (see Chapter 5). The only direction of spatial variation is the x-direction which is also the direction of wave propagation. The magnetized beam-plasma systems consists of a Maxwellian bulk of electrons, complemented by bulk Maxwellian ions when they are mobile, and a cold relativistic electron beam. The system length is $256\lambda_{De}$, where λ_{De} is the Debye length of the electrons at $\omega_{pe}t = 0$. The number of beam electrons and bulk electrons is 256 and 2304 respectively, so that the beam to background density ratio is $n_b/n_p = 1/9$. The speed of light is set at $c = 9v_e$, where v_e is the initial thermal velocity of the electrons and the initial beam electrons momentum is $p_b = 7.63mv_e$ where m is the electron mass. This yields a relativistic factor for the beam electrons $\gamma = 1.31$ at $\omega_{pe}t = 0$. When there is a dc electric field, it is applied to particles with momentum larger than the cut-off momentum $p_\ell = 6.0mv_e$. The strength of the dc field E_{dc} is set equal to $\tilde{E}_{dc}$ $(= eE_{dc}/m\omega_{pe}^2\Delta) = 0.2$, where ω_{pe} is the plasma frequency, e is the charge of one electron and Δ is the unit grid spacing equal to the initial electronic Debye length. This dc field is above the Dreicer field.[6] When the ions are mobile, the ion-to-electron mass ratio is $M_i/m_e = 10$ and the ion-to-electron temperature ratio is $T_i/T_e = 1$.

The constant magnetic field, along which the beam propagates and the dc field is applied, is such that the electron cyclotron frequency $\omega_{ce} = \omega_{pe}$. This static magnetic field is tilted in the x-y-plane at various angles θ with respect to the x-axis. This allows for quasi-two-dimensional effects because we can effectively have two projections of the wavevector, one along the field and one perpendicular to it as illustrated in Fig. 12.1.

The temporal evolution of plasma waves and radiation energy under the influence of a dc electric field is considered with various tilt angles of the magnetic field, dc electric field and beam propagation directions (all coinciding) with respect to the wavevector lying along the x-axis. The results without the dc field will be discussed first to shed light on the more complex cases with the dc field.

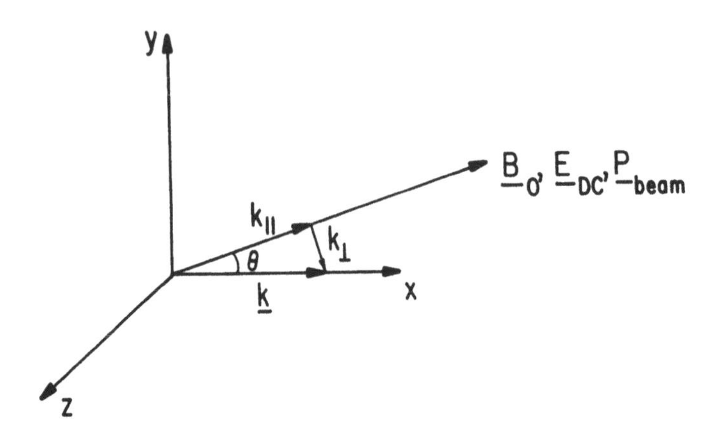

FIGURE 12.1 The direction of the external magnetic and dc electric fields and the direction of wavenumber vectors in the model (Ref. 5)

Beam-plasma Interaction Without dc Field

A beam of electrons causes a beam-plasma instability.[3,4] Electrostatic waves excited grow in time first linearly and then nonlinearly until the wave amplitude becomes large enough to trap[7] electrons or the decelerated phase velocity of the waves merges ("crashes") with the bulk distribution.[8,4] In general, the former happens when the beam is weak (low density), while the latter takes place when it is strong. With the dc field set equal to zero, simulations[5] show a definite change in the character of the beam-plasma instability as the tilt angle θ is increased. The simulation results are summarized in Fig. 12.2. The growth rate for electrostatic modes decreases[5] approximately with the angle θ as

$$\gamma_g(\theta) \sim \gamma_g(0) \cos^2 \theta \ , \tag{12.1}$$

where $\gamma_g(0)$ is the growth rate for $\theta = 0°$. The growth rate increases from 0 at $\theta = 0°$, thereafter attaining a value equivalent to the electrostatic growth rate and following its decrease with tilt angle θ. The simulation finds[5] that the tendency is for the electrostatic modes to dominate for $\theta \lesssim 45°$ and for the magnetic modes to take over for $\theta \gtrsim 45°$. Up to 45°, shorter wavelengths or larger wavenumbers grow. This can be understood as follows. In the electrostatic approximation the dispersion relation of the plasma without beam may be written[3] as

$$1 - \omega_{pe}^2 \cos^2 \theta / \omega^2 - \omega_{pe}^2 \sin^2 \theta / (\omega^2 - \omega_{ce}^2) = 0 \ , \tag{12.2}$$

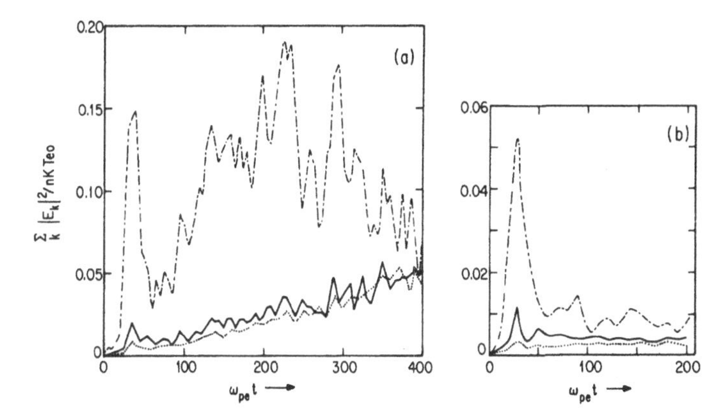

FIGURE 12.2 The field energy growth of the beam-plasma instability: (a) with d.c. field and (b) without it. The dash-dot curves are electrostatic waves and others are electromagnetic wave energies (Ref. 5)

with normal mode solutions

$$\omega^2 = 1/2(\omega_{pe}^2 + \omega_{ce}^2) \pm 1/2 \left[(\omega_{pe}^2 + \omega_{ce}^2)^2 - 4\omega_{pe}^2\omega_{ce}^2 \cos^2\theta\right]^{1/2} . \tag{12.3}$$

The upper root with the plus sign will be referred to as ω_1 and the lower one as ω_2 hereafter. Their electromagnetic counterpart can also be identified as an upper hybrid-type branch for the former and a whistler-type branch for the latter. For $\theta \lesssim 45°$ the coupling is between the beam mode with frequency $\omega \simeq kv_b \cos\theta$ and the upper hybrid branch with asymptotic frequency $\omega \simeq \omega_1$. For $\theta > 45°$, the coupling occurs between the beam mode again with frequency $\omega \approx kv_b \cos\theta$ and the whistler branch with asymptotic frequency $\omega \approx \omega_2$. According to linear theory, the wavevector of the excited mode is approximately given by

$$\begin{aligned} k_{1\alpha} &\sim \omega_1/v_b \cos\theta, \quad \theta \lesssim 45° , \\ k_{2\alpha} &\sim \omega_2/v_b \cos\theta, \quad \theta > 45° . \end{aligned} \tag{12.4}$$

These linear theory estimates agree with the simulation results.[5] The simulations show that saturation of the beam-plasma instability occurs.[7]

Beam-plasma Interaction With dc Field

When a dc electric field is applied to the beam-plasma system, the beam particles (or higher energy particles) have preferential acceleration because

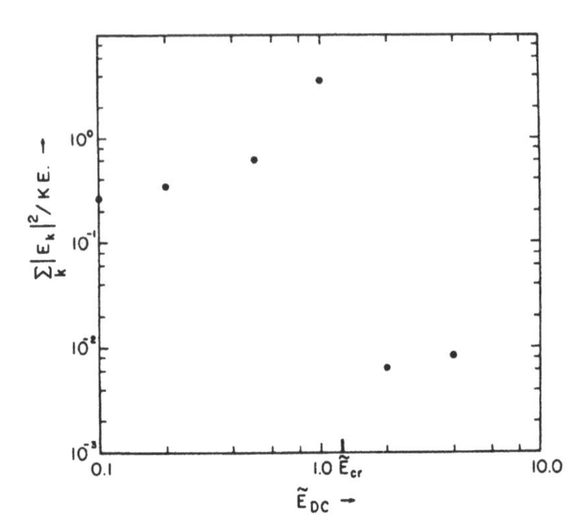

FIGURE 12.3 The excited electric field energy vs. the applied dc electric field E_{dc}. There is a collective threshold E_{cr} (Ref. 4)

Coulomb collisions are less frequent for these than less energetic particles. As a result, the wave electric field that decelerates the beam particles competes with the dc field and it takes more time to saturate them without the field. A major difference with the dc field is that (i) at saturation the amplitude of the wave is much higher than without it, (ii) after saturation of the initial beam-plasma instability (the beam-plasma phase), the wave energy continues to grow, attaining levels much higher (the runaway phase) than during the beam plasma phase. The process of wave energy enhancement for $\theta > 0°$ in Fig. 12.2(a) is illustrated (the title angle $\theta = 15°$) with $\tilde{E}_{dc} = 0.2$ and in Fig. 12.2(b) the same evolution without the dc field.

When the dc field is increased, the saturated amplitude of the wave is also increased. (See Fig. 12.3.) In fact, it increases rapidly as $E \rightarrow E_c$, a critical dc field. However, interestingly, when $E > E_c$, the beam-plasma system shows no instability as shown in Fig. 12.3. All energies remain at the thermal level. Beyond a collective threshold dc field, itself greater than the collisional Dreicer field, $\tilde{E}_c = \left[\gamma_s(1+s)^{-5/2}\right]^{1/2}$ with $s = (n_b/2n_p)^{1/2}(v_b/c)^2$, the beam would just runaway without any interaction with the background plasma.[4] The threshold dc field E_c corresponded to the maximum electrostatic field amplitude achieved by the waves during the beam-plasma phase. The reason for this pure runaway beam is that for such a dc field, the force imparted to the beam by the dc field is greater than the maximum drag or trapping force the waves can supply.

We have seen that with an applied dc electric field the beam-plasma interaction and the saturated amplitude of electric and magnetic fields are much enhanced. This phenomenon may be called driven enhanced radiation. This may play a role in a variety of astrophysical plasmas such as a mechanism of auroral kilometric radiation[9] and decametric and other radiation from solar coronal loop plasmas. Likewise this may be useful to enhance the gain of radiation output in a free electron laser, which will be discussed below.

Wave Spectra

We now consider what kind of waves are excited. The dispersion relations obtained from temporal autocorrelations (Chapter 11) of the various components of the longitudinal and transverse fields (E_x or electrostatic; E_y, E_z, B_y or electromagnetic) over the whole length of the simulations are presented. The dispersion relations are displayed as plots of the frequency, normalized to ω_{pe}, as a function of the mode number $M = kL_x/2\pi$.

When the beam is emerged in a plasma without $E_{dc}(E_{dc} = 0)$, the spectrum of waves as shown in Fig. 12.4 are excited. These various excited waves can be understood as follows. The simulation dispersion relations are first compared with a solution of the full electromagnetic dispersion relation without beam[2] yielding the normal modes of the system, upon which we superimpose the beam mode at $\omega = kv_b \cos\theta$ and the beam cyclotron resonances at $\omega = kv_b \cos\theta \pm \omega_{ce}/\gamma$ (the normal modes dispersion relation). This analysis should give a dispersion relation in the limit of infinitesimally dilute beam density. This first comparison is carried out for both electrostatic and electromagnetic fields. As a second approximation, the electrostatic simulation dispersion relation is compared with a solution of the electrostatic dispersion relation with beam given in Ref. 2 (the beam dispersion relation).

In terms of the polarizations of the normal modes in the full electromagnetic dispersion relation, the right-circularly polarized (R) and left-circularly polarized (L) modes are primarily defined from the parallel propagation characteristics of the waves, while the ordinary (O) and extraordinary (X) modes are primarily defined from their perpendicular propagation characteristics. For a case with a general tilt angle $\theta (0 < \theta < 90°)$, however, modes may be defined in both ways. The cut-off frequencies for the R waves and the L waves are

$$\omega_{\pm}^{c} = \frac{1}{2}\left[\left(\omega_{ce}^2 + 4\omega_{pe}^2\right)^{1/2} \pm \omega_{ce}\right] , \tag{12.5}$$

where ω_{+}^{c} is the R wave cut-off and ω_{-}^{c} is the L wave cut-off. A typical simplified dispersion relation for these modes is

$$c^2 k_{\parallel}^2/\omega^2 = 1 - \left(\omega_{pe}^2/\omega^2\right)/(1 \pm \omega_{ce}/\omega) . \tag{12.6}$$

On the other hand, the frequencies of the O wave and the X wave may be

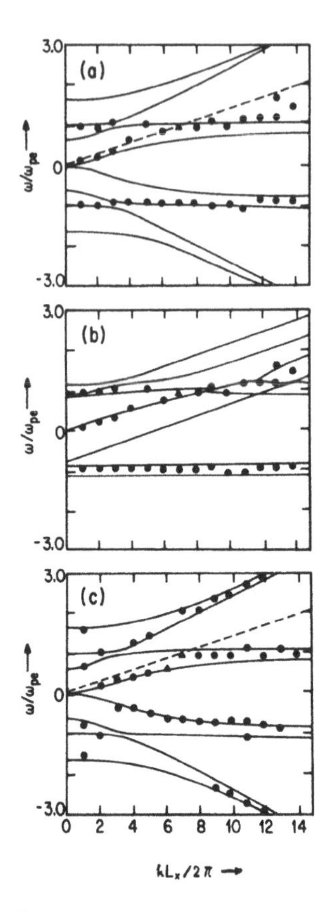

FIGURE 12.4 The wave dispersion relation without E_{dc}. The dots are simulation and curves are theory by (a) electrostatic dispersion without beam (beam is drawn in by a broken line), (b) electrostatic dispersion with beam, (c) electromagnetic dispersion without beam (Ref. 5)

characterized by the ω_1 and ω_2 frequencies of Eq. (12.3) respectively. In the limit of $\theta \to \pi/2$, ω_1 tends to the upper hybrid frequency $\omega_h = \left(\omega_{pe}^2 + \omega_{ce}^2\right)^{1/2}$ and ω_2 tends to the lower hybrid frequency if ion effects are included (right now no ion terms are included and $\omega_2 \to 0$ as $\theta \to \pi/2$). The typical dispersion relations for the O and X waves are respectively

$$c^2k_\perp^2/\omega^2 = 1 - \omega_{pe}^2/\omega_{ce}^2 , \tag{12.7}$$

$$c^2k_\perp^2/\omega^2 = 1 - \left[(\omega_{pe}^2/\omega^2)(\omega^2 - \omega_{pe}^2)/(\omega^2 - \omega_h^2)\right] . \tag{12.8}$$

For an arbitrary tilt angle θ, modes with a smaller wavenumber are more easily classified as an R or L wave. In Fig. 12.4(c), for instance, at $k \to 0$ the highest frequency is the R wave cut-off frequency ω_+^c, the second one is ω_{pe} and the third one is ω_-^c. In the lower half of the ω-plane the order is reversed. At $k \to \infty$, the upper two branches (X and O modes in that order) tend to $\omega = kc$, the next one to ω_1, and the last one to ω_2. Similarly in the lower half of the ω-plane the order is reversed. For intermediate values of k, modes contain both features (R or L and X or O). A whistler-like mode may be seen in the lower branches of Fig. 12.4(c) between mode numbers 1 and 6. As θ increases, the characteristics of the waves can be less identifiable as an R or L wave and more as an O or X wave.

Let us now discuss cases with the dc electric field applied ($\tilde{E}_{dc} = 0.2$) in Fig. 12.5. Note that compared with Fig. 12.4 we now see many new modes are strongly excited. Modes with frequency $\omega > \omega_{pe}$ can radiate out of the system.

Figures 12.5(a), (b) and (c) depict the dispersion relation with the dc field for $\theta = 15°$. Compare the electrostatic simulation dispersion relation with the theoretical normal mode with the beam mode at $\omega \simeq kv_f \cos\theta$, with $v_f = 8.54v_e$, superimposed on it [Fig. 12.5(a)] and with the beam dispersion relation [Fig. 12.5(b)]. We see that the beam branch (the runaway branch) is strongly excited with dc field for all the wavenumbers plotted. A harmonic branch appears clearly at $\omega \sim 2\omega_{pe}$ and the start of one at $\omega \sim 3\omega_{pe}$ to which the runaway branch strongly couples. The electrostatic mode with the highest spectral intensity is mode 5 with a measured frequency slightly smaller than ω_{pe}. Figure 12.5(a) indicates that the excitation of mode 5 corresponds to a coupling of the runaway branch with the ω_1 branch so that $\omega \sim kv_f \cos\theta \lesssim \omega_{pe}$. In summary, the beam branch and harmonics of ω_{pe} are strongly excited under E_{dc}.

B. Free Electron Laser

Sources of tunable coherent high power electromagnetic radiation are highly desirable of the application in many areas. In recent years, extensive efforts have been devoted to finding new mechanisms for the production of electromagnetic radiation by a relativistic electron beam interacting with a spatially modulated magnetic field. For example, CO_2 laser light amplification by this[10] and the generation of a large amount of microwave power[11] may be attributed to the process called the free electron laser. When a relativistic electron beam is injected through a helical static magnetic field, the generation of electromagnetic radiation is observed. The physical situation is shown schematically in Fig. 12.6 where λ_0 is the wavelength of the helical magnetic field. From a kinematic argument similar to the listener of the whistle from an incoming locomotive, the lasing wavelength is shortened by the Doppler shift to $\lambda = \lambda_0(c/v_0 - 1) \simeq \lambda_0/2\gamma^2$. In computer experiments,[12] it has been found that as much as 35% of the beam energy converted to radiation, of which as

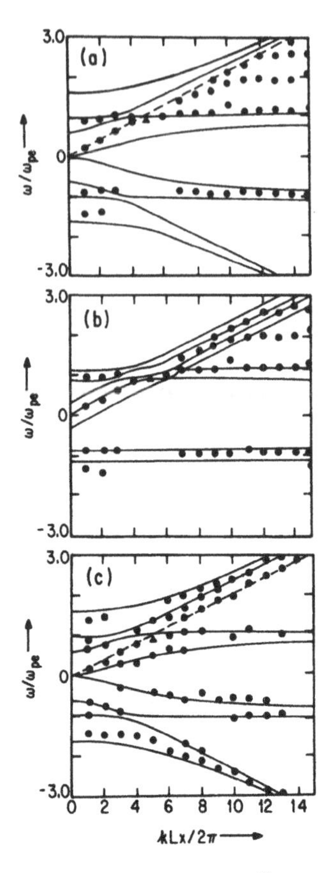

FIGURE 12.5 The wave dispersion relation with $\tilde{E}_{dc} = 0.2$. (a), (b), (c) are plotted likewise as in Fig. 12.5 (Ref. 5)

much as 90% was in the most rapidly growing mode.

The linear analysis shows the dispersion relation

$$\left[(\omega-(k+k_0)v_0)^2-\frac{1}{\gamma_0^3}\left(1+3(k+k_0)^2\right)\right]\times\left(\omega^2-k^2c^2-\frac{1}{\gamma_0}\right)$$

$$=\frac{\omega_{ce}^2(k+k_0)^2}{2\gamma_0^5k_0^2}\,, \tag{12.9}$$

where k_0, v_0, and γ_0 are the ripple (or wiggler) wavenumber, the beam velocity and energy at the beginning. The dispersion relation indicates that the plasma oscillations are coupled to the electromagnetic waves through the rippled magnetic field. Since the slow wave of plasma oscillation is a negative

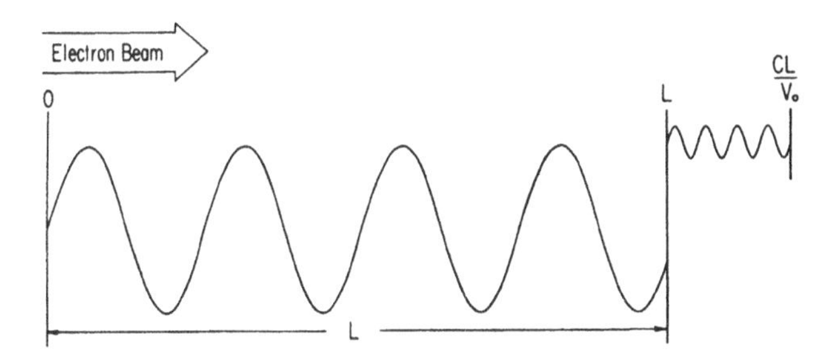

FIGURE 12.6 The wavenumber simplification in the free electron laser by the Doppler shift due to the incoming electron beam (Ref. 12)

energy wave,[13] we expect that its interaction with the electromagnetic wave will lead to instability. The unstable regions are the intercepts of the slow wave and the electromagnetic wave as shown. The approximate solution is[12]

$$\omega = T + \frac{1}{2}\left[-\mu \pm i(Q/TL - \mu^2)^{1/2}\right] , \tag{12.10}$$

where

$$T = (k^2c^2 + 1/\gamma_0)^{1/2} , \; L = \gamma_0^{-3/2}\left[1 + 3(k+k_0)^2\right]^{1/2} ,$$

$$D = (k+k_0)v_0 \quad , \quad \mu = T + L - D .$$

Equation (12.10) shows that for a given mismatch in frequencies, instability will occur if the strength of the rippled magnetic field is greater than a critical value. The threshold value of the magnetic field is given by

$$\Omega_e = (2TL\gamma_0^5)^{1/2}\mu k_0/(k+k_0) \tag{12.11}$$

and the growth rate is

$$\omega_i = \frac{1}{2}(Q/TL - \mu^2)^{1/2} . \tag{12.12}$$

For $\gamma_0 \gg 1$, $\omega_i \simeq \Omega_e/2(k_0c)^{1/2}\gamma^{3/4}$. The ratio of the electromagnetic and the electrostatic wave energies for large γ is

$$W_{em}/W_{es} \simeq 4k_0c\gamma^{7/2} , \tag{12.13}$$

therefore, most of the energy loss of the electron beam should be converted into radiation.

Using the fully relativistic electromagnetic particle code,[14] (see Chapter 5) many cases with different parameters are run to investigate the dependence

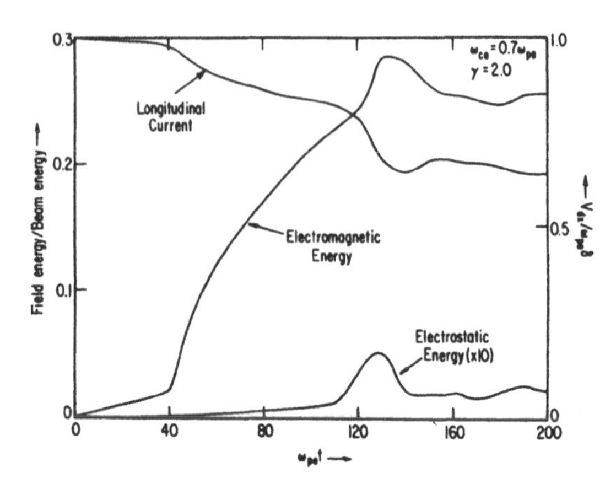

FIGURE 12.7 Temporal evolution of field energies in the free electron laser process (Ref. 12)

of the instability and its saturation level.[12] The system size is chosen[12] to be 256 grid lengths, and 2560 particles are followed in time. The normalized velocity of light is chosen to be 15. The static rippled or wiggler field was such that $k_0 c = 2.2\omega_{pe}$. The particles initially have a drift velocity corresponding to $\gamma = 2.0$.

Simulation results[12] show that the wavenumber matching condition, $k_p = k_{em} + k_0$, is satisfied between the unstable modes in both the electromagnetic and the electrostatic spectrum. The electromagnetic waves are coupled to the plasma oscillations through the rippled or wiggler magnetic field. The time evolution of the electromagnetic and the electrostatic energy, as well as the longitudinal current, are plotted in Fig. 12.7. At the time of saturation, the current was decreased by 36% while roughly 30% of the beam energy has been converted into radiation; of this, half is in the most unstable electromagnetic spectrum. In the computer experiments,[12] as the initial unstable spectrum grows to high amplitude, longer wavelength modes become unstable. The later process is believed to be caused by nonlinear mode coupling of the initial unstable waves. The phase space diagram shows that the electron beam has been slowed down considerably. The difference in beam energy goes into electromagnetic radiation.

The maximum growth rates are plotted versus the beam energy γ for three different strengths of the rippled magnetic field in Fig. 12.8. The solid curves are the maximum growth rates as predicted by the theory. The simulation data agreed well with the theory. The percentage of beam energy deposited in the initial unstable spectrum is given as a function of γ in Fig. 12.9 for three different strengths of the rippled magnetic field. The efficiency of energy conversion is found to decrease for increasing γ. Besides being a monotonically

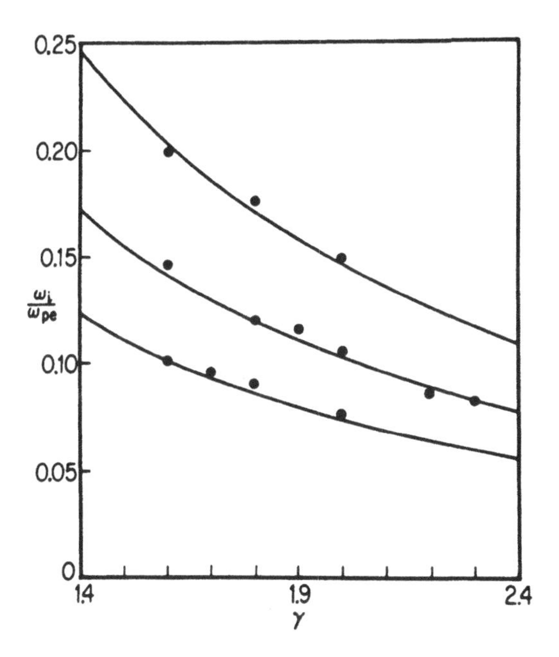

FIGURE 12.8 The growth rate of the pumped EM wave as a function of the electron beam energy γ for three different wiggler separations (Ref. 12)

decreasing function of γ, it is also sensitive to the wavelength of the rippled magnetic field. Cases with $\gamma = 1.8$ and $\gamma = 2.0$ for a ripple such that now $k_0 c \simeq 1.5\omega_{pe}$ are studied. The energy of the electromagnetic spectrum concentrates in the most unstable mode. For the case of $\gamma = 1.8$, the most rapidly growing mode (mode 9) has nearly 90% of the total electromagnetic energy. Therefore, the process could become efficient in amplifying light wave with a particular frequency by choosing the appropriate wavelength of the rippled magentic field. Since the lasing wavelength is roughly given by $\lambda_0/2\gamma^2$, where λ_0 is the wavelength of the ripple, the frequency of the radiation is reduced when we use a rippled magnetic field with a longer wavelength. In the case with $\gamma = 2.3$, radiation with frequencies up to $14\omega_{pe}$ has been observed in the simulation, and the generation of even higher frequency electromagnetic waves is possible by increasing γ.

Of considerable importance is the saturation level since this determines the potential efficiency of radiation production. The simulations indicate that saturation is due to electron trapping in the electrostatic wave which is generated.

The dispersion relation given by Eq. (12.10) indicates that the instability

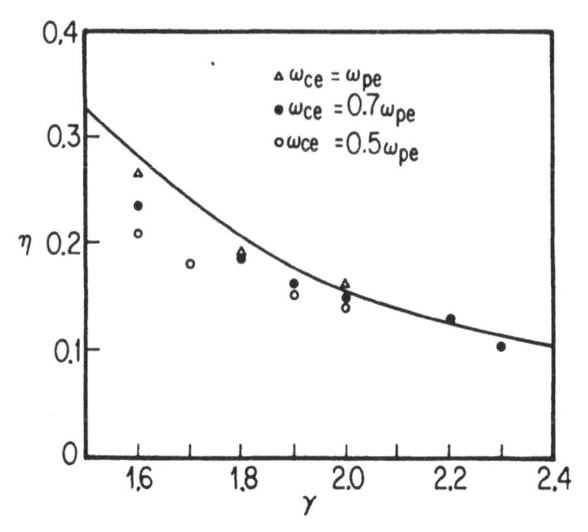

FIGURE 12.9 The lasing efficiency as a function of the beam energy γ (Ref. 12)

arises from the coupling between the electromagnetic and electrostatic plasma waves through the rippled magnetic field. As the wave grows and the beam loses energy during the instability, it will eventually reach an amplitude at which it can trap most of the electrons in the beam. After the trapping process sets in, the instability eventually saturates. At this time the electron beam is, on the average, slowed down to the phase velocity of the beam mode. Then, the change in energy of the electron beam is given by

$$\Delta W = m_0 c^2 (\gamma_0 - \gamma_{ph}) \tag{12.14}$$

where γ_0 applies to the initial beam velocity, while γ_{ph} is the γ corresponding to the phase velocity of the wave. Assuming all the energy loss from the electron beam converted into electromagnetic radiation, the efficiency, η, is then

$$\eta = |\Delta W|/(\gamma_0 - 1) m_0 c^2 = (\gamma_0 - \gamma_{ph})/(\gamma_0 - 1) \; . \tag{12.15}$$

Figure 12.9 illustrates its dependence on the relativistic γ. For large γ,

$$\eta \simeq (2 k_0 c \gamma^{3/2})^{-1} \; . \tag{12.16}$$

The simulation[12] has shown the nonlinear saturation mechanism and the efficiency of the beam energy conversion into radiation.

12.2 Laser Plasma Accelerators

A reverse process of the free electron laser is particle acceleration by electromagnetic and plasma waves.

An applied electric field of 0.1 GeV/cm (rf or dc) corresponds to 1eV/Å and thus severely modifies the electron wavefunction within the atom, leading to sparking, breakdown, etc. Thus when we try to accelerate particles to high energies by applying a high voltage, the structure for holding or molding accelerating fields tends to break down. Since the particles accelerated are at such a high energy that they travel with a velocity very close to the speed of light c, the phase velocity of the accelerating structure (electrostatic or electromagnetic) must also be very close to the speed of light c. A possible way to overcome this difficulty is to try to accelerate particles using the waves of an (already broken down) plasma.[15,16]

There have been many attempts to gain net acceleration by electromagnetic waves, which may be categorized into two types: the virtual photon approach and the real photon approach.[15] Consider Fig. 12.10(a). In order to obtain an electric field component parallel to the wave propagation $k_{\parallel}$, some (metal) reflector is used. The wave has $E_{\parallel}$ component, but unfortunately the phase velocity of the wave is larger than the speed of light $\omega/k_{\parallel} > \omega/k = c$, thus no coupling occurs. One may confine the electromagnetic field by two conductors in a waveguide [Fig. 12.10(b)] instead of the semi-infinite case in Fig. 12.10(a). The characteristics of wave phase velocity of a waveguide are well-known [see Fig. 12.10(c)]. The phase velocity is always larger than c: $v_{ph} = c(1-(\omega_c/\omega)^2)^{-1/2}$ where ω_c is the cutoff frequency for the waveguide. An accelerator such as linacs alleviates this problem by implementing a periodic structure in the waveguide (irises). A periodically rippled waveguide introduces the so-called Brillouin effect into the phase velocity characteristics. The Brillouin diagram for the frequency vs. wavenumber in the ripple waveguide is shown in Fig. 12.10(d). Here sections of phase velocity less than c are realized. There are other ways to make phase velocity less than c, such as by using a dielectric coating. All these techniques may be collectively called a technique for slow-wave structure. Particles can now *surf* on such field crests and obtain a net acceleration. The intensity of the fields is limited by materials considerations such as electric breakdowns, and almost invariably, the localized high electric field results at and near the slow-wave structure, making the breakdown easier there. It may be possible to create a slow-wave structure in the form of a plasma waveguide (or plasma "optical fiber").[17] A narrow waveguide with rippled surface accentuated either by a plasma or by plasma and magnetic fields provides a slow-wave structure in which the particles may be able to surf on the electromagnetic wave. This structure could be supplied by a crystal channel for injected hard X-rays as well.[18]

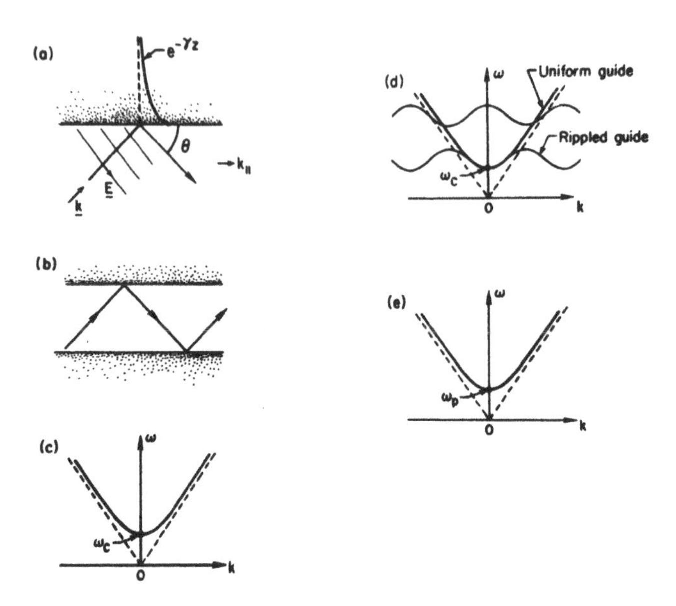

FIGURE 12.10 The electrostatic wave propagation in semi-infinite (a) and waveguide (b) geometry, and plasma (e) and their corresponding dispersion relation (c) and (e). The effect of slow-wave structure is indicated in (d). (Ref. 16)

An alternate approach is to nonlinearly excite a longitudinal wave of plasma oscillations induced by two beating electromagnetic waves (laser lights).[15]

The basic mechanism of particle acceleration in this approach is as follows. The two injected laser beams induce plasmons (or a Langmuir wave) through the forward Raman scattering process. This may be regarded as optical mixing.[19] The resultant large-amplitude plasma wave with phase velocity very close to the speed of light grows and is sustained by the laser lights. It grows until the amplitude becomes relativistic, i.e., the quivering velocity of the electrostatic field becomes on the order of c.

A. Beat-wave accelerator

Consider two large-amplitude traveling electromagnetic waves, (ω_0, k_0) and (ω_1, k_1), injected into an underdense plasma, which induce a plasma wave $(\omega_p, k_0 - k_1)$ through the beating of two electromagnetic waves if the frequency separation of the two electromagnetic waves is equal to the plasma frequency:

$$\omega_0 - \omega_1 = \omega_p \ , \tag{12.17}$$

$$k_0 - k_1 = k_p \ , \tag{12.18}$$

where k_p is the wavenumber of the plasma wave. The beat of the two electromagnetic waves gives rise to a nonlinear ponderomotive force, which sets off the plasma oscillations. It may be possible to achieve the objective through the forward Raman instability,[20] i.e., the second electromagnetic wave (ω_1, k_1) grows from a thermal noise. Rosenbluth et al., have discussed plasma heating via the beat of the electromagnetic wave.[21–23] It is important that the plasma be sufficiently underdense so that ω_0 is much larger than ω_p. This will ensure that the phase velocity of the plasma wave v_p is very close to the speed of light:

$$v_p = \frac{\omega_p}{k_p} = \frac{\omega_0 - \omega_1}{k_0 - k_1} \tag{12.19}$$

In the limit of $\omega_p/\omega_0 \ll 1$, Eq. (12.19) yields

$$v_p = \frac{\omega_p}{k_p} = \lim_{\omega_p/\omega_0 \to 0} \frac{\omega_0 - \omega_1}{k_0 - k_1} = v_g^{\mathrm{EM}} = c\left(1 - \frac{\omega_p^2}{\omega_0^2}\right)^{1/2} . \tag{12.20}$$

Suppose that ω_0 is not very much larger than ω_p, then the phase velocity of the plasma wave is much less than c. The resultant plasma wave can quickly trap electrons and saturate. In the course of interaction the nonlinear effects may change the phase velocity of the plasma wave. Thus, in the case where ω_p/ω_0 is not small, the interaction of light waves and plasma is strong, and the light waves suffer strong feedback from the plasma. (The light waves may be called "plastic" or "soft" in this case.)

This mechanism calls for a highly underdense plasma $\omega_p/\omega_0 \ll 1$, which guarantees that the plasma wave remain reinforced or "regulated" by the two beating laser beams

$$\omega_0 = k_0 c/\sqrt{1 - \omega_p^2/\omega_0^2} \sim k_0 c \ ,$$

$$\omega_1 = k_1 c/\sqrt{1 - \omega_p^2/\omega_1^2} \sim k_1 c \ .$$

(The light waves may be called "hard" in this case.) Since the phase velocity v_p is very close to c for $\omega_p/\omega_0 \ll 1$, the electron trapping and, therefore, saturation, occur only when the electrostatic wave grows up to an amplitude so large that it becomes relativistic. The other important consequence of $\omega_p/\omega_0 \ll 1$ is that particles will be in phase with the plasma wave for a long time and achieve a large amount of acceleration, because, again, the phase velocity $v_p \sim c$ and the particles would not exceed v_p easily.

Let us consider the energy gain of an electron trapped in the electrostatic wave with phase velocity $v_p = \omega_p/k_p$. We go to the rest frame of the photon-induced longitudinal wave (plasma wave). Since the wave has the approximate phase velocity given in Eq. (12.20), $\beta = v_p/c$ and $\gamma = \omega_0/\omega_p$. Note that this frame is also the rest frame for the photons in the plasma: in this frame the photons have no momentum and the photon wavenumber is zero. The Lorentz transformations of the momentum four-vectors for the photons and the plasmons (the plasma wave) are

$$\begin{pmatrix} \gamma & i\beta\gamma \\ -i\beta\gamma & \gamma \end{pmatrix} \begin{pmatrix} k_0 \\ i\omega_0/c \end{pmatrix} = \begin{pmatrix} 0 \\ i\omega_p/c \end{pmatrix} ;$$

$$\begin{pmatrix} \gamma & i\beta\gamma \\ -i\beta\gamma & \gamma \end{pmatrix} \begin{pmatrix} k_p \\ i\omega_p/c \end{pmatrix} = \begin{pmatrix} k_p/\gamma \\ 0 \end{pmatrix} \tag{12.21}$$

where the right-hand side refers to the rest frame quantities with respect to the plasma wave ($k_p^{\text{wave}} = k_p/\gamma$), k_0 is the photon wavenumber in the laboratory frame, and the well-known dispersion relation for the photon in a plasma $\omega_0 = (\omega_p^2 + k_0^2c^2)^{1/2}$ is invoked. Equation (12.22) is reminiscent of the relation between the meson and the massless (vacuum) photon:[24] Eq. (12.22) indicates that the photon in the plasma (dressed photon) has rest mass ω_p/c, because the electromagnetic interaction shielded by plasma electrons can reach only the collisionless skin depth c/ω_p in the plasma. This is just as the nuclear force reaches the inverse of the meson mass, and Yukawa[24] predicted the meson energy as $\omega = (c^2/a^2 + k^2c^2)^{1/2}$, where a is the nuclear interaction radius. Compare:

Meson	**Photon in Plasmas**	**Vacuum Photon**
(interaction length a)	(interaction length c/ω_p)	(Interaction length ∞)
$\omega = \left(\frac{c^2}{a^2} + k^2c^2\right)^{1/2}$	$\omega = \left(\omega_p^2 + k^2c^2\right)^{1/2}$	$\omega = kc$

At the same time, the Lorentz transformation gives the longitudinal electric field associated with the plasmon as invariant ($E_L^{\text{wave}} = E_L$). We note here that the plasma frequency really plays the role of the rest mass, as the plasma frequency ω_p is invariant under the Lorentz transformation. An alternative to this is to Lorentz-transform the plasma density $n' = \gamma n$ and the electron mass $m' = \gamma m$. In doing so, the plasma frequency $\omega_p = (4\pi e^2 n/m)^{1/2} = (4\pi e^2 n'/m')^{1/2}$ remains invariant.

The electrostatic wave amplitude can be evaluated as follows.[15] When most of electrons are bunched as a result of the beat electromagnetic waves, we may estimate the electrostatic field by assuming that most of the electrons give rise to this field:

$$\nabla \cdot \mathbf{E} = -4\pi e n \quad \text{or} \quad k_p E_L = -4\pi e n \ , \tag{12.22}$$

where n is the electron density. The maximum electrostatic field is, therefore,

$$E_L = m\omega_p c/e \ . \tag{12.23}$$

Rosenbluth and Liu[21] studied the nonlinear saturation due to the relativistic mass and subsequent detuning and obtained the saturated electric field at $\Delta\omega = \omega_0 - \omega_1 = \omega_p$ as

$$E_L = \frac{m\omega_p c}{e}\left(\frac{16}{3}\frac{eE_0}{m\omega_0 c}\frac{eE_1}{m\omega_1 c}\right)^{1/3} \ . \tag{12.24}$$

The electric potential due to the plasma wave evaluated in the laboratory frame is

$$e\varphi = e\int_0^\lambda E_L dx = e\frac{m\omega_p c}{e}\left(\frac{c}{\omega_p}\right) = mc^2 \ . \tag{12.25}$$

Going to the wave frame, we obtain the potential in the wave frame

$$e\varphi^{\text{wave}} = \gamma e\varphi = \gamma mc^2 \ . \tag{12.26}$$

This energy in the wave frame corresponds to the laboratory energy by the Lorentz transformation

$$\begin{pmatrix} \gamma & -i\beta\gamma \\ i\beta\gamma & \gamma \end{pmatrix}\begin{pmatrix} \gamma\beta mc \\ i\gamma mc^2 \end{pmatrix} = \begin{pmatrix} 2\gamma^2\beta mc \\ imc\gamma^2(1+\beta^2) \end{pmatrix} , \tag{12.27}$$

where the right-hand side refers to the laboratory frame quantities. Thus we obtain the maximum energy electrons can achieve by the plasma wave trapping as[15]

$$W^{\max} = \gamma^{\max} mc^2 = 2\gamma^2 mc^2 = 2\left(\frac{\omega_0}{\omega_p}\right)^2 mc^2 \ . \tag{12.28}$$

The time to reach energies of Eq. (12.28) may be given by

$$t_a \simeq W^{\max}/ceE_L = 2\left(\frac{\omega_0}{\omega_p}\right)^2 /\omega_p \tag{12.29}$$

and the length of acceleration to reach the Eq. (12.29) energy as

$$\ell_a \simeq 2\omega_0^2 c/\omega_p^3 \ . \tag{12.30}$$

To demonstrate the above mechanism for electron acceleration, computer simulations have been carried out employing a $1\frac{2}{2} - D$ (one spatial and three velocity and field dimensions) fully self-consistent relativistic electromagnetic code,[14] as discussed in Chapter 5. Two parallel electromagnetic waves (ω_0, k_0) and (ω_1, k_1) are imposed on an initially uniform thermal electron plasma. The

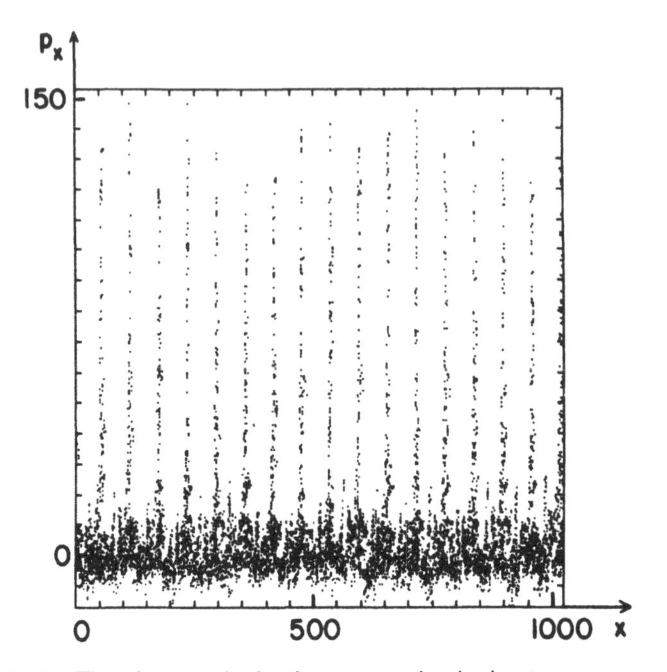

FIGURE 12.11 The electron plot in phase space by the beat wave

direction of the photon propagation as well as the allowed spatial variation is taken as the x-direction. The system length is $L_x = 10240$, the speed of light $c = 10\omega_p\Delta$, the photon wavenumber $k_0 = 2\pi \times 68/1024\Delta$, the number of electrons 1024Δ, and the particle size 1Δ with a Gaussian shape, and the ions are fixed and uniform. The thermal velocity $v_e = 1\omega_p\Delta$. The photon frequencies are taken as $\omega_0 = 4.29\omega_p$ and $\omega_1 = 3.29\omega_p$, while the amplitudes are $v_i \equiv eE_i/m\omega_i = c(i = 0 \text{ or } 1)$.

Figure 12.11 shows the phase space of electrons accelerated by the beat plasma wave $k_p \simeq \omega_p/c$. High energy electrons are seen as highly elongated tails in every ridge of each length of the resonantly excited electron plasma wave. The horizontally stretched arms in Fig. 12.11 are separated by a length $\lambda = 2\pi/k_p$. Extensive numerical works[25–27] have been carried out to confirm the acceleration mechanism of Tajima and Dawson.

As the laser beam becomes more intense, the accelerated electrons become more numerous and they are more energetic. When the laser beam is ultrarelativistic (i.e., $eE_0/m\omega_0 c > 1$) and the laser wavepacket is localized, such a wavepacket exerts a large ponderomotive force on the plasma and can create a local vacuum.[28] The intense electromagnetic wave pulse pushes the plasma forward and expands the region of plasma-plowed area (vacuum or very low density plasma). Since the expanding electromagnetic pulse acts like a piston that reflects incoming particles, we can evaluate how much momentum transfer takes place during the process. Equating the electromagnetic pressure (piston pressure) to the momentum exchange by electrons which are pushed by the piston, we obtain

$$P = \frac{E_0^2}{8\pi} = nv_g p \simeq nv_g \gamma mc \ , \tag{12.31}$$

where v_g is the velocity of the electromagnetic wave front. From Eq. (12.31), the particle energy γ^{ave} is calculated as

$$\gamma^{\text{ave}} \simeq \left(\frac{c}{v_g}\right)\left(\frac{\omega_0}{\omega_p}\right)^2\left(\frac{eE_0}{m\omega_0 c}\right)^2 , \tag{12.32}$$

which agrees with Eq. (12.28) in scaling when $eE_0/m\omega_0 c = 1$. Simulations performed[28] show this parametric dependence. The obtained energy scaling[28] in fact shows $\gamma^{\text{max}} \propto (eE_0/m\omega_0 c)^2$.

In the original paper by Tajima and Dawson[15] they pointed out that the excitation of an intense plasma wave can be accomplished also by an appropriately pulsed electromagnetic beam. In place of the photon pulse one can immerse an electron beam pulse. This has been recently called a wake-field accelerator.[29]

B. Plasma Fiber Accelerator

It is of considerable interest and concern to see whether intense laser light can be propagated over a long distance without deteriorating the laser beam quality too much. If the laser light deteriorates too quickly or refracts over a short distance, we have to reshape the laser light or inject a fresh laser light, which would amount to a complicated engineering headache and too much higher power consumption for the lasers.

We address here a few crucial questions associated with the acceleration of particles to high energies with the present concept. The first problem is *transverse* deterioration: the defocusing of the laser light due to the laser optics as well as to the plasma nonlinear effects on the laser light. The self-trapping effect could overcome the defocusing problem. The second problem is the *longitudinal* deterioration of acceleration: the dephasing between the accelerating electrostatic plasma wave and the particles being accelerated. We discuss the concept of the plasma fiber accelerator and triple soliton to cope with problems of both transverse and longitudinal detuning.

Let us consider the self-trapping and self-focusing problem. For general reference see Akhmanov et al.[30] The laser light has a focal length associated with it, the Rayleigh length $z_R = \pi w_0^2/\lambda_\ell$ where λ_ℓ is the laser wavelength and w_0 the focal waist. Fortunately, when the laser light is sufficiently intense, self-trapping of the light beam can take place at a certain threshold laser power.[31]

The self-focusing has been observed in the simulation of Refs. 26 and 27. When the laser power is at the threshold, the laser beam propagates without defocusing, overcoming the natural tendency of spreading over the Rayleigh length. The idea of the plasma fiber accelerator[17] is introduced to tackle two problems simultaneously: (i) the longitudinal phase mismatch between the plasma wave and the particles; (ii) the tendency of laser light to spread in the transverse direction. The reader is referred to Refs. 16 and 17 for a more detailed discussion.

C. Triple Soliton Accelerator

The pump power depletion in the ultra-high energy accelerator is a severe problem. Its severity may be appreciated by considering the desired luminosity $\mathcal{L}$. Since the relevant cross-section σ decreases inversely proportional to the square of the center-of-mass energy $E_{\rm cm}$, the luminosity $\mathcal{L}$ has to increase as $\mathcal{L} \propto E_{cm}^2$, (e.g., $\mathcal{L} = 10^{34} cm^{-2}\,\sec^{-1}$ at 10TeV e^+e^-),[16] in order to keep the number of relevant events constant in a given time, and thus the necessary power supplied to an accelerator $P \propto E_{cm}^2$. In the beat-wave accelerator,[15] however, the pump depletion problem is uniquely important and possibly quite damaging. This is because the accelerating electrostatic beat wave propagates with group velocity close to zero, i.e., $3(v_e/c)v_e$ with v_e being the electron thermal velocity, while the laser beams propagate approximately with the speed of light c. Thus the laser beams that enter the fresh plasma expend their energy in creating the plasma wave that is left out in the laboratory (plasma rest frame) behind the pump beams. Thus the beat wave accelerator would suffer greatly from this heavy pump depletion in addition to the already severe pump power requirement.

The idea of wakeless triple soliton acceleration is introduced[32] in order to overcome this difficulty. The essence of this idea is to use the self-induced transparency due to the plasma nonlinearity and deliberately shaped optical beams, to drive a triple soliton accelerating structure without a wake. A phenomenon similar to this was discovered in nonlinear crystals.[33] A triple soliton is a "vector soliton" composed of two transverse electromagnetic potentials and one electrostatic potential, while the conventional plasma soliton is a "scalar soliton" composed of one electrostatic potential.[34]

The coupled three-wave equations for the beat wave excitation are given by

$$\left(\frac{\partial}{\partial t} + v_{g_0}\frac{\partial}{\partial x}\right)\phi_0 = \beta_0\phi_1\phi_p^* \,, \tag{12.33}$$

$$\left(\frac{\partial}{\partial t} + v_{g_1}\frac{\partial}{\partial x}\right)\phi_1 = -\beta_1\phi_0\phi_p^* \,, \tag{12.34}$$

$$\left[\frac{\partial}{\partial t} + v_{gp}\frac{\partial}{\partial x} - i\Delta\omega - \frac{3i}{16}\omega_p|\phi_p|^2\right]\phi_p = -\beta_p\phi_0\phi_1^* , \qquad (12.35)$$

where $\beta_0 = \omega_p^2/2\omega_0$, $\beta_1 = \omega_p^2/2\omega_1$, $\beta_p = \omega_p/2$, and v_{gj} is the group velocities of electromagnetic and electrostatic waves. The frequency mismatch is $\Delta\omega = \omega_0 - \omega_1 - \omega_p$. Note that the fourth term on the left-hand side of Eq. (12.35) is the nonlinear frequency shift associated with the electron relativistic mass correction[21] which limits the amplitude of the ideal plasma wave growth to $eE_p/m\omega_p = (16/3)^{1/3}(\phi_0\phi_1)^{1/3}$. The mode coupling equations (12.33)-(12.35) conserve two action integrals and the interaction Hamiltonian.[35] This integrable system contains soliton solutions.[36]

In Eqs. (12.33)-(12.35) the complex amplitudes $\phi_j(x,t)$'s are slowly varying functions of x and t such that $eE_j/mc\omega_j = v_j/c = \frac{1}{2}\phi_j(x,t)\exp[i(k_jx - \omega_jt)] + \text{c.c.}$ ($j = 0, 1$, or p). Looking for a stationary structure $\phi_j(\xi = x - \lambda t)$ traveling with speed λ, we have

$$d\phi_0/d\xi = -\tilde{\beta}\phi_1\phi_p , \qquad (12.36)$$

$$d\phi_1/d\xi = \tilde{\beta}_1\phi_0\phi_p^* , \qquad (12.37)$$

$$\left(d/d\xi + i\Delta\tilde{\omega} + i\tilde{\Lambda}\right)\phi_p = \tilde{\beta}_p\phi_0\phi_1^* , \qquad (12.38)$$

where $\tilde{\Lambda} = (3\omega_p/16)|\phi_p|^2/(\lambda - v_{gp})$, $\tilde{\beta}_j = \beta_j/(\lambda - v_{gj})$ ($j = 0, 1$ or p) and $\Delta\tilde{\omega} = \Delta\omega/(\lambda - v_{gp})$.

When the relativistic nonlinear frequency shift $\tilde{\Lambda}$ is neglected and $\Delta\tilde{\omega} = 0$, the exact solutions to Eqs. (12.36)-(12.38) are $\phi_0^2 = a^2 sn^2[-\sqrt{\tilde{\beta}\tilde{\beta}_p}\xi, a/b]$, $\phi_1^2 = b^2 - \phi_0^2$ and $\phi_p^2 = \left(\tilde{\beta}_p/\tilde{\beta}\right) a^2 \left[1 - \phi_0^2/a^2\right]$. Here, we approximated $\tilde{\beta}_0 = \tilde{\beta}_1 = \tilde{\beta}$ and $v_{g0} = v_{g1} = v_g$ for $\omega_0 \gg \omega_p$. When $a = b$, we have the triple soliton solution,

$$\phi_0 = -a \tanh\ q(x - \lambda t) , \qquad (12.39)$$

$$\phi_1 = a \text{ sech } q(x - \lambda t) , \qquad (12.40)$$

$$\phi_p = a_p \text{ sech } q(x - \lambda t) , \qquad (12.41)$$

where $q = \sqrt{\tilde{\beta}\tilde{\beta}_p}$ and the soliton velocity λ is given by

$$\lambda = \left(v_g - v_{gp}\beta a_p^2/\beta_p a^2\right) \Big/ \left[1 - (\beta a_p^2/\beta_p a^2)\right] . \qquad (12.42)$$

Note that the electromagnetic and electrostatic envelopes have the same characteristic wavenumber q and phase velocity λ, hence the name of the triple soliton. The speed λ of the triple soliton is a function of a and β, which can be tuned to be the speed of light when $\frac{a_p}{a} = \frac{1}{\sqrt{2}}(\frac{\omega_p}{\omega_0})$. Physically, the three envelopes phase lock such that phase $(\phi_0\phi_1^*) = 0$ in $x > \lambda t$ (the pulse front) and the amplitude of the higher frequency pulse decreases in time as the lower side-bands ϕ_1 and ϕ_p are excited through (12.37) and (12.38). Around the pulse center ($x \simeq \lambda t$) the phase of the pump wave ϕ_0 jumps by π and thus the inverse decay process begins. Namely, ϕ_0 increases and the energy in ϕ_p is restored back to ϕ_0.

Kinetic effects on the trisoliton can be investigated by particle simulation. Figure 12.12 shows the triple soliton structure and particle plots in phase space from particle simulation.[32] The structure at time $t = 12.5\omega_p^{-1}$ in Fig. 12.12(b) and (c) after the two tailored laser beams are injected resembles the anticipated profiles Eqs. (12.39)-(12.42). Note also the substantial electron forward momentum created by the accelerating electrostatic field [Fig. 12.12(d)]. At this time the electron plasma wave is localized around $x = 100\Delta$. At $t = 50\omega_p^{-1}$, the plasma wave is around $x = 500\Delta$ as seen in Fig. 12.12(e). This corresponds to the velocity of the electrostatic wave of $\sim 0.9c$, much faster than the group velocity of the plasma wave. However, in this simulation the plasma wavepacket speed is still slightly less than c; thus the wake creation is much reduced but remains. The difference of the simulation result from the ideal case Eq. (12.41) may be due to the effective dissipation of the plasma wave from the acceleration of the thermal electrons shown in Fig. 12.12(e). In addition, it may be the phase mismatch occurring from the accelerated electrons, or the non-ideal start of the electrostatic wave amplitude, ϕ_p, in the particle simulation.[32]

12.3 Ion Cyclotron Resonance Heating of a Plasma

Shining radiation on a plasma is expected to excite various oscillations and to heat a plasma. In particular, the ion cyclotron resonance heating (ICRH) has been regarded as one of the important heating methods of magnetized plasmas capable of heating ions directly. Heating a plasma with two (or more) species of ions has particular applications[37] in fusion. This mechanism may also be in operation in the ^{3}He-rich solar wind during some strong solar flares.[38] In this heating mechanism, a propagating fast Alfvén magnetosonic wave in the ion cyclotron frequency range interacts with a magnetized plasma with two ion species. In a typical confinement plasma when the magnetosonic wave is launched from the side of the weak magnetic field, it encounters the minority cyclotron resonance before the ion–ion hybrid resonance, thus heating the minority ions.[37] When the magnetosonic wave is launched from the high magnetic field side, the ion–ion hybrid resonance is met before the minority

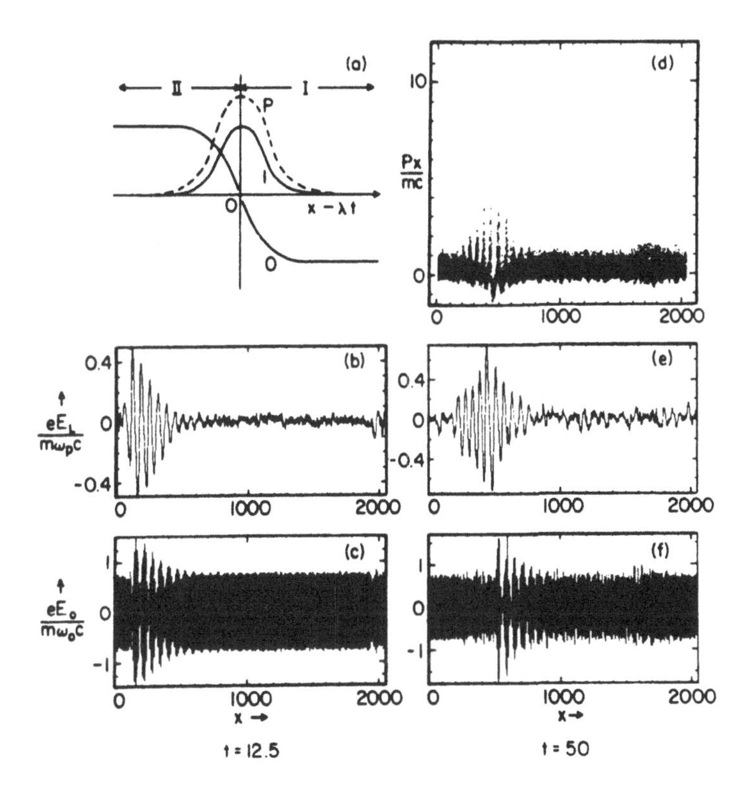

FIGURE 12.12 Triple soliton in the plasma beat wave system. Electrostatic fields (b), (e); electromagnetic fields (c), (f) and phase space plot (d). (b) and (c) early and (e) and (f) at later time (Ref. 32)

ion cyclotron. In the cold plasma theory, the wave must be tunneled to the cyclotron resonance layer; with the inclusion of the finite Larmor radius effects, the wave is mainly mode-converted to a short wavelength electrostatic wave,[39–41] eventually dumping its energy into the plasma. For the alternative case with the lighter of the ions in the majority, the above sequence of resonances is reversed. On the surface of the solar photosphere, hydrogens are the majority and heliums (4) are the minority. Intense flares may trigger the magnetosonic waves (see Chapter 14), whose frequency may fall onto the hybrid frequency of H and ^{4}He. Since the cyclotron frequency of ^{3}He can be in the neighborhood of the hybrid frequency, intense preferential heating of ^{3}He may develop. It is possible to have a simultaneous second harmonic heating of heavier elements such as Fe in their highly ionized states.

Experimental results[42] on tokamak plasmas with the antenna placed on the high field side have shown that localized ion heating takes place closer to the hybrid than the minority cyclotron layer and that this heating persists even when the cyclotron layer is placed outside the plasma. The energy transfer to the ions cannot be attributed to conventional Landau damping in view of negligible $k_{\|}$ and the fact that the difference between the ion hybrid frequency ω_{IH} and the minority cyclotron frequency Ω_{β} is typically of order $\frac{1}{10}\Omega_{\beta}$ for minority concentration 10–20%. Nonlinear wave particle interaction, stochastic as well as coherent, can be the heating mechanism. It is necessary to delineate this situation. Numerical calculations on ion–ion hybrid heating have been carried out[43] by integrating the differential equation for wave propagation in an inhomogeneous medium. The inclusion of self-consistent ion dynamics is indispensable in opening all the channels for energy absorption including nonlinear particle trapping and turbulent diffusion caused by the stirring of fluctuations.

The standard electromagnetic particle code (see Chapter 5) retaining full electron dynamics has been used to describe wave propagation in the high-frequency regime, as for example the study of the upper hybrid wave propagation and mode conversion.[44] However, inclusion of the full electron dynamics makes the code too costly and too noisy to run on the ion cyclotron period time scales. Numerical stability considerations limit the simulation time step to the electron characteristic periods, much shorter than the relevant time scales for the ion motion. The need to suppress the noise levels also arises during low-frequency, small phase velocity runs as the typical electric fields of the launched waves are smaller than those in case of high-frequency, large phase velocity waves. This is seen from the scaling $E/B_0 = (v_{ph}/c)(B/B_0)$ given that B/B_0 is limited to a few percent for linearity and that $v_{ph}/c \sim v_A/c < 10^{-1}$ for usual tokamak parameters. It is beneficial to reduce the unnecessary electron noise in the simulation while keeping the full ion dynamics.

Therefore, there has been motivation for generating numerical algorithms suppressing the electron dynamics to the minimum required, depending on the situation under consideration. One approach is to utilize a simplified model with the exact electron motion reduced by time-averaging the guiding center motion while keeping the full ion dynamics, as described by the drift-kinetic equations (see Chapter 7). A second approach may be to treat electrons as a massless, cold neutralizing dielectric fluid while ions are treated as particles. Thus, fluid-particle "hybrid" codes have been developed. (see Chapter 8) The applicability of the electrostatic model is limited to one-dimensional cases and for short wavelength low phase velocity waves, such as ion Bernstein modes near cyclotron harmonics.

In the general case of wave propagation across a magnetized plasma, such as a fast magnetosonic mode, both the electrostatic and the electromagnetic field components, coupled through the external B-field, must be retained. The need to include the electron response to the transverse electric field makes the electrostatic shielding method hardly adequate even in one dimension. Full treatment of the fields is especially important during mode conversion in order to simulate the progressive change in the wave character, from mainly electrostatic to mainly transverse and vice versa. Electromagnetic hybrid codes such as the ones developed by Byers et al.[45] and Hewett et al.[46] have a common characteristic in the above algorithms in that they employ the quasineutrality assumption $n_e \approx n_i = n_0$ and the resulting generalized Ohm's law in order to compute the electron charge and current density source terms from the ion charge and current density. The displacement current is omitted from Maxwell's equations, justified for $v_{\text{phase}} \ll c$, and the wave equation for the transverse electromagnetic components becomes an elliptic equation(magnetoinductive model).

The first approach is adopted here following the electron guiding center motion[47] in the direction perpendicular to the external magnetic field and retaining the full set of Maxwell's equations. For the electron motion parallel to the magnetic field, the electrostatic screening method[48] may be extended to include the effects of small but finite $k_{\parallel}$ without increasing the dimensionality of the code.

A one-dimensional slab of a hot plasma is simulated[43] embedded in a magnetic field $\mathbf{B}_0$ along the z-direction and varying along the x-direction according to $B(x) = b_0/(a+x)$. The pressure and density are uniform across x and no zeroth order current runs along the magnetic field. Thus our system satisfies the trivial magnetohydrodynamic (MHD) equilibrium $\nabla P = j \times B = 0$. The algorithm following the fields and particles are described in Chapter 5.

The plasma is excited by a current driven along the y-direction through an infinite antenna sheet on the yz-plane placed at distance $x = D$ from the boundary. The periodic boundary conditions are employed with electromagnetic wave absorbing boundaries (Section 5.3) in order to avoid penetration for the waves launched near the one edge from the other edge.

A fast magnetosonic wave of frequency ω propagating nearly perpendicular to the magnetic field in a two-ion species plasma will undergo partial mode conversion into a short wavelength electrostatic mode near the layer x_{mc} defined by

$$\omega = \omega_{IH}(x_{mc}) \equiv \Omega_\alpha \left[\left(\frac{\omega_\beta^2}{\omega_\alpha^2} + \frac{\Omega_\beta^2}{\Omega_\alpha^2} \right) \Big/ \left(1 + \frac{\omega_\beta^2}{\omega_\alpha^2} \right) \right]^{1/2} \tag{12.43}$$

with ω_α, ω_β the plasma frequencies, and Ω_α, Ω_β the cyclotron frequencies of the majority and minority ion species, respectively. See Fig. 12.13. Note that the ion-hybrid frequency $\omega_{IH} \rightarrow \Omega_\beta$ as the minority concentration $n_\beta \rightarrow 0$. The local dispersion relation shows the existence of two branches $k_i^2(x,\omega)(i = 1,2)$ of equal real parts near the mode conversion zone[43]; further away from resonance, these branches could be identified as the fast magnetosonic mode and the minority ion Bernstein mode.

As the WKB approximation equation fails near the mode conversion region, a solution of the full wave propagation equation is required to study

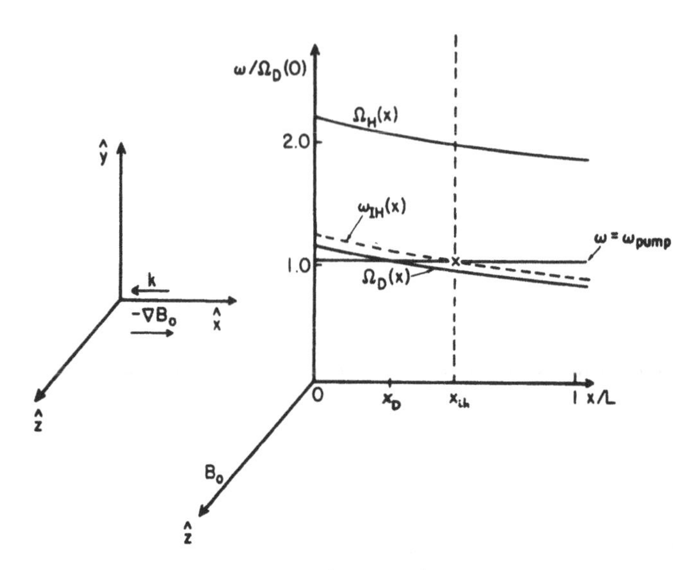

FIGURE 12.13 Cyclotron resonances (Ω_H, Ω_D) and two-ion hybrid resonance ω_{IH} for a nonuniform magnetic field (Ref. 43)

the effect of mode conversion. The results of the extensive analytic work by Swanson,[39,41] as well as by Jacquinot,[40] including finite $k_{\|}$, Landau damping and toroidal effects, can be formulated in terms of a transfer matrix, connecting the coefficients between the two asymptotic wave solutions at $x = \pm\infty$ (with x measured from the resonance $x = 0$),

$$e^{+iz} + Re^{-iz} \overset{-\infty}{\longleftarrow} \quad \overset{+\infty}{\longrightarrow} Te^{iz} + Ce^{i\frac{2}{3}\lambda|z|^{3/2}} \tag{12.44}$$

where evanescent terms have been omitted in Eq. (12.44). In the simple case with $k_{\|} = 0$ and linear profile for the plasma parameters, the reflection, transmission, and mode conversion coefficients are given by

$$R = 0 \quad , \quad T = e^{-\eta}, \quad C = -\left(1 - e^{-2\eta}\right) \tag{12.45}$$

for incidence from the high field side and

$$R = -\left(1 - e^{-2\eta}\right), \quad T = e^{-\eta}, \quad C = -e^{-\eta}\left(1 - e^{-2\eta}\right) \tag{12.46}$$

for incidence from the low field side where λ and η depend on the plasma parameters. In particular, η is roughly proportional to the relative minority concentration n_β/n_α and was shown to be much larger than 1 for typical thermonuclear plasmas.[39] Then examination of the coefficients in Eqs. (12.45) and (12.46) shows a major difference between the two cases; as for high field side incidence, the incoming wave is primarily mode converted to the slow electrostatic mode, while for low field side incidence it is primarily reflected away from the mode conversion region.

Energy conservation yields the energy fraction extracted from the incident fast magnetosonic as

$$\Delta E = 1 - |R|^2 - |T|^2 = \Delta E_{\text{abs}} + \Delta E_{mc} \ . \tag{12.47}$$

ΔE_{abs} is the fraction of energy absorbed via the combined minority ion cyclotron and the majority first cyclotron harmonic Landau damping as for small minority concentration of hydrogen in deuterium $\omega_{IH} \cong \Omega_H \cong 2\Omega_D$. ΔE_{mc} is the fraction of energy carried away by the electrostatic mode generated by conversion, $\Delta E_{mc} = |C|^2 \left(1 - e^{-2\eta}\right)^{-1}$. Since the energy in the electrostatic branch cannot leave the plasma and it is eventually absorbed by electrons, the mode conversion and local ion cyclotron absorption are the two crucial aspects in determining energy deposition during ion-hybrid heating.

There is a considerable difference in local absorption, depending on whether the wave is incident from the high or from the low field side. In the case of incidence from the low field side, the wave first encounters the minority cyclotron resonance before the two-ion hybrid resonance. Direct minority heating occurs through the cyclotron resonance due to the left-hand circularly polarized component E_+ in the magnetosonic mode.[37] In the opposite case of incidence from the high field side the wave encounters the two-ion hybrid resonance first and must be tunneled to the minority cyclotron regime, being at the same time mode-converted to an electrostatic mode. Nevertheless, direct ion heating has been observed experimentally[42] near the two-ion hybrid layer even for cases with the minority cyclotron resonance located outside the plasma. This heating cannot be accounted for by the linear theory demanding $(\Omega - \omega)/k_{\|} \leq v_{\text{min}}$ for resonance and suggests the existence of nonlinear wave particle interactions for energy transfer.

Simulation can focus on the nonlinear effects on heating, with wave propagation with $k_{\|} = 0$. Then local energy absorption due to Landau damping cannot take place. However, nonlinear wave particle interaction provides additional heating mechanisms and strong energy absorption is actually observed.[43] There can be two channels for energy transfer from the wave to the particles.

A series of runs have been performed using the code in Ref. 43 (also see Chapter 8) to investigate numerically the physics of the ion–ion hybrid mode conversion for wave propagation perpendicular to the magnetic field in a two-ion species plasma with deuteron (D) as majority and hydrogen (H) as minority.

For wave incidence from the low field side, it is arranged to place the mode conversion layer at $x_{mc} \cong (1/2)L$. The time span simulated was typically 10 to 20 deuteron cyclotron periods, enough for wave propagation from boundary to boundary. The antenna current is limited by the requirement $B_z/B_0 \lesssim 5\%$. As

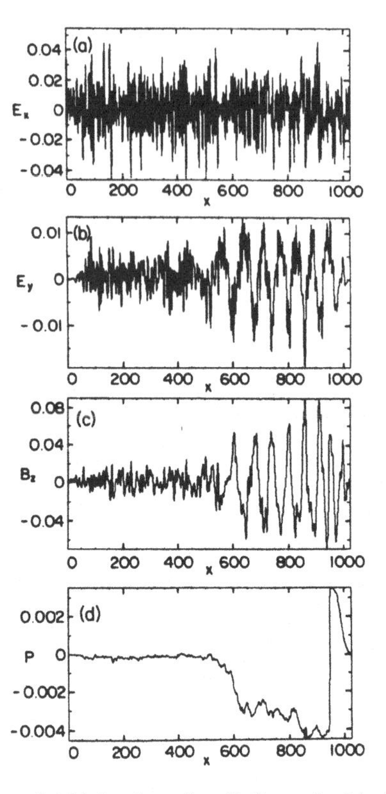

FIGURE 12.14 The low-field injection of radiation, electric fields (a), (b), magnetic field (c), and Poynting flux (d) (Ref. 43)

the electric field components scale as $E_x \sim E_y \cong \frac{v_A}{c} B_z$, the wave electrostatic field is moderately above the electrostatic thermal fluctuation level for runs with high $\omega_D/\Omega_D \cong c/v_A$ ratios. Results obtained for $n_H/n_D = 1/4$ are shown in Figs. 12.14.

The sudden reduction in the electromagnetic energy flux (the Poynting vector) that occurs near the middle of the plasma slab, Fig. 12.14(d), is an indication of the partial mode conversion of the incoming electromagnetic energy associated with the fast magnetosonic mode into the electrostatic branch of the ion–ion hybrid mode. This reduction in electromagnetic energy flux cannot be attributed to any strong localized absorption by particles as it is checked from the phase space plots Figs. 12.15(a)–(d): Therefore, the reduction in the Poynting vector can be used to estimate the percentage of the mode conversion taking place near the ion–ion hybrid resonance layer. This method of estimate yields a surprisingly high conversion coefficient of 90% or higher for a wide range of densities between $\omega_D/\Omega_D = 3.33$ and $\omega_D/\Omega_D \cong 40$.

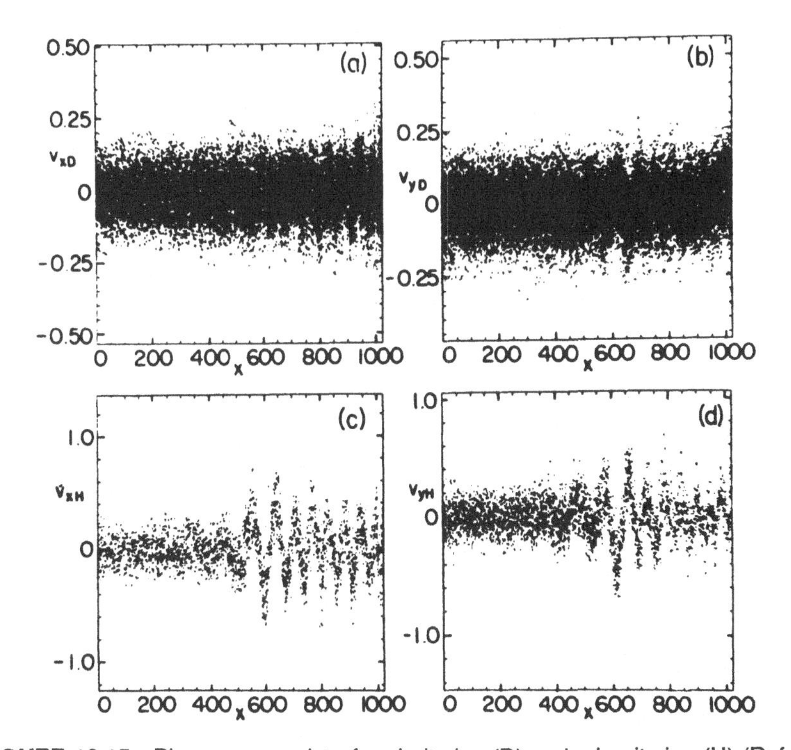

FIGURE 12.15 Phase space plot of majority ion (D) and minority ion (H) (Ref. 43)

Finite Larmor radius effects combined with the inhomogeneity place the mode conversion layer before x_{mc} that is derived by the WKB approximation. An analogous effect is observed during upper hybrid mode conversion simulation and is explained analytically by the inclusion of spatial derivatives of odd order in the wave propagation differential equation.

In this chapter we have surveyed applications of the numerical techniques developed in the preceding chapters. We have looked at three representative problems of interaction between radiation and a plasma, (i) the beam-plasma instability and subsequent radiation, including the free electron laser, (ii) particle acceleration by radiation and plasma coupling, and (iii) particle heating in a magnetized plasma by radiation. These problems involve the radiation time scale and thus the computational code employed has to resolve short time scales and least coarse-grained, following almost all microscopic physics. In problem (iii) the radiation is in time scale so that electron time scales have been coarse-grained.

Problems

1. Apply a dc electric field to the free electron laser system to enhance the power of radiation. Find an optimal strength of the field.
2. In a plasma fiber by a judicious choice of the fiber radius and laser modes (such as the injection angles of two lasers), one can tune the beat plasma wave so that it has phase velocity c. Derive this condition for the radius.
3. By choosing the parameters in Eq. (12.42) one can make the soliton velocity λ larger or equal to c. Discuss why this does not violate the special theory of relativity.
4. The triple soliton comes in three types (Ref. 35). Discuss the consequences on the boundary and/or initial conditions by the choice of the type of soliton.
5. Make a case why a strong solar flare may be able to trigger a process of Alfvén waves that pick up and accelerate elements such as ^{3}He and highly ionized Fe etc.
6. In relativistic dynamics how does the trapping process of Ref. 7 change?

References

1. H. Alfvén, Nature **150**, 405 (1942).
2. B.B. Godfrey, W.R. Shanahan, and L.E. Thode, Phys. Fluids **18**, 346 (1975).
3. A.B. Mikhailovskii, *Theory of Plasma Instabilities*, vol. 1 (Consultant Bureau, New York, 1974) p. 159.
4. J.N. Leboeuf and T. Tajima, Phys. Fluids **22**, 1485 (1979).
5. M. Thaker, J.N. Leboeuf, and T. Tajima, UCLA PPG preprint 547 (1981).
6. H. Dreicer, Phys. Rev. **115**, 238 (1959).
7. T.M. O'Neil, Phys. Fluids **8**, 2255 (1965); J.M. Dawson, Phys. Rev. 113, 383 (1959).
8. S. Kainer, J.M. Dawson, R. Shanny, and T.P. Coffey, Phys. Fluids **15**, 493 (1972).
9. D.A. Gurnett, J. Geophys. Res. **79**, 4227 (1974).
10. L.R. Elias, W.M. Fairbank, J.M. Madey, H.A. Schwettman, and T.I. Smith, Phys. Rev. Lett. **36**, 717 (1976).

11. M. Friedman and M. Herndon, Phys. Rev. Lett. **28**, 210 (1971).

12. T. Kwan and J.M. Dawson, A.T. Lin, Phys. Fluids **20**, 581 (1977).

13. B. Coppi, M.N. Rosenbluth, and R.N. Sudan, Ann. Phys. **55**, 207 (1969).

14. A.T. Lin, J.M. Dawson, and H. Okuda, Phys. Fluids **17**, 1995 (1974).

15. T. Tajima and J.M. Dawson, Phys. Rev. Lett. **43**, 267 (1979).

16. T. Tajima, *Laser and Particle Beams* **3**, 351 (1985).

17. T. Tajima, *Proc. of 12th Int. Conf. High Energy Accel.*, eds. F. T. Cole and R. Donaldson (Fermi National Accel. Lab., Batavia, 1983) p. 470.

18. T. Tajima and M. Cavenago, Phys. Rev. Lett. **59**, 1440 (1987).

19. N. Kroll, A. Ron, and N. Rostoker, Phys. Rev. Lett. **13**, 83 (1964).

20. C. Joshi, T. Tajima, J.M. Dawson, H.A. Baldis, and N.A. Ebrahim, Phys. Rev. Lett. **47**, 1285 (1981).

21. M.N. Rosenbluth and C.S. Liu, Phys. Rev. Lett. **29**, 701 (1972).

22. B.I. Cohen, A.N. Kaufman, and K.M. Watson, Phys. Rev. Lett. **29**, 581 (1972).

23. R. Galvao and T. Tajima, IFSR **110** (1985).

24. H. Yukawa, Proc. Math. Phys. Soc. Jpn. **17**, 48 (1935).

25. D.J. Sullivan and B.B. Godfrey, IEEE Trans. Nucl. Sci. NS-28, 3395 (1981).

26. C. Joshi, W.B. Mori, T. Katsouleas, J.M. Dawson, J.M. Kindel, D.W. Forslund, Nature **311**, 525 (1984).

27. D.W. Forslund, J.M. Kindel, W.B. Mori, and J.M. Dawson, Phys. Rev. Lett. **54**, 558 (1985).

28. M. Ashour-Abdalla, J.N. Leboeuf, T. Tajima, J.M. Dawson, and C.F. Kennel, Phys. Rev. **A23**, 1906 (1981).

29. P.S. Chen, J.M. Dawson, R.W. Huff, and T. Katsouleas, Phys. Rev. Lett. **54**, 693 (1985).

30. S.A. Akhmanov, A.P. Sukhorukov, and R.V. Khokhlov, Sov. Phys. Uspekhi **93**, 609 (1968).

31. G. Schmidt and W. Horton, Comments Plasma Phys. Controlled Fusion **E9**, 85 (1985); D.C. Barnes, T. Kurki-Suonio, and T. Tajima, IEEE Trans. Plas. Sci. **PS-15**, 154 (1987).

32. K. Mima, T Ohsuga, H. Takabe, K. Nishihara, T. Tajima, E. Zaidman, and W. Horton, Phys. Rev. Lett. **57**, 1421 (1968).

33. S.L. McCall and E.L. Hahn, Phys. Rev. **183**, 457 (1969).

34. V.E. Zakharov, Sov. Phys. JETP **35**, 908 (1972).

35. C.J. McKinstrie and D.F. Dubois, Phys. Rev. Lett. **57**, 2022 (1986).

36. K. Nozaki and T. Taniuti, J. Phys. Soc. Jpn. **34**, 796 (1973).

37. T.H. Stix, Nucl. Fusion **15**, 737 (1975).

38. C.Y. Fan, G. Gloeckler, and D. Hoverstadt, Space Sci. Rev. **38**, 143 (1984).

39. D.G. Swanson, Phys. Rev. Lett. **36**, 316 (1976).

40. J. Jacquinot, B.D. McVey, and J.E. Scharer, Phys. Rev. Lett. **39**, 88 (1976).

41. Y.C. Ngan and D.G. Swanson, Phys. Fluids **20**, 1920 (1977).

42. TFR Group, in *Plasma Physics and Controlled Nuclear Fusion Research* (IAEA, Vienna, 1981) vol. 2, p. 75.

43. S. Riyopoulos and T. Tajima, Phys. Fluids **29**, 4161 (1986).

44. A.T. Lin, C.C. Lin, and J.M. Dawson, Phys. Fluids **25**, 646 (1982).

45. J. Byers, B.I. Cohen, W.C. Condit, and J.D. Hanson, J. Comp. Phys. **27**, 363 (1978).

46. D.W. Hewett and C.W. Nielson, J. Comp. Phys. **29**, 219 (1978).

47. W.W. Lee and H. Okuda, J. Comp. Phys. **36**, 870 (1976).

48. H. Okuda, J.M. Dawson, A.T. Lin, and C.C. Lin, Phys. Fluids **21**, 476 (1978).

13

DRIFT WAVES AND PLASMA TURBULENCE

A plasma is known to exhibit instabilities if it is spatially nonuniform or inhomogeneous.[1] This can be easily understood. The plasma in a spatially nonuniform state, retaining a higher free energy, tries to release it by modulating itself to form an energy release channel much more rapidly than the otherwise collisional (or thermal) process. For controlled thermonuclear fusion the plasma has to be isolated perhaps by surrounding magnetic fields and thus the plasma is necessarily (or forced to be) nonuniform. The plasma in space and astrophysical settings has been increasingly found to have naturally a quite nonuniform structure, such as a bundle of filaments or a sponge-like structure.[2–4] Because of the relevance of the way in which the plasma attains a nonuniform state, either naturally or by force, the plasma very often exhibits the instabilities associated with plasma nonuniformities.

The properties of these instabilities vary, of course, depending upon the nature of the plasma nonuniformity and other conditions, and the properties of ensuing turbulence driven by these instabilities vary as well. However, the natural scale length of plasma nonuniformity is generally much larger than the microscopic scales such as the Debye length and the collisionless skindepth.

This sets the typical time scale much longer than the microscopic ones such as those considered in the previous chapter. These instabilities can be kinetic, fluid, or in-between. Drift waves, for example, driven by a density or a temperature gradient are essentially kinetic modes, athough many of them can be described by the fluid equations, particularly those involving collisions. The Rayleigh instability due to the temperature gradient is a typical example.[5] The instabilities of these hydrodynamic or magnetohydrodynamic origins have a macroscopic time scale and can be even slower than the kinetic processes. In terms of hierarchy, the present processes are situated between the microscopic processes as considered in Chapter 12 and the macroprocesses such as MHD (see Chapter 6 and Chapter 14) and (even longer) transport processes.

The drift wave instability has a double personality, kinetic and fluid-like. It is driven by the presence of a density (or temperature) gradient across the magnetic field that permeates the plasma. The name of this instability arose because particles in a nonuniform plasma drift across the magnetic field.[6,7] Analysis of the stability of the simplest possible case (a slab plasma with a density gradient, and without magnetic shear or collision effects) shows that the density perturbation perpendicular to both the magnetic field direction and the density gradient direction develops and grows unstably.[6–10] In fact, this mode is always unstable no matter how small the density gradient is and sometimes it is referred to as the universal instability. It is also unstable with collisions in a wide range of parameters. To the lowest approximation the magnetized electron motion and ion motion are characterized by the $\mathbf{E} \times \mathbf{B}$ drift (see Chapter 7) and they are identical in magnitude and direction. There arises, therefore, no momentum or energy exchange. When wave-particle interactions (Landau damping), finite ion Larmor radius effects such as the polarization drift (see Chapter 7), or collisional effects are taken into consideration, the electron and ion $\mathbf{E} \times \mathbf{B}$ drifts cease to be identical and become out of phase. This turns out to allow particles (typically electrons) to give up energy to the waves and to let them grow; this is an applicable physical picture for both collisional and collisionless cases.

The history of the theory of the universal mode (collisionless drift waves) is intriguing. It is a course of zig-zags between stability and instability in drift waves, but during these zig-zags, the theory has increased in sophistication and depth. The drift waves are "universally" unstable if there is no magnetic shear.[1] The "universal" instability was generally considered to be stabilized by the presence of a magnetic shear. Yet Krall and Rosenbluth's[7] theory predicted a restricted class of absolute instability. Rutherford and Frieman's[11] theory found that a wavepacket can be convectively unstable, which may be stabilized by sufficiently strong magnetic shear. Here the absolute instability refers to an instability that grows exponentially in time at a fixed spatial point, while the convective instability does not. It grows exponentially in time only on a frame moving with a certain velocity. Pearlstein and Berk[12] introduced outgoing wave boundary conditions into the mode analysis of Ref. 7 and found that propagating modes can be unstable, which may be stabilized by strong enough magnetic shear. According to Pearlstein and Berk[12] under the shear the energy input into the drift wave from electrons (the destabilizing effect) through electron Landau resonance can be offset by the energy outflow from the drift wave to ions (the stabilizing effect) through ion Landau resonance.

In 1978 Ross and Mahajan[13] and Tsang et al.[14] finally (and correctly) found that the drift wave eigenmodes of Pearlstein and Berk are in fact *always* absolutely stable in sheared slab; the stability arose from the inclusion of the nonresonant electron contribution in the vicinity of the mode rational surface, which had been neglected in previous work. All the above works have been linear theoretical approaches. In 1979, however, Hirshman and Molvig[15] showed that by including electron nonlinearity under certain conditions, the "universal" drift wave can be absolutely unstable. On the other hand, Chang and Chen[16] in 1980 argued, based on a linear theory, that the inclusion of toroidal geometry and trapped particle effects leads to absolute instability of drift waves. We see that the theory is becoming increasingly more correct and sophisticated in time. The above historical sketch of drift wave stability indicates the subtlety of the physics of drift waves and the difficulty of theoretical study of the drift waves and the associated anomalous transport of electrons and their heat.[17]

Nonuniformities and inhomogeneities of other types in a plasma lead to other types of plasma instabilities and often to a turbulent plasma. The nonuniformity of plasma flow (shear flow) gives rise to the Kelvin-Helmholtz instability[5] — this instability may develop in a neutral fluid or in a magnetized plasma. The nonuniformity of temperature causes the Rayleigh convection instability[5]: again this can occur either in a neutral fluid or in a magnetized plasma. In the following we discuss three examples of plasma nonuniformities and associated instabilities and their subsequent consequences such as turbulence: the drift wave instability, the Kelvin-Helmholtz instability, and the Rayleigh instability.

13.1 Drift Wave Instabilities

As we reviewed in the introduction of this Chapter, the traditional analytical approach has had difficulty in completely describing the stability of drift waves and their nonlinear properties, although many authors[18] suspected the chief reason for the so-called anomalous transport (anomalous compared to the neoclassical transport theory[19,20]) is that the drift wave instability induces a high level of plasma fluctuations or turbulence which lead to wave-enhanced transport. Computational simulation has been carried out for many years[21–26] in an attempt to understand this complicated and difficult physical process,

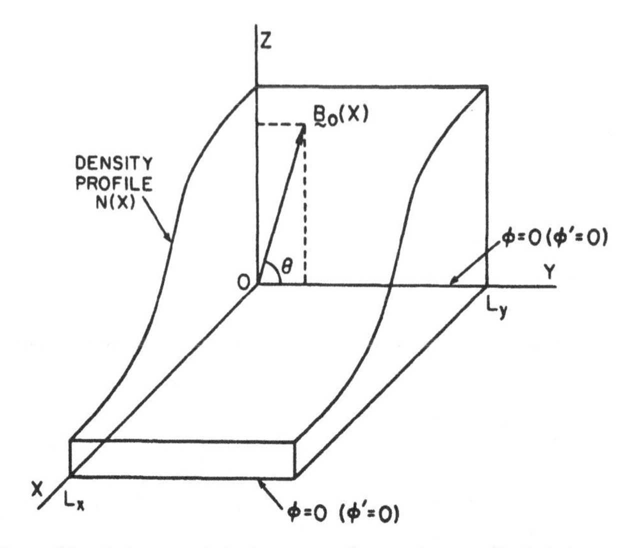

FIGURE 13.1 Simulation model of a nonuniform plasma (Ref. 36)

and like the theory, the simulation has become increasingly more sophisticated over the years. Most complete and self-consistent simulations of the "universal" mode so far have been based on the particle simulation approach, although numerous other simulations have been carried out as well. The main techniques of these particle simulations are described in Chapters 2, 4 and 7 and were pioneered by Okuda, Lee and Cheng.[27–29] The other approaches include those spearheaded by Waltz, Hasegawa and others. This approach is based on a reduced, less formidable set of equations that have been derived from the original set of equations under certain plausible assumptions and closure ansatz. A typical set of equations is that of the so-called Hasegawa-Mima equation[30,31] and their numerical simulation. These approaches have been quite successful in seeing some nonlinear developments and nearly stationary states.[32–35] However, these approaches invariably assumed a certain *ad hoc* mechanism of instability itself. For this reason, we shall not discuss this latter approach here.

Lee et al.'s work[22–24] in a shearless slab model found that the instability saturates when the density profile is flattened by the wave convection of particles (the quasilinear flattening). When the magnetic shear is introduced, however, care must be taken in the simulation in order to realize the condition of theory[12–14] that treats the outgoing wave condition. The configuration of the simulation is schematically shown in Fig. 13.1. The drift wave develops along the y-direction with a localized eigenfunction structure in the

x-direction. The electron resonance point ($x = x_e \cong \omega_* L_s/k_y v_e$, where the electron diamagnetic frequency $\omega_{*e} = k_y cT_e/eBL_n$ with T_e being the electron temperature, v_e the electron thermal velocity, L_n the density scale length, and L_s the shear scale length) neighborhood supplies energy to the wave and the ion resonance point $x = x_i \cong \omega_{*i} L_s/k_y v_i$, where $\omega_{*i} = k_y cT_i/eBL_n = \omega_{*e}$, if $T_i = T_e$, and v_i the ion thermal velocity) neighborhood extracts energy from the wave. The overall balance of the energy and its flow between these two regions determines the stable[13,14] eigenfunction structure in the sheared slab geometry. Thus, if the simulation system is too small to accommodate the ion resonance point neighborhood within it, the energy sink for drift waves ceases to exist and the system fails to exhibit the localized stable eigenfunction structure the theory predicts. In order to realize this eigenmode, it is necessary for the simulation region to include the ion resonance within it. This was first done by Sydora et al.[36]

In the simulation of Sydora et al.,[36] the plasma is confined between two boundaries $x = 0$ and $x = L_x$. In the y-direction, the system is periodic with length L_y. A strong confining sheared magnetic field is externally imposed with

$$\mathbf{B} = B_0 \{\hat{z} + [(x - x_0)/L_s]\,\hat{y}\} \;, \tag{13.1}$$

where x_0 defines the location of the rational surface. Therefore the wavenumber along the magnetic field is expressed as

$$k_{\|} = \mathbf{k} \cdot \mathbf{B}/|\mathbf{B}| \simeq k_y \left[(x - x_0)/L_s\right] \;, \tag{13.2}$$

with $k_y = 2\pi m/L_y\Delta$, $m = 0, \pm 1, \ldots, \pm L_y/2$. The plasma is assumed to have initially an exponential density profile

$$n(x) = n_0 k L_x \{\exp(-\kappa x)/\left[1 - \exp(-\kappa L_x)\right]\} \;.$$

This profile gives a constant density gradient scale length $L_n[\equiv 1/\kappa$, with $\kappa = -n'(x)/n]$, although no temperature gradient is imposed initially.

The boundary condition imposed on the electrostatic potential at the endpoints $x = 0$ and $x = L_x$ are of two types. The first requires $\phi(0) = 0 = \phi(L_x)$ which allows for odd parity modes with respect to the endpoints in the simulation. The second type used is $\partial\phi/\partial x(x = 0) = 0 = \partial\phi/\partial x(x = L_x)$ which corresponds to even parity modes. Therefore the eigenmode parity, with respect to the rational surface, is determined by its position relative to the endpoints. For example, if the rational surface is placed at $x = 0$ and odd parity modes are imposed with respect to the boundary points, then the eigenmodes are odd with respect to the rational surface.

The particle and field solvers are as discussed in Chapter 7 (Refs. 27 and 28). The particles are specularly reflected at the walls in such a way as to eliminate the sheath currents which may occur.[37] Reflection of particles implies that there is no mechanism for net energy loss in the system. The system size used for the strong shear simulations is $L_x \times L_y = 64\Delta \times 32\Delta$. The average number density is $n_0 = 16/\Delta^2$, ion to electron mass ratio $m_i/m_e = 100$, temperature ratio $T_i/T_e = 1$, and electron Debye length $\lambda_{De} = 2.5\Delta$. The ion Larmor radius is $\rho_i = 2.5\Delta$ where $\rho_i = v_i/\Omega_i$ and $v_i = (T_i/m_i)^{1/2}$ with Ω_i the ion cyclotron frequency, $\Omega_e/\omega_{pe} = 10$, and time step $\omega_{pe}\Delta t = 2 - 4$. The density gradient scale length is $L_n = 1/\kappa = 14.3\Delta$ and $\kappa = 0.07\Delta^{-1}$. The shear scale length is varied from $L_s = 400\Delta(\omega_{pe}\Delta t = 4)$ to $L_s = 200\Delta(\omega_{pe}\Delta t = 2)$. The rational surface position, x_0, has been placed either at $x_0 = 0$ or at $x_0 = L_x/2 = 32\Delta$. The system described above supports discrete wavenumbers $k_y\rho_i = 0.49m$, $m = 0, \pm1, \ldots, \pm L_y/2$, where m is the mode number in the y-direction and $\omega_*/\omega_{pe} = 0.0086m$. The simulations have been carried out for the duration of $\omega_* t = 70(m = 1)$.

The diagnostics involve the determination of the frequencies of potential fluctuations for a fixed $k_y\rho_i$ at various positions in the x-direction. They are obtained from the power spectra of the potential by the maximum entropy correlation technique.[38] (see Chapter 11) The equation for the power spectral intensity is given by

$$I_{k_y}(x,\omega) = \int_{-\infty}^{\infty} d\tau A_{k_y}(x,\tau)\exp(i\omega\tau) \; , \tag{13.3}$$

where the correlation function is

$$A_{k_y}(x,\tau) = \lim_{T\to\infty} \frac{1}{T}\int_0^T dt\tilde{\phi}_{k_y}(x,t)\tilde{\phi}^*_{k_y}(x,t+\tau) \tag{13.4}$$

for a lag time τ. The spatial structure of the fluctuations at a particular frequency is best analyzed by an interferogram technique[39] which is a correlation of the plasma potential at a particular $k_y\rho_i$ for a fixed frequency, ω_0, determined by the power spectra. The following quantity is computed

$$C_{k_y}(x,\tau) = \frac{1}{T}\int_0^T \tilde{\phi}_{k_y}(x,t+\tau)\sin(\omega_0 t)dt \; , \tag{13.5}$$

where τ is a variable lag time and T the total time of integration.

Here let us briefly review the linear theory of drift waves. The governing equation for potential fluctuations of the form $\tilde{\phi} = \tilde{\phi}(x)\exp\left[i(k_y y - \omega t)\right]$ can be written as

$$\left[1 + \eta + (1 - \omega_*/\omega)\xi_e Z(\xi_e) + (\eta + \omega_*\omega)\xi_i Z(\xi_i)I_0(b)e^{-b}\right]\tilde{\phi} = 0 \; , \tag{13.6}$$

where

$$\eta = T_e/T_i, \quad \xi_\alpha = \omega L_s/\sqrt{2}k_y(x - x_0)v_\alpha, \; b = (k_x^2 + k_y^2)\rho_i^2 \; ,$$

and where Z is the plasma dispersion function[40] and I_0 is the modified Bessel function ($\alpha = e$ or i). For $k_x\rho_i \ll 1$ we can replace k_x by $-i\partial/\partial x$ in Eq. (13.6) to obtain the eigenmode equation[41]

$$\tilde{\phi}'' + Q(\tilde{x}, \omega)\tilde{\phi} = 0 \ , \tag{13.7}$$

where

$$Q(\tilde{x}, \omega) = \eta - k_y^2 \rho_s^2 + \frac{(1+\eta)(\omega/\omega_*) + (\omega/\omega_* - 1)\xi_e Z(\xi_e)}{(\omega/\omega_* + 1/\eta)\xi_i Z(\xi_i)} \ ,$$

with $\rho_s = (T_e/T_i)^{1/2}\rho_i$ and $\tilde{x} = x/\rho_s$. The prime denotes differentiation with respect to $\tilde{x}$.

The electron and ion resonance locations, x_e and x_i, are determined from the condition $\xi_\alpha = 1$, so that $x_e = \omega L_s/\sqrt{2}k_y v_e$ and $x_i = \omega L_s/\sqrt{2}k_y v_i$ where ω refers to the normal mode frequency. Equation (13.7) has been solved using a standard shooting code[42] method with the simulation parameters. The eigenfrequencies and damping rates for various values of $k_y\rho_s$ are shown in Fig. 13.2. The real frequencies for odd and even parity were almost the same numerically, and therefore only the values for the even parity are shown. In all cases linear theory predicts stable, linearly damped modes for odd and even parity. It is also necessary to solve the local dispersion relation and it is simply obtained by setting $k_x^2 = 0$ in Eq. (13.7).[36]

A. Simulation with Strong Shear in Two Dimensions

Simulation results with strong magnetic shear are discussed. The modes in this case are expected to be well localized. The even parity mode is the ordinary universal mode of drift waves. The odd mode parity is often neglected because it is a heavily damped mode as is seen in Fig. 13.2. A value of $L_s/L_n = 28$ is used, and with $\omega_*/\omega_{pe} = 0.0086$ the ion resonance location is at $x_i(\omega_*) = 49.5\Delta$, which gives $x_i(\omega) < x_i(\omega_*) < L_x$. $\partial\phi/\partial x = 0$ boundary conditions are used at $x = 0$ and $x = L_x$. With rational surface at $x = 0$ only even modes given by $\tilde{\phi} = \sum_n \phi_n \cos(n\pi x/L_x)$ are allowed in the simulation. According to linear theory, these modes are stable and are less heavily damped than the odd parity, with which the observed frequencies agree with the local dispersion relation [$k_x^2 = 0$ in Eq. (13.7)] (see Fig. 13.3). The time evolution of the electrostatic energy shows no observable increase in the energy of the individual discrete modes, as expected. The electrostatic energy is dominantly in the $m = 0$ and $m = 1$ modes because finite particle size as well as shielding effects enter for $m > 1$. The interferogram of the potential fluctuations at the measured frequency for mode $k_y\rho_s = 0.49$, is displayed in Fig. 13.4 along with the shooting code[42] theory results. Good agreement is found for both the real and imaginary parts of the potential structure. The results from this case indicate that the eigenmode structure established is stable and, therefore, in agreement with the linear theory of the universal mode.[13,14] The long time evolution represents the steady state sustained by energy balance between the

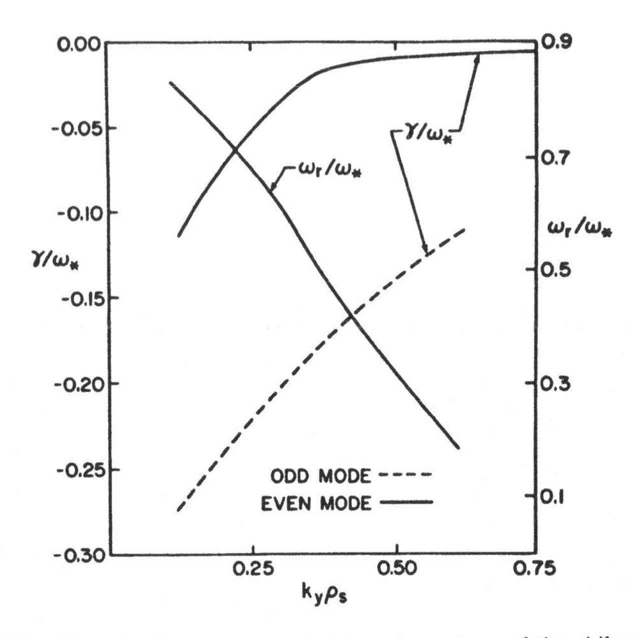

FIGURE 13.2 The eigenfrequency ω and damping rate γ of the drift waves as a function of wavenumbers (Ref. 36)

"Cherenkov emission" of electron kinetic energy into the wave at the electron resonance and shear damping of the wave into parallel ion kinetic energy at the ion Landau resonance. Because of this energy exchange from electrons to waves to ions, there is a net energy flow even though the eigenmodes are stable. In fact, we observe that the electron momentum in the y-direction, near $x = x_e$, gives up a finite value in the electron diamagnetic drift direction.

To estimate the momentum transfer quantitatively, the quasilinear theory is used and, assuming that resonant diffusion is dominant, the diffusion coefficient is given by

$$D_x = \left(\frac{c}{B}\right)^2 \sum_k k_y^2 |\phi_k|^2 \pi \delta(k_\parallel v_\parallel - \omega_k) \ , \tag{13.8}$$

using the fact that $\gamma \ll \omega$ and $\omega < \omega_*$. The evolution of the average distribution is governed by

$$\frac{\partial f_0}{\partial t} = \frac{\partial}{\partial x}\left(D_x \frac{\partial f_0}{\partial x}\right) \ , \tag{13.9}$$

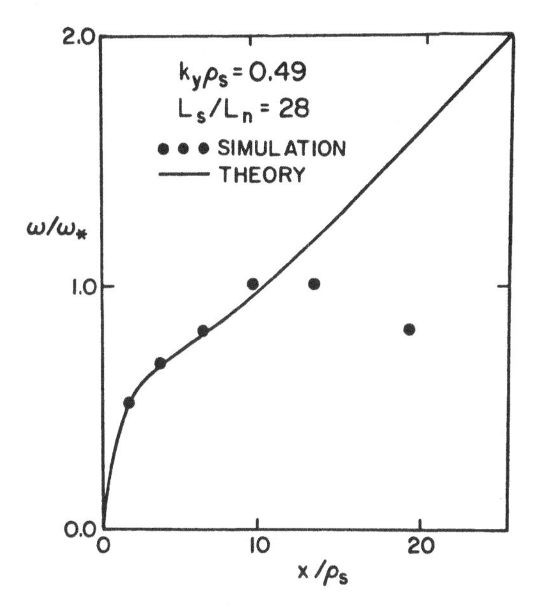

FIGURE 13.3 The local dispersion relation, theory and simulation (Ref. 36)

and the rate of change of momentum is

$$\frac{d}{dt}\int f_0 v_{\parallel} dv_{\parallel} \approx \pi \sum_k \left|\frac{e\phi_k}{T_e}\right|^2 \frac{\omega_k}{k_{\parallel}^2}\omega_*^2 f_0\left(\frac{\omega_k}{k_{\parallel}}\right) . \tag{13.10}$$

For fluctuation levels of $|e\phi_k/T_e| \sim 10^{-3} - 10^{-2}$, we obtain[36]

$$\frac{d\left\langle v_{\parallel}\right\rangle / v_e}{d(\omega_{pe}t)} \sim 2\times 10^{-6} ,$$

and compare with the observed value of 10^{-6}. The ion momentum in the y-direction, near $x = x_i$, acquires a finite value in the direction of the electron diamagnetic drift as a result of the wave-particle interaction. Equation (13.10) also gives agreement within a factor of two for the rate of change of momentum in the ions.[36] The energy transport and momentum transport in the x-direction is, therefore, nonzero even in the absence of instability (i.e., beyond the collisional contributions).

B. Three-dimensional Simulation and Drift Wave Turbulence

If the linear drift wave eigenfunction in a sheared magnetic field is stable as we discussed above, why and how is the fusion plasma such as a plasma in

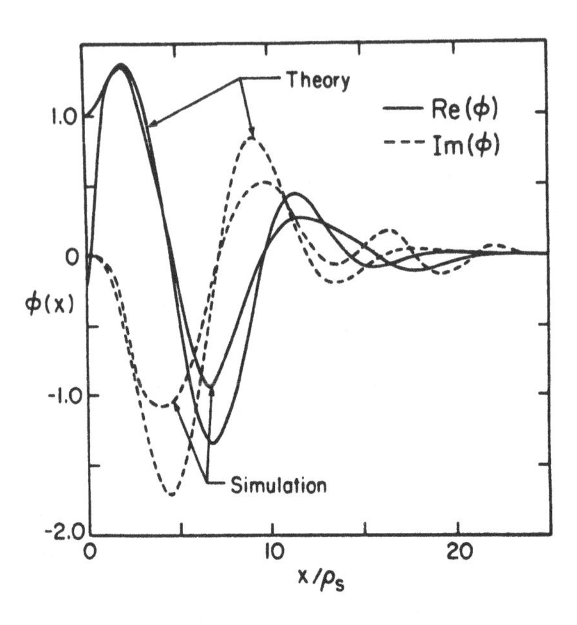

FIGURE 13.4 The spatial structure of drift wave eigenmode with strong magnetic shear, theory and simulation (Ref. 36)

a tokamak believed to sustain a high level of electrostatic fluctuations of the drift wave origin?[18] One possibility suggested by Hirshman and Molvig[15] is that the electron orbit under turbulent drift waves that overlap each other becomes stochastic, modifies the linear electron (adiabatic) response, and thus destabilizes the drift waves. This theory is based on strongly turbulent drift waves and subsequent nonlinear electron dynamics. This nonlinear instability should take place in a three-dimensional system in which many different drift waves whose rational surfaces x_{mn} for mode (m, n) are closely packed and the drift waves around them are "overlapped."[43] (island overlap, see also Chapter 15). Such an effect can not be realized within the realm of two spatial dimensions. A three-dimensional simulation becomes necessary to simulate this physics. The other possibility was suggested by Cheng et al.[16] or others, who proposed that the toroidicity of the tokamak plasma, couples many poloidal modes, that facilitates the overriding of the ion Landau damping. This theory predicts the linear instability of a toroidicity-induced drift wave that did not exist without the toroidal effects such as toroidal geometry and trapped particles.

In the following, we summarize the idea of the electron-nonlinearity induced instability[15] of drift waves. The electron equations of motion in the guiding-center approximation (see Chapter 7) in a sheared magnetic field are

$$\dot{x} = cE_y/B \tag{13.11}$$

$$\dot{y} = -cE_x/B + v_{\parallel}x/L_s \tag{13.12}$$

$$\dot{z} = v_{\parallel} = \text{const.} \tag{13.13}$$

in the neighborhood of a mode rational surface ($x = 0$), i.e., $|x|/L_s \ll 1$.

Letting $x(0) = x_0$, $y(0) = y_0 + v_{\parallel}x_0 t/L_s$ be the unperturbed orbits for $\mathbf{E} = 0$ and $\delta x = x - x(0)$, $\delta y = y - y(0)$, the equations for the perturbed orbits become

$$\delta\dot{x} = \frac{-c}{B}\frac{\partial\phi}{\partial\delta y} , \tag{13.14}$$

$$\delta\dot{y} = \frac{c}{B}\frac{\partial\phi}{\partial\delta x} - v_{\parallel}\frac{\delta x}{L_s} , \tag{13.15}$$

and have been studied by Molvig et al.[43] This set of equations can be treated dynamically as in Section 15.3, or by mapping as in Section 15.4, when ϕ is given. The first term on the right-hand side of Eq. (13.14) and Eq. (13.15) represents the radial diffusion from the background of fine scale fluctuations (thermal noise) seen by the particles. The second term in Eq. (13.15) gives the free streaming contribution along the magnetic field. When the potential is given by $\phi(\mathbf{x}, t) = \sum_{m,n} \phi_{mn}(x) \exp[i(k_y y + k_z z - \omega_{mn} t)]$, resonances occur when

$$k_{\parallel}(x_0)v_{\parallel} - \omega_{mn} = 0 . \tag{13.16}$$

Around each resonant position, x_0, located at

$$k_{\parallel} = k_z \frac{k_y x_0}{L_s} = 0 , \tag{13.17}$$

there is an island in $(\delta x, \delta y)$ space, in which the particles are trapped[44] by the wave, if the islands are well separated. The island width can be estimated by assuming

$$\delta\dot{x} = \frac{c}{B}k_y\bar{\phi}\sin(k_y\delta y) \simeq \frac{c}{B}\bar{\phi}k_y^2\delta y , \tag{13.18}$$

and substituting into Eq. (13.15) gives

$$\delta\ddot{y} + \frac{c}{B}\bar{\phi}k_y^2\frac{v_{\parallel}}{L_s}\delta y = 0 , \tag{13.19}$$

where $\bar{\phi}$ is a root mean square value of the local potential. The width[44] of the resonant island is given by the trapping width

$$\delta x_t = 4\left[\frac{c}{B}\frac{L_s}{v_{\parallel}}\bar{\phi}\right]^{1/2} . \tag{13.20}$$

From Eq. (13.17) the rational surfaces are located at

$$x_{mn} = \frac{-L_y L_s}{L_z}\frac{n}{m} \tag{13.21}$$

with spacing

$$\Delta x_{mn} \simeq \frac{L_y L_s}{L_z}\frac{n}{m(m+1)} , \tag{13.22}$$

as an upper bound where n is held fixed and m varies.

When $\delta x_t > \Delta x_{mn}$,[43] the island overlap condition of Chirikov et al.[45] is satisfied and resonance overlap occurs, resulting in stochastic particle orbit behavior. This means two particles moving along nearly identical trajectories will separate exponentially according to the general formula

$$\frac{\Delta x(t)}{\Delta x(0)} \propto \exp[t/\tau_c] , \tag{13.23}$$

where the characteristic time scale of exponentiation, τ_c, is the approximate inverse of the Kolmogorov entropy[46] and $\Delta x(0)$ is the initial separation. On a time scale large compared to τ_c, the net effect of stochastic behavior is diffusion of the electron orbits.

The behavior of electrons in a strongly turbulent plasma due to drift waves embedded in a sheared magnetic field may be described by the drift-kinetic equation (Chapter 7) for electrons

$$\frac{\partial f_e}{\partial t} + \boldsymbol{\nabla}\cdot(\hat{n}vf_e) + \boldsymbol{\nabla}\cdot\left(\frac{c}{B}\mathbf{E}\times\hat{n}f_e\right) - \frac{|e|}{m_e}E_{\|}\frac{\partial f_e}{\partial v_{\|}} = 0 , \tag{13.24}$$

where $\mathbf{B} = B\hat{n}$ and $\hat{n} = \hat{z} + (x/L_s)\hat{y}$. If we separate the distribution function into its average (f_0) and fluctuating ($\hat{f}_0$) parts the equation for the fluctuating part of the electron distribution is

$$\left[\frac{\partial}{\partial t} + v\hat{n}\cdot\nabla + (\hat{n}\times\nabla\phi)\cdot\nabla\right]\hat{f}_e$$

$$= \frac{-|e|}{T_e} f_0 \left[v_*(\hat{n}\times\hat{x})\cdot\nabla - v\hat{n}\cdot\nabla\right]\hat{\phi} , \tag{13.25}$$

where the carats on f_e and ϕ indicate perturbed quantities. In this equation only the $\mathbf{E}\times\mathbf{B}$ nonlinearity is kept and $\omega_* = k_y v_{*e}$, with $v_{*e} = v_e^2/L_n\Omega_e$.

Several simplifying assumptions are made in order to proceed with the solution of this equation. First, since stochastic electron diffusion is perpendicular to the magnetic field, the electron $\mathbf{E}\times\mathbf{B}$ nonlinearity is approximated by an x-space diffusion operator, $D\partial^2/\partial x^2$. The coefficient, D, replaced all nonlinearities in the problem[15] and is itself a functional of the potential fluctuation spectrum,

$$D = \sum_{\omega,k} G_{k,\omega} k_y^2 \left\langle \left| \hat{\phi}_{k,\omega} \right|^2 \right\rangle \tag{13.26}$$

where $G_{k,\omega}$ is the propagator of the particle motion. In this sense the term "strong turbulence" here refers to the orbital nonlinearities of electron guiding-centers, but note that other zeroth order quantities are not strongly modified. This means that this type of strong turbulence preserves a certain structure of the plasma such as the gyromotion and ion ballistic motion, while it becomes strongly chaotic in its electron guiding-center motion and its low frequency band of spectrum in the vicinity of $\omega = \omega_*$ in Eq. (13.26). Thus, when we refer to strong turbulence, the strongly chaotic behavior takes place at a particular frequency band for a medium like a plasma which has multiple layers of time hierarchy. For example, the electron guiding centers may behave chaotically responding to the drift waves with frequency $\omega \sim \omega_*$, while the high frequency response of electrons to plasma waves may be perfectly laminar.

Second, the only nonlinear terms retained in the expression for $(\hat{f}_e)_k$ are ones explicitly proportional to $\hat{\phi}_k$. This implies that we keep only the part of $\hat{f}_k$ which is phase-coherent with $\hat{\phi}_k$.

Formally, we integrate the equation for $\hat{f}_e$ along the nonlinear orbits. The drift-kinetic equation for the perturbed electron distribution function is

$$\left[-i(\omega - k_\parallel v_\parallel) - \frac{D\partial^2}{\partial x^2}\right] \hat{f}_e = \frac{-i|e|}{T_e} f_0 \left[\omega_* - k_\parallel v_\parallel - \frac{iD\partial^2}{\partial x^2}\right] \hat{\phi} \ . \tag{13.27}$$

The last term on the right-hand side of Eq. (13.27) assures that diffusion does not affect the adiabatic response[49] of $\hat{f}_e$ (see Chapter 7). By rewriting the nonadiabatic electron response

$$\hat{h}_e = \hat{f}_e - \frac{|e|}{T_e} \hat{\phi} f_0 \ , \tag{13.28}$$

the drift-kinetic equation becomes (Chapter 7)

$$\left[-i(\omega - k_\parallel v_\parallel) - D\frac{\partial^2}{\partial x^2}\right] \hat{h}_e(x) = i(\omega - \omega_*)\frac{|e|}{T_e} f_0 \hat{\phi}(x) \ . \tag{13.29}$$

The integral operation solution to the nonadiabatic diffusive electron response[15] is

$$\hat{h}_e(x; v_\parallel) = \int_0^\infty d\tau \int_{-\infty}^\infty dx' \exp\left[\frac{i(\omega - k_\parallel v_\parallel)\tau - \frac{1}{3}(k_\parallel' v_\parallel)^2 D\tau^3}{\sqrt{4\pi D\tau}}\right]$$

$$\cdot \exp\left[\frac{-(x - x' - iDk_\parallel' v_\parallel \tau^2)}{4D\tau}\right]$$

$$\cdot \left[i(\omega - \omega_*) \frac{|e| f_0}{T_e} \hat{\phi}(x') \right] . \tag{13.30}$$

To solve the integral equation, we note that

$$\hat{h}_e(x, v_\parallel) = \int dx' G(x, x', v_\parallel) \hat{\phi}(x') . \tag{13.31}$$

The function G is a Green function of $(x - x')$ peaked with a width x_c. $x_c = \omega_c / k'_\parallel v_e$ is the correlation length for the radial diffusion. The scale length of $\hat{\phi}(x)$ is x_T, the approximate width of the eigenfunction. For $x_c < x_T$, we can Taylor expand $\hat{\phi}(x')$ around $x = x'$, to give

$$\hat{h}_e(x, v_\parallel) = \int d(\delta x) G(x, x + \delta x, v_\parallel) \left[\hat{\phi}(x) + \delta x \frac{d\hat{\phi}}{dx} + \frac{(\delta x)^2}{2} \frac{d^2 \hat{\phi}}{dx^2} + \cdots \right] , \tag{13.32}$$

where $\delta x = x - x'$. The nonadiabatic (see Chapters 7 and 9) perturbed electron density response is

$$n_e^{NA} = \int dv_\parallel \hat{h}_e(x, v_\parallel) \tag{13.33}$$

and the total perturbed density response[47,48] becomes

$$\hat{n}_e = \frac{|e| n_0 \hat{\phi}}{T_e} + \hat{n}_e^{NA} = \frac{|e| n_0 \hat{\phi}}{T_e} + \left[C_e^{(0)} + C_e^{(1)} \frac{d}{dx} + C_e^{(2)} \frac{d^2}{dx^2} \right] \hat{\phi} , \tag{13.34}$$

where the coefficients, C_e, may be found in Refs. 47 and 48. The perturbed ion density becomes

$$\hat{n}_i = \frac{-|e| \hat{n}_0}{T_i} + \left[C_i^{(0)} + C_i^{(2)} \frac{d^2}{dx^2} \right] \hat{\phi} , \tag{13.35}$$

where, again, C_i's may be found in Refs. 47 and 48.

The renormalized electron density response is coupled to the linear ion response through the quasi-neutrality condition. This gives the eigenmode equation (13.7) with the (electron) nonlinearly modified "potential" Q defined as

$$Q(x) = \frac{\left[1 + \tau \frac{\omega}{\omega_*} + \xi_i Z(\xi_i) \left[1 + \tau \frac{\omega}{\omega_*} \right] \Gamma_0(b) + i \left[\frac{\omega}{\omega_*} - 1 \right] \frac{\omega}{\omega_c} I^{(0)} \right]}{\left[\xi_i Z(\xi_i) \left[\frac{\omega}{\omega_*} + \frac{1}{\tau} \right] (\Gamma_0 - \Gamma_1) + 3i \left[\frac{\omega}{\omega_*} - 1 \right] \frac{\omega}{\omega_c} x_c^2 I^{(2)} \right]} , \tag{13.36}$$

and $\tilde{x} = x/\rho_s$, $\tau = T_e/T_i$. The boundary conditions are the same as in the linear eigenmode problem[12]; outgoing waves.

The simulations are carried out[49] in slab geometry with the inhomogeneous exponential density profile. The y and z-directions are homogeneous and periodic. A normal mode expansion of the fields and charge density is employed in the z-direction.[28] (See Chapter 7). Otherwise, the electron and ion dynamics are identical to the two-dimensional simulation discussed earlier in this section. The value $x_0 = L_x/2$ is chosen in Eq. (13.1) and the rational surfaces are concentrated in the middle of the simulation domain. The boundary condition on the electrostatic potential is $\phi(0) = 0 = \phi(L_x)$ and the eigenmode parity is dominantly even with respect to the rational surfaces. The rational surface positions are located at $x_{mn} = x_o \pm (n/m)(L_s L_y/L_z)$, where $k_y = 2\pi m/L_y$ and $k_z = 2\pi n/L_z$. Typical simulation parameters used are: $L_x \times L_y \times L_z = 12\rho_s \times 6\rho_s \times 1200\rho_s$, $\tau = T_e/T_i = 1$, $m_i/m_e = 500$, $L_s/L_n = 14$, $L_n/\Delta = 16$, $v_e\omega_{pe}\delta = 2.6$, $\Omega_{ce}/\omega_{pe} = 11$, $k_y\rho_s = 1.1m$, $n_0 = 16$ particles/cell, $\Omega_{ci}\Delta t = 0.07$, $m = 0, \pm 1, \ldots, \pm 16$, $n = 0, \pm 1, \ldots, \pm 5$, and the finite particle size is given by $a_x = 1.5\Delta$, $a_y = \Delta$, and $a_z = 270\Delta$. The parameter definitions are $T_e = m_e v_e^2$, $\rho_s = \sqrt{\tau} v_i/\Omega_{ci}$, $\Omega_{ci} = eB/m_i c$.

The analysis and interpretation of the simulation results consists of two parts. First, particle orbits and diffusive electron behavior are investigated. Second, the wave potential fluctuations are temporally and spatially analyzed in order to determine their stability.[50]

Test electrons are selected in the vicinity of the mode rational surfaces. The positions and velocities are stored at each time step in order to obtain the spatial diffusivity of the particles. To determine the Chirikov condition[45] for stochasticity at the simulation thermal level in the mode rational surface region, the island width (determined by the particular drift wave eigenmode amplitude at the electron Landau resonances) is compared with the mode rational surface separation. In the neighborhood of the electron resonance, from Eq. (13.20), the island width is approximately

$$\delta x_t/\rho_s \simeq (4/\rho_s)\left(L_s \left|\phi_{mn}\right| / B v_e\right)^{1/2} . \tag{13.37}$$

A crude estimate of the rational surface spacing is from Eq. (13.22)

$$\Delta x_{mn}/\rho_s \simeq 0.21 n/[m(m+1)] . \tag{13.38}$$

From the fluctuation levels measured in the simulation,[50] $10^{-3} \lesssim e\phi_{mn}/T_e \lesssim 10^{-1}$, the island overlap condition is well satisfied ($\delta x_t > \Delta x_{mn}$) and the electron orbits are stochastic as demonstrated in Fig. 13.5.

The diffusion coefficient, D, for test particles may be measured by use of

$$D = \lim_{t\to\infty} \sum_{i=1}^{N} \frac{(\Delta X_i)^2}{2Nt} , \tag{13.39}$$

where ΔX_i is the change in position of the guiding center for the ith particle at time t and where N is the number of test particles. The guiding-center displacement of test electrons in the x-direction as a function of time, initially located in the range $5 \lesssim x/\rho_s \lesssim 7$, is illustrated in Fig. 13.6. The value of

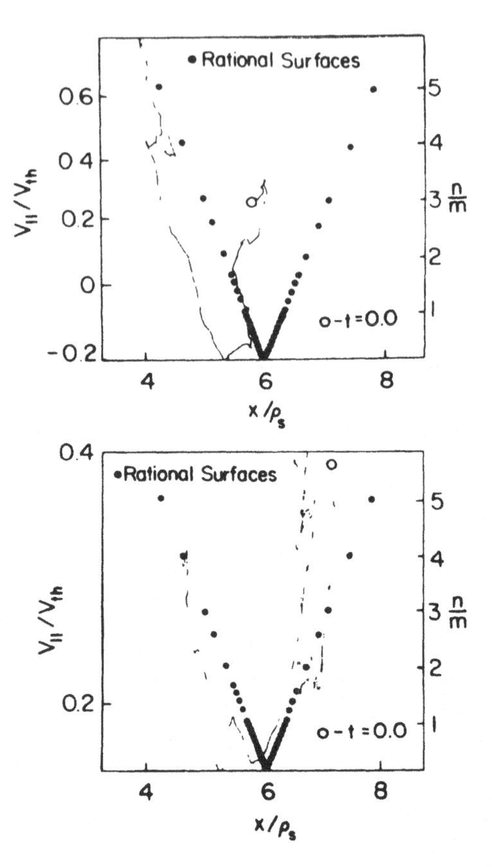

FIGURE 13.5 Stochastic meandering of electron orbits due to overlap of drift waves at different rational rational surfaces (Ref. 50)

the test particle diffusion coefficient ($D = 0.006\rho_s^2\Omega_i$) is verified by two independent calculations. One estimate is based on the potential fluctuations measured in the simulation[50] on the assumption that resonant diffusion dominates. Another estimate is made from the exponential divergence of neighboring orbits (Lyapunov exponent[46]) shown in Fig. 13.6(b), (c) where the relation between the parallel correlation length and diffusion coefficient is determined. It is evident from Fig. 13.6(a) that the diffusion coefficient decreases beyond $\Omega_i t = 132$. This indicates that the diffusion process is time-dependent and is related to the saturation of the potential fluctuations.

For the case of resonance overlap, the stochastic particle orbit behavior resulted in the net effect of electron-orbit diffusion transverse to the magnetic field. The electrons decorrelate from the drift wave resonances at a rate[15] of

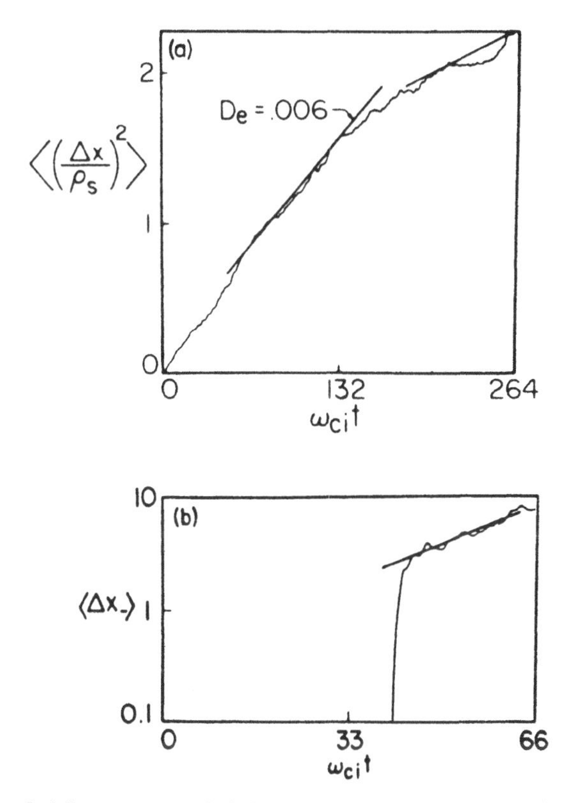

FIGURE 13.6 Guiding center radial displacement of electrons in time and the Lyapunov exponent (Ref. 49)

$\omega_c = [(k'_{\|} v_e)^2 D]^{1/3}$, where $k'_{\|} = k_y/L_s$. At short wavelengths ($k_y \rho_s \gtrsim 1$), the linearly stable modes can be nonlinearly destabilized for $\omega_c \gtrsim \omega$, where ω is the linear drift wave eigenmode frequency.[50]

For the case of multiple rational surface with $k_y \rho_s \gtrsim 1$, approximately a 30 percent increase in the amplitude above the thermal fluctuation level was observed in the simulation[50] and it occurs over several drift periods. The growth rates of the fluctuations correspond closely to the predictions derived from the nonlinear eigenmode equation. The variation of the mode in the x-direction agrees qualitatively with the wave function obtained from the eigenmode equation. The growth rates and real frequencies for various mode numbers, as well as comparisons with the analytic model are shown

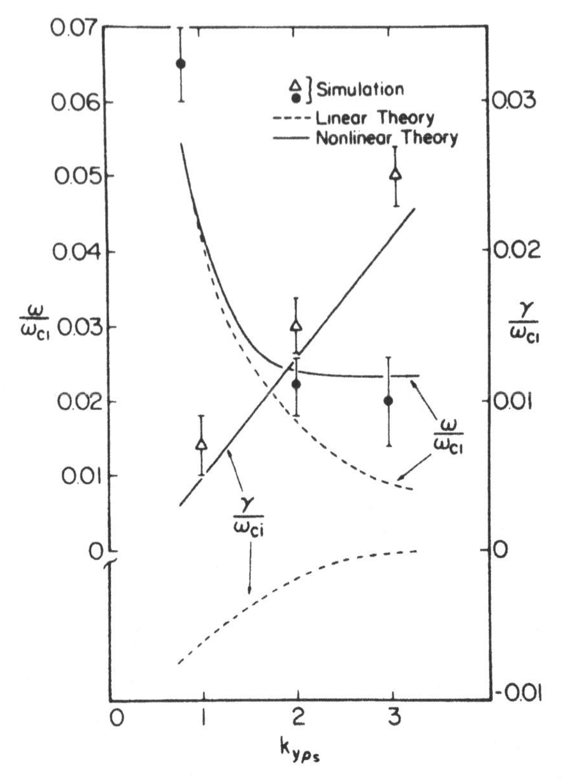

FIGURE 13.7 Eigenfrequencies and growth rates of the nonlinearly destabilized drift wave, theory and simulation (Ref. 50)

in Fig. 13.7. The eigenfunctions, $\hat{\phi}(\tilde{x})$, and eigenvalues, $\omega = \omega_r + i\gamma$, are determined by varying the parameters $k_y\rho_s$, τ, L_s/L_n, m_i/m_e, and ω_c/ω_*, where ω_c/ω_* measures the strength of the background fluctuations causing the electron diffusion in x.

The enhanced fluctuation level is large enough to cause relaxation of the equilibrium density profile in the resonance overlap region. The decrease in the electron diffusion coefficient illustrated in Fig. 13.6(a) coincides with the saturation time of the longest-wavelength modes, $(m, n) = (\pm 1, n)$. The increased amplitude of the potential fluctuations is sufficient to push particles out of the resonance overlap region. This changes the diffusivity of the particles, which in turn affects the mode stability. The enhanced diffusion out of the resonance overlap region manifests itself as a local density-profile relaxation. The time for an electron to pass through the finite resonance region diffusively corresponds roughly to the initial departure of the diffusion coefficient from the constant slope.

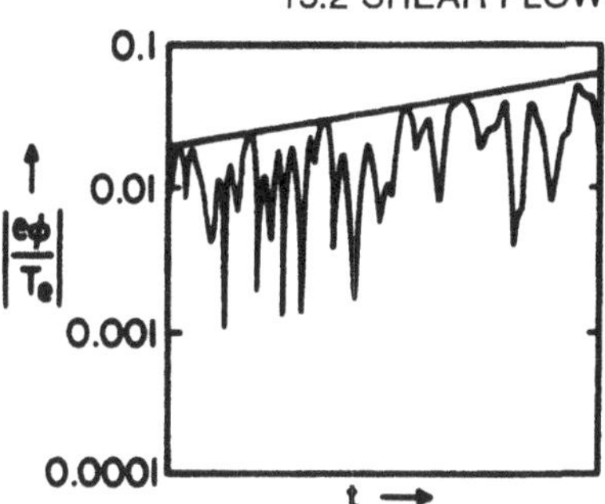

FIGURE 13.8 The amplitude of electric potential grows exponentially in time for a mode ($m = 8$, $n = 9$) that would be stable in slab geometry (Ref. 52)

C. Toroidal Simulation of Drift Waves

Simulation of the toroidicity-induced drift instability[16,51] has also been carried out[52] based on the toroidal particle code (see Chapter 10). In this mechanism the destabilization of drift waves is the annihilation of the ion shear damping by the poloidal mode coupling and the periodicity of the mode structure in the poloidal angle. The poloidal coupling and periodicity produce the band structure and alias coupling characteristic of the periodic system, we considered in Chapter 4 (Fig. 4.4). As the drift wave radially propagates out, it rotates along the poloidal direction because of the magnetic shear. When it goes around the poloidal angle, it copules to different poloidal aliases (which are equivalent to the radial aliases). Thus as the wave propagates outward, it can get to a different branch (or alias) of the drift wave centered at a different rational surface before the original wave suffers strong enough ion shear Landau damping. This is the mechanism of the toroidicity induced drift wave destabilization. The simulation[52] shows this mechanism of destabilization (Fig. 13.8) and the band structure of modes (Fig. 13.9).

13.2 Shear Flow Instability

The Kelvin-Helmholtz instability arises at the interface of two relatively sliding fluids or plasma.[5] This instability is believed to play a major role in creating turbulence and mixing fluids in such diverse arenas as laboratory plasmas with an injected flow of particles, fusion plasmas, oceanography, the magnetosphere, and astrophysical circumstances. One example is the auroral oval region where hot magnetospheric ions are injected into ionospheric plasma, with hot ions causing charge separation.[53] Another example is an ignited plasma in the field reverse configuration (FRC),[54] in which charged fusion products such as alpha particles or protons cause the frictional trans-

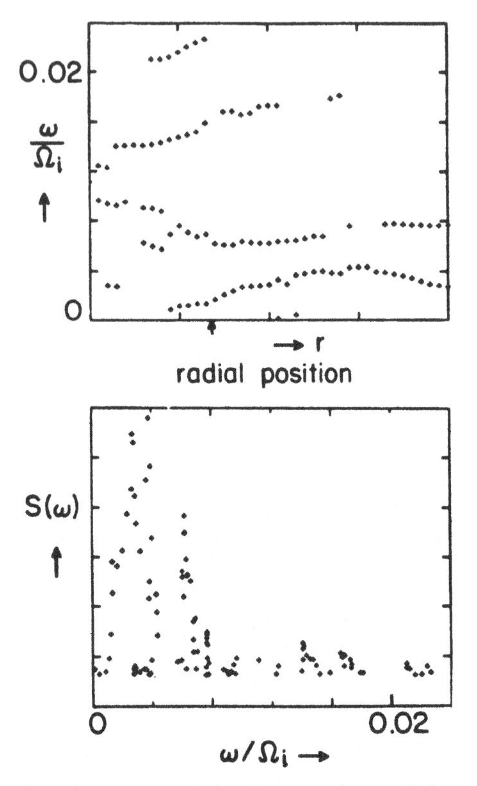

FIGURE 13.9 The band structure of the frequencies and the spectral intensity as a function of radial position (Ref. 52)

port of electrons in the radial direction.[55] The resulting crossed electric field $E_0(x)\hat{x}$ and magnetic field $B_0\hat{z}$ produces a sheared plasma $\mathbf{E} \times \mathbf{B}$ flow given by $v_y(x) = -cE_x/B_0$. In this sense the charge separation and subsequent shear flow instability may arise as a general signature of ignited confined plasmas in the future.

Byers,[56] Levy and Hockney,[57] Lindl,[58] and Horton et al.[59] studied the basically electrostatical Kelvin-Helmholtz instability. When the mode is purely or almost perpendicular to the magnetic field ($k_\parallel v_e < \gamma$, where $k_\parallel$ is the parallel wavenumber, v_e the electron thermal velocity, and γ the growth rate or typical frequency of the mode), electrons cannot short out charge separation and electrostatic fields develop. This situation may be simulated by employing the explicit electrostatic particle code as discussed in Chapters 4 and 7, or the implicit electrostatic particle code in Chapter 9. In the latter one can deal with a much longer time scale and slower instability.

When the mode is not purely perpendicular ($k_{\|} v_e > \gamma$), on the other hand, electrons short out charge separation and the quasineutrality holds, allowing the Boltzman distribution of electrons. This situation may be simulated by the fluid description (Chapter 6) or by the hybrid model of plasma (Chapter 8). Christiansen[60] modeled an inviscid and incompresssible fluid by employing point vortices interacting through a stream function satisfying Poisson's equation for an ordinary fluid. Tajima and Leboeuf[61] used the MHD particle model (see Chapter 6) to study the instability in an ordinary fluid as well as in an MHD fluid. Anomalous viscosity may result from the Kelvin-Helmholtz instability.

The equations of motion for a fluid element obeying magnetohydrodynamics are as discussed in Chapter 6 and Ref. 61. In the present runs,[61] the density ρ is a fixed constant. An isothermal equation of state in two dimensions is used so that the pressure is defined as $p = Tn$. Equation (6.76) is pushed with a viscous drag term, and in finite difference form reads

$$v_j^{n+1} = v_j^n - F^{n+1/2}\Delta t - \nu(v_j^n - v_g^n) \ , \tag{13.40}$$

where F regroups the magnetic and fluid forces and v_g^n is the fluid velocity defined as a grid quantity (see Chapter 6). Here, ν is chosen to be 1, so that we are simulating fluid elements for which no multistream is allowed within the element. Since we are dealing with single-fluid dynamics, if two differently directed particle flows appear in a single cell, the fluid velocity can then be zero and lose its meaning. To be physical, such multistreaming should be avoided in single-fluid dynamics. The choice of $\nu = 1$ represents an instantaneous adjustment of all particles in a given cell to the fluid velocity while the choice of $0 \neq \nu < 1$ signifies a slower adjustment. Note that the Kelvin-Helmholtz instability without a magnetic field is relevant for collisional cases ($\nu \neq 0$). With a magnetic field, the choice of ν does not affect the results significantly (in practice). An initially laminar flow sandwiched by stationary ones is set up in the x-direction.[61] For the present runs, particles are initially at rest except for the middle layer which is initially given a flow velocity v_f, usually exceeding the speed of sound c_s in the medium. Both the flowing fluid and the stationary fluid have the same density and the discontinuity in velocity is sharp at the interfaces in order to allow for strong velocity shear. To start up the Kelvin-Helmholtz instability, a small perturbation of the form $y(j) = \epsilon \sin[k_x x(j)]$ with $\epsilon = 0.01$ is given at $t = 0$ to each jth particle of the flowing fluid, with $k_x = 2\pi/L_x\Delta$, i.e., the fundamental wavelength is excited. The magnetic field, if there is one, is taken along the direction of flow. The field strength is the same for both stationary and flowing parts of the fluid. The system size is typically $32\Delta \times 32\Delta$, Δ being the unit grid spacing, with 16 particles per cell. Typical runs extend to $t = 200 c_s^{-1}\Delta$.

The Kelvin-Helmholtz instability is incurred, in principle, by an infinitesimal shear in velocity when no external magnetic field is applied. However, the nonlinear evolution of the instability differs radically when the velocity difference is subsonic and when it is supersonic. The slow subsonic case with the charge non-neutrality will be discussed in more detail later.

In the subsonic case, with $v_f = 0.3c_s$, small vortices formed at the contact surface between flowing and stationary fluids slowly propagate inward, an indication of the linear growth of the Kelvin-Helmholtz instability. The linear theory for the mode on the interface of two semi-infinite fluids, one flowing with velocity v_f relative to the stationary fluid, yields the dispersion relation on the frame of the stationary fluid as[5]

$$\omega = \frac{1}{2} k_x v_f (1 + i) , \tag{13.41}$$

where k_x is the wavenumber along the flow direction. If the vortices generated by the instability are small enough in the early development, the two interfaces in the simulation may be treated independently and each interface is supposed to obey the dispersion relation of Eq. (13.41). Analysis of the simulation results by correlation techniques performed on the Fourier components of the fluid velocity yields a growth rate for the instability in agreement with Eq. (13.41). For $v_f = 0.3c_s$ and a perturbation at the fundamental wavelength ($k_x = 0.196\Delta^{-1}$), theory predicts $\gamma = 2.95 \times 10^{-2} c_s \Delta^{-1}$, while simulation yields $\gamma \sim 0.03 c_s/\Delta$.[61] As time progresses, the tendency is for the two fluids to intermix. Saturation occurs at $t \sim 300 c_s^{-1}$ due to the viscous drag which tends to equilibrate the velocities of the two fluid layers.

In the supersonic case, seeds of vortices rapidly evolve into large vortices which subsequently rotate around each other and intertwine. These vortices survive for a long time ($t \sim 200 c_s^{-1} \Delta$) and provide the system with a marked configuration. Figure 13.10 depicts the $x - y$ space pattern for the streaming particles at equivalent times for flows with $M_s = 1.5$ [Fig. 13.10(a)] and $M_s = 2.5$ [Fig. 13.10(b)]. A sharp break is apparent in both frames of Fig. 13.10. The angle of the break with respect to the x axis decreases with increasing Mach number. Typical angles of 42° for $M_s = 1.5$ and 24° for $M_s = 2.5$ are measured. Such a structure in the nonlinear stage of the supersonic Kelvin-Helmholtz instability may be understood as follows: In the laboratory frame, early on the instability gives rise to a series of vortices, flowing roughly with the streaming layer; these vortices, now protruding into the (originally) stationary layers, trigger shocks into the stationary layers. The shock angle created by an obstacle in a supersonic flow (this time, these seed vortices) is given[61] by

$$\alpha_s = \sin^{-1}(1/M_s) . \tag{13.42}$$

As the vortices grow in size, Fig. 13.10 shows that the arms of the vortices preferentially stretch along the shock characteristics. The supersonic Kelvin-Helmholtz instability thus evolves into large parallelogram-structured vortices delimited by the shock characteristics. One notices that within the large vortices there appear smaller parallelogram-structured vortices that are rotating

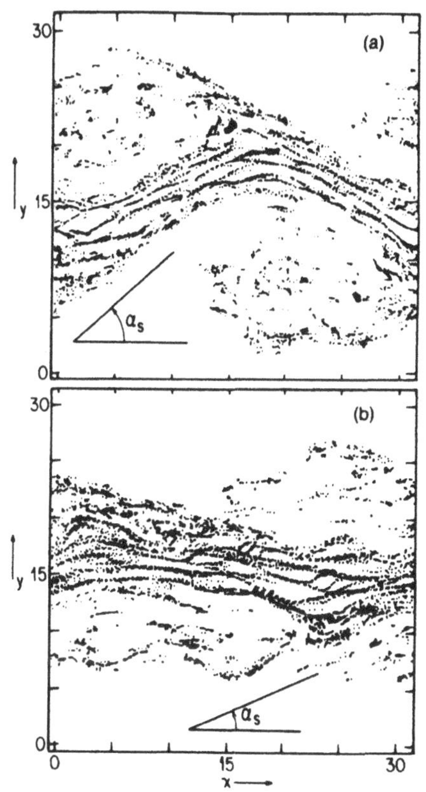

FIGURE 13.10 Supersonic Kelvin-Helmholtz instability and vortices augmented by shock fronts (Ref. 61)

and winding tighter and tighter. Later on, because of the strong drag created by such large-scale vortices, the two relatively sliding fluids become more and more intermingled, creating sound turbulence in both the x and y-directions. The shock-created vortex structure may provide a mechanism for an anomalously large mixing length for the supersonic interface of fluids. Such a pronounced shock structure may help interpret some astrophysical jets.[62]

Next, we discuss a simulation assuming nonquasineutrality in the parallel directions in a low frequency regime (i.e., perpendicular quasineutrality). For strongly magnetized plasmas the Kelvin-Helmholtz growth rate is slow compared to the ion cyclotron frequency $|dv_E/dx| \ll \Omega_i = eB/m_ic$ and the wavelengths are long compared to the ion Larmor radius $a \gg \rho_i = v_i/\Omega_i = c(m_iT_i)^{1/2}/eB$. In order to investigate this situation the technique discussed in Chapter 9, the implicit electrostatic particle code, is employed.

The simulations are performed in a box of width L_x and length L_y parallel to the reversed flow $v_y(x)$.[59] The system is periodic in y and has conducting wall boundaries at $x = 0$ and $x = L_x$ with $v_x(0) = v_x(L_x) = 0$. In the reference case the simulation parameters are $a = 10\Delta_x$, $\omega_{pe}\Delta t = 200$, $L_x/\Delta_x = L_y/\Delta_y = 64$ and the number of particles per cell $N_e = N_i = 9$ per cell. The mesh size in the y-direction is either $\Delta_y/\Delta_x = 1$ or 2. The particles are taken as gaussian clouds with size $a_x = 3\Delta_x$ and $a_y = 3\Delta_y$. The time decentering parameters of the Lorentz force are $\gamma_i = \gamma_e = 0.1$ (see Chapter 9). We wish to study the low frequency $\omega \sim v_0/a \ll \Omega_i$ regime of the Kelvin-Helmholtz instability for realistic ion-to-electron mass ratios, where a is the velocity shear gradient length. We take $m_i/m_e = 1600$ and the magnetic field strength $\Omega_e/\omega_{pe} = 10 - 80$. The space scales L and Δ_x are to range over the longest to the shortest wavelength of interest which requires $L > a \gg \Delta_x$. The time step is determined by $v_{\max}\Delta t < \Delta \equiv \min(\Delta_x, \Delta_y)$ the Courant condition. We limit the maximum flow velocities to be such that the maximum frequency $v_{\max}/\Delta$ in the dynamics is $\omega_{\max} = \pi v_{\max}/\Delta \ll \Omega_i$. The fastest large space scale dynamics is then on the frequency $\omega_k \sim kv_{\max} = 2\pi m v_{\max}/L = m\Omega_i(2\pi\Delta/10L) \sim m\Omega_i/100$. The reference time step is $\Delta t\omega_{pe} = 200$. In the problem with $B_y/B_0 = 0$ or $k_\parallel = 0$ the distribution of particles in $v_\parallel$ has no effect on the dynamics of the instability.

A time series is shown in Figs. 13.11. The initial flow is perturbed by a small amplitude (1%) $m = 3$ perturbation in the initial x position $x = x_0 + \delta\cos(k_s y)$. After a transient the exponential growth of the unstable mode dominates in frames (b)–(c) of Fig. 13.11 giving the contours of the electrostatic potential $\varphi(x, y, t)$. In Fig. 13.12 we show the particle motion in the x-y plane perpendicular to the magnetic field. (only those particles with initial velocities $v_y < 0$). We first see tongues of particles moving across the symmetry line from $v_x = cE_y/B$. Subsequently, these tongues undergo $\mathbf{E} \times \mathbf{B}$ trapping into vortical motions in the x-y-plane. The particle orbits can become strongly stochastic while the potential $\varphi(x, y, t)$ is relatively smooth and slowly varying. At late time ($t > 10^2 a/v_0$) the flow is dominated by a few large, $R > (1 - 5)$, nearly stationary vortices.[59] From the longest simulations ($t_f \sim 10^3 a/v_0$) it appears that the long-time limit is a single large vortex with characteristic radius $R \sim L_x/2$.

13.3 Heat Convection Instability

Under gravity a fluid takes a stratified equilibrium in which the higher it is, the lighter (or more dilute) the fluid becomes. When the fluid is heated from the bottom (or equivalently heat flux is passed through the bottom cross-section), the heat propagates through the fluid via conduction if the heat flux is small enough for a given heat conductivity field. If the heat flux that is injected into the fluid from below is increased beyond a certain critical

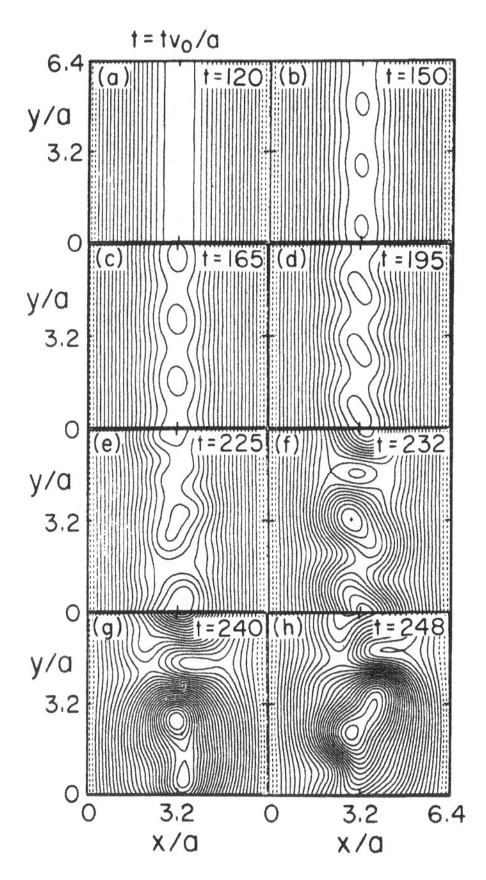

FIGURE 13.11 The electric potential in the shear flow instability at various times, where the reversal of flow occurs at the midpoint in x (Ref. 59)

value, the heat conduction alone is not sufficient for this amount of heat flux. Then, convective motion of the fluid begins. The Rayleigh-Benard convection instability[5] in an incompressible fluid is an example. The convective instability in a compressible fluid such as the solar convection in the convection zone takes place if the following Schwarzschild criterion[63] is satisfied:

$$\rho^* < \rho(z + dz) , \tag{13.43}$$

where $\rho(z+dz)$ is the density at height $z+dz$ and ρ^* is the density of the fluid at $z+dz$ if we adiabatically raise the fluid $\rho(z)$ from z to $z+dz$. This criterion is sometimes called the superadiabatic temperature (or density) gradient. If this is fulfilled, convection ensues. There have been a number of numerical simulations[64–66] on this topic.

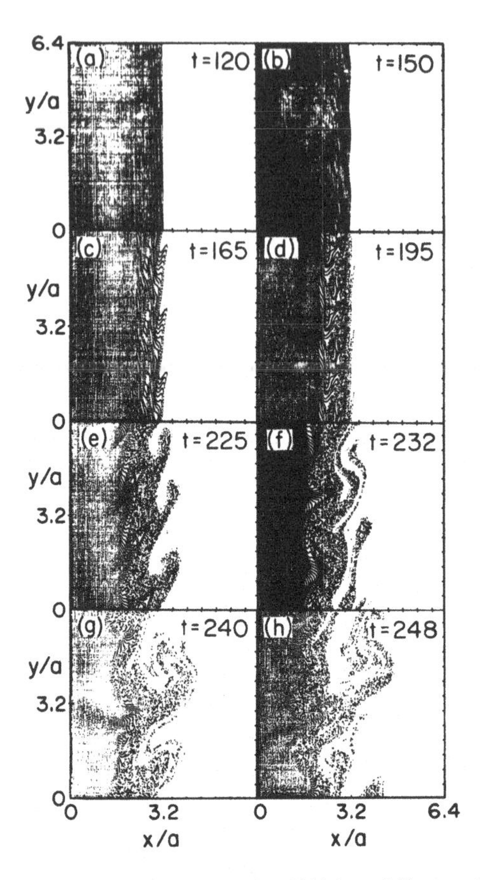

FIGURE 13.12 Particle positions during the K-H instability evolution at the same times as in Fig. 13.11 (Ref. 59)

A set of questions relevant to compressible convection concerns the possible locations in the hydrodynamic parameter space where dynamic behavior exists and the physical mechanisms that cause such behavior. The relevant dimensionless parameters are[5]: the Prandtl number $\sigma = \nu/\kappa$, the Rayleigh number $R = g\alpha\bar{\beta}d^4/\kappa\nu$, and the Reynolds number $R_e = Lv/\nu$, where L is a typical dimension of the system, v a typical velocity, ν the kinematic viscosity, g the gravitational acceleration, d the depth of the layer, $\bar{\beta} = d\ell nT/dz$, α the coefficient of volume expansion, and κ the thermal conductivity. The convection zone is defined by the following parameter values. The ratio of the reference atmosphere superadiabatic temperature gradient to the temperature gradient is defined as Δ, the ratio of the bottom to the top densities as X, and

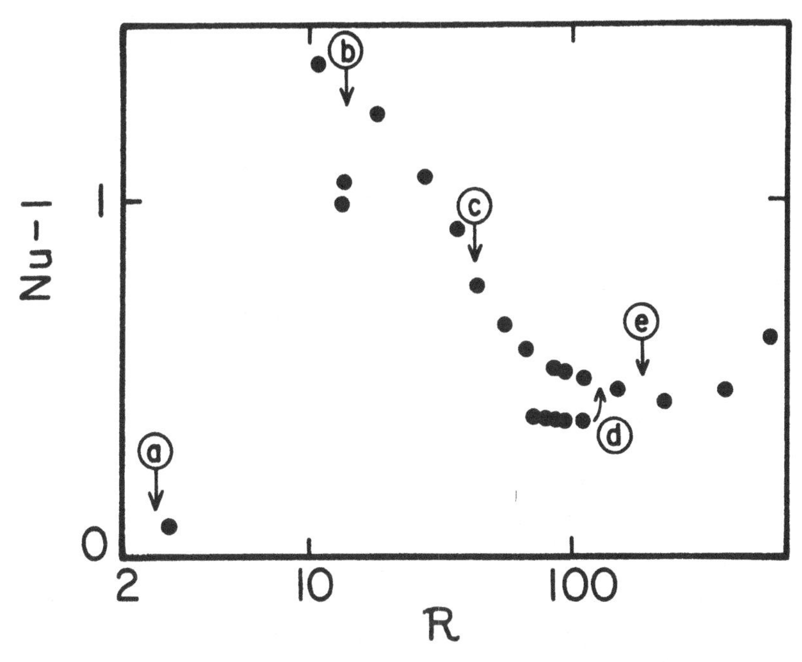

FIGURE 13.13 Nu (the Nusselt number)-1 vs. the normalized Rayleigh number. Simulation shows transitions at a, b, c, d, and e (Ref. 67)

the aspect ratio (width to height) as A: Ginet's study[67] took the following values:

$$\Delta = 4.0 \times 10^{-3}, \quad \chi = 22.4, \quad A = 1 , \tag{13.44}$$

where the choice of the relatively small superadiabatic temperature gradient, Δ, ensures the validity of the anelastic approximation,[68] in which the elastic restoring force of the pressure is dropped so that no sound wave time scale shows up. For the present description[67] the Rayleigh number is given by

$$R = \frac{3.52 \times 10^4}{\sigma} . \tag{13.45}$$

The Prandtl number is used as the "order parameter."[69] The minimum critical Rayleigh number for the onset of the convective instability can be estimated from the calculations of Gough et al.[70]

$$R_c(\min) = 1.26 \times 10^3 . \tag{13.46}$$

The normalized Rayleigh number, $\mathcal{R} = R/R_c(\min)$ is

$$\mathcal{R} = \frac{27.9}{\sigma} . \tag{13.47}$$

In the simulation by Ginet and Sudan[67] sound waves are filtered out by dropping $\partial\rho/\partial t$ in the continuity equation; the thermodynamic quantities such as pressure are linearized about the static reference atmosphere; however, the advective nonlinearity is retained. In the following, their computer simulation[67] is presented.

The Prandtl number regime that will be investigated is naturally bounded above by the critical Prandtl number corresponding to the onset of the convective instability, $\sigma_c = 9.8$. It is the numerical resolution that determines the lower bound on the Prandtl number and with a 33×33 grid the lowest value considered will be $\sigma = 0.05$. In this range the normalized Rayleigh number [Eq. (13.47)] varies from 2.85 ($\sigma = 9.8$) to 558 ($\sigma = 0.05$).

A plot of the time averaged Nusselt number, Nu, versus the normalized Rayleigh number $\mathcal{R}$ (Fig. 13.13), serves as a useful roadmap to guide us through a discussion of the results. The time averaged Nusselt number is a measure of the convective efficiency which is defined as

$$\mathrm{Nu} = 1 + \partial_z T/\bar{\beta} \ , \tag{13.48}$$

where $\bar{\beta}$ is the superadiabatic temperature gradient wavenumber. Thus the Nusselt number (minus one) as defined is an indicator of the degree of convective transport present.

There are five observed transitions[70] between states of qualitatively different temporal behavior which will be denoted as transition a, b, c, d, and e in order of their appearance as $\mathcal{R}$ is increased from a minimum value of 2.8. In the following discussions we will refer to different regimes of solutions that exist between transitions and label the regime with the letters of the bounding transitions. For example, the set of solutions that exist between transitions a and b are in the regime a-b. Ginet's simulations[67] have also revealed the non-uniqueness of long-time anelastic solutions which start from different initial conditions. Two different branches have been observed which we have denoted Branch A (most points in Fig. 13.13) and Branch B [the points left to letter (d) in Fig. 13.13]. In the following we pick a couple of transitions for example.

1. Regime a-b

Transition (a) which occurs at $\mathcal{R} = 2.85$ is the onset of stationary convection. Below $\mathcal{R} = 2.85$ the convection zone is conductively stable and no motion is possible. On the unstable side of the transition a single branch of solutions is found (Branch A) in the regime a-b. Solutions in this regime are stationary, after some initial transient behavior, and are exemplified by the temporal plots of the temperature at the midpoint T, the net energy flux F, and the integrated vorticity Ω that are shown in Fig. 13.14 for the case $\mathcal{R} = 11$. The characteristics of the flow in this regime are similar to those encountered in stationary two-dimensional Rayleigh-Benard convection. The integrated momentum vorticity, Ω, is indicative of a convective cell composed of two counter

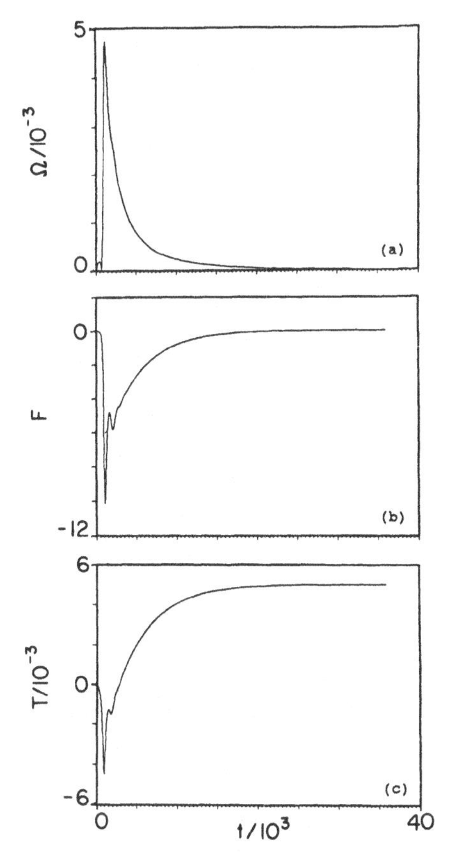

FIGURE 13.14 The temporal behavior of the temperature T, energy flux F, vorticity Ω in the weakly unstable regime ($R = 11$) (Ref. 67)

rotating rolls. Efficient convection means that entropy can be convected upwards at a rate faster than a conductive or viscous process, would diffuse the perturbations. A cellular flow pattern would alter the initial vertical entropy distribution by transporting entropy from high value regions to low value regions in the upwards plume and vice versa in the downwards plumes thus tending to flatten out the total average entropy profile. This is a self-limiting process, if the entropy gradient becomes too positive then the convective instability will be shut off as dictated by the Schwarzschild criteria.[63] The observed increase in the Nusselt number between $\mathcal{R} = 2.85$ and $\mathcal{R} = 11$ (Fig. 13.13) is consistent with the Landau-Hopf[71,72] type of bifurcation that characterizes the onset of convection at transition (a). Near the onset, the magnitude of the

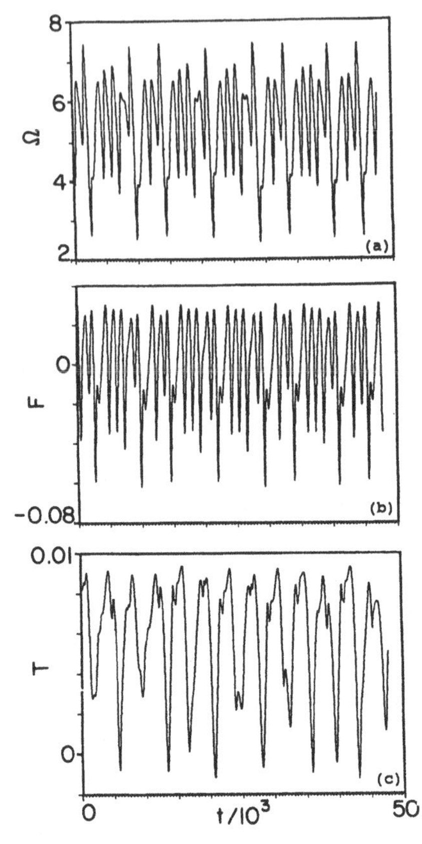

FIGURE 13.15 The aperiodic oscillatory behavior of T, F, and Ω in the more unstable regime ($\mathcal{R} = 13.6$) (Ref. 67)

dynamic variables will increase as the square root of the distance because of the continuous variation in the growth rate with the Rayleigh number.

2. Regime b-c

At a Rayleigh number of about 14 there appears to be a transition from the stationary flow of regime a-b to the temporally periodic flow which is characteristic of regime b-c. In Fig. 13.13 the transition at (b) is indicated by a dip in the Nusselt number measured for runs done at $\mathcal{R} = 14$ and $\mathcal{R} = 13.6$.

Temporal plots of Ω, F, and T at $\mathcal{R} = 13.6$ which are characteristic of transition (b) are shown in Fig. 13.15. Unlike the stationary behavior in regime

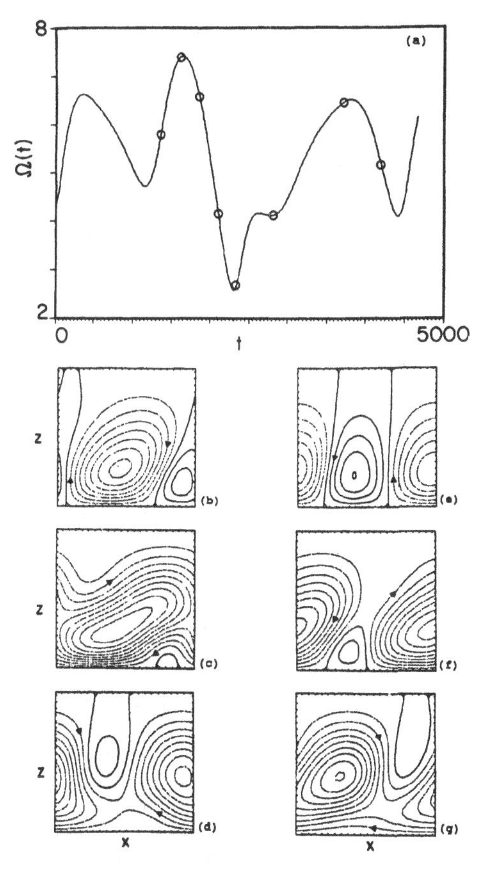

FIGURE 13.16 The temporal variations of the vorticity Ω and the contours of stream function φ at the various points in the $\Omega - t$ diagram (Ref. 67)

a-b or the periodicity in regime b-c, the time series in Fig. 13.15 become aperiodic. One can prove this by determining the largest Lyapunov exponent[73,74] characteristic of the system. The Lyapunov exponent measures the rate of divergence (or convergence) of nearby trajectories in a system's phase space, as we discussed in Section 13.1.

The temporal variation in the flow pattern at a fixed Rayleigh number solution near transition b is striking. In Fig. 13.16 we show a temporal plot of Ω for the $\mathcal{R} = 13.6$ run as shown, covering one of the roughly defined cycles. Snapshots of the streamlines defined by contours of constant stream function φ are illustrated in Fig. 13.16. It is possible to construct a phase space for this system of many degrees of freedom from the time series of a single dynamic variable by the method of time delays. In Fig. 13.17 we show two-dimensional

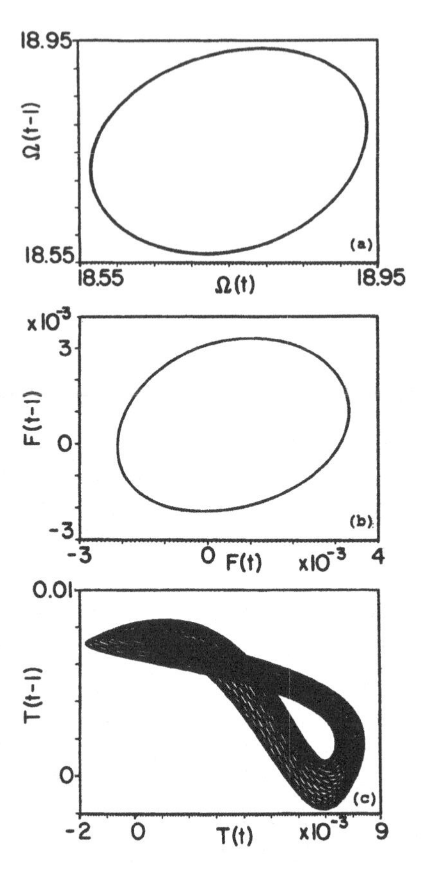

FIGURE 13.17 The Lorentz plot (phase space plot) of time series of vorticity Ω, energy flux F, and temperature T at $\mathcal{R} = 13.6$ (Ref. 67)

phase plots for Ω, F, and T. The periodic motion of Ω and F is inferred from the observation of elliptical phase orbits (or limit cycles). The trajectory in the T phase space indicates quasiperiodic motion on a torus, and a Fourier analysis of the temporal T data has shown one fundamental frequency and several harmonics. Periodic behavior is also consistent with the measurements of the largest Lyapunov exponent from a time series of F.

The spatial structure of the Lorentz plot of the flow through one period is depicted in Fig. 13.17.

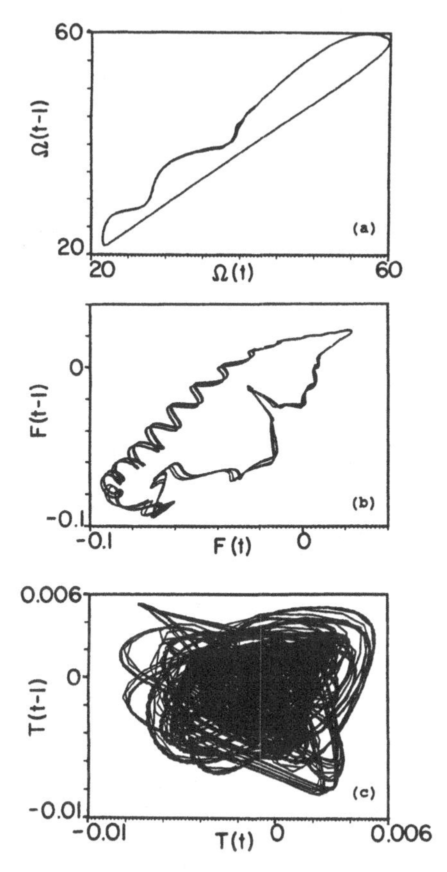

FIGURE 13.18 The Lorentz plot of time series of Ω, F, and T at $\mathcal{R} = 149$ (Ref. 67)

3. Regime c-e

The periodic temporal behavior characteristic of regime b-c in Branch A disappears[67] when the Rayleigh number is raised above a critical value somewhere between $\mathcal{R} = 37$ and $\mathcal{R} = 45$. A new class of solutions appears whose temporal behavior can best be described as quasiperiodic. The transition between the periodic and quasiperiodic regime bears the signature of a Landau-Hopf bifurcation. When we further increase the Rayleigh number beyond $\mathcal{R} = 100$, the system shows less regular oscillations. We show the phase space solutions for the simulation at $\mathcal{R} = 149$ in Fig. 13.18 which reveals a sharp kinking of the two-dimensional torus in F space that might be a precursor to the transition at (d). A nearly chaotic behavior is anticipated, which commences in the regime beyond the point (e). Further discussions on transition beyond (e) and on Branch B are left to Ref. 67.

As we reviewed in this chapter and in Chapter 1, understanding of some of the most fascinating and challenging problems, the problem of instability and turbulence induced by it, has tremendously increased by the numerical computation. It will continue to guide and inspire the physicist.

Problems

1. Explain why the toroidal effects weaken or eliminate the magnetic shear damping of the drift wave.[75]
2. Trace the history of the drift wave theory and simulation.
3. Even though the "universal" drift wave is absolutely stable with magnetic shear, it can be convectively unstable. Explain why this can be possible.
4. The introduction of an additional dimension adds a break of symmetry and/or the addition of degrees of freedom. Why is the three-dimensional theory/simulation of drift waves so different from that in two dimensions?
5. In the nonuniform direction the customary method of diagnosis by the Fourier analysis is often ill-suited. List a set of alternative approaches to diagnosis in such nonuniform systems.
6. When islands of the adjacent drift wave eigenmodes at different rational surfaces overlap, electron orbits become stochastic. Molvig et al.[43] assumed a normal distribution. Discuss deviations from this, such as kurtosis in the orbital stochasticity.
7. In fully developed turbulence higher and higher harmonics are generated. It thus becomes quite challenging to both theory and simulation to carry out satisfactory analysis for a very high Reynolds number. Discuss various approaches to fully developed strong turbulence, such as the direct interaction approximation,[76] the renormalization group technique,[77,80] renormalized perturbation expansion,[78] Lagrangian renormalized approximation,[79] and decimation under symmetry.

References

1. for example, A.B. Mikhailovskii, *Theory of Plasma Instabilities*, vol. 2 (Consultant Bureau, New York, 1974).
2. H. Alfvén, *Cosmic Plasmas* (Reidel, Dordrecht, 1981)
3. T.D. Tarbell, A.M. Title, and S.A. Schoolman, Ap. J. **229**, 387 (1979).

4. W. Calvert, Geophys. Rev. Lett. **8**, 919 (1981).
5. S. Chandrasekhar, *Hydrodynamic and Hydromagnetic Stability* (Clarendon, Oxford, 1961).
6. L.I. Rudakov and R.Z. Sagdeev, Zhur. Eksptl. Teoret. Fiz. **37**, 1337 (1959) [Sov. Phys. JETP **10**, 952 (1960)].
7. N.A. Krall and M.N. Rosenbluth, Phys. Fluids **6**, 254 (1963).
8. A.A. Galeev, V.N. Oraevskii and R.Z. Sagdeev, Zhur. Eksptl. Teoret. Fiz. **44**, 903 (1963) [Sov. Phys. JETP **17**, 615 (1963)].
9. A.B. Mikhailovskii and L.I. Rudakov, Zhur. Eksptl. Teoret. Fiz. **44**, 912 (1963) [Sov. Phys. JETP **17**, 621 (1963)].
10. A.B. Mikhailovskii and A.V. Timofeev, Zhur. Eksptl. Teoret. **44**, 919 (1963) [Sov. Phys. JETP **17**, 626 (1963)].
11. P.H. Rutherford and E.A. Frieman, Phys. Fluids **10**, 1007 (1967).
12. L.D. Pearlstein and H.L. Berk, Phys. Rev. Lett. **23**, 220 (1969).
13. D.W. Ross and S.M. Mahajan, Phys. Rev. Lett. **40**, 324 (1978).
14. K.T. Tsang, P.J. Catto, J.C. Whitson, and J. Smith, Phys. Rev. Lett. **40**, 327 (1978).
15. S.P. Hirshman and K. Molvig, Phys. Rev. Lett. **42**, 648 (1979).
16. C.Z. Cheng and L. Chen, Phys. Fluids **23**, 1770 (1980); J.B. Taylor, L. Chen, M.S. Chance, and C.Z. Cheng, Nucl. Fusion **20**, 901 (1980); J.W. Conner, R.J. Hastie, and J.B. Taylor, Plasma Phys. **22**, 757 (1980).
17. B.B. Kadomtsev, *Plasma Turbulence* (Academic, New York, 1965).
18. P.C. Liewer, Nucl. Fusion **25**, 543 (1985).
19. A.A. Galeev and R.Z. Sagdeev, Zh. Eksptl. Teoret. Fiz. **53**, 348 (1967) [Sov. Phys. JETP **26**, 233 (1968)].
20. F. Hinton, M.N. Rosenbluth and R.D.Hazeltine, Phys. Fluids. **15**, 116 (1972).
21. H. Okuda and J.M. Dawson, Phys. Fluids **16**, 408 (1973).
22. W.W. Lee and H. Okuda, Phys. Rev. Lett. **36**, 870 (1976).
23. W.W. Lee, Y.Y. Kuo and H. Okuda, Phys. Fluids **21**, 617 (1978).
24. W.W. Lee, W.M. Nevins, H. Okuda, and R.B. White, Phys. Rev. Lett. **43**, 347 (1979).
25. C.Z. Cheng and H. Okuda, Phys. Rev. Lett. **38**, 708 (1977).
26. S. Tokuda, T. Kamimura, and H. Ito, J. Phys. Soc. Jpn. **48**, 1722 (1980).
27. W.W. Lee and H. Okuda, J. Comp. Phys. **26**, 139 (1978).
28. C.Z. Cheng and H. Okuda, J. Comp. Phys. **25**, 133 (1977).

29. H. Okuda, W.W. Lee and C.Z. Cheng, Comp. Phys. Comm. **17**, 233 (1979).

30. T. Hatori and Y. Terashima, J. Phys. Soc. Jpn. **42**, 1010 (1977).

31. A. Hasegawa and K. Mima, Phys. Fluids **21**, 87 (1978).

32. R.E. Waltz, Phys. Fluids **26**, 169 (1983).

33. W.W. Lee, J.A. Krommes, C.R. Oberman, and R.A. Smith, Phys. Fluids **27**, 2652 (1984).

34. A. Hasegawa and M. Wakatani, Phys. Rev. Lett. **50**, 682 (1983).

35. N. Bekki and Y. Kaneda, Phys. Rev. Lett. **57**, 2176 (1986).

36. R.D. Sydora, J.N. Leboeuf and T. Tajima, Phys. Fluids **28**, 528 (1985).

37. H. Naitou, S. Tokuda and T. Kamimura, J. Comp. Phys. **33**, 86 (1979).

38. J.P. Burg, Ph.D. dissertation (Stanford University, Palo Alto, Ca., 1975).

39. V.K. Decyk, G.J. Morales and J.M. Dawson, Phys. Fluids **23**, 826 (1980).

40. B.D. Fried and S.D. Conte, *The Plasma Dispersion Function* (Academic, New York, 1961).

41. W.M. Nevins and L. Chen, Phys. Fluids **23**, 1973 (1980).

42. N.T. Gladd and W. Horton, Phys. Fluids **16**, 879 (1973); D.W. Ross and S.M. Mahajan, Phys. Fluids **22**, 294 (1979).

43. K. Molvig, J.P. Freiberg, R. Potok, S.P. Hirshman, J.C. Whitson, and T. Tajima, *Long Time Prediction in Dynamics*, ed. by W. Horton, L. Reichl and V. Szebeheli (Wiley, New York, 1983) p. 319.

44. T.M. O'Neil, Phys. Fluid **8**, 2255 (1965); L.M. Al'tshul' and V.I. Karpman, Zhur. Eksptl. Teoret. Fiz. **49**, 515 (1965) [Sov. Phys. JETP **22**, 361 (1960)].

45. G.M. Zaslavskii and B.V. Chirikov, Sov. Phys. Uspekhi **14**, 549 (1972).

46. for example, A.J. Lichtenberg and M.A. Lieberman, *Regular and Stochastic Motion* (Springer Verlag, New York, 1983).

47. W.M. Tang, Nucl. Fusion **18**, 1089 (1978).

48. C.O. Beasley, W.I. van Rij and K. Molvig, Phys. Fluids **28**, 271 (1985).

49. R.D. Sydora, Ph.D. Dissertation (The University of Texas, Austin, 1985).

50. R.D. Sydora, J.N. Leboeuf, D.R. Thayer, P.H. Diamond, and T. Tajima, Phys. Rev. Lett. **57**, 3269 (1986).

51. T.J. Schep and M. Venema, Plas. Phys. Control. Fus. **27**, 653 (1985); R.J. Hastie, K.W. Hesketh, and J.B. Taylor, Nucl. Fusion **19**, 1223 (1979).

52. M.J. LeBrun and T. Tajima, to be published.

53. J.S. Wagner, J.R. Kan, S.-I. Akasofu, T. Tajima, J.N. Leboeuf, and J.M. Dawson, *Physics of Auroral Arc Formation*, ed. by S.-I. Akasofu and J.R. Kan (American Geophysical Union, Washington, D.C., 1981) p. 304.

54. J.M. Finn and R.N. Sudan, Nucl. Fusion, **22**, 1443 (1982).

55. J.H. Hammer and H.L. Berk, Nucl. Fusion **22**, 89 (1982); A. Reiman and R.N. Sudan, Comments on Plasma Phys. Contr. Fusion **5**, 167 (1979).

56. J.A. Byers, Phys. Fluids **9**, 1038 (1966).

57. R.H. Levy and R.W. Hockney, Phys. Fluids **11**, 766 (1968).

58. J.D. Lindl, Ph.D. dissertation (Princeton University, Princeton, 1969) unpublished.

59. W. Horton, T. Tajima, and T. Kamimura, Phys. Fluids **30**, 3485 (1987).

60. J.P. Christiansen, J. Comp. Phys. **13**, 363 (1973).

61. T. Tajima and J.N. Leboeuf, Phys. Fluids **23**, 884 (1980).

62. M.L. Norman, L. Smarr and K.H.A. Winkler, *Numerical Astrophysics* eds. J.M. Centrella, J.M. LeBlanc, and R.L. Bowers (Jone and Bartlett Publ., Boston, 1985) p. 88.

63. M. Schwarzschild, *Structure and Evolution of the Stars* (Dover, New York, 1958) p. 44.

64. E. Graham, J. Fluid Mech. **70**, 689 (1975).

65. N.E. Hurlburt, J. Toomre, J.M. Massaguer, Ap. J. **282**, 557 (1984).

66. K.L. Charr, S. Sofia,C.L. Wolff, Ap. J. **263**, 935 (1982).

67. G.P. Ginet, Ph.D dissertation (Cornell University, 1987); also G.P. Ginet and R.N. Sudan, Phys. Fluids **30**, 1667 (1987).

68. P.A. Gilman and G.A. Glatzmaier, Ap. J. Sup. Ser. **45**, 335 (1981); G.A. Glatzmaier and P.A. Gilman, *ibid*, p. 381, *ibid*

69. L.P. Kadanoff, W. Götze, D. Hamblen, R. Hecht, E.A.S. Lewis, V.V. Palcianskas, M. Rayl, J.B. Swift, D. Aspnes, and J. Kane, Rev. Mod. Phys. **39**, 395 (1967).

70. P.O. Gough, D.R. Moore, F.A. Spiegel, and N.O. Weiss, Ap. J. **206**, 536 (1976). p. 351, *ibid* **47**, 108.

71. L.D. Landau and E.M. Lifshitz, *Fluid Mechanics* (Pergamon, Oxford, 1959).

72. B.L. Hao, ed. *Chaos* (World Press, Singapore, 1984).

73. N.H. Packard, J.P. Crutchfield, J.D. Farmer, and R.S. Shaw, Phys. Rev. Lett. **45**, 712 (1980).

74. A. Wolf, J.B. Swift, H.L. Swinney, and J.A. Vastano, Physica **16D**, 285 (1985).

75. J.B. Taylor, *Plasma Physics and Controlled Nuclear Fusion Research*, (IAEA, Vienna, 1977) vol. 2, p. 323.

76. R.H. Kraichnan, J. Fluid Mech. **5**, 497 (1959).

77. V. Yakhot and S.A. Orszag, J. Scient. Comp. **1**, 3 (1986).

78. R.H. Kraichnan, J. Math. Phys. **2**, 124 (1961).

79. Y. Kaneda, J. Fluid Mech. **107**, 701 (1981).

80. R.N. Sudan, in *From Particles to Plasmas*, J.W. Van Dam ed. (Addison-Wesley, Redwood City, 1989) p.273.

14

MAGNETIC RECONNECTION

When there is a strong magnetic field present in a plasma, and the plasma is said to be magnetized, a number of new phenomena which include frozen-in magnetic flux discussed in Chapter 6 and the gyromotion discussed in Chapter 7 occur. The magnetic field in a plasma and the plasma displacement perpendicular to it obey the same convective equation[1] in a perfectly conducting plasma. The equation of motion for an electron component of the fluid in the magnetic field under the massless electron approximation is

$$\mathbf{E} + \mathbf{v}_e \times \mathbf{B}/c = 0 , \tag{14.1}$$

where $\mathbf{v}_e$ is the electron fluid velocity and the inertia term in $d\mathbf{v}_e/dt$ is neglected since it is proportional to the electron mass. The pressure effect is dropped here for simplicity (see Chapter 6). If $\mathbf{v}_e$ represents electron velocity, this equation reduces to the zeroth order electron guiding-center equation (see Chapter 7). By the non-relativistic Lorentz transformation of electric field on the frame moving with the velocity $\mathbf{v}_e$ the above equation is put in the form

$$\mathbf{E} + \mathbf{v}_e \times \mathbf{B}/c = \mathbf{E}' = 0 . \tag{14.2}$$

This indicates that the electric field in the frame of reference moving with the fluid is zero and that the fluid experiences no electric acceleration. This is a manifestation of vanishing electric resistivity in the conducting fluid (plasma)

$$E' = \eta J' = 0 \tag{14.3}$$

which constitutes the ideal magnetohydrodynamic condition in which the plasma and the field move together. For example, in the shear Alfvén wave the fluid and field vibrate together perpendicular to the magnetic field (and the propagation) direction.[2] The time scale τ_A of the ideal MHD motion such as in the Alfvén oscillations is determined by the length a over which such perturbation travels divided by the Alfvén velocity v_A: $\tau_A = a/v_A$.

A mathematical possibility in which a partial detachment of the plasma from the magnetic field lines takes place, may be envisioned by having a non-vanishing resistivity $\eta \neq 0$ which is often called Ohm's law

$$\mathbf{E} + \mathbf{v}_e \times \mathbf{B}/c = \eta \mathbf{J} \ . \tag{14.4}$$

This yields a second (and usually much longer) time scale of the resistive time τ_r and is characterized by the second term of Faraday's equation

$$\frac{\partial \mathbf{B}}{\partial t} = \mathbf{\nabla} \times (\mathbf{v}_e \times \mathbf{B}) + \eta \nabla^2 \mathbf{B} \ . \tag{14.5}$$

The temporal change of flux Φ through any closed contour that moves with the material motion $\mathbf{v}_e$ is expressed as (see Fig. 14.1)

$$\begin{aligned} \frac{d\Phi}{dt} &= \int \frac{\partial \mathbf{B}}{\partial t} \cdot d\mathbf{S} + \oint \mathbf{v}_e \times d\ell \cdot \mathbf{B} = \int \left[\frac{\partial \mathbf{B}}{\partial t} - \nabla \times (\mathbf{v}_e \times \mathbf{B}) \right] \cdot d\mathbf{S} \\ &= \eta \int \nabla^2 \mathbf{B} \cdot d\mathbf{S} \ . \end{aligned} \tag{14.6}$$

Or we have

$$\frac{d\Phi}{dt} = \eta \nabla^2 \Phi \ . \tag{14.7}$$

If $\eta = 0$ (perfectly conducting fluid or the ideal MHD fluid) and the fluid motion is characterized by Eq. (14.1), the right-hand side of Eq. (14.7) vanishes and the flux change through the moving fluid is zero i.e., the magnetic flux is tied to the fluid. On the other hand, when $\eta \neq 0$ (resistive MHD fluid), the magnetic flux diffuses out of the contour according to Eq. (14.6). Thus the resistive (diffusion) time is given by $\tau_s = a^2/\eta$ with a being the characteristic length of diffusion.

In the model resistive MHD equations the parameter that governs the system besides the Alfvén time is the ratio of the resistive time to the Alfvén time, $S = \tau_r/\tau_A$, the so-called magnetic Reynolds number or the Lundquist number. The ideal MHD system is dissipationless, while the resistive MHD system is dissipative in that the magnetic energy is converted into kinetic

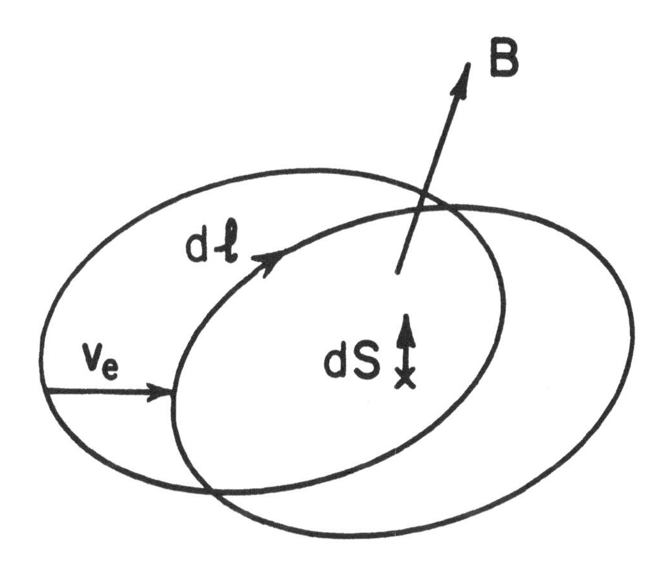

FIGURE 14.1 The frozen field line and the magnetic flux

energy via the resistivity. In many applications the magnetic Reynolds number is quite large, perhaps 10^7 in a present day tokamak plasma and $10^{10} \sim 10^{12}$ in a solar coronal loop plasma. In these problems it has been either observed or postulated that the magnetic energy configuration changes its topology accompanied by the magnetic field line reconnection which involves the resistive process, and the magnetic energy is converted into kinetic energy in a time much shorter than the resistive time. It is particularly challenging to plasma physics theory that in some instances in the solar flares' impulsive phase the energy conversion from magnetic to kinetic forms seems to take place in a matter of Alfvén time. This is to say that such a process is some $10^{10} - 10^{12}$ times faster than the resistive time. A similarly abrupt energy release is observed in the laboratory, e.g., in the tokamak plasma disruption and sawteeth oscillations, which seem to take place in only a matter of the Alfvén time scale. The theoretical community has continued to quest for a resolution of this spectacularly fast magnetic reconnection process for more than three decades.

In a situation where the magnetic energy is converted into kinetic energy and the magnetic field line configuration changes via the resistive (diffusive) process it is convenient to separate the magnetic fields into two categories:

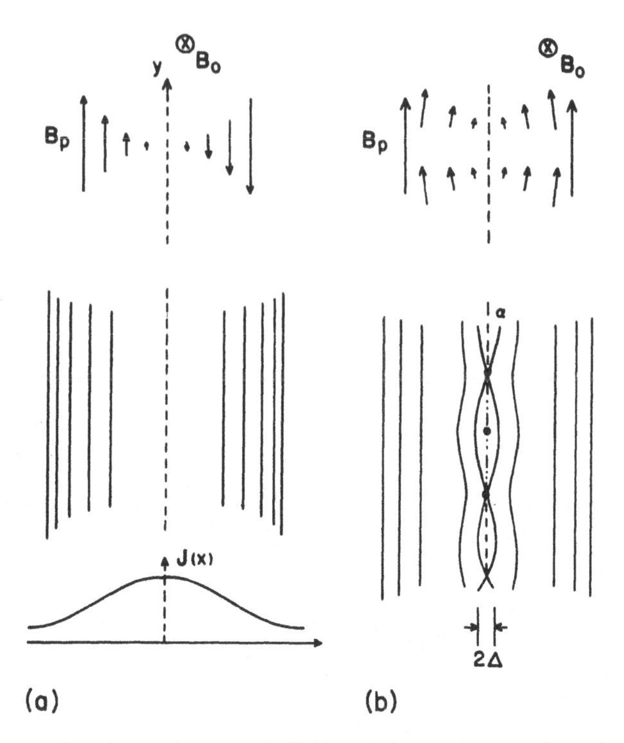

FIGURE 14.2 The sheared magnetic field and the tearing configuration

(i) the potential magnetic field $\mathbf{B}_0$ which is expressible as $\mathbf{B}_0 = \nabla y \times \nabla \psi_0$, where ψ_0 is the external flux (or often toroidal flux in toroidal geometry) given externally and is not ordinarily coupled with the plasma in a time scale of interest and y is a coordinate (Fig. 14.2); (ii) the magnetic field $\mathbf{B}_p = \nabla z \times \nabla A_z$ which is created by the plasma current $\mathbf{J} = c\nabla \times \mathbf{B}_p/4\pi$ and exhibits shear, i.e., $\mathbf{B}_p$ becomes a function of the coordinate x that is perpendicular to y and z (Fig. 14.2). The latter field is often called the poloidal field as well, since $\mathbf{B}_p$ is perpendicular to $\mathbf{B}_0$ which is often called the toroidal field in cylindrical or toroidal geometry. The sheared magnetic field $\mathbf{B}_p$ may accompany the null line or the field reverse layer on which $\mathbf{B}_p$ vanishes and changes sign on one side from the other. This line is very sensitive to magnetic and thus plasma perturbation, since even an infinitesimally small amount of magnetic perturbation in the $x - y$ plane induces a topological change in the magnetic field line. This is called reconnection of magnetic field lines, as the originally unconnected antiparallel field lines are now connected after the perturbation. It has been shown, moreover, it is often here that the instability of magnetic perturbation and the energy conversion from the magnetic to the kinetic takes place. As a result of reconnection, the new magnetic configuration exhibits X-points and O-points, the latter of which are also called magnetic islands. Furth, Killeen and Rosenbluth[3] found the tearing instability that tears the poloidal magnetic field lines as discussed above and grows with growth rate $\gamma = C\tau_A^{-1}S^{-3/5}$, where the constant $C = [\Gamma(1/4)a\Delta'/\pi\Gamma(3/4)]^{4/5}$. Here a is the plasma current channel width, Γ the gamma function, and Δ' is the jump in the derivative of perturbed magnetic flux A_z. This growth time γ^{-1} corresponds to a time scale in between the Alfvén time, τ_A, and the resistive time, τ_r, thus much faster than τ_s itself. In this linear theory the angle of the X-point is infinitesimally small.

Earlier Sweet[4] and Parker[5] obtained a steady-state solution whose time scale of reconnection is characterized by $\tau_{\text{SP}}^{-1} = \eta^{1/2}(v_*/L)^{1/2}1/a \propto S^{-1/2}$, where a is the thickness of the sheet current channel and L its length and the X-point angle is characterized by a/L. Petschek[6] obtained a steady-state solution which has a much larger X-point angle than that assumed by Sweet and Parker. The time scale of reconnection by the Petschek process is independent of resistivity, $\tau_p^{-1} \propto S^0$. The Sweet-Parker process and the Petschek process are, however, based on the steady-state assumption. There has been a considerable amount of research continuing Furth et al.'s work in the fusion plasma physics community with ramifications primarily on the nonlinear evolution of tearing instabilities and on the influence of toroidal geometry on the tearing modes. This effort was eloquently reviewed by White.[7] In what follows, kinetic tearing instabilities and magnetic reconnection in astrophysical settings will be discussed.

14.1 Collisionless Tearing Instabilities

In a collisionless plasma the collision-induced resistivity and Eq. (14.4) are inadequate to describe the tearing modes[8] that result in a bunching or filamentation of the plasma current $\mathbf{J}$ along the y-direction. Within the central layer, the field lines slip with respect to the plasma, producing an induced electric field in the z-direction which accelerates the electrons along the local magnetic field. This perturbed current in the z-direction, $\tilde{J}_z$ results in a localized region $|x| \leq \Delta \ll a$. In the outer region ($|x| > \Delta$), the field lines remain frozen in the plasma. In the tearing instability the released magnetic energy in the outer region is converted into the internal kinetic energy of the plasma in the central layer. The model system for simulation of this instability is given by the following. The magnetic field in a slab plasma is

$$\mathbf{B}_0 = B_z\hat{z} + B_y(x)\hat{y} \quad \text{with} \quad |B_z| \gg |B_y| \,. \tag{14.8}$$

Restating the above description of the modes, the collisionless tearing modes can be treated as follows: (1) The perturbed current is primarily in the direction of the external magnetic field; (2) The electron response is much larger than the ion response in the tearing layer; (3) In the outer region the induced electrostatic potential $\tilde{\phi}$ is negligible, while in the inner region magnetic flux fluctuation $\tilde{\psi}$ is relatively small. The corresponding initial current profile used in the simulation by Katanuma and Kamimura[9] is

$$J_z(x) = -en_0 v_d \exp\left[-(\ell n2)(x-x_0)^2/a^2\right] \ ,$$

where v_d is the electron drift velocity in the z-direction. The initial equilibrium for the plasma is given by the electron distribution function

$$f(x,v) = \frac{1}{\left[(2\pi)^{1/2}v_e\right]^3} \exp\left\{-\frac{1}{2v_e^2}\left[v^2 - \left(\frac{P_z}{m} - v_{0z}\right)^2 - \left(\frac{P_z}{m}\right)^2\right]\right\} \ ,$$

where the canonical momentum P_z is given by $P_z = mv_z - eA_z(x)/c$.

The linear theory is considered by making an infinitesimal perturbation to the above equilibrium, which yields a drift-kinetic equation for the electron perturbed distribution $\tilde{f}$

$$\left(\frac{\partial}{\partial t} + v_\parallel \frac{B_y}{B_z}\frac{\partial}{\partial y} + v_\parallel \frac{\partial}{\partial z}\right)\tilde{f} = -v_\parallel \frac{\tilde{B}_x}{B_z}\frac{\partial}{\partial x} f_0 + \frac{e}{m}\tilde{E}_\parallel \frac{\partial}{\partial v_\parallel} f_0 \ , \tag{14.9}$$

where the perturbed quantities have a tilde. The magnetic field $\tilde{\mathbf{B}}$ and the electric field $\tilde{E}_\parallel$ are related by

$$\tilde{\mathbf{B}} = \nabla \times \tilde{A}_z \hat{z} \ , \tag{14.10}$$

and

$$\tilde{E}_\parallel = -\frac{1}{c}\frac{\partial}{\partial t}\tilde{A}_z \ , \tag{14.11}$$

where Ampere's law (with the displacement current neglected) gives the relation between the perturbed plasma current (determined by $\tilde{f}$) and $\tilde{A}_z$

$$\nabla^2 \tilde{A}_z = -\frac{4\pi n_0 e}{c}\int d\mathbf{v} v_\parallel \tilde{f} \ . \tag{14.12}$$

Equations (14.9)-(14.12) yield the linear dispersion relation

$$\begin{aligned} \left(\frac{\partial^2}{\partial x^2} - k_y^2\right)\tilde{A}_z \ = \ & \frac{1}{c^2\lambda_{De}^2}\left[\omega\left(\left\langle\left\langle v_\parallel^2\right\rangle\right\rangle - v_{0z}\left\langle\left\langle v_\parallel\right\rangle\right\rangle\right) - \frac{k_y}{\Omega_e}\left(\left\langle\left\langle v_\parallel^3\right\rangle\right\rangle\right)\right. \\ & \left. - v_{0z}\left\langle\left\langle v_\parallel^2\right\rangle\right\rangle \frac{\partial}{\partial x} v_{0z}\right]\tilde{A}_z \ , \end{aligned} \tag{14.13}$$

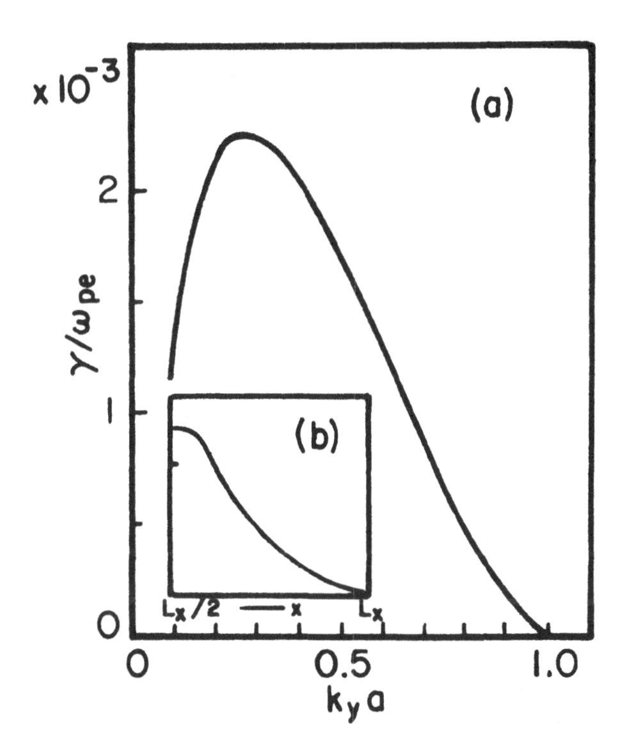

FIGURE 14.3 The linear growth rate (a) and the structure of the eigenfunction (b) of the collisionless tearing instability (Ref. 9)

with $\left\langle\left\langle v_{\|}^{n}\right\rangle\right\rangle \equiv \int d\mathbf{v} f_0/(\omega - k_{\|}v_{\|})$. Equation (14.13) reduces to the case of the collisionless Drake-Lee result[8] on one hand and to the case of Furth et al.[3] The growth rate of tearing modes in the regime of the simulation parameters described below, however, does not quite fit these cases because the parameter δ' in these theories is computed based on the boundary layer theory while in the simulation the tearing layer is not completely isolated from the outer layer, but the transition from one to the other is gradual. Thus it is necessary to resort to the numerical shooting code calculation to obtain an accurate growth rate. Figure 14.3 depicts the result for the parameters used in the simulation. The initial density profile is in pressure equilibrium

$$n(x)(T_i + T_e) + B_y^2(x)/8\pi = \text{const.}\,. \tag{14.14}$$

The particle guiding-centers are initially loaded according to the density profile. The system is periodic in the y-direction and is bounded by two conducting walls at $x = 0$ and $x = L_x$ in the x direction. Particles carry three

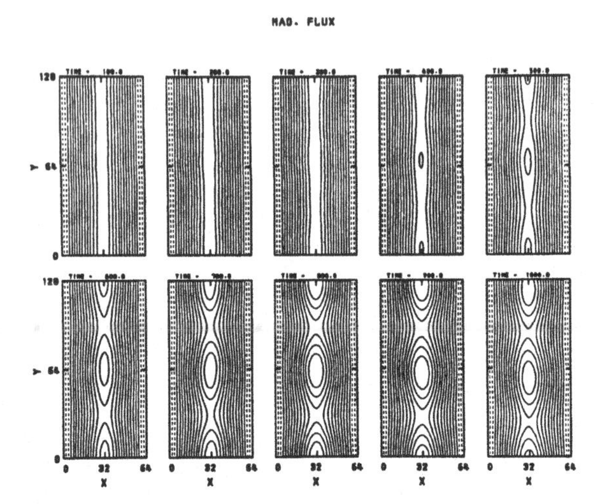

FIGURE 14.4 Contours of the magnetic flux A_z in time from the upper left corner to the lower right corner during tearing instability (Ref. 9)

components of velocity (v_x, v_y, v_z), and two spatial coordinates (x, y) $(2 - \frac{1}{2}$ dimensional model). The particles which hit the walls are reflected according to the method designated (I) in Ref. 10. This produces neither macroscopic plasma flows nor density perturbations near the walls.

The investigation has been done on a magnetoinductive particle code with full electron and ion dynamics[9] (and earlier by Ref. 11). The parameters used[9] are: the number of particles (electrons and ions, respectively) $N_e = N_i = 17424$, the size of grid $L_x = 64\Delta$, $L_y = 64\Delta$, the electron cyclotron frequency $\Omega_e = 1.5\omega_{pe}$, the speed of light $c = 10v_e$, the Debye length $\lambda_{De} = \Delta$, the ion-to-electron mass ratio $m_i/m_e = 16$, the ion and electron temperatures $T_i = T_e$, and the electron drift velocity in the z-direction $v_{dz} = -v_e$, where v_e is the electron thermal velocity, Δ the unit mesh size, and ω_{pe} the electron plasma frequency. Then, the ion Larmor radius is 2.7Δ, while the electron Larmor radius is 0.67Δ. The ion cylcotron frequency $\Omega_i = 0.094\omega_{pe}$.

Figure 14.4 shows the evolution of tearing modes and their saturation. In this case the current width $a = 5.1\Delta$, for which the mode $k_y = \pi/32\Delta$ is unstable. The temporary evolution of the perturbed current exhibits an exponential growth in the linear regime with the theoretical linear growth rate written in by a solid line. The observed growth[9] of the tearing mode was in good agreement with the theory. In Fig. 14.5 the island width is measured in time.

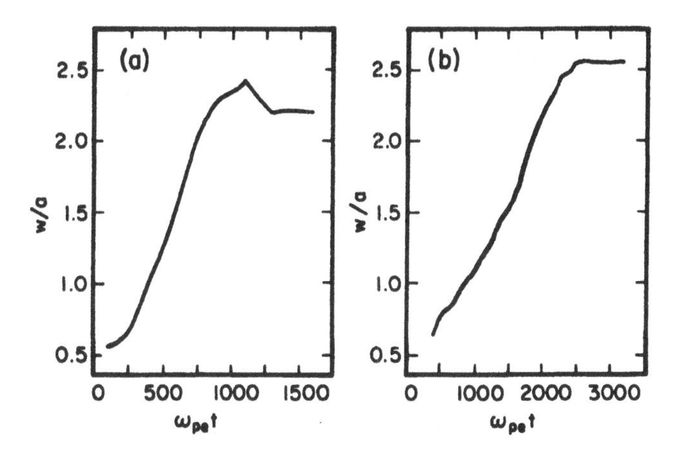

FIGURE 14.5 Temporal evolution of island width w during the tearing instability (Ref. 9). Results from the explicit particle simulation

The above simulation took typically 20,000 time steps to reach the saturation stage. By employing the implicit electromagnetic particle code[12] it has become possible to simulate essentially the same physics with much fewer time steps. The parameters employed in the implicit code (see Chapter 9) are identical to the case with the explicit code except for the timestep related parameters. The decentering parameters for electrons and ions are chosen as $\gamma_e = \gamma_i = 0.1$. In the decentered algorithm described in Chapter 9 the cyclotron motion remains damped out, since electrons and ions are initially loaded with zero perpendicular kinetic energy. The initial density profile of particles with uniform parallel temperature balances the shear magnetic pressure $B_y^2(x)/8\pi$. In these runs a typical time step chosen is $\Delta t = 5\omega_{pe}^{-1}$, about two orders of magnitude larger than the equivalent explicit calculation as described earlier. The series of snapshots of magnetic field lines obtained from various corresponding times of this simulation are in fact indistinguishable[12] from the explicit simulation.[9] The logarithmic amplitude of the unstable mode plotted against time indicates a measured growth rate of $\gamma \sim 1 \times 10^{-3}\omega_{pe}$.[12] The theory based on Eq. (14.13) solved numerically for the simulation parameters by the shooting method[13] yields a growth rate $\gamma \sim 1.3 \times 10^{-3}\omega_{pe}$, while the Drake-Lee formula[8]

$$\gamma = k_y v_e \Delta'/2\pi^{1/2}(\omega_{pe}/c)^2/L_s \,, \tag{14.15}$$

with L_s being the shear length and a rough estimate of $\Delta' \simeq 1/a$ produces $\gamma \sim 2 \times 10^{-3}\omega_{pe}$. Figure 14.6 shows the growth of the magnetic island size that is characterized by the distance between the O-point (center of island)

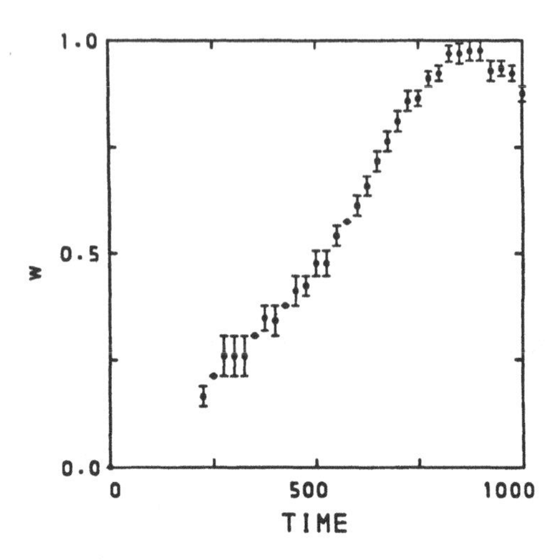

FIGURE 14.6 Evolution of the width of island duirng the tearing instability (Ref. 12). Result from the implicit particle simulation

to the separatrix in the x-direction. According to Drake and Lee[8] the size of island w grows exponentially in the linear regime of collisionless tearing modes and then increases linearly in time in the regime[14] similar to the Rutherford regime, which is expressed as

$$w = \frac{1}{16G}\left(\frac{c}{\omega_{pe}}\right)^2 \Delta' \nu t \ , \tag{14.16}$$

where the geometrical factor $G \simeq 0.41$ and ν is the collision frequency.

The collision frequency in a simulated plasma is in general quite different from a real plasma (see Chapter 2). The collision frequency in a two-dimensional plasma is given by $\nu = F\omega_p/16N_D$ with $N_D = n\lambda_D^2$ [Eq. (2.41)], where the collision-reducing factor due to the finite size particles is $F = \sigma_f/\sigma_p$ found in Fig. 2.3.[15] For the present case the computational collision frequency is estimated to be $\nu \sim 1 \times 10^{-3}\omega_{pe}$. Thus $w/a \sim 0.8 \times 10^{-3} t\omega_{pe}$, which is in good agreement with the Rutherford regime theory.

The linear tearing instability and its nonlinear development into the Rutherford regime have been demonstrated in both explicit and implicit particle simulations which generated indistinguishably similar results with the difference being a much larger time step (and thus much faster computation) for the implicit code.

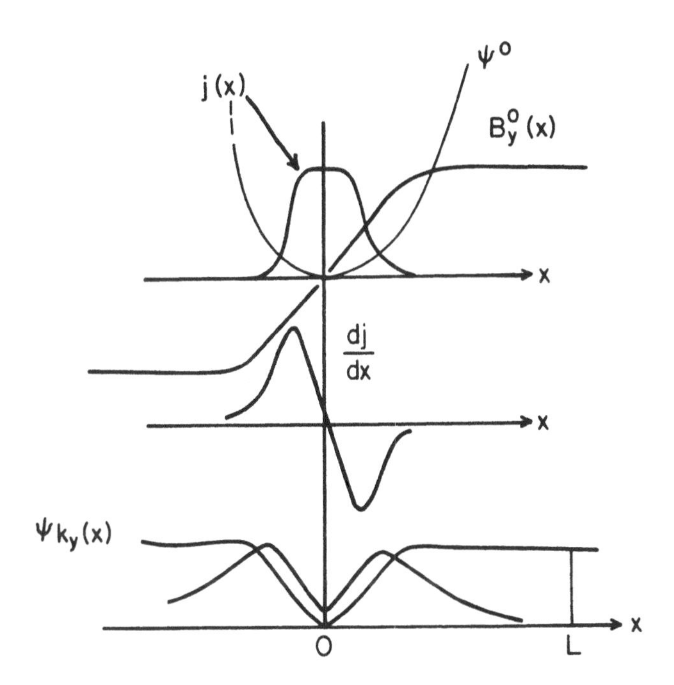

FIGURE 14.7 The reversed field $B_y(x)$ and its associated profile of current etc. The eigenfunctions of spontaneous and driven reconnection $\psi_{k_y}(x)$ (Ref. 16)

14.2 Linear Theory of Driven Reconnection

In the previous section, the spontaneous reconnection due to the tearing instability has been surveyed. Furth et al.'s collisional tearing instability[3] is based on the MHD theory and the boundary layer theory, which generates the jump in the derivative of the perturbed vector potential Δ'. The peculiar entrance of the quantity Δ' into the theory of reconnection in the linear MHD theory is due to the fact that the (ideal) MHD lacks any spatial scale and that the collisional (resistive) MHD with a high magnetic Reynolds number can be treated as the boundary layer theory. Thus in the global MHD equation, the scale length for the resistive layer does not appear and its effect is tucked into Δ'. We saw in the previous section, however, that in the kinetic model a more exact treatment may be required (the shooting code analysis), since there are specific scale lengths entering such as the collisionless skin depth and the Debye length. In the spontaneous reconnection (i.e., the tearing instability) the eigenmode of the perturbed vector potential behaves as the asymptotically decaying function $e^{-\alpha x}$. In contrast to this, if there is a driving agent at $x = x_0$ that forces the MHD fluid to move at a specific perturbation (k_y: given), a linear combination of two solutions that correspond to the asymptotic functions of $e^{-\alpha x}$ and $e^{+\alpha x}$ can be constructured . Within a linear theoretical framework Horton and Tajima[16] examined this problem. To see how the transition from the spontaneous to driven reconnection takes place in linear theory, let us first examine the two-dimensional incompressible MHD case. Ohm's law $\mathbf{E} + \frac{1}{c}\mathbf{v} \times \mathbf{B} = \eta \mathbf{J}$ and the fluid momentum equation $\rho(\partial \mathbf{v}/\partial t + \mathbf{v} \cdot \nabla v) = \frac{1}{c} J \times B - \nabla P$ under these conditions $\nabla \cdot \mathbf{v} = 0$, $\nabla \cdot \mathbf{B} = 0$, $\rho =$ const. may be cased into the following equations

$$\frac{\partial}{\partial t}\psi = \frac{c}{B_z}[\varphi, \psi] + \eta \nabla^2 \psi \ , \tag{14.17}$$

and

$$\nabla_\perp^2 \frac{\partial}{\partial t}\varphi = \frac{v_A^2}{cB_z}[\psi, \nabla_\perp^2 \psi] = \frac{c}{B_z}[\varphi \ , \nabla_\perp^2 \varphi] \ , \tag{14.18}$$

where [ ] is the Poisson bracket $[A, B] = \partial_x A \partial_y B - \partial_y A \partial_x B$, $\mathbf{v} = c\hat{z} \times \nabla \psi / B_z$, and $\nabla_\perp$ is the in-plane gradient. In the MHD region

$$E_z = \frac{1}{2}(v_y B_x - v_x B_y) \tag{14.19}$$

and

$$\mathbf{v} \cdot \nabla J_z = 0 \ . \tag{14.20}$$

Expanding $\phi(x, y, t) = \phi_k(x, t)\sin(ky)$ and $\psi(x, y, t) = \psi(x) + \psi_k(x, t)\cos(ky)$, Eqs. (14.17) and (14.18) are written as

$$\frac{\partial \psi_k}{\partial t}\cos(ky) = kB_y(x)\varphi_k(x)\cos(ky) \tag{14.21}$$

$$\frac{\partial^2 \psi_k(x)}{\partial x^2} + \left(-k^2 - \frac{\partial^3 \psi/\partial x^2}{\partial \psi/\partial x}\right)\psi_k(x) = 0 \ . \tag{14.22}$$

When the equilibrium current $J_z(x)$ is given by

$$J_z(x) = \frac{4\pi}{c} B_y^0 / a \, \mathrm{sech}^2(x/a) \ , \tag{14.23}$$

an exact solution for Eq. (14.21) is

$$\psi_k(x) = C_1 \left[k_y a + \tanh\left(\frac{x}{a}\right)\right] e^{-k_y x} + C_2 \left[k_y a - \tanh\left(\frac{x}{a}\right)\right] e^{+k_y x} \ . \tag{14.24}$$

For an arbitrary boundary condition at $x = L$ $d\psi_k(L)/dx = \sigma \psi_k(L)$ we obtain for

$R \equiv -C_2/C_1$ as

$$R = R(k_y L, L/a, \sigma) = -\frac{C_2}{C_1}$$

$$= e^{-2k_y L} \left\{ \frac{\sec h^2(L/a) + k_y + \sigma)\,[k_y, a + \tanh(L/a)]}{-\sec h^2(L/a) + (\sigma - k_y)\,[\tanh(L/a) - k_y a]} \right\} \quad (14.25)$$

The value of the outer solution at $x = 0$ (at the boundary layer) is

$$\psi_k(0) = C_1(1 - R)k_y a\ . \quad (14.26)$$

For example, if we force the fluid to push in at a certain portion of y coordinate (the normal flow at $x = \pm L$), the boundary condition $\mathbf{v}(L, y) = v_x \cos(ky)\hat{x}$ yields

$$\sigma = \frac{1}{\psi_k}\frac{\partial \psi_k}{\partial x} = 1\ . \quad (14.27)$$

With Eq. (14.27), Eq. (14.25) provides a connection from the boundary condition at $x = \pm L$ (flow) to the tearing layer. See Fig. 14.7. For the spontaneous tearing the asymptotic behavior of eigenfunction $\psi_k(x)$ is an exponentially decaying one. As we integrate Eq. (14.22) from $x = \infty$ toward $x = 0$, ψ_k at $x = 0+$ take certain values with $\psi_k(0) \neq 0$ and $\psi_k'(0) \neq 0$. Since $\psi_k(x)$ is an even function, $[\psi_k'(0+) - \psi_k'(0-)]\,/\psi_k(0) \neq 0$, which is called Δ' in Ref. 3. When we start from the forced boundary condition at $x = L$ and march toward $x = 0$, $\psi_k(0)$ vanishes under appropriate conditions for k_y, L and a. Figure 14.8 shows the boundary on which spontaneous reconnection changes over to driven reconnection.

Based on this, let us consider the driven collisionless reconnection. Ampere's law $\nabla^2\psi = \frac{4\pi}{c}J_\parallel$ yields

$$\frac{\partial^2 \psi_k}{\partial x^2} = \frac{4\pi n e^2}{mc}\left\langle \frac{1}{-i(\omega - k_\parallel v_\parallel)} \right\rangle E_\parallel + \frac{4\pi}{c}J_{\text{ion}}\ , \quad (14.28)$$

where $E_\parallel = -k_\parallel \phi + \frac{\gamma}{c}\psi$ and J_{ion} is the ion contribution to the current. Let x_t be the tearing layer width $x_t = \gamma/|k_\parallel' v_e|$. Then within $|x| < x_t$, $E_\parallel \simeq \gamma\psi_k(x)/c \neq 0$, while $|x| > x_t$, $E_\parallel \simeq 0$ and ideal MHD applies. In order to calculate the growth rate, we integrate Eq. (14.28) over the tearing layer $(-x_t, x_t)$

$$\int_{-x_t}^{x_t} dx \frac{\partial^2 \psi_k}{\partial x^2} = 2\psi_k'(x_t) \cong \frac{\omega_{pe}^2}{c^2}\int_{-x_t}^{x_t} dx \frac{\psi_k(x)}{\gamma + ik_\parallel(x)v_e}$$

$$= \frac{\omega_{pe}^2}{c^2}\gamma\pi \int \delta[k_\parallel(x)v_e]\psi_k(x)dx\ , \quad (14.29)$$

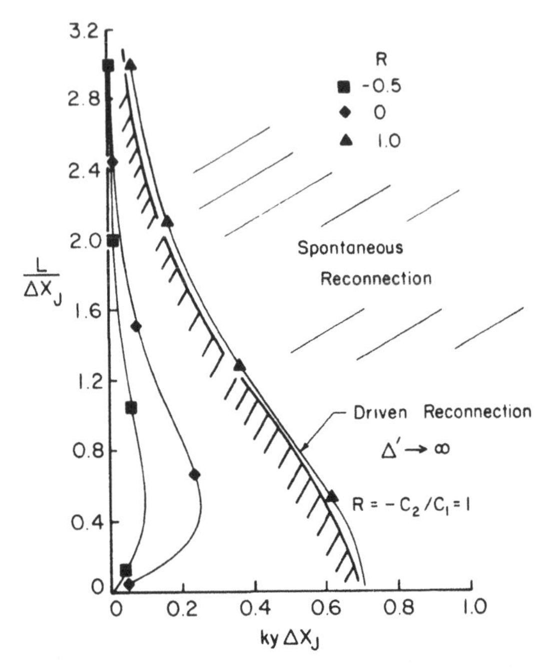

FIGURE 14.8 Transition from spontaneous to driven reconnection in the parameter space of the width of the system L and the wavelength k_y (Δx_J is the current width) (Ref. 16)

which yields

$$\gamma \simeq \left(\frac{1}{\pi}\right)\frac{c^2}{\omega_{pe}^2}\left|k_{\|}'\right| v_e \left[\frac{\psi_k'(0+) - \psi_k'(0-)}{\psi_k(0)}\right] . \tag{14.30}$$

This is essentially the Drake-Lee result [Eq. (14.15)] except for a numerical factor of $\pi^{1/2}/2$. However, when the perturbed magnetic flux $\psi_k(x)$ vanishes at $x = 0$ (i.e., $\psi_k \propto |x|$), Eq. (14.29) now changes

$$2\psi_k'(x_t) = \frac{\gamma\omega_{pe}^2}{c^2}\int_{-x_t}^{x_t} dx \frac{|x|\psi_k'(0)}{\gamma + ik_{\|}(x)v_e} = \gamma^2\frac{\omega_p^2}{c^2}\frac{1}{(k_{\|}'v_e)^2}\psi_k'(0)\ell n\left|\frac{k_{\|}'^2 v_e^2 x_t^2 + \gamma^2}{\gamma^2}\right| . \tag{14.31}$$

The linear growth rate for driven reconnection in a collisionless plasma becomes

$$\gamma \simeq \left(\frac{2}{\ell n 2}\right)^{1/2}\frac{c|k_{\|}'|v_e}{\omega_{pe}} . \tag{14.32}$$

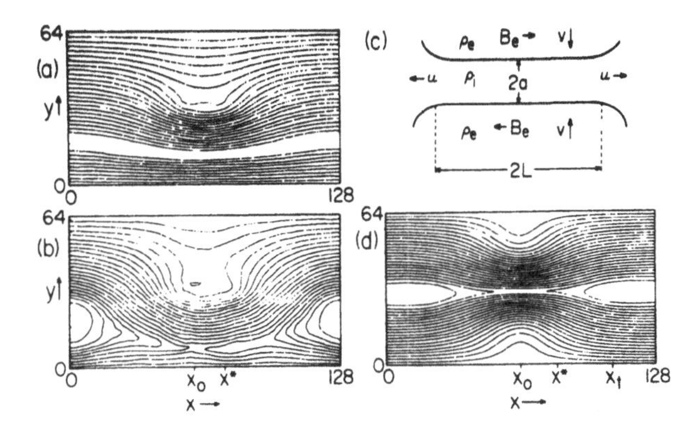

FIGURE 14.9 Pinching in a reversed field configuration. (a), (b) on one-sided pinch, (d) symmetric pinching, (c) a schematics (Ref. 19)

As opposed to the growth rate for spontaneous reconnection, the growth rate for driven reconnection does not involve the quantity Δ'. A similar effect appears in the $m = 1$ tearing mode growth rate.

14.3 Fast Reconnection

It seems unrealistic to expect a linear process to play a major role when the forcing at a boundary causes the wavefunction $\psi_k(x)$ to go beyond zero at the tearing layer (see Fig. 14.7). In many astrophysical and geophysical situations, external disturbances trigger the process of magnetic reconnection, in which the process seems often outright nonlinear and the time scale is many orders of magnitude shorter than the resistive time scale.

Two types of steady-state reconnection solutions have been known; one by Sweet[4] and Parker[5] and the other by Petschek.[6] These are relevant for the start of driven reconnection. Figure 14.2 shows the geometry of reconnection for the slower process of the two. Consider pressure equilibrium $P(y) + B^2(y)/8\pi = \text{const.}$. As the plasma is externally pinched, the magnetic field lines as well as the plasma slab in the neutral sheet with higher density are pinched inward [Fig. 14.9]. As the pressure increases by pinching, this excess pressure causes flow along the field lines. This flow causes the contraction of the neutral layer, whose thickness decreases exponentially in time.[17] Within ideal MHD the plasma layer develops into a singular current sheet, in a quasistationary state, always out of equilibrium.[18] For nonideal MHD (resistive plasmas), however, the layer width becomes stabilized as field lines begin reconnecting at such a rate that the perpendicular inflow of particles into the current sheet due to the field-line annihilation matches the plasma end loss along the magnetic field lines due to parallel pressure drop. The inflow is governed by magnetic diffusion due to resistivity in the layer. This is a relatively slow process as described by Sweet and Parker. Computer simulation has shown dynamical transition and development of various reconnection configuration.[19]

The in-the-plane magnetic flux ψ (poloidal magnetic flux) and out-of-the plane (along the z axis) magnetic field B_z (toroidal magnetic field) are described by

$$\frac{\partial \psi}{\partial t} + \mathbf{v} \cdot \boldsymbol{\nabla} \psi = \eta \nabla^2 \psi , \tag{14.33}$$

$$\frac{\partial B_z}{\partial t} + \boldsymbol{\nabla} \cdot \mathbf{v} B_z = \eta \nabla^2 B_z , \tag{14.34}$$

where $\mathbf{B} = \mathbf{B}_\perp + B_z \hat{z}$ and $\mathbf{B}_\perp = \nabla\psi \times \hat{z}$; and v_z is neglected. For the initial configuration the flux function is assumed to be linearly increasing in y, $\psi = B_e|y|$, on each side of the exteriors of the current sheet located at $y = 0$; this is equivalent to the assumption of a uniform magnetic field of magnitude B_e. In the exterior region the diffusion is negligible and the flux velocity is determined by the fluid in the y-direction:

$$v = \dot{\psi}/B_e . \tag{14.35}$$

Inside of the current sheet with half-width a, the perpendicular velocity v is zero and the diffusion becomes important:

$$\dot{\psi} = \eta \nabla^2 \psi \simeq \eta B_e / a . \tag{14.36}$$

The balance of the perpendicular inflow with the parallel outflow gives

$$\rho_e v L = \rho_i u a , \tag{14.37}$$

where subscripts e and i refer to the external and internal quantities with respect to the current sheet. Equations (14.35) and (14.37) given

$$\dot{\psi} = B_e u \rho_i a / \rho_e L , \tag{14.38}$$

while Eqs. (14.36) and (14.37) yield

$$a = \eta^{1/2} (\rho_e L / \rho_i u)^{1/2} . \tag{14.39}$$

When the plasma is compressible (with relatively weak or no B_z magnetic field), the peak (interior) density ρ_i with respect to ρ_e (exterior density) is determined by the perpendicular pressure balance

$$\mathcal{P}_i = (P + B_z^2/8\pi)_i = \mathcal{P}_e = (P + B_z^2/8\pi + B_{e\perp}^2/8\pi)_e \ , \tag{14.40}$$

where the left-hand side refers to the quantities in the interior and the right-hand to the exterior. Equation (14.40) is replaced by a dynamical equation when reconnection is more rapid. We also use an adiabatic law for P, i.e., P/ρ^γ = const. . The plasma slab develops a diffuse profile as it becomes thinner as shown by Eq. (14.36). The flow velocity u is determined by the drop in the perpendicular pressure along x, and is $v_{Ae}(\rho_e/\rho_i)^{1/2}$, where v_{Ae} is the Alfvén velocity in the exterior region. This is given by

$$v = S^{-1/2} v_{Ae}(\rho_i/\rho_e)^{1/2} \ . \tag{14.41}$$

Sweet[4] and Parker[5] gave an incompressible version of Eq. (14.41) when $\rho_i = \rho_e$.

If the plasma is compressible, however, the poloidal flux reconnected in the Sweet-Parker phase (the process we just looked at) piles up in the current sheet as time goes on. The sheet then becomes tapered, shortening the effective fluid exhaust length L.[17] It is at this stage where an enhancement in the reconnection rate is observed in the simulation[19] [see Fig. 14.10]. A simplified model of this stage may be depicted as in Fig. 14.10. The trapped reconnected flux has a tapered structure with opening angle α, where α is small enough compared with unity that the initial pressure balance and ρ_i/ρ_e are not significantly modified. As a new flux tube reconnects at $x = 0$, its plasma pressure goes up to $\mathcal{P}_i = (P + B_z^2/8\pi)_i$ over width a. At distance $L_* = a/\alpha$ along the x axis, this flux tube is located at $a < y < 2a$ outside the diffusive area, with a magnetic field B_e and pressure $\mathcal{P}_e$. Since no magnetic pressure gradient exists along the field line (with angle α from x axis), at $x = L_*$ the parallel pressure satisfies $\rho_i u^2/2 = \mathcal{P}_i - \mathcal{P}_e$, therefore, the plasma in that flux tube flows with a velocity $u = v_A(\rho_i/\rho_e)^{1/2}$ along the x direction. The angle α is determined by $\alpha = B_y/B_e$ [see Fig. 14.10], where B_y is the average magnetic field along y in the diffusion layer, given by $B_y = \psi/L_t$, where ψ is the total reconnected flux and L_t the length along the x direction traveled by flux ψ. With $L_t = L_* + ut$, we obtain

$$L_* = ut\psi_c/(\psi - \psi_c) \ , \tag{14.42}$$

where $\psi_c \equiv aB_e$. In the evaluation of L_t it is assumed that the flux in the current sheet is carried instantly over a distance L^* where diffusion is predominant, and then remains trapped with the fluid which moves with velocity u along x. Once L^* becomes shorter than L, which happens when $\psi > \psi_c$ or $t > t_0 \simeq L/u$, the substitution of Eq. (14.42) into Eq. (14.38) yields $t\dot{\psi} - (\rho_i/\rho_e)\psi = -\psi_c$ or

$$\psi = \psi_0(t/t_0)^{\rho_i/\rho_e} + (\rho_e/\rho_i)\psi_c \ . \tag{14.43}$$

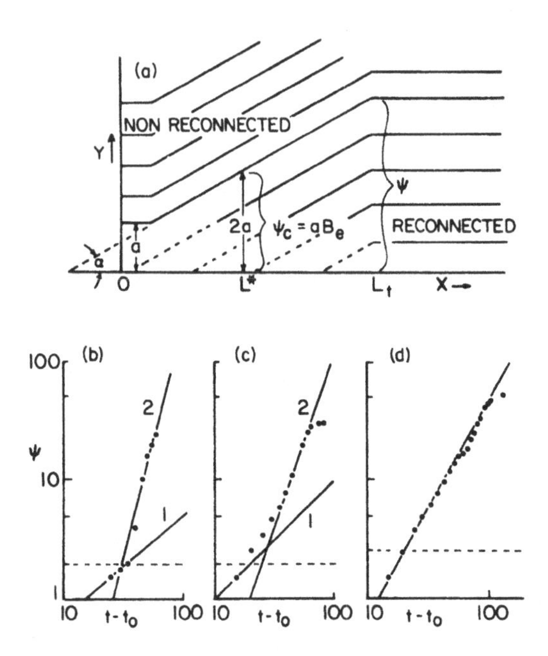

FIGURE 14.10 Schematic geometry of fast second stage of reconnection (a) and examples of fast temporal evolution of reconnected flux (b-d) (Ref. 19)

If the plasma is compressible, the reconnection rate becomes much faster than that in the Sweet-Parker phase after t_0 with a shrinking current sheet L^*. There will be a sharp increase in the reconnection velocity given by $v = \dot{\psi}/B_e \propto t^{\delta}$, with $\delta = \rho_i/\rho_e - 1$, and this stage is called the *second phase*. If the plasma is incompressible (ρ_i/ρ_e), the Sweet-Parker phase lasts beyond t_0. Correspondence to relations in Eqs. (14.42) and (14.43) can be found in simulation of Figs. 14.9(a) and 14.9(b) for shrinking L^* and in Figs. 14.10(b) and 14.10(c) for two (or more) phases of $\psi(t)$. The exponent obtained from Fig. 14.10(b) for the second phase is 4.0 with $\rho_i/\rho_e \simeq 4.0$ in the simulation, and the exponent from Fig. 14.10(c) is 2.9 with $\rho_i/\rho_e \simeq 3.4$; both cases are in good agreement with Eq. (14.43).

The geometry of the system is important both for the second phase and for the eventual saturation of reconnection. When the current layer is pinched from both sides so that it remains straight during the reconnection process, a different exponent for the time dependence for the reconnected flux is observed in the second phase. Figure 14.9(d) shows the flux lines in this case. The rate of reconnection in this case is given in Fig. 14.10(d): $\psi \simeq (t - t_0)^m$ where $m \simeq 2$.[19]

Many other simulation investigations are consistent with this theory and simulation. Sato and Hayashi's simulation[20] (their Fig. 1) shows that fast reconnection sets in when L^* becomes the length of their system, consistent with the present theory of the second phase. Park's work[21] notes that the incompressible case stays in the Sweet-Parker phase all the way.

14.4 Coalescence Instability

In order to convert magnetic energy into kinetic energy rapidly and by a substantial amount, it seems necessary that the bulk of the available magnetic energy must participate in the conversion process. The resistive heating in the X-point alone is too meager because the available magnetic energy at the X-point is small by itself. On the other hand, the ideal MHD instabilities such as the kink instability and the coalescence instability are the processes that involve the bulk current redistribution in a matter of the Alfvén time scale. In the previous section, we considered basic processes of fast reconnection and in the present section we consider nonlinear driven reconnection triggered as a secondary process by a primary instability. It is instructive to pick the coalescence instability as the primary instability because (i) although it is an ideal MHD instability in the linear sense, it would not nonlinearly evolve if there were no resistive (non-ideal MHD) effect; (ii) it can involve a large amount of conversion of magnetic to kinetic energies in a short time; (iii) it is essentially a two-dimensional instability, thus more amenable to thorough analysis of the fundamental processes of the instability. The coalescence instability starts from the Fadeev equilibrium,[22] which is characterized by the current localization parameter ϵ_c: The equilibrium toroidal current (in the z-direction) is given as $J_z = B_{ox} k(1 - \epsilon_c^2)(\cosh ky + \epsilon_c \cos kx)^{-2}$. The parameter ϵ_c varies from 0 to 1 with small ϵ_c corresponding to a weak localization and ϵ_c close to unity corresponding to a peaked localization; in the limit where $\epsilon_c \to 1$ the current distribution becomes the delta function.

When ϵ_c is as small as 0.3, the rate of reconnection was that of Sweet-Parker.[23] When ϵ_c is larger than 0.3 but smaller than 0.8, the reconnection rate experiences two phases.[23] This emergence of two phases is similar to the case of the driven reconnection,[19] which was examined in the previous Sec. 14.3. The intensity of coalescence and the rate of subsequent reconnection are controlled by just one parameter, the current localization (ϵ_c). In this problem there is no ambiguity as to the nature of the driver in contrast to the reconnection driven by external boundary conditions. For the case where ϵ_c=0.7, the second phase showed the reconnected flux ψ increasing as t^α with $1 < \alpha$. This indicates that the more the current localizes, the faster the reconnection becomes.

This leads to a question: Can the reconnected flux ψ increase explosively as $(t_0 - t)^{-\alpha}(\alpha > 0)$ triggered by the coalescence instability? The answer from computer simulation is: yes it can, when ϵ_c is further increased. The computational results obtained by an electromagnetic particle simulation (see Chapter 5) based on the same parameters as reported in Ref. 24 and on a set-up similar to Ref. 25.

Figure 14.11 displays the time history of various field and particle quantities observed in our simulation in which, after the initial transient, the phase of coalescence of two magnetic islands commences. It is seen in Figs. 14.11(a)-(c) that around $t = 27$ the magnetic and electrostatic field energies shoot up explosively as well as the ion temperature in the direction of coalescence (the x-direction). It is also seen in Fig. 14.11(a)-(c) that (i) after the explosive increase of the field energies and temperature, this overshooting results in synchronous amplitude oscillations of all these quantities with the period being approximately the compressional Alfvén period; and (ii) superimposed on these overall amplitude oscillations is a distinct double-peak structure in the electrostatic field energy and the ion temperature. Although we are interested in analyzing the entire episode of the run, including the initial phase and the post-explosive phase, we focus particularly on the explosive phase of the coalescence. We replot Figs. 14.11(a)-(c) into Figs. 14.11(d)-(f) to find the way in which these quantities increase toward the catastrophic point. It is found from Figs. 14.11(d)-(f) that (i) the magnetic energy explodes as $(t_0 - t)^{-8/3}$; (ii) the electrostatic energy explodes as $(t_0 - t)^{-4}$; and (iii) the ion temperature in the coalescing direction explodes as $(t_0 - t)^{-8/3}$ until saturation due to overshooting sets in, where t_0 is the explosion time measured here to be $t_0 \sim 27$ in this run. See Table 14.1.

It is very interesting to observe the existence of an explosive process (or instability) and its indices of explosion (the exponent to the time) that governs the explosive magnetic process (the magnetic collapse).[24]

The following investigation of the coalescence of magnetic islands is through extensive numerical simulations of both MHD and kinetic types. Analysis of simulation results leads to a heuristic model of the explosive coalescence and further a more complete theory of this process. The results from these two different models are consistent in basic points, but are complementary in many detailed aspects. In general, kinetic models allow more physical effects at the sacrifice of cramped time and space scales; MHD models are more realistic in scales at the sacrifice of neglecting kinetic effects.

The kinetic simulation model here is the electromagnetic particle code (Chapter 5) with the configuration of the plasma and magnetic fields being that of Refs. 24 and 25. The plasma density is initially uniform in the x- and y-directions and the z-direction is the ignorable direction. Fields are solved with periodic boundary conditions in the x- and y-directions. The sheared

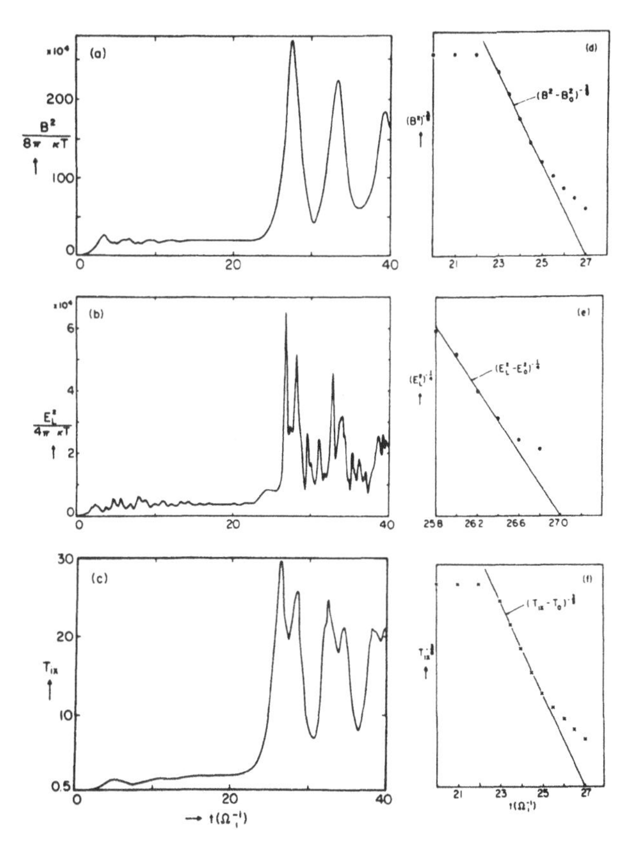

FIGURE 14.11 The magnetic (a), (d) and electric (b), (e) energies and ion temperature (c), (f), as a function of time during the explosive coalescence process (Ref. 24)

magnetic fields are generated by the externally imposed sheet currents J_z at $y = 0$ and L_y, where L_y is the length of the periodic box in the y-direction. The sheet currents are turned with a ramp function profile in time. The excess of the plasma return current or lack of it for the uniform component (the wavenumber $k = 0$) is compensated by the displacement current term alone, since the term $\nabla \times B$ vanishes for $k = 0$. The process of island generation from this configuration was discussed in detail in Ref. 25. The later process of coalescence of generated islands is focused here. A uniform external (toroidal) magnetic field B_z is applied with various chosen strengths. Typical parameters in the code are: the numbers of grid points in the x- and y-directions are $L_x/\Delta = 128$ and $L_y/\Delta = 32$, the number of electrons (and that of ions) 16384, the speed of light $c = 4\omega_{pe}\Delta$, the thermal velocities of electrons in the

Indices of Explosion [exponents to the $1/(t_0 - t)$] During Coalescence

		$\Omega_{et} = 0$ $L_x \times L_y$ $= 128 \times 32$ (see Ref. 26) [NB: $L_x \times L_y$ $= 256 \times 32$ many islands]	$\Omega_{et} = 0.2\omega_{pe}$ $L_x \times L_y$ $= 128 \times 32$	$\Omega_{et} = 1.0$ $L_x \times L_y$ $= 128 \times 32$	$\Omega_{et} = 2.0$ $L_x \times L_y$ $= 128 \times 32$ No formation of islands
Magnetic Energy B^2	(S)	8/3	8/3	2	
	(T)	8/3	8/3	2*	N/A
Electrostatic Energy E_L^2	(S)	4	4		
	(T)	4	4	4	N/A
Ion Temperature in x-direction T_{ix}	(S)	8/3	8/3	3	
	(T)	8/3	8/3	2	N/A
Explosive Time t_0	(S)	$24.3\Omega_i^{-1}$	$27\Omega_i^{-1}$	$19\Omega_i^{-1}$	N/A
Compressional Alfvén Oscillation Period τ_{os}		$6.3\Omega_i^{-1}$	$6.0\Omega_i^{-1}$	$8.8\Omega_i^{-1}$	N/A

* incompressibility is assumed.
Derivation from observation might be due to plasma rotation in $\Omega_{et} = 1$ case.
S is for simulation results and T for theory.

TABLE 14.1 Summary of simulation and subsequent theory on explosive coalescence

x-, y-, and z-directions $v_e = 1\omega_{pe}\Delta$, the electron-to-ion mass ratio $m_e/m_i = 1/10$, the electron-to-ion temperature ratio $T_e/T_i = 2.0$, the (poloidal) sheared magnetic field B_x at $y = 0$ and L_y is such that $eB_x/m_e c = 0.77\omega_{pe}$, and the size of particles $a = 1\Delta$. The toroidal field B_z is varied with $eB_z/mc\omega_{pe}$ ranging from 0 to 0.2, 1, and 2.

In these parameters the poloidal Larmor radius at the external current sheets for electrons and ions are $\rho_{pe} = 1.2\lambda_{De}$ and $\rho_{pi} = 5.3\lambda_{De}$, the poloidal cyclotron frequencies for electrons and ions $\Omega_e = 0.77\omega_{pe}$ and $\Omega_i = 0.077\omega_{pe}$, and the poloidal Alfvén velocity $v_{Ap} = 1.22v_e$. These numbers change accordingly when there is an imposed toroidal magnetic field B_z. Because of the nature of the particle code and electromagnetic interactions retained, the temporal and spatial scales of simulation are compressed by using an unrealistically large electron-to-ion mass ratio and small grid. However, it is noted that the main time scales we are interested in are those of the Alfvén time and the electron time scales are sufficiently isolated from this. The chief purpose of this simulation is not to reproduce laboratory plasma behavior but to extract some fundamental underlying processes and try to understand them. Although the fundamental physics emerges in the kinetic simulation, many parameters are strained in this model such as an unrealistically large mass ratio 1/10. On the other hand, the MHD model dispenses many of these processes and thus it is unnecessary to strain many parameters.

The MHD simulation model used is the MHD particle code[26] with $2\frac{1}{2}$ dimensions. The configuration of the plasma and magnetic fields is that of Ref. 23 based on the initial conditions of Fadeev et al.'s equilibrium.[23] The MHD particle code is robust in applications to problems even with strong turbulence, flows, convections, and density depression, which is helpful because the present problem involves fast (explosive) reconnection, strong density depression and compression, and strong flows. The magnetic induction equation is advanced by the Lax-Wendroff method. The plasma is originally uniform in density and temperature contained by metallic (conducting) walls at $y = 0$ and L_y. Here typical parameters are: $L_x/\Delta = 128$ and $L_y/\Delta = 64$, the number of fluid particles 32768, the poloidal magnetic fields B_x at $y = 0$ and L_y are such that the (poloidal) Alfvén velocity $v_{Ap} = 3.5c_s$, the adiabatic constant $\gamma = 2$, and the size of particles $a = 1.0\Delta$, where c_s is the sound speed and Δ is the unit grid length. The current localization parameter ϵ_c is varied from the value $\epsilon_c = 0.3$ to 0.85. The Alfvén transit times across the y-direction and the x-direction are $\tau_{Ay} = 18.3\Delta/c_s$ and $\tau_{Ax} = 36.5\Delta/c_s$, respectively. The typical magnetic Reynolds number is $R_m \cong 10^4$. As is well known, the ideal MHD dynamics does not contain any characteristic length, except for the system's overall length; in the present case it is either L_y or the island width. For example, the collisionless skin depth c/ω_{pe} and the Debye length vanish. Therefore, in contrast with the kinetic model discussed earlier, the spatial scales are not compressed. Similarly the relevant time scales are the Alfvén time and the much larger resistive time. However, the MHD model largely lacks the kinetic effects such as the Landau and cyclotron dampings, particle acceleration, finite Larmor radius effects, etc. Thus the study by the MHD model is complementary to that by the kinetic model mentioned earlier.

Results from the electromagnetic particle model shall now be discussed.[24] Figure 14.12 presents the time history of various field quantities and temperatures in the course of the early formation and the coalescence process with the toroidal field being such that $eB_z/m_ec \equiv \Omega_{et} = 0.2\omega_{pe}$. In Fig. 14.12 both the magnetic field energy and the ion temperature in the direction of coalescence (x) show that a long relatively dormant period ($t \cong 3 - 22$) sets in, followed by a stage ($t \cong 22 - 27$) of rapid and huge increases in these oscillatory quantities. After the rapid increase ($t \geq 27$) amplitude oscillations ensue in the coalescence due to overshooting. All the other quantities shown in Fig. 14.12(a)-(e) closely follow the pattern of Fig. 14.12(a) with their characteristic events simultaneously occurring. The amplitude oscillations of the temperatures (T_{ix} and T_{iz} as well as T_{ex} and T_{ez}) and the electrostatic field energy have a structure of marked double peaks. The valley of the double-peak

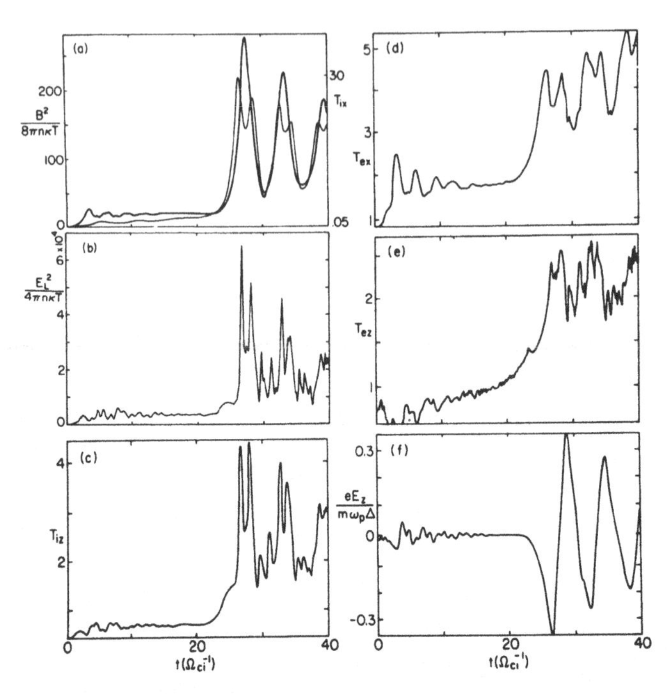

FIGURE 14.12 Temporal evolution of various quantities during and after the explosive coalescence (Ref. 24)

structure coincides with the peak of the magnetic field energy amplitude. The quiescent period ($3 < t \leq 20$) corresponds to the stage where the formed islands slowly attract each other. The rapid explosive rise ($t \geq 20$) marks the commencement of the explosive coalescence. The following stage of amplitude oscillations correspond to the pulsations of coalesced islands. The induced electric field E_z increases explosively when there is rapid flux reconnection during the explosive coalescence and then oscillates as the magnetic flux in the coalesced island is compressed and decompressed.

Let us study the structure of the plasma and fields shortly before and after the completion of island coalescence. Figure 14.13 exhibits the spatial structure of the magnetic field lines, In the toroidal current density J_z, the plasma electron density, the electron flow, the ion flow, and the electric field at $t = 24\Omega_i^{-1}$. In Fig. 14.13 the island near $x = 55\Delta$, $y = 16\Delta$ is rapidly approaching the other island at $x = 0$ and $y = 0$ pinned by the sheet current gaps. The former island is accelerated by the intense magnetic fields behind it. The density of electrons is sharply peaked just behind the center of the island because of the acceleration [Fig. 14.13]. Electrons flow mainly along

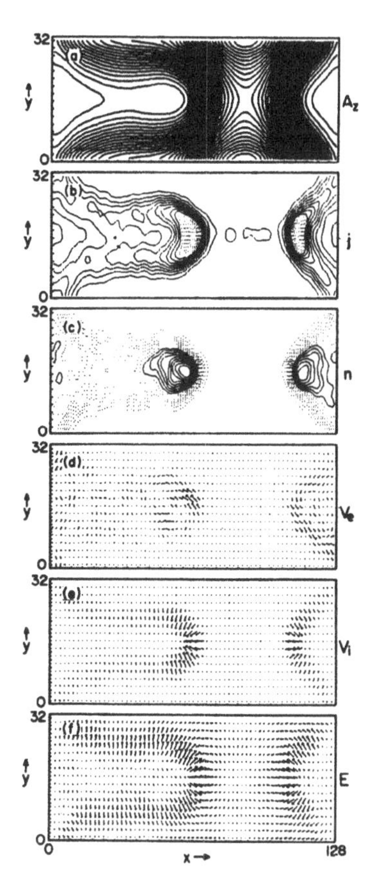

FIGURE 14.13 Spatial pattern of various field variables during the explosive coalescence (Ref. 24)

the field lines [Fig. 14.13], while ions which are left behind by the electrons try to catch up with them [Fig. 14.13]. This sets up an electrostatic field pattern with arrows pointing from the inside to the outside of the island [Fig. 14.13(f)].

The particle acceleration in the high energy tail of ions and electrons can be studied. The tail formation is due to a combination of localized electrostatic field acceleration across the poloidal magnetic field and magnetic acceleration ($v_{ph} \times B$ acceleration).[24,27] The difference of motions between ions and electrons around the time of Fig. 14.13 causes a strong localized shock-like electrostatic field, E_L, whose phase propagates with a phase velocity of the structure $v_{ph} = v_x$. Here the mechanism of $v_{ph} \times B$ acceleration[27,28] is that

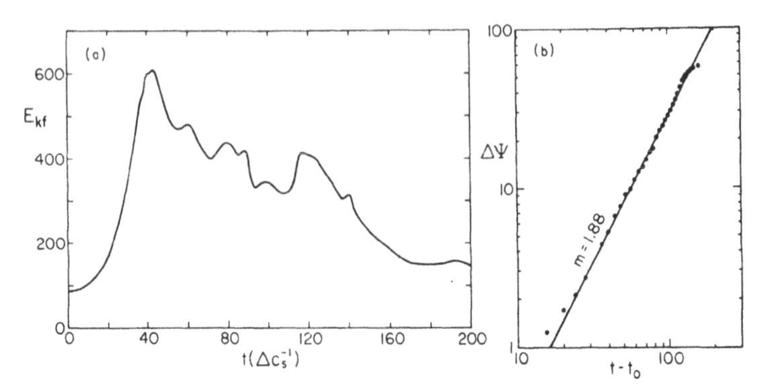

FIGURE 14.14 Temporal evolution of kinetic energy (a) and reconnected flux (b) when the coalescence is rapid ($\epsilon_c = 0.7$) (Ref. 24)

an electrostatic wave propagating with phase velocity v_{ph} can add energy to a particle that is propagating obliquely to the wave propagation direction by combining the electric acceleration and $v \times B$ acceleration when the particle is trapped in the wave. This $v_{ph} \times B$ acceleration causes the formation of high energy particles in the toroidal direction and accelerates ions and electrons to relativistic energies in opposite directions along the toroidal magnetic field.

Results from the magnetohydrodynamic particle models are presented hence. For the case with $\epsilon_c = 0.3$ the reconnected flux increases $\Delta\psi \propto t^m$ with $m = 1$.[23] Figure 14.14 displays the case with $\epsilon_c = 0.7$. The reconnected flux ψ increases rapidly with $\Delta\psi \propto t^m (m \sim 1.9)$.[23] Figure 14.15 shows the kinetic energy and the reconnected flux upon coalescence as a function of time for the case with $\epsilon_c = 0.85$. A theoretical curve $(t_0 - t)^{-4/3}$ is superimposed on the simulation result. During the phase of the rapid increase of reconnected flux ($t \cong 50 - 90\Delta c_s^{-1}$) the simulation result matches reasonably with the theoretical curve. Beyond $t = 90\Delta c_s^{-1}$ the increase begins to be mitigated due to a saturation effect (the flux depletion).

It is clear that as ϵ_c increases, the process of reconnection becomes faster, changing from the Sweet[4] and Parker[5] rate to the faster[19] rate and then to the explosive rate.[24] It is also noted that the explosive increase of reconnected flux during the coalescence is observed in the MHD simulation as well as in the kinetic simulation discussed earlier. The structure and its evolution of the plasma and magnetic fields during the coalescence are now shown. Figure 14.16 presents the pattern of the plasma and fields of the case $\epsilon_c = 0.85$, where we see faster and explosive reconnection corresponding to Fig. 14.15. In frames of Figs. 14.16(a)-(d) ($t = 50$) one sees the coalescence behavior before it becomes explosive. Although, in Figs. 14.16(a) and (b), in particular, one can detect some deviation from the Sweet-Parker type for $\epsilon_c = 0.3$; it is qualitatively similar to the $\epsilon_c = 0.3$ case and the $\epsilon_c = 0.7$ case[23] at this stage. In Figs. 14.16(e) and (f) ($t = 75$), we now see significant deviations in pattern

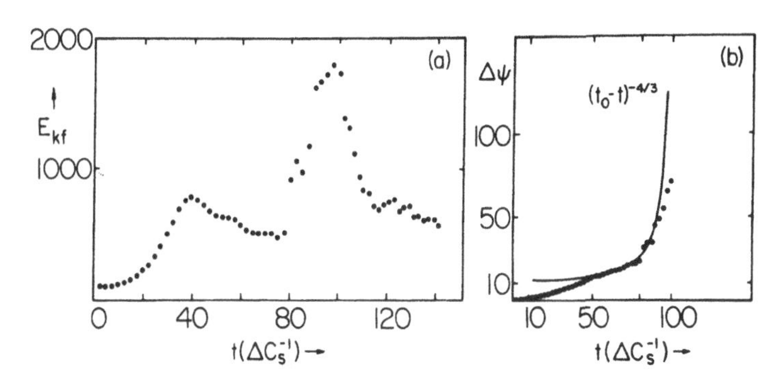

FIGURE 14.15 Temporal evolution of kinetic energy (a) and reconnected flux (b) when the coalescence is explosive ($\epsilon_c = 0.85$) (Ref. 24)

Coalescence and Current Peakedness (ϵ_c)

ϵ_c	0	0.3	0.7	0.85
Process	Sheetpinch tearing instability[3]	Slow Coalescence Sweet-Parker process[4,5]	Fast Coalescence Ref. 19	Explosive Coalescence Ref. 24
Recon. flux $\Delta\psi$	$e^{\gamma_L t}$ $\gamma_L \propto \eta^{3/5}$	$\eta^{1/2}t$	$\eta^{1/2}t^m$ $(m \geq 1)$	$\eta^0/(t_0 - t)^{4/3}$

TABLE 14.2 The current peakedness parameter (ϵ_c) controls the physical process and its speed of the coalescence instability

from the cases[23] with less ϵ_c. A much wider reconnection angle than the previous times is observed[24] in Fig. 14.16(e). From these observations it can be argued that the widening of the reconnection angle has to be accompanied by fast or explosive coalescence. In Table 14.2 we summarize the results.

14.5 Theory of Explosive Coalescence and Comparison with Simulation

Some of the observed features of the explosive coalescence detailed in the above stimulate theoretical investigation and can be explained by theory. The

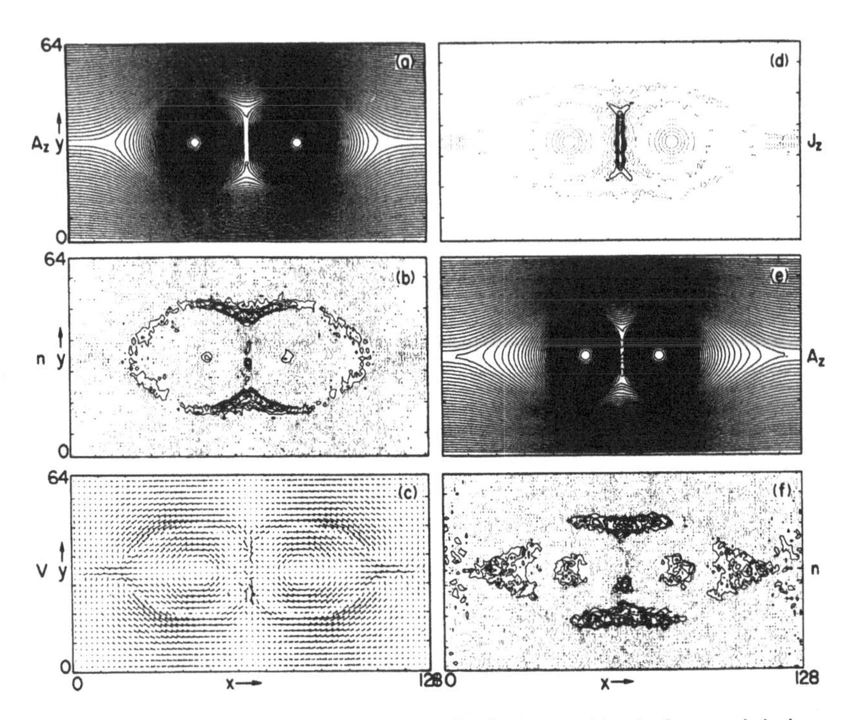

FIGURE 14.16 Spatial pattern of various physical quantities before and during explosive coalescence

geometry of magnetic fields is exemplified here by Fig. 14.16. We are primarily concerned with the plasma sheet region (in the neighborhood of $x = 64\Delta$ and $y \sim 20\Delta - 42\Delta$) where the physics is nearly one-dimensional, that is, the variation of quantities in the y-direction is much less than that in the x-direction. We employed Eqs. (14.33), (14.34), (14.37) and (14.40) in Sec. 14.3. Equation (14.43) successfully extracted the main physics of the fast reconnection and explained the driven reconnection simulation results[19] as well as the coalescence simulation results with $\epsilon_c \leq 0.7$.[23]

This set of equations Eqs. (14.33)-(14.37) and (14.40) can be extended to describe the explosive reconnection. In the explosive process, the local equilibrium, as assumed in the above, can no longer be maintained, but the dynamical pressure equation or the equation of motion as it is has to be employed. In place of Eq. (14.40) we have

$$\frac{\partial v_x}{\partial t} = \frac{1}{8\pi\rho}\frac{\partial}{\partial x}B_y^2 - \frac{1}{2}\frac{\partial}{\partial x}v_x^2 \; . \tag{14.44}$$

where the predominant one-dimensionality of the problem in the x-direction has been taken into account and the plasma internal pressure p is dropped compared with the dynamical pressure $\frac{1}{2}v_x^2$.

The current sheet of nearly one-dimensional structure is formed[24] in the explosive stage of the coalescence instability. As the coalescence proceeds further, the magnetic field structure approaches an X-type (Petschek type[6]) configuration. It is assumed that $\frac{\partial}{\partial x} \gg \frac{\partial}{\partial y}, \frac{\partial}{\partial z}$, in which x is the direction of coalescence, while y is the direction of poloidal magnetic field line and z is the direction of plasma current. The external plasma dynamics of the explosive stage is treated as a one-dimensional problem.

We start from the two-fluid model equations of plasma and the Maxwell equations, neglecting the displacement current. The adiabatic law of states for both electrons and ions is assumed. The basic equations read as follows:

$$\frac{\partial n_j}{\partial t} + \nabla \cdot (n_j \mathbf{v}_j) = 0 \tag{14.45}$$

$$m_j n_j \frac{d\mathbf{v}_j}{dt} = n_j e_j \left(\mathbf{E} + \frac{\mathbf{v}_j}{c} \times \mathbf{B}\right) - \nabla p_j \ , \tag{14.46}$$

$$\nabla \times \mathbf{B} = \frac{4\pi}{c} \sum_j n_j e_j \mathbf{v}_j \ , \tag{14.47}$$

$$\nabla \cdot \mathbf{E} = 4\pi \sum_j n_j e_j \ , \tag{14.48}$$

$$\nabla \times \mathbf{E} = -\frac{1}{c}\frac{\partial \mathbf{B}}{\partial t} \ , \tag{14.49}$$

$$\frac{\partial p_j}{\partial t} + \mathbf{v}_j \cdot \nabla p_j + \gamma p_j \ \mathrm{div}\ v_j = 0, \tag{14.50}$$

where j denotes the species of particles and γ is the ratio of heat capacities.

During explosive coalescence, there is no specific scale length. The scale length characterizing the current sheet varies continuously in time without deformation of global structure of current sheet. That is, the relations (laws) that govern the explosive coalescence themselves are invariant under a changing time scale. This was the manifestation of the presence of self-similarity in the system during explosive coalescence. This may be called universality in time, as opposed to the conventional universality in space such as in Kadanoff's spin block problem[29] and in Kolmogorov's turbulence spectrum.[30] A similar situation also arises in the general theory of relativity in which the scale factor a plays a role in the Hubble expansion of the universe. Such a physical situation may be best described by self-similar solutions in which scale factors vary continuously.

Scale factors $a(t)$ and $b(t)$ are introduced as follows,[24]

$$v_{ex} = \frac{\dot{a}}{a} x \,, \tag{14.51}$$

$$v_{ix} = \frac{\dot{b}}{b} x \,, \tag{14.52}$$

where a dot represents the time derivative. The ansatz imposed here is that the velocities are linear in x, which implies that particles flow in the opposite direction around the center of current sheet, $x = 0$. The scale factors a and b will be determined from the above basic equations. From the continuity equations of electrons and ions, Eq. (14.45), we obtain

$$n_e = n_0/a \,, \tag{14.53}$$

$$n_i = n_0/b \,, \tag{14.54}$$

where n_0 is a constant. Equations (14.53) and (14.54) show that the densities of ions and electrons are nearly homogeneous in space and vary only in time during coalescence. The self-similar solutions obtained here are local solutions in space whose properties are dominated by the physical process near the current sheet. We therefore neglect the higher order terms in space proportional to x^3 and higher hereafter. The current J_z in the sheet is nearly constant. This means that as n is nearly constant, v_z is also approximately constant in space. Neglecting the term with x^3 in Eq. (14.48), we obtain

$$\frac{B_0(t)}{\lambda} = \frac{4\pi e n_0}{c} \left(\frac{v_{iz}^{(0)}}{b} - \frac{v_{ez}^{(0)}}{a} \right) \,, \tag{14.55}$$

where it is assumed the magnetic-field B_y varies as $B_y = B_0(t)\frac{x}{\lambda}$, where λ is the magnetic field scale length. This ansatz is consistent with the assumption that the sheet current is (nearly) constant in space.

From the y-component of Eq. (14.49) and the z-component of equation of motion for electrons Eq. (14.46) we obtain

$$\dot{B}_0 = 2c \frac{E_{z1}}{\lambda} \,, \tag{14.56}$$

$$E_{z1} \frac{x^2}{\lambda^2} + \frac{\dot{a}}{a} \frac{B_0(t)}{\lambda c} x^2 = 0 \,, \tag{14.57}$$

$$\frac{\partial v_{ez}^{(0)}}{\partial t} = -\frac{e}{m_e} E_{z0} \,, \tag{14.58}$$

$$E_z = E_{z0}(t) + E_{zl}(t) \frac{x^2}{\lambda^2} \,. \tag{14.59}$$

From Eqs. (14.56), (14.58) and Problem 3, $v_{ez}^{(0)}$ and E_z are given by

$$v_{ez}^{(0)} = -\frac{cB_{00}b}{4\pi e n_0 \lambda a^2 \left(\frac{b}{a} + \frac{m_e}{m_i}\right)}, \tag{14.60}$$

$$E_z = -\frac{B_{00}\dot{a}x^2}{ca^3\lambda} + \frac{m_e c B_{00}}{4\pi n_0 e^2 \lambda}\frac{d}{dt}\left(\frac{b}{a^2\left(\frac{b}{a} + \frac{m_e}{m_i}\right)}\right). \tag{14.61}$$

Under the assumption that the electrostatic field E_x varies like $E_x = E_0(t)x/\lambda$, Poisson's equation (14.48) becomes

$$E_0 = 4\pi e n_0 \lambda \left(\frac{1}{b} - \frac{1}{a}\right). \tag{14.62}$$

When the plasma is quasi-neutral $n_i = n_e$, i.e., $a = b$, small terms of the order of m_e/m_i are neglected. Equations (14.47)-(14.62) are utilized to yield the basic equation for $a(t)$

$$\ddot{a} = -\frac{v_A^2}{\lambda^2 a^2} + \frac{c_s^2}{\lambda^2 a^\gamma}, \tag{14.63}$$

where

$$v_A^2 = \frac{B_{00}^2}{4\pi n_0(m_i + m_e)} \quad \text{and} \quad c_s^2 = \frac{(P_{0e} + P_{0i})}{(m_e + m_i)n_0}. \tag{14.64}$$

In Eq. (14.63) the first term of the RHS corresponds to the $J \times B$ term and is the term that drives magnetic compression (collapse). The second term corresponds to the pressure gradient term and may eventually be able to balance the magnetic collapse when $\gamma = 3$. The condition $\gamma = 3$ means that the plasma compression takes place in a nearly one-dimensional fashion so that the degree of freedom of the system f becomes unity. When $\gamma = 3$, we obtain

$$\ddot{a} = -\frac{v_A^2}{\lambda^2 a^2} + \frac{c_s^2}{\lambda^2 a^3}. \tag{14.65}$$

when $\gamma = 2 (f = 2)$, on the other hand, we obtain

$$\ddot{a} = -\frac{(v_A^2 - c_s^2)}{\lambda^2 a^2}. \tag{14.66}$$

Once the behavior of the scale factor $a(t)$ is determined from the above equations, various physical quantities are obtained as follows, in the quasi-neutral plasmas, and neglecting the electron mass ($\frac{m_e}{m_i} \to 0$),

$$B_y = \frac{B_{00}}{a^2}\frac{x}{\lambda} \tag{14.67}$$

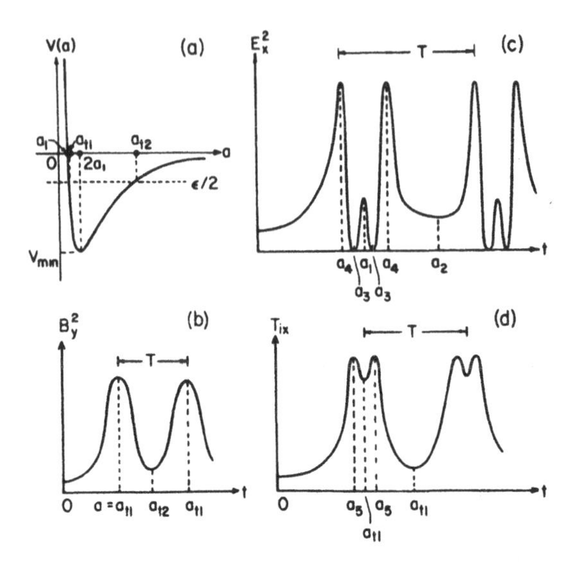

FIGURE 14.17 Schematic effective potential and amplitude oscillations of various physical quantities (Ref. 24)

$$E_x = \left(-\frac{m_i}{e_\lambda}\frac{v_A^2}{a^3} + \frac{P_{0e}}{e_\lambda a^4 n_0}\right)\frac{x}{\lambda} \tag{14.68}$$

$$E_z = -\frac{B_{00}\dot{a}x^2}{ca^3\lambda} - \frac{B_{00}m_e c\dot{a}}{4\pi n_0 e^2\lambda a^2} \tag{14.69}$$

$$v_{ez} = -\frac{cB_{00}}{4\pi e n_0 \lambda a} \tag{14.70}$$

$$v_{ix} = v_{ex} = \frac{\dot{a}}{a}x \tag{14.71}$$

$$n_i = n_e = \frac{n_0}{a} \tag{14.72}$$

where the electrostatic field E_x in the quasi-neutral plasmas is determined from the equation of motions for ions, not from Poisson's equations. From Eqs. (14.67) and (14.68) in the explosive phase ($a \to 0$) the electrostatic field [$E_x \propto (a^{-3} + a^{-4})$] grows more rapidly than the magnetic field ($B_y \propto a^{-2}$) does.

Now we investigate the global time behavior of coalescence by making use of the first integral of Eq. (14.65). Equation (14.65) is rewritten as

$$\ddot{a} = -\frac{\partial V(a)}{\partial a} , \tag{14.73}$$

where the effective (Sagdeev) potential $V(a)$ is given by

$$V(a) = -\frac{v_A^2}{\lambda^2 a} + \frac{c_s^2}{2\lambda^2 a^2} , \tag{14.74}$$

where the first term may be reminiscent of the "gravitational attractive potential" while the second of the "centrifugal repulsive potential." The schematic graph of the effective potential is drawn in Fig. 14.17. The value a which satisfies $V(a_1) = 0$ is given by

$$a_1 = \frac{1}{2}\frac{c_s^2}{v_A^2} . \tag{14.75}$$

The minimum of the potential, $V_{\rm min}$, obtained from $\partial V/\partial a = 0$ is

$$V_{\rm min} = \frac{-v_A^4}{2\lambda^2 c_s^2} , \tag{14.76}$$

at $a = 2a_1 = c_s^2/v_A^2$. The potential becomes deeper when the ratio of the kinetic to magnetic energy densities β decreases. This means that the driving force $J \times B$ is dominant compared with the pressure term. The first integral of Eq. (14.73) is given by

$$\dot{a}^2 = \frac{2v_A^2}{\lambda^2 a} - \frac{c_s^2}{\lambda^2 a^2} + \mathcal{E} , \tag{14.77}$$

where $\mathcal{E}$ is the initial (Sagdeev) "energy" (dimension: $1/\text{time}^2$) in space of stretching factor a.

$$\mathcal{E} = \dot{a}_0^2 - \frac{2v_A^2}{\lambda^2 a_0} + \frac{c_s^2}{\lambda^2 a_0^2} . \tag{14.78}$$

As seen from Fig. 14.17, the explosive magnetic compression corresponds that the scale factor $a(t)$ rapidly changes in time by orders of magnitude and nearly vanishes. This may be called an explosive magnetic collapse. Such an explosive collapse can be realized (i) when the effective potential has a sharp and deep potential well and this means that $\beta = c_s^2/v_A^2$ is very small; and (ii) when the initial total energy $\mathcal{E}/2$ is nearly zero. On the other hand, when $\mathcal{E}/2$ is close to $-V_{\rm min}$, we have oscillations near the potential minimum and no explosive collapse.

The time history of various physical quantities is examined based on the qualitative time behavior of $a(t)$ derivable from the effective potential $V(a)$. The magnetic field energy is proportional to B_y^2, which is given by

$$B_y^2 = \frac{B_{00}^2}{a^4}\left(\frac{x}{\lambda}\right)^2 . \tag{14.79}$$

If the scale factor a becomes smaller, B_y^2 must increase. The maximum is given by

$$\frac{\partial B_y^2}{\partial t} = 0 \ , \tag{14.80}$$

which yields $\dot{a} = 0$, namely $a = a_{t1}$. After the maximum, B_y^2 decreases again and reaches minimum at $a = a_{t2}$. The oscillatory behavior of the magnetic field energy is schematically drawn in Fig. 14.17.

The electrostatic field E_x is given by Eq. (14.68). The time history of the electrostatic field energy, which is propositional to E_x^2, is analyzed by investigating

$$\frac{\partial E_x^2}{\partial t} = 0 \ .$$

This condition is equivalent to

$$\mathcal{E}_0 = 0 \ , \tag{14.81}$$

or

$$\frac{\partial \mathcal{E}_0}{\partial t} = 0, \tag{14.82}$$

where $\mathcal{E}_0(t) = -\frac{m_i}{e\lambda}\frac{v_A^2}{a^3} + \frac{P_{0e}}{e\lambda n_0 a^4}$. The first condition $\mathcal{E}_0 = 0$ occurs at $a = a_3 = P_{0e}/m_i n_0 v_A^2 \cong c_s^2/v_A^2$. The second condition $\frac{\partial \mathcal{E}_0}{\partial t}$ gives two conditions, namely

- $\dot{a} = 0$, which occurs at $a = a_{t1}, a_{t2}$
- $a = a_4 = \frac{4}{3}\frac{P_{0e}}{m_i n_0 v_A^2} \cong \frac{4}{3}\frac{c_s^2}{v_A^2} = \frac{4}{3}\beta$.

The above considerations give us the schematic time history of the electrostatic field energy E_x^2 as drawn in Fig. 14.17(b). Figure 14.17(b) indicates a triple-peak structure in the electrostatic field energy. When the plasma β is small, a_3 and a_{t1} are close. In this case, the triple-peak structure in the electrostatic field energy would become a double-peak structure. The maximum value of the electrostatic field, $E_{\max}$, achieved at $a = a_4$ is given by

$$E_{\max} = \frac{1}{4}\left(\frac{3}{4}\right)^3 \frac{m_i}{e\lambda}\frac{v_A^8}{c_s^6}\frac{x}{\lambda} \cong 0.1\frac{m_i}{e\lambda}\frac{v_A^2}{\beta^3}\frac{x}{\lambda} \ . \tag{14.83}$$

The induced electric field E_z is given by Eq. (14.69), which shows that E_z becomes zero, when $\dot{a} = 0$. E_z changes its sign around $\dot{a} = 0$ because $\dot{a} = 0$ is the point where the magnetic field reaches a maximum or a minimum.

Next, the time behavior of ion temperature T_{ix} is examined. In the early stage of coalescence, the plasma should be adiabatically compressed. However, as the magnetic field energy increases near the peak and approaches the peak, the ion flow energy becomes dominant over the thermal energy. From the consideration that v_x^2 gives a maximum or a minimum, namely $\frac{\partial v_x^2}{\partial t} = 0$, we find two conditions for the extrema; (i) $v_x = 0$, which gives $\dot{a}$, (ii) $\frac{\partial v_x}{\partial t} = 0$, which gives

$$a\ddot{a} = \dot{a}^2 \ . \tag{14.84}$$

When the explosive coalescence takes place ($\mathcal{E} = 0$), we estimate the condition (14.88) as

$$a = a_5 \simeq \frac{2}{3}\frac{c_s^2}{v_A^2} = \frac{2}{3}\beta \ .$$

After $a = a_5$, the kinetic energy must decrease, which means that the plasma is in the state of colliding phase. The above considerations give us the schematic time history of the ion temperature, which is shown in Fig. 14.17(c). Figure 14.17(c) shows a double-peak structure in the ion temperature. In the limit of quasi-neutrality, we can estimate the dominant term governing the explosive phase where the adiabatic compression is predominant.

The explosive phase of the coalescence is investigated in more detail in a case where we can neglect the effect of plasma pressure; it acts only as a saturation mechanism. In the explosive phase, therefore, we can neglect the second term on the right-hand side of Eq. (14.63);

$$\ddot{a} = -\frac{v_A^2}{\lambda^2 a^2} \ . \tag{14.85}$$

The solution of Eq. (14.78) with small ε is given by

$$a(t) \cong \left(\frac{9}{2}\right)^{1/2} \left(\frac{v_A}{\lambda}\right)^{2/3} (t_0 - t)^{2/3} + \mathcal{O}(\varepsilon) \ , \tag{14.86}$$

where the order of ε is neglected and t_0 is the explosion time. Once the solution $a(t)$ is given by Eq. (14.86), the various physical quantities behave as follows, which is valid in the explosive phase of the coalescence;

$$v_x = v_{ix} = v_{ex} = -\frac{2}{3}\frac{x}{(t_0 - t)} \ , \tag{14.87}$$

$$n = n_i = n_e = \left(\frac{2}{9}\right)^{1/3} \frac{\lambda^{2/3} n_0}{v_A^{2/3}(t_0 - t)^{2/3}} \ , \tag{14.88}$$

$$E_x = -\frac{2}{9}\frac{m_i}{e}\frac{x}{(t_0 - t)^2} , \tag{14.89}$$

$$B_y = \left(\frac{2}{9}\right)^{2/3} \frac{B_{00}\lambda^{1/3} x}{v_A^{4/3}(t_0 - t)^{4/3}} \ , \tag{14.90}$$

$$E_z = \frac{2}{3}\left(\frac{2}{9}\right)^{2/3} \frac{B_{00}\lambda^{1/3} x^2}{v_A^{4/3} c(t_0 - t)^{7/3}}$$

$$+\frac{2}{3}\left(\frac{2}{9}\right)^{1/3}\frac{B_{00}c}{\omega_{pe}^2\lambda^{1/3}v_A^{2/3}(t_0-t)^{5/3}}\,. \qquad (14.91)$$

Let us compare the theoretical results discussed here for the explosive phase with the computer simulation results.[24] The global structure of the magnetic field energy, electrostatic field energy, and ion temperature in the x-direction observed in the simulation is well explained by the theoretical model obtained here. Especially, the double-peak structure in the ion temperature and the triple-peak structure in the electrostatic field energy are also observed in the simulation (see Fig. 14.12). Table 14.1 summarizes the results of the comparison of the explosion indices between the theory and the collisionless simulation.

In Table 14.1 we show the index m of explosiveness [the exponent to the time $(t_0-t)^{-m}$], which shows a good agreement between simulation and theory in the electrostatic energy. For the ion temperature we find its explosiveness as

$$T \sim \frac{1}{a^4} \sim \frac{1}{(t_0-t)^{8/3}} \qquad (14.92)$$

when $\gamma = 3$. In Fig. 14.11, the electrostatic field energy, magnetic field energy, and ion temperature are well explained by the one-dimensional model of the explosive collapse.

The explosive coalescence is a process of magnetic collapse, in which the magnetic and electrostatic field energies and ion temperatures explode toward the explosive time t_0 as shown in Table 14.1 and Fig. 14.11. Single-peak, double-peak, and triple-peak structures of magnetic, temperature, and electrostatic energy, respectively, are observed in the simulation as overshoot amplitude oscillations.

We summarize the temporal behavior of reconnected flux for various values of the current peakedness parameter ϵ_c in Table 14.2. At $\epsilon_c = 0$ no chain of islands exists initially and the system is unstable against the linear tearing instability, which grows exponentially with the linear growth rate γ_L. At a small value of $\epsilon_c = 0.3$ the chain of islands is unstable against the coalescence mode, which grows exponentially in the early linear regime with a linear growth rate that is independent of the resistivity η. In its later nonlinear regime flux grows linearly in time with the coefficient proportional to $\eta^{1/2}$. At a moderately large value of $\epsilon_c = 0.7$ the nonlinear regime exhibits a faster time dependence of $t^m (m \geq 1)$. Its coefficient is approximately proportional to $\eta^{1/2}$. At a still larger value of $\epsilon_c = 0.85$ reconnected flux can go explosively as we have discussed.

The explosive process we described in this discussion possesses a common property found in much wider and general areas of physics. That is, the explosive phase of the coalescence (until it saturates or bounces back) lacks any characteristic time scale, a form of temporal universality. This type of phenomena occurs in a wide range of plasma physics.

One may compare the tokamak major disruption with the present explosive coalescence in two dimensions, which is a result of a current-current interaction. The explosive process we find in the coalescence instability is characterized by singular functions of time with a pole (or a branch point) at the explosion time $(t_0 - t)^{-m}$. The tokamak major disruption has been modeled[31] as nonlinear evolution of unstable tearing. (mainly the $m = 2$, $n = 1$ mode and $m = 3$, $n = 2$) and their destabilization of other beat modes such as the $m = 5$, $n = 3$ mode and its coupling to higher m and n's. It is interesting to extract from their numerical simulation[32] that the toroidal voltage approximately scales as $(t_0 - t)^{-2}$ toward the disruption time t_0 to our best fit (Fig. 5 of Ref. 32). The toroidal voltage is related to E_z in this chapter, which has power exponents of 7/3 and 5/3 according to Eq. (14.91), close to 2. Furthermore, it is of interest to note that we are unable to make a good fit of type $(t_0 - t)^{-m}$ with the kinetic energy growth in Fig. 3 of Ref. 32: Rather, the exponent γ itself explodes as $(t_0 - t)^{-2}$ to our best fit, suggesting an approximate functional form of $\exp[(t_0 - t)^{-1}]$ for the kinetic energy. That is, we may conclude from this study of Fig. 3 of Ref. 32 that the singularity of the temporal explosion of the major disruption is characterized by an essential singularity. In this model the electrostatic potential ϕ and the magnetic flux ψ obey equations of quadratic nonlinearity. With single helicity calculations it is not possible to construct temporal functions with an essential singularity. However, with multiple helicities interacting with each other it may be possible to argue that a strongly developed turbulence which is established by a cascade of many higher order beat modes gives rise to an essential singularity in the temporal behavior: secondary processes yielding $(\varphi, \psi) \propto (t_0 - t)^{-2}$, tertiary processes (beats of beats) $\propto (t_0 - t)^{-3}$, quartic processes $\propto (t_0 - t)^{-4}, \ldots$. One sums up the entire energy as $\sum_{n=0}^{\infty} C_n (t_0 - t)^{-n}$, where C_n's are coefficients; the analytic property of such a function may be generically related to that of $\exp[(t_0 - t)^{-1}]$. It may be argued, on the other hand, that the toroidal voltage is not a result of a multiplicative process but an additive one, thus resulting in the temporal behavior of a pole type as in two dimensions.

It may be of interest to compare the present process of the magnetic collapse with other physical processes. An obvious counterpart to this is the electrostatic collapse of plasma waves[33] in which the process proceeds explosively in two or three dimensions. In the problem of the long-time tail of the temporal autocorrelation function of fluctuations of fluid turbulence etc., it has been known that the autocorrelation function does not have its own proper time scale, but it has a long tail and thus there is no characteristic time scale. It exhibits a self-similarity in time scales. Examples of such behavior include (i) the so-called $1/f$ noise problem,[34] where the frequency spectrum contains an infrared as well as ultraviolet divergence in frequency (f) and (ii) the anomalous fluctuations in the neighborhood of the thermal equilibrium phase transition.[35] The intermittency of turbulence again manifests a self-similar character in the frequency domain,[36] just like the $1/f$ noise. These examples show a kind of temporal universality. On the other hand, spatial counterparts to these may be examples of the critical opalescence in phase transition and the Kolomogorov universal spectrum of a fully developed turbulence. Kadanoff's[29] spin-block model of phase transition yields the spin correlation function $g(r)$ going like $(r - r_0)^{-\alpha}$. On the other hand the Kolmogorov universal spectrum[30] of fully developed turbulence yields the wavenumber space energy spectrum $E(k)$ going like $k^{-\alpha}$. Possible analogues among these were suggested in Table 1.1. Also consult Ref. 36 on some of these analogues.

14.6 Current Loop Coalescence Model of Solar Flares

The nonlinear coalescence instability of current carrying loops heretofore discussed seems to be taking place in the magnetotail [37] and possibly related to the magnetospheric substorm process. It, moreover, explains many of the characteristics of solar flares such as their impulsive nature, simultaneous heating and high-energy particle acceleration, and amplitude oscillations of electromagnetic emission as well as the characteristic development of microwave images obtained during a flare.[24] The main characteristics of the explosive coalescence are: (i) a large amount of impulsive increases in the kinetic energies of electrons and ions, (ii) simultaneous heating and acceleration of electrons and ions in high and low energy ranges, (iii) ensuing quasi-periodic amplitude oscillations in fields and particle quantities, (iv) the double peak (and triple peak) structures in these oscillations, and (v) the characteristic break in energy spectra of electrons and ions. A single pair of currents as well as multiple currents may participate in the coalescence process, yielding varieties of phenomena. These physical properties seem to underlie some impulsive solar flares. In particular, double sub-peak structures in the quasi-periodic oscillations found in the time profiles of two solar flares on 1980 June 7 and 1982 November 26 are well explained in terms of the coalescence instability of two current loops. This interpretation is supported by the observations of two microwave sources and their interaction in the November 26 flare. In the following these observations of solar flares are discussed in light of the theory of nonlinear coalescence.

The solar flare phenomenon is a manifestation of an explosive release process of energy stored in the lower corona, involving plasma heating up to $\sim 5 \times 10^7$K, acceleration of charged particles up to the order of the rest mass energy of electrons and ions, and production of electromagnetic radiation from radio to γ-ray wavelengths. For a previous summary of solar flares, see Refs. 38, 39 and 40. Since the Solar Maximum Mission (SMM) and Hinotori satellites were launched, it has become clear from observations of hard X-rays and γ-rays that the electrons with energies up to MeV and ions with energies up to GeV are simultaneously accelerated within a second during the impulsive phase of a solar flare.[41] In a particular flare (03: 12UT of 1980 June 7) γ-rays showed quasi-periodic amplitude oscillations which closely correlated with the quasi-periodicity in both microwave bursts and hard X-ray bursts.[42–45]

Direct observations in soft X-rays[46] showed that in the active regions there exist multiple coronal loops. These loops may carry plasma currents. The interaction of coronal loops is an important physical process for energy release in the solar corona. Indeed many recent observations ranging from H_α and radio[47–49] to hard X-ray[50] emissions provide evidence for such a physical process at work in solar flares.

In order to explain the rapid quasi-periodic particle acceleration of both electrons and ions observed in the 1980 June 7 flare, a likely mechanism for the impulsive energy release in solar flares is the current loop coalescence instability discussed in the previous section. During the coalescence of two current loops, magnetic energy stored by the plasma current is explosively transformed to plasma heating as well as to production of high energy particles within an Alfvén transit time across the current loop (which is about $1 \sim 10$ seconds for appropriate radius of the loop) through the magnetic reconnection process. Furthermore, the energy release is achieved in a quasi-periodic fashion whose periodicity depends on plasma parameters such as the plasma beta ratio (β), the ratio B_P/B_t between the poloidal (B_P: produced by the loop current) and the toroidal (B_t: potential field) components of the magnetic field, as well as the colliding velocity of two current loops that is determined mostly from its initial total plasma loop current profile. The plasma is heated up to ~ 60 times its initial temperature. At the same time, electrons and ions are accelerated simultaneously by the transverse electrostatic field which is produced during the explosive coalescence process.

In the following a comparison of the above simulation/theory with observations of two solar flares are presented: the 1980 June 7 event and the 1982 November 26 event. This should serve to provide an example to indicate the insight obtained from the cooperative study between simulation and observation. Both flares showed quasi-periodic amplitude oscillations with double sub-peak structure in hard X-ray and microwave time profiles. Since the two events vary widely in duration, source size, source height, etc., they provide a stringent test for examining the validity of our model of particle acceleration in solar flares in terms of the coalescence instability.

A. Explosive Coalescence — the 1980 June 7 Flare

The 1980 June 7 solar flare at 0312 UT (Fig. 14.18) has been investigated by many authors.[42–45,51] Some essential points from these observations follow:

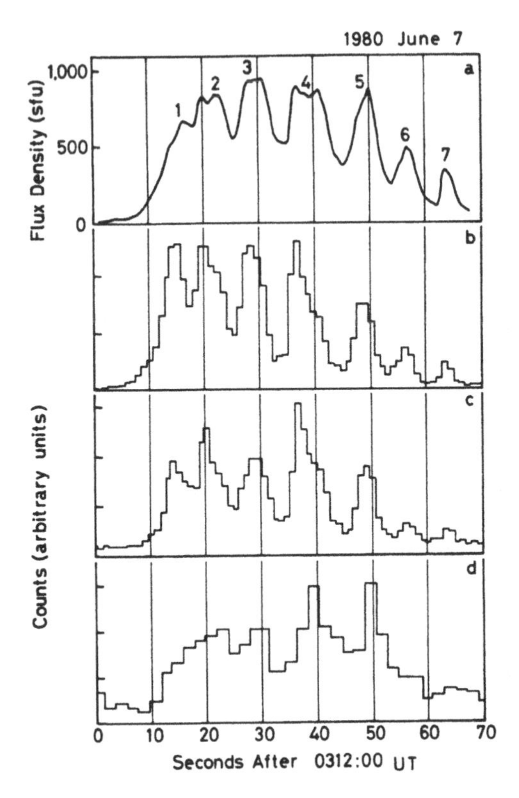

FIGURE 14.18 Electromagnetic signals observed from the 1980 June 7 flare (Ref. 24)

1. The flare was composed of seven successive pulses with a quasi-periodicity of about 8 seconds. Each of the pulses in hard X-rays, prompt γ-ray lines, and microwaves was almost synchronous and similar in shape.
2. Several microwave pulses consisted of double sub-peaks as vividly seen especially in the second and fourth pulses in Fig. 14.18(a). The first sub-peak coincided with the peak of the corresponding hard X-ray pulse [Fig. 14.18(b) and (c)], while the second sub-peak coincides with the peak of the corresponding γ-ray pulse [Fig. 14.18(d)] and with the small hump in hard X-ray time profiles.
3. The starting times of hard X-rays, prompt γ-ray lines, and microwaves coincide within ± 2.2 seconds. Therefore, the acceleration of electrons (up to several MeV) and ions (up to several tens of MeV/nucleon) must have

begun almost simultaneously. The time scales of the accelerations are less than 4 seconds.

4. The height of the microwave source was estimated to within 10 arcsec above the photosphere (Hα flare: N12°, W64°). The source had a small size of less than 5 arcsec in the east-west direction and showed no motion.
5. According to the Hα photographs taken at the Peking Observatory, the flaring region had two loops or two arcades of loops that appear to be in contact with each other, one stretching in the east-west direction and the other in the north-south.

With all the above-mentioned characteristics of this flare it is hard to argue that the coalescence of two current loops has nothing to do with an essential release mechanism of this flare. Indeed, the observed time history shown in Fig. 14.18 resembles those obtained from computer simulation for the case in which two parallel loops have sufficient electric currents so that they attract each other fast enough (in about one Alfvén transit time).

The simulated time history of the electron temperature is shown in Fig. 14.12. Clearly seen is a quasi-periodic oscillation, the period of which is about one Alfvén transit time ($8\Omega_i^{-1}$). The observed period ($\sim$ 8s) is close to one Alfvén transit time, while a theoretical Alfvén transit time is $\sim$4s.

Figure 14.12 also shows that the electron temperature oscillation is characterized by a prominent double sub-peak structure. The double sub-peaks occur just before and after each peak in the magnetic field intensity. As discussed in Section 14.5, this double sub-peak structure should be directly related to an overshooting phenomenon of the colliding current blobs.

A similar time history with double sub-peak structure is obtained for the kinetic energy of high-energy tail electrons and protons as well as for the proton temperature. The acceleration of the high energy-tail particles is due to a combination of the localized electrostatic acceleration and the magnetic $\mathbf{J} \times \mathbf{B}$ acceleration. If we recall that these processes accompany the local plasma compression/decompression associated with the overshooting, the surprising resemblance of the time profile of the microwave emissions caused by high-energy tail electrons [Fig. 14.18(a)] with that of Fig. 14.12(a) may be explained.

We discuss next the energy spectra of electrons and protons, after the explosive coalescence of two current loops. The energy spectrum obtained from the simulation[24] consists of three components: (a) a background thermal component due to the adiabatic heating, (b) an intermediate component due to inductive electric field, (c) a high energy component due to $v_{ph} \times B$ acceleration. In order to compare the spectra from the simulation with those observed, the spectra in simulation are shown in Fig. 14.19. For electrons, as is seen in Fig. 14.19(a), the intermediate non-thermal component has a power-law index of $\gamma \simeq 2$ near the peak, while near the valley the spectrum becomes softer. The global structure of the electron energy spectrum appears

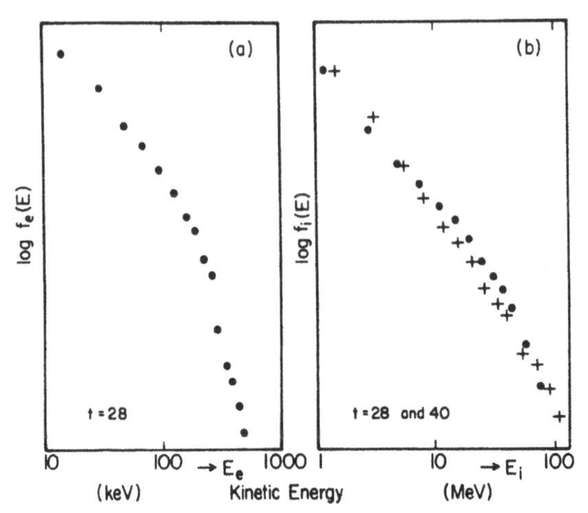

FIGURE 14.19 High energy spectra of electrons (a) and protons (b) from simulation (Ref. 24)

to be consistent with the observations.[42,44]

The spectrum of ions is shown in Fig. 14.19(b) for two instances. The one is just after the coalescence ($T = 28\Omega_i^{-1}$ with crosses) and the other is at $T = 40\Omega_i^{-1}$ (with dots). As seen in the figure, the spectrum becomes harder with increasing energy up to several tens of MeV. The energy spectra of the intermediate and high energy components of protons appears to be consistent with the energy spectrum of observed γ-rays.

B. Slow Coalescence — the 1982 November 26 Flare

We briefly outline the characteristics of the 1982 November 26 flare (Fig. 14.20). This event had a much longer duration than the event on 1980 June 7, about 20 minutes as opposed to about 1 minute. The microwave observations were made with the 17-GHz interferometer at Nobeyama, Japan, and the hard X-ray observation with the Hard X-ray Burst Spectrometer (HXRBS) on SMM. The main characteristics of this flare were:

1. The microwave burst was composed of three successive peaks with a quasi-periodicity of about 6 minutes as indicated by numbers 1-3 in Fig. 14.20(a).
2. Each of the microwave peaks consisted of two sub-peaks.
3. The microwave and hard X-ray emissions started almost simultaneously (within 10 seconds).
4. The microwave source was composed of two sources, one at a height of $\sim 10^4$ km above the photosphere and the other at $\sim 3 \times 10^4$ km.

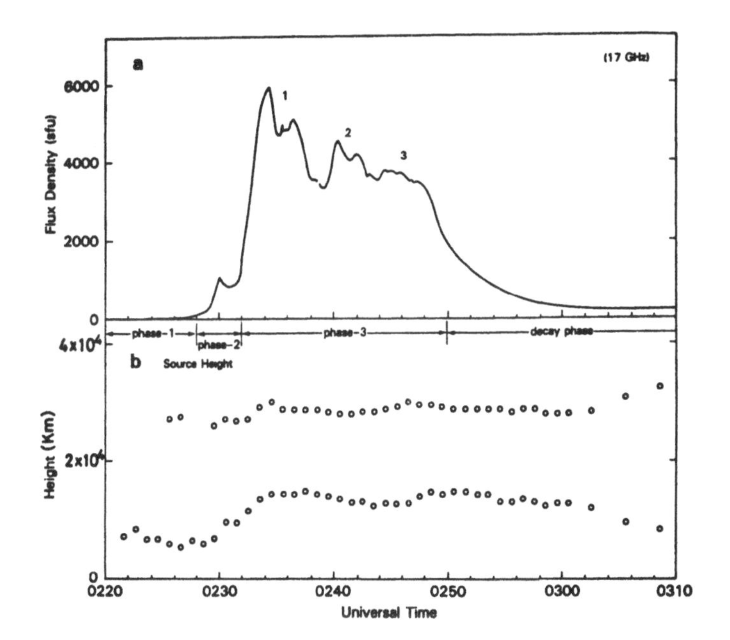

FIGURE 14.20 Microwave signals from the 1982 November 26 flare (Ref. 24)

Figure 14.20(b) shows the height of the two microwave sources as a function of time. In the pre-burst phase (phase 1: 0220-0228 UT), the upper source appeared at a height of $\sim 2.9 \times 10^4$ km above the photosphere and the lower one at $\sim 0.7 \times 10^4$ km. In phase 2, the lower source rose at a velocity of $\sim 30\ \mathrm{km\ s^{-1}}$. The main phase (phase 3) started when the lower source reached a height of $\sim 1.5 \times 10^4$ km implying that the two loops collided with each other at this time. In fact, a small up-and-down motion of the lower source was observed in the main phase. The oscillation period and its peak-to-peak amplitude of the up-and-down motion were ~ 14 min and $\sim 2 \times 10^3$ km (significantly larger than the observational error) respectively. After the main phase, the lower source began to go down towards its previous position. On the other hand, the upper source rose gradually, though it remained at almost the same height until the decay phase started (see Fig. 14.20).

The observational facts summarized above, especially the collision of the two microwave sources and the small up-and-down motion of the lower source, imply that the current-loop coalescence takes place. However, the slow variation together with the small peak-to-valley ratio of the time profile requires a slightly different physical condition of the coalescence than that of the 1980 June 7 event. When two parallel loops have insufficient electric currents or are not well separated and hence the attracting force between them is weaker than that of the previous case, reconnection of poloidal magnetic fields during loop coalescence becomes slower.

The observed plasma kinetic energy oscillation in Fig. 14.14(a) in Section 14.5 exhibits a structure similar to the microwave time profile of the 1982 November 26 flare as shown in Fig. 14.20(a). The source size of the November 26 flare is about 10 times larger than that of the June 7 flare. The calculated period of 5 Alfvén time of the oscillation in Fig. 14.14(a) corresponds to 200 sec. This period is close to the observed period of about 6 minutes. Note also that in this case the flow velocity is much below the Alfvén velocity in agreement with the observational fact that the 40 km/s colliding velocity of the lower loop is much smaller than the Alfvén velocity of $\sim 10^3$ km/s.

The current loop coalescence in Section 14.5 seems to provide a vital key to understanding several important features of solar flares as follows: (i) the sudden explosive development in time profiles, (ii) simultaneous acceleration of electrons and protons up to several times their rest mass energy, (iii) plasma heating due to adiabatic compression up to several times $10^{7\circ}$K, and (iv) quasi-periodic pulses of microwave, X-ray, and γ-ray radiations and also double-peak structure in each pulse. With regard to (ii), it is noteworthy that the energy spectrum of accelerated electrons is consistent with observations. Probably in impulsive flares, such as the 1980 June 7 flare, compact current loops develop and interact with each other in the lower corona, while in gradual flares such as the 1982 November 26 flare, large current loops which reach several 10^4 km in height, develop and interact with each other.

14.7 Reconnection-Driven Oscillations in Dwarf Nova Disks

Plasmas of astrophysical disks of various kinds are believed to be immersed with magnetic fields with differential Keplerian rotation. An example is a disk associated with dwarf novae. A common phenomenon observed in dwarf novae systems is the presence of quasi-periodic oscillations (QPO's). These oscillations are usually observed during the outburst period,[52,53] but may also be present during the quiescence period. QPO's typically have coherence times of a few cycles (2 to ~15) and are characterized by a broad peak in their power spectrum. The periods of the oscillations imply a rotational origin of the disk of a dwarf nova or vertical oscillations of the disk (the two frequencies are comparable for non-self-gravitating thin disks). Some observed periods are 31.5s (SS Cyg), ~ 50s (RU Peg), 70-146s (U Gem), 82-147s (KT Per), 23-253s (VW-Hyi), and 351s (GK Per). QPO's are also seen in soft x-rays in SS Cyg.[64] These latter oscillations have a period of ~ 9s but are otherwise very similar to the optical QPO's; i.e., phase incoherence with period stability.

Reconnecting magnetic fields may drive oscillations at preferred radii.[54] The reconnection in the present model occurs roughly once every rotational period, although not at the same azimuth. The energy released in each reconnection episode can drive local oscillations through the formation of magnetic islands. The oscillations damp through whatever viscosity source is present, and through the centrifugal action of the sheared disk. Phase incoherency in this model is naturally provided by the stochasticity of the reconnection events. Period stability arises from the reconnection occurring at a fairly definite radius.

Three components contribute to the magnetic fields in the accretion disk. The first of these is the dwarf nova, which may have magnetic fields of a rotating dipole. The second consists of external magnetic fields, including magnetic fields of a possible accompanying binary and interstellar magnetic fields. The third is the spontaneous generation of magnetic fields in the differentially rotating disk.

The following three elements play decisive roles in determining magnetic fields in the disk. First, the accretion disk traps (freezes in) the magnetic fields, which are to be churned by the differential rotation of the disk. Second, the dipole-like magnetic fields originating from the dwarf nova are rotating with the stellar angular velocity and are immersed in the disk plasmas as well as in the plasma wind from the nova. Third, the nova emanates the plasma wind, which possibly includes well collimated jets. It is noted that the wind that leaves the surface of the nova with a supersonic speed escapes the gravity field, so that the wind angular velocity $\omega_w(r)$ is primarily determined by its initial angular velocity and the angular momentum conservation, and that $\omega_w(r) \propto r^{-2}$. On the other hand, the matter in the disk executes the Keplerian motion with angular velocity ω_d and obeys the Kepler's law $\omega_d(r) \propto r^{-3/2}$. Therefore, the plasma wind spirals away from the dwarf nova with much less angular velocity than that of the disk. Thus, in general, the angular velocity of the dwarf nova (and its dipole field), that of the disk and that of its wind are all different. Typically, they are in this order.

To the first and crudest approximation, it is expected that the magnetic fields in the disk, in the dwarf nova, and in the wind, are frozen in. Since the emanating dipole magnetic fields from the dwarf nova are rapidly rotating and they end up immersed in the wind and the disk plasmas, the fields become spirals by rapidly twisting, as is seen in the solar magnetic fields and the galactic magnetic fields. The magnetic fields emanate from the rapidly spinning nova and are frozen in the disk that is spinning less rapidly than the star, but yet more so than the wind. The magnetic field lines in the disk are strongly twisted in the azimuthal direction and become primarily oriented in that direction. The other important feature that characterizes the magnetic fields in the disk is the differential rotation. Due to this, the basically azimuthal fields are churned in the disk to form a more complex structure.

In the following model a simpler uniform magnetic field immersed in the

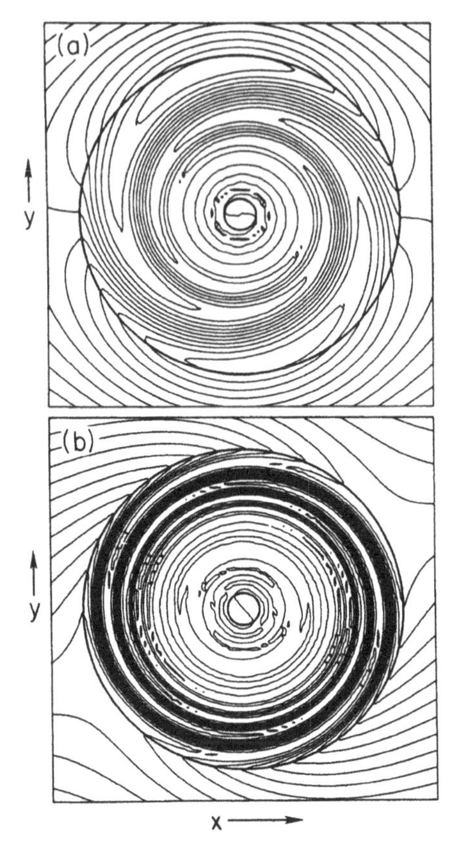

FIGURE 14.21 Magnetic field lines in differentially rotating Keplerian disks (a) $R_m = 1000$ and (b) $R_m = 10,000$ (Ref. 54)

disk is implemented. The magnetic field geometry in an accretion disk is determined by the source and by the shearing motion in the disk. Consider two heuristic problems: a uniform field in the presence of (i) a rigid, rotating conductor; and (ii) a conductor in Keplerian differential rotation. These idealizations are deficient in two important respects. First, the field is resolved only in two dimensions; i.e., only the in-disk components are considered. Second, the field is uniform at infinity and is eventually mostly expelled from the disk. The fields that are important in real accretion disks are anchored in the compact star and are continuously replenished. Given these two major assumptions, it is possible to solve the posed problems – exactly for the rigid conductor and numerically in the differentially rotating conductor.

The problem of a rigid, rotating cylindrical conductor immersed in a uniform magnetic field oriented orthogonally to the axis of the cylinder has been solved exactly.[55]The problem solved is the motion of a conductor with finite resistivity in a field which is initially everywhere uniform. The displacement current and the magnetic fields generated by free charges are ignored. In this approximation the induction equation is given by Eq. (6.20), where $\mathbf{v} = \boldsymbol{\omega} \times \mathbf{r}$ (if one neglects the response motion of the disk to the magnetic fields) is the local velocity, and $\boldsymbol{\omega}$ is the rotation frequency.

We analyze the induction equation where the rotation law is Keplerian,[54] i.e., $\omega \sim r^{-3/2}$. The initial condition of a uniform field threading the disk was imposed, but since the rotation law is singular at the origin, it was necessary to provide a special prescription near the center. It is chosen to enforce rigid rotation at $r_1 = 0.1b$ with $\omega(r < r_1) = \omega(r_1)$.

The Keplerian rotation is not essentially different from the rigid rotation solved by Parker.[55] There are at most two islands present at any one time and there is a final epoch of island formation followed by the onset of a steady state. The Keplerian rotator has a magnetic Reynolds number R_m that decreases inwards with radius, since $R_m \sim r^2\omega \sim r^{1/2}$. Unlike the rigid rotation, the inner portion of the disk comes into steady state before the outer portions; the time for steady state is proportional to a positive power of R_m. Consequently a Keplerian disk that has the same value of R_m at its outer boundary as the corresponding rigid conductor will reach steady state earlier.

In Fig. 14.21 we show the magnetic field lines in the disk for $R_m = 1000$ (a) and 10^4 (b). The field structure is similar to that discussed by Sofue at al.[56] In Fig. 14.22 we show the loci of $|B| = 0$ in the Keplerian rotation at $R_m = 10^4$. The slow progression of the islands inwards to the center is evident.

Reconnection is an inevitable consequence of the wrap-up and shearing of magnetic field lines in an accretion disk. Determining the exact topology at the field will require detailed three-dimensional magnetohydrodynamic (MHD) simulations, and then it will be possible only to compute for a relatively small R_m. Several general features of the reconnection, however, are suggested by this simulation. Reconnection will occur every period with the formation of a few islands even in the regime of tight wrapping. The magnetic field will probably be organized into bands of flux. Reconnection events naturally lead to the formation of plasma blobs in the islands and to the formation of shocks.[6] Blobs of plasma have already been implicated as being the underlying structure in dwarf novae oscillations. Amplitude fluctuations and phase incoherency are not unexpected in reconnection driven oscillations, although detailed calculations must be performed if this is to be demonstrated conclusively. If the field source is stochastically generated by a dynamo in the compact star, and then wrapped up by the disk, amplitude and phase incoherency would be natural outcomes.

We have considered several examples of computational studies of magnetic field line reconnection. This is a phenomenon which involves a delicate mixture of two physics, the (ideal) MHD dynamics and the resistive (or dissipative) process. The spatial characteristics of this problem is that it presents

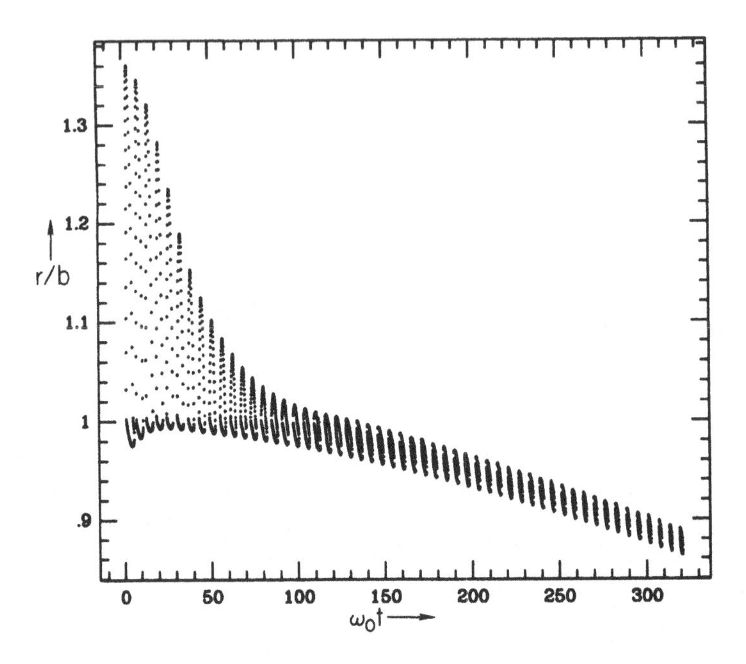

FIGURE 14.22 Radial position of loci of neutral points in the disk plasma (Ref. 54)

boundary layer where a rapid variation of fields and fluid velocity takes place. Its temporal characteristics are a mixture of the above two time scales, the emergence of a new time scale that is in between the two. Because of this complex physics in time and space, the physics of reconnection will continue to attract the attention of plasma physicists, particularly that of computational plasma physicists.

Problems

1. Compare Eq. (14.32) with Galeev's ion tearing instability.[57]

2. Discuss the behavior of solution when the boundary conditions at $x - L$ and $x = -L$ are not identical with respect to y in the forced reconnection calculation Eq. (14.27).

3. From Eqs. (14.56)-(14.59) and (14.46) derive: (i) $B_0(t) = B_{00}/a^2$ where

B_{00} is constant; (ii) $\partial v_{iz}^{(0)}/\partial t = eE_{z0}/m_i$ and $v_{iz}^{(0)} = -m_e v_{ez}^{(0)}/m_i$; (iii) Eqs. (14.60) and (14.61).

4. Derive Eq. (14.63).
5. Derive equations of motion for a and b when the quasineutrality is not satisfied and m_e/m_i are not negligible.
6. Instead of a loop, consider an arcade of magnetic fields whose feet are sheared by the photospheric motion.[58] How much energy is accumulated, when and how fast is it released, what are the magnetic topologies during and after the energy accumulation phase?
7. In the force-free twist of a loop Taylor's states[59] have discrete energy levels, resembling hydrogen atom's energy levels. On the other hand, in the force-free shear of an arcade Taylor's states have continuous energy levels. Specify these energy levels and discuss the reason for discrete and continuous levels.
8. Discuss the process of coalescence of cosmic strings.[60]

References

1. L.D. Landau and E.M. Lifshitz, *Electrodynamics of Continuous Media* translated by J.B. Sykes and J.S. Bell; (Pergamon, New York, 1960) Chapter 7.
2. H. Alfvén and C.G. Fälthammer, *Cosmical Electrodynamics* (Oxford University Press, Oxford, 1963) p. 80.
3. H.P. Furth, J. Killeen, and M.N. Rosenbluth, Phys. Fluids **6**, 459 (1963).
4. P.A. Sweet, in *Electromagnetic Phenomena in Cosmic Physics* (Cambridge University Press, London, 1958), p. 123.
5. E.N. Parker, J. Geophys. Res. **62**, 509 (1957).
6. H.E. Petschek, *Proc. AAS-NASA Symposium on Physics of Solar Flares* (NASA SP-50, Washington, D.C., 1964) p. 425.
7. R.B. White, in *Handbook of Plasma Physics*, ed. by M.N. Rosenbluth and R.Z. Sagdeev (North-Holland, Amsterdam, 1983) vol. 1, R.B. White, Rev. Mod. Phys. **58**, 183 (1986).
8. J.F. Drake and Y.C. Lee, Phys. Fluids **20**, 1341 (1977).
9. I. Katanuma and T. Kamimura, Phys. Fluids **38**, 2500 (1980).
10. H. Naito, S. Tokuda and T. Kamimura, J. Comp. Phys. **33**, 86 (1979).
11. D.O. Dickman, R.L. Morse, and C.W. Nielson, Phys. Fluids **12**, 1708 (1969).

12. T. Kamimura, E. Montalvo, D.C. Barnes, J.N. Leboeuf, and T. Tajima, submitted to J. Comp. Phys.

13. R.B. White, J. Comp. Phys. **31**, 409 (1983).

14. P.H. Rutherford, Phys. Fluids **16**, 1903 (1973).

15. C.K. Birdsall, A.B. Langdon, and H. Okuda, in *Methods of Computational Physics*, ed. by B. Alders, S. Fernbacher and M. Rotenberg (Academic, New York, 1970) vol. 9, p. 241.

16. W. Horton and T. Tajima, J. Geophys. Res. **93**, 2741 (1988).

17. F. Brunel, T. Tajima, J.N. Leboeuf, and J.M. Dawson, Phys. Rev. Lett. **44**, 1494 (1980); F. Brunel and T. Tajima, Phys. Fluids **26**, 535 (1983).

18. S.I. Syrovatskii, Zh. Eksp.Teor. Fiz. **60**, 1727 (1971) [Sov. Phys. JETP **33**, 933 (1971)].

19. F. Brunel, T. Tajima, and J.M. Dawson, Phys. Rev. Lett. **49**, 323 (1982).

20. T. Sato and T. Hayashi, Phys. Fluids **22**, 1189 (1979).

21. W. Park, Bull. Am. Phys. Soc. **26**, 845 (1981).

22. V.M. Fadeev, I.F. Kvartskhava, and N.N. Komarov, Nucl. Fusion **5**, 202 (1965).

23. A. Bhattacharjee, F. Brunel and T. Tajima, Phys. Fluids **26**, 3332 (1983).

24. T. Tajima, F. Brunel, and J-I. Sakai, Ap. J. **245**, L45 (1982); T. Tajima, J.-I. Sakai, H. Nakajima, T. Kosugi, F. Brunel, and M.R. Kundu, Ap. J. **321** 1031 (1987).

25. J.N. Leboeuf, T. Tajima, and J.M. Dawson, Phys. Fluids **25**, 784 (1982).

26. F. Brunel, J.N. Leboeuf, T. Tajima, J.M. Dawson, M. Makino, and T. Kamimura, J. Comp. Phys. **43**, 268 (1981).

27. R.Z. Sagdeev and V.D. Shapiro, JETP Lett. **17**, 279 (1973).

28. J.M. Dawson, V.K. Decyk, R.W. Huff, I. Jaeckert, T. Katsouleas, J.N. Leboeuf, B. Lemberge, R.M. Martinez, Y. Ohsawa, and S.T. Ratliff, Phys. Rev. Lett. **50**, 1455 (1983).

29. L.P. Kadanoff, W. Götze, D. Hamblen, R. Hecht, E.A.S. Lewis, V.V. Palcianskas, M. Rayl, J. Swift, D. Aspnes, and J. Kane, Rev. Mod. Phys. **39**, 395 (1967).

30. A.N. Kolmogorov, Comput. Ren. Acad. Sci. USSR **30**, 301 (1941) [Sov. Phys. - Uspekhi **10**, 734 (1968)].

31. B. Carreras, M.N. Rosenbluth, and R. Hicks, Phys. Rev. Lett. **46**, 1131 (1981).

32. P.H. Diamond, R.D. Hazeltine, Z.A. An, B.A. Carreras, and H.R. Hicks, Phys. Fluids **27**, 1449 (1984).

33. V.E. Zakharov, Sov. Phys. JETP **35**, 968 (1972).

34. P. Dutta and P.M. Horn, Rev. Mod. Phys. **53**, 497 (1981).

35. K. Tomita in *Chaos and Statistical Methods* ed. by Y. Kuramoto (Springer, Berlin, 1984) p. 10.

36. H.A. Rose and P.L. Sulem, J. Phys. (Paris) **39**, 441 (1978).

37. R.C. Elphic, C.A. Cathell, S.J. Bame, and C.T. Russell, Geophys. Res. Lett. **13**, 648 (1986); J.A. Slavin, E.J. Smith, B.T. Tsurutani, D.G. Sibeck, D.N. Baker, J.T. Gosling, E.W. Hones, and F.L. Scarf, Geophys. Res. Lett. **11**, 657 (1984).

38. Z. Svestka, *Solar Flares* (Reidel, Dordrecht, 1976).

39. P.A. Sturrock, ed. *Solar Flares: Monograph from Skylab Solar Workshop II* (Colorado Associated Univ. Press, Boulder, 1980).

40. E.R. Priest, *Solar Magnetohydrodynamics* (Reidel, Dordrecht, 1982).

41. E.L. Chupp, Ann. Rev. Astron. Astrophys. **22**, 359 (1984).

42. S.R. Kane, K. Kai, T. Kosugi, S. Enome, P.B. Landecker, and D.L. McKenzie, Ap. J. **271**, 376 (1983).

43. D.J. Forrest and E.L. Chupp, Nature **305**, 291 (1983).

44. A.L. Kiplinger, B.R. Dennis, K.J. Frost, and L.E. Orwig, Ap. J. **273** 783 (1983).

45. H. Nakajima, T. Kosugi, and S. Enome, Nature **305** 292 (1983).

46. R. Howard and Z. Svestka, Solar Phys. **54**, 65 (1977).

47. M.R. Kundu, E.J. Schmahl, T. Velusamy, and L. Vlahos, Astron. Astrophys. **108**, 188 (1982).

48. R.F. Wilson and K.R. Lang, Ap. J. **279**, 427 (1984).

49. H. Nakajima, T. Tajima, F. Brunel, and J. Sakai, In *Proc. Course and Workshop on Plasma Astrophysics* (Varenna, Italy, ESA, 1984) p. 193.

50. M.E. Machado, A.M. Hernandez, M.G. Rovira, C.V. Sneibrun, Adv. Space Res. Vol. 4, No. 7, 91 (1984).

51. D.J. Forrest, E.L. Chupp, C. Rippin, E. Rieger, J.M. Ryan, et al., in *Proc. 17th Int. Cosmic Ray Conf.* (Centre de'Etudes Nucleares de Saclay, Paris, 1981) Vol. 10, p. 5.

52. J. Patterson, E.L. Robinson, and R.E. Nather, Ap. J. **214**, 144 (1977).

53. E.L. Robinson and B. Warner, Ap. J. **277**, 250 (1984).

54. T. Tajima and D. Gilden, Ap. J. **320**, 741 (1987).

55. R.L. Parker, Proc. Roy. Soc. **A291**, 60 (1966).

56. Y. Sofue, M. Fujimoto, and R. Widebinski, Ann. Rev. Astron. Ap. **24**, 459 (1986).

57. A.A. Galeev, Space Sci. Rev. **23**, 411 (1979).

58. N. Bekki, T. Tajima, J. Van Dam, Z. Mikic, D.C. Barnes, and D.D. Schnack, to be published in Phys. Rev. Lett.

59. J.B. Taylor, Phys. Rev. Lett. **33**, 1139 (1974).

60. A. Vilenkin, Phys. Rep. **121**, 265 (1985); T.W. B. Kibble, Phys. Rep. **67**, 183 (1980).

15

TRANSPORT

The slowest time scale of a many-body system such as a plasma is generally the transport time scale. If we coarse-grain a plasma over the fast time scales, such as the electron time scales (gyro period, plasma period), the ion time scales (ion plasma period, ion gyro period, ion acoustic period), the magnetohydrodynamic time scales (fast Alfvén period, shear Alfvén period, the resistive MHD time scales, and the slow kinetic instability time scales (drift wave period etc.), we are left with a nearly stationary plasma which evolves slowly, due to collisions and diffusion. Nearly all wave phenomena are smoothed out and the slow collisional or diffusive processes determine the eventual slow evolution of the plasma state, such as the confinement of the plasma. The transport process is naturally described by the Boltzmann equation

$$\frac{\partial f_a}{\partial t} + \mathbf{v} \cdot \frac{\partial f_a}{\partial \mathbf{r}} + \frac{\mathbf{F}}{m_a} \cdot \frac{\partial f_a}{\partial \mathbf{v}} = \left(\frac{\partial f_a}{\partial t}\right)_c + S_a + L_a \ ,$$

where the subscript a refers to the species, S and L are the source and sink of particles, and $(\partial f_a/\partial t)_c$ is the collision term. From kinetic theory (see Chapter 1) Landau,[1] and Balescu-Lenard[2] derived the collision term Eq. (1.35).

The transport phenomena are complex, however. Often transport studies cannot be isolated in the transport time scale alone. These phenomena arise from stochastic processes of the system or from deterministic processes of the system, where an example of a stochastic process is Brownian motion, and an example of a deterministic process is the orbit of a planet around the sun. It has been recognized, however, that more and more systems that have been considered deterministic are actually stochastic if they are examined in a long-range time scale. When we consider transport phenomena, certain phenomena are simply stochastic processes, while others are governed by a dynamic process but exhibit essentially stochastic behavior. Such a difference may be schematized in Fig. 15.1. The deterministic processes in dynamical phenomena are common problems that do not belong in transport study. One of the most challenging questions in the transport problem of plasma physics and, in fact, in all fields of nonlinear dynamics is why and how an essentially deterministic system acquires stochasticity. For example, why and how can an essentially collisionless plasma sometimes exhibit a rapid dissipation of its internal energy as observed in solar flares and tokamak disruption (the problem of anomalous disruption)? (see Chapter 14.) Why and how can a collisionless plasma show anomalously fast transport of energy across the magnetic fields (the problem of anomalous transport)? Often interplay occurs among different hierarchies of plasma dynamics corresponding to different time scales (and spatial scales), which leads to such anomalies in transport.

15.1 Monte-Carlo Method

For stochastic phenomena, the Monte Carlo method is the most commonly employed method. This method was developed by Los Alamos scientists in the 1940's, including J. von Neumann, S. Ulam, R. Richtmyer, E. Fermi, and N. Metropolis. An excellent historical perspective is found in the recent issue of Los Alamos Science.[3] A famous paper was published in 1953.[4] Some recent developments in condensed matter physics may be found in Ref. 5. Some of those in particle physics may be found in Ref. 6. The basic idea is to evaluate a weighted integral

$$I = \int_0^1 f(x)p(x)dx \ , \tag{15.1}$$

by generating a set of random numbers ξ_i. Here $p(x)$ is the weighting probability function and $f(x)$ is an arbitrary function that is weighted. If $\{\xi_i\}$ are a uniform series of N random numbers generated in [0,1], $p(x) = 1$ if $x \epsilon [0, 1]$ and 0 otherwise. The expectation value of $f(\xi)$ is

$$I = E[f(\xi)] = \int_0^1 f(x)p(x)dx \cong \frac{1}{N}\sum_{i=1}^{N} f(\xi_i) \ . \tag{15.2}$$

We expect from "the large number theorem" that the approximation in Eq. (15.2) becomes progressively better with the statistical uncertainty proportional to $1/\sqrt{N}$, as N is increased. When the weighting function $p(x)$ is an

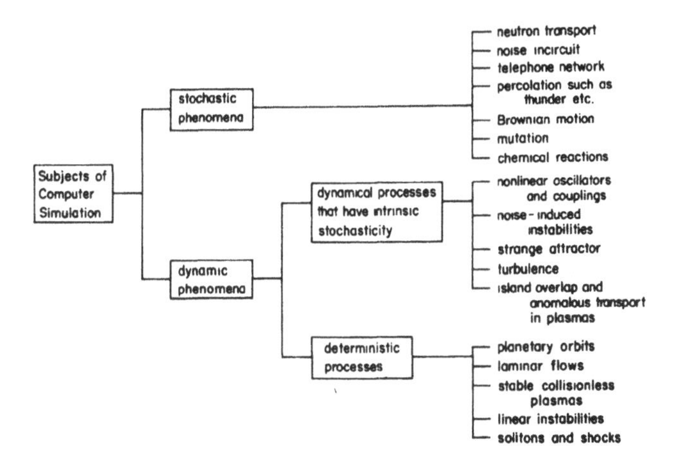

FIGURE 15.1 Some examples of subjects of computer simulation both stochastic and dynamical

exponential function, it often appears in statistical physics (partition functions) and particle physics (Feynman path integrals). In plasma physics, a Gaussian function and an exponential function often appears. We leave the reader to explore this powerful method for himself, as this method is not a dynamical method and out of the scope of this book.

15.2 Fokker-Planck Model

A standard model of solving the transport problem is the Fokker-Planck equation. The method of solution is either solving the phase space distribution itself or solving for the moment equations of the distribution function. Here we discuss the former.

It can be shown that under certain conditions the type of kinetic theoretical approach is equivalent to the Brownian motion approach,[7] in which a test particle feels fluctuating disturbance and executes a random relaxation process. The behavior of the Brownian motion may be described by the Fokker-Planck equation.[8] The details of these two approaches and relations may be found in Refs. 9 and 10. A Fokker-Planck equation is often truncated at the second order of a phase space step $\Delta X \equiv \Delta\mathbf{x}\Delta\mathbf{v}$. For the problem of velocity space transport (i.e., spatially uniform plasma) the Fokker-Planck equation can be written as

$$\left(\frac{\partial f_a}{\partial t}\right)_c = -\Gamma_a \frac{\partial}{\partial \mathbf{v}} \cdot \left(f_a \frac{\partial h_a}{\partial \mathbf{v}}\right) + \frac{\Gamma_a}{2} \frac{\partial^2}{\partial \mathbf{v} \partial \mathbf{v}} : \left(f_a \frac{\partial^2 g_a}{\partial \mathbf{v} \partial \mathbf{v}}\right) , \tag{15.3}$$

where $\Gamma_a = 4\pi Z_a^2 e^4 / m_a^2$,

$$g_a = \sum \left(\frac{Z_b}{Z_a}\right)^2 \ell n \Lambda_{ab} \int f_a(v') |\mathbf{v} - \mathbf{v}'| dv' , \tag{15.4}$$

$$h_a = \sum_b \frac{m_a + m_b}{m_b} \left(\frac{Z_b}{Z_a}\right)^2 \ell n \Lambda_{ab} \int f_b(v') \frac{1}{|\mathbf{v} - \mathbf{v}'|} d\mathbf{v}' . \tag{15.5}$$

A Fokker-Planck equation for physical space coordinates can be treated in a similar fashion. Equations (15.4) and (15.5) are sometimes called the Rosenbluth potentials. Here[11]

$$\ell n \Lambda_{ab} = \ell n \left\{ \left(\frac{m_a m_b}{m_a + m_b}\right) \frac{2\alpha c \lambda_D}{e^2} \max \left(\frac{2E}{m}\right)_{a,b}^{1/2} \right\} - \frac{1}{2} , \tag{15.6}$$

with $\alpha = e^2/\hbar c$, λ_D the Debye length, E the mean energy of particles of species a or b on the normalization is such that $n_a = \int f_a(\mathbf{v}) d\mathbf{v}$.

Since the collision term contains velocity derivatives of f_a multiplied by velocity moments over f_a, Eq. (15.3) is a nonlinear, partial integro-differential equation in seven independent variables. We choose a spherical coordinate system for velocity space (with $\theta = 0$ corresponding to the direction along a magnetic field line) and a cylindrical coordinate system for physical space (z along the magnetic axis).

Let us discuss a typical numerical model.[11] Here we assume that the "Rosenbluth potentials," given by Eqs. (15.4) and (15.5), are isotropic, i.e.,

$$\frac{\partial g_a}{\partial \mu} = \frac{\partial h_a}{\partial \mu} = 0 . \tag{15.7}$$

With this assumption, Eq. (15.3) becomes

$$\frac{1}{\Gamma_a} \frac{\partial f_a}{\partial t} = -\frac{1}{v^2} \frac{\partial}{\partial v} \left(f_a v^2 \frac{\partial h_a}{\partial v}\right) + \frac{1}{2v^2} \frac{\partial^2}{\partial v^2} \left(f_a v^2 \frac{\partial^2 g_a}{\partial v^2}\right)$$

$$+ \frac{1}{2v^3} \frac{\Gamma g_a}{\partial v} \left[(1 - \mu^2) \frac{\partial^2 f_a}{\partial f_a} \partial \mu^2 - 4\mu \frac{\partial f_a}{\partial \mu} - 2 f_a\right]$$

$$- \frac{1}{v^2} \frac{\partial}{\partial v} \left(f_a \frac{\partial g_a}{\partial v}\right) + \frac{1}{v^3} \frac{\partial g_a}{\partial v} \left(\mu \frac{\partial f_a}{\partial \mu} + f_a\right) . \tag{15.8}$$

Equation (15.8) is separable, and if we let

$$f_a(v,\mu,t) = U_a(v,t)M_a(\mu) \ , \tag{15.9}$$

then Eq. (15.8) yields an eigenvalue problem for $M_a(\mu)$

$$(1-\mu^2)\frac{d^2M_a}{d\mu^2} - 2\mu\frac{dM_a}{d\mu} + \Lambda_a M_a = 0 \ , \tag{15.10}$$

which is Legendre's equation. In the full velocity space we have $-1 \le \mu \le 1$; hence the solutions of Eq. (15.10) are Legendre polynomials, and the lowest mode corresponds to an isotropic distribution function. In this case Eq. (15.10) is solved numerically as an eigenvalue problem with $M_a(\mu) = 0$ at the end-points. For each eigenvalue Λ_a of Eq. (15.10) we have an equation of the form

$$\frac{1}{\Gamma_a}\frac{\partial f_a}{\partial t} = -\frac{1}{v^2}\frac{\partial}{\partial v}\left[f_a\left(v^2\frac{\partial h_a}{\partial v}\right)\right] + \frac{1}{2v^2}\frac{\partial^2}{\partial v^2}\left(v^2 f_a \frac{\partial^2 g_a}{\partial v^2}\right)$$
$$-\frac{\Lambda_a}{2v^3}\frac{\partial g_a}{\partial v} f_a \ . \tag{15.11}$$

We consider only the lowest eigenvalue Λ_a. Equation (15.11) can be cast in the form

$$\frac{\partial f_a}{\partial t} = \frac{1}{v^2}\frac{\partial}{\partial v}\left[\alpha_a f_a + \beta_a \frac{\partial f_a}{\partial v}\right] - \frac{\gamma_a}{v^2} f_a \ . \tag{15.12}$$

Equation (15.12) is a system of coupled partial differential equations. For each particle species, a, we have an equation corresponding to Λ_a, the eigenvalue of (15.10). Hence, we have one equation for each species. In solving the system given by Eq. (15.12), a vector equation of the form

$$\frac{\partial F}{\partial t} = \frac{1}{v^2}\frac{\partial G}{\partial v} - \frac{C}{v^2}F + D \ , \tag{15.13}$$

is considered where

$$F = \begin{bmatrix} f_1 \\ \vdots \\ f_p \end{bmatrix} \ ; \quad G = AF + B\frac{\partial F}{\partial v} \ , \tag{15.14}$$

and A, B, C are diagonal $p \times p$ matrices and D is the source vector. Without source (and loss) terms, Eq. (15.13) becomes

$$\frac{\partial F}{\partial t} = \frac{1}{v^2}\frac{\partial G}{\partial v} \ . \tag{15.15}$$

This equation is in conservation form (divergence of a flux), which is consistent with the correct boundary conditions.

Equation (15.15) is solved by finite-difference methods. On the domain $0 \leq v \leq v_j$, $t \geq 0$, we have a finite-difference mesh denoted by v_j, $j = 0, \ldots, J$ and t_n, $n = 0, 1, 2, \ldots$. The v spacing is variable and we define $\Delta v_{j+1/2} = v_{j+1} - v_j$, $\Delta v_{j-1/2} = v_j - v_{j-1}$, $\Delta v_j = \frac{1}{2}(v_{j+1} - v_{j-1})$. We approximate Eq. (15.15) by the following implicit difference equation.[11]

$$\frac{F_j^{n+1} - F_j^n}{\Delta t} = \rho \left[\frac{1}{v_j^2} \frac{G_{j+1/2}^{n+1} - G_{j-1/2}^{n+1}}{\Delta v_j} - \frac{C_j^{n+1}}{v_j^2} F_j^{n+1} + D_j^{n+1} \right]$$

$$+(1-\rho) \left[\frac{1}{v_j^2} \frac{G_{j+1/2}^n - G_{j-1/2}^n}{\Delta v_j} - \frac{C_j^n}{v_j^2} F_j^n + D_j^n \right] , \qquad (15.16)$$

where

$$G_{j+1/2}^n = \frac{1}{2} A_{j+1/2}^n (F_{j+1}^n + F_j^n) + B_{j+1/2}^n \frac{F_{j+1}^n - F_j^n}{\Delta v_{j+1/2}} , \qquad (15.17)$$

$$G_{j-1/2}^n = \frac{1}{2} A_{j-1/2}^n (F_j^n + F_{j-1}^n) + B_{j-1/2}^n \frac{F_j^n - F_{j-1}^n}{\Delta v_{j-1/2}} . \qquad (15.18)$$

For numerical stability we must have $\frac{1}{2} \leq \rho \leq 1$; usually $\rho = 1$ is taken. Without source and loss terms, i.e., $C_j^n = D_j^n = 0$ for all j and n, we have

$$\sum_{j=1}^{J-1} \left(\frac{F_j^{n+1} - F_j^n}{\Delta t} \right) v_j^2 \Delta v_j = 0 \qquad (15.19)$$

for all n, independent of the mesh spacing, as long as $G_{1/2}^n = G_{J-1/2}^n = 0$, for all n. This condition is the boundary condition, thus this difference scheme[11] rigorously conserves particle density in the absence of source and loss terms.

In order to solve the difference equations given by Eq. (15.16), we can put it in the linear algebraic system:

$$-\alpha_j^{n+1} F_{j+1}^{n+1} + \beta_j^{n+1} F_j^{n+1} - \gamma_j^{n+1} F_{j-1}^{n+1} = \delta_j^n , \qquad (15.20)$$

where $j = 1, \ldots, J$ and α, β, γ, and δ are defined in Ref. 11. In order to linearize the system of Eq. (15.20), we extrapolate the α, β, γ defined above to the new time step, $n+1$. The method used to solve Eq. (15.20) is the standard tridiagonal method given by Richtmyer and Morton.[12] A generalization of the present method to cases for spatially nonuniform systems is detailed in Ref. 11.

15.3 Particle Transport for Energetic Particles

When the configuration is geometrically complex and/or particles are energetic enough that approximations such as Eqs. (15.3)-(15.5) are not justified, the previous Fokker-Planck method to determine the transport becomes too complicated or sometimes no longer valid. We discuss methods to treat such problems in this section and the next.

The distribution of energetic charged particles generated by fusion reactions such as d$-^3$He reactions $[^3\mathrm{He(d,p)}\alpha]$ in a field reversed configuration[13] (FRC) is an example of this type of situation. The Larmor radius of the fusion particles can be as large as the radius of the plasma, since the toroidal magnetic field is absent in this configuration. A fraction of the charged fusion products escapes directly, while the others are confined to form a directed particle flow parallel to the plasma current.[13] The transport process can be studied by a dynamical simulation of plasma particles. The integration of particle dynamics is carried out a la Chapter 4 and Chapter 10, combined with appropriate initial conditions, which often consist of the Monte Carlo method (see Section 15.1) to generate an appropriate distribution *ab initio*.[4] In this sense this method distinguishes itself from the Monte Carlo method discussed in Section 15.1, in that it resorts to the Monte Carlo method only to determine the initial condition, and the dynamics is determined by the particle simulation method.

There is a fundamental identity between the time integrated solution of an initial value problem and a steady-state problem. Consider the following dynamical kinetic equation

$$\frac{\partial F}{\partial t} + \mathcal{L}F = 0 \ , \tag{15.21}$$

where $\mathcal{L}$ is a kinetic operator independent of time that includes phase space flow (the Vlasov equation), collisions (Fokker-Planck equation) and sinks through the boundary conditions on $\mathcal{L}$. Suppose at $t = 0$ the distribution function is determined by an initial condition $F = S_0(r, v)$ and at $t = \infty$, $F = 0$ (i.e., all particles are lost by the system). Then integrating this equation from $t = 0$ to infinity yields

$$\mathcal{L}G = S_0(r, v) \ , \tag{15.22}$$

where $G = \int_0^\infty F dt$. Note that this time averaged equation is the steady-state kinetic equation for a source $S_0(r, v)$. This observation allows us to calculate by a dynamical simulation method steady-state quantities from an initial set of data. In particular, if $S_0(r, v)$ is the distribution of fusion products created at birth, we can use $S_0(r, v)$ as an initial distribution and calculate the steady-state distribution, as well as any steady-state moment of the distribution.

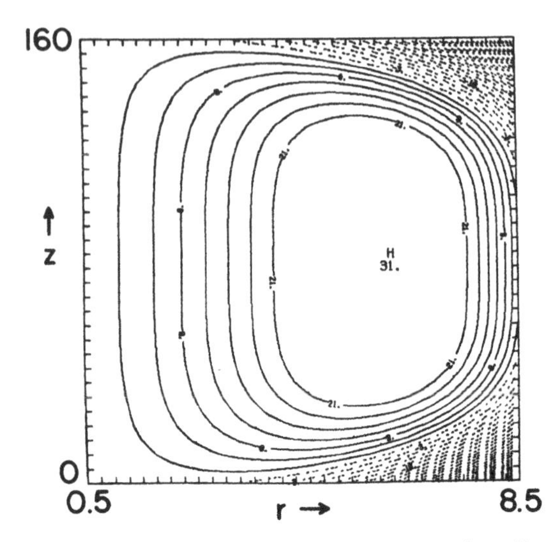

FIGURE 15.2 The poloidal flux contour of the field reversed configuration (Ref. 14)

The particle simulation is performed with a toroidal particle code[15] (see Chapter 10). The particle dynamics is evaluated in the coordinates $r, \theta, z, p_r, p_\theta$ and p_z. An equilibrium for an axially extended constant pressure gradient ($\partial P/\partial\psi \equiv$ pressure, $\psi \equiv$ magnetic flux) is chosen, which is of the form[16]

$$\psi = -\frac{b\tilde{r}^2}{8}\left\{1-\tilde{r}^2-\left(\frac{1}{1+\frac{6}{\epsilon^2}+\frac{4}{\epsilon^4}}\right)\left[\frac{1}{2}\left(\tilde{r}^2-1\right)\left(\tilde{r}^2-2\right) -6\tilde{z}^2\left(\tilde{r}^2-1\right)+4\tilde{z}^4\right]\right\}, \tag{15.23}$$

and the magnetic field $\mathbf{B}$ is given by

$$\mathbf{B} = \hat{\theta} \times \nabla\psi/r$$

where $\tilde{r} = r/a$, $\tilde{z} = z/z_0$, $\epsilon = a/z_0$, $4b$ is the axially magnetic field at the separatrix, a is the radius of the separatrix at $z = 0$. Poloidal flux ψ is defined such that $\psi = 0$ on the separatrix. We assume the plasma is isothermal, so the pressure is of the form

$$P(\psi) = n_0(\psi)T_0 \equiv C\psi \ . \tag{15.24}$$

The equation of motion for individual particles is

$$\frac{d\mathbf{v}}{dt} = \frac{q}{m}\mathbf{v} \times \mathbf{B}(\mathbf{r}) - \nu_s\mathbf{v} \ , \tag{15.25}$$

where ν_s is the slowdown drag rate due to collisions, which is a good approximation if particles we consider are of high energy. Electric field effects are neglected, and we remove particles if they reach the separatrix, beyond which ψ becomes negative. The steady-state current is obtained by accumulating in time

$$J_\theta(r,z) = \frac{1}{T}\int_0^T dt \sum_j v_{\theta_j}(t) = \frac{1}{T}\int_0^T dt v_\theta F(v_\theta, p_\theta, v_z, r, z, t) , \qquad (15.26)$$

where T is the length of simulation. Figure 15.2 shows the flux contours according to Eq. (15.23). Here $2z_0 = 160$ and $a = 8.5$ in the normalized code units. Then the magnetic field is relatively weak so that $\alpha \equiv a\Omega_0/v_0$ (where a is the radius of the separatrix, Ω_0 the Larmor frequency, and v_0 the initial speed) is approximately 5, and particle orbits are largely of the betatron type.[17] The parameter α means the normalized radius of the plasma with respect to the gyroradius as appropriately defined, and an example of which is shown in Fig. 15.3(a), where dots indicate particle position at regular time intervals in (r,z) coordinates. With increasing magnetic fields and α somewhat greater than 10, particle orbits may bifurcate into two radial regions surrounding an inaccessible region. In the three-dimensional problem of the present simulation, we typically observe the situation shown in Fig. 15.3(b). Figures 15.3(a) and (b) are obtained in runs with $\nu_s = 0$. Figure 15.3(c) shows the same particle under the same conditions (but with slowdown collisions).

The following issues are of importance:

(i) Current profile: The radial current profile has a double peak in strong field cases ($\alpha \gtrsim 10$).

(ii) Particle density profile: Although the current may vanish near the middle, the density is high in the central region, indicating that there are equal numbers of positively and negatively moving particles.

(iii) Particle orbits: Orbits that are bifurcated in the strong field case vs. orbits that do not. We investigate the extent in which particles with $p_\theta - av_0 < 0$ can be contained as this condition conflicts with the orbit ergodicity.

(iv) Properties of confined particles: Do confined particles satisfy

$$p_\theta - av_0 > 0 \qquad (\text{when} \quad \nu_s = 0), \qquad \text{and}$$

$$p_\theta(t) - av(t) < 0 \qquad (\text{when} \quad \nu_s \neq 0)?$$

(v) Ergodicity: Are orbits regular, chaotic, or somewhere in between? For a completely chaotic orbit, one obtains a constant density (x,r) profile in the accessible region of phase space, while if KAM (Kolmogorov-Arnold-Mozer)[18] barrier exists, even chaotic orbits do not fall in the $r - z$ density space uniformly. Only regular orbits give rise to curves on the surface of section plots.

Concerning these issues simulation[14] gives the following results:

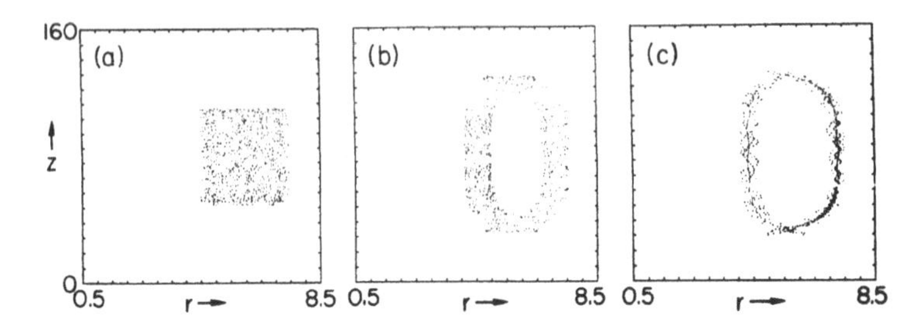

FIGURE 15.3 Surface of section of orbits of confined particles in FRC (Ref. 14)

(i) Current profile: A hollow distribution begins to emerge[14] as α exceeds ~ 10. The high current region encircles the low current (or zero) current region in r, z contours.[14]

(ii) Typical particle orbits are shown in Fig. 15.3.

(iii) Confinement properties: Particles with $p_\theta < 0$ are immediately lost from the system. We find selective confinement of fusion particles. As α becomes smaller, so does the fraction of confined fusion protons. All unconfined particles (for case $\nu_s \neq 0$) satisfy $p_\theta(t) - av(t) < 0$ as they must from dynamical considerations. Not all confined particles satisfy $p - av_0 > 0$ (case $\nu_s = 0$). However, all confined particles satisfy $p_\theta(t) - av(t) > 0$ for $\nu_s \neq 0$. The stochasticity properties of the orbits of confined particles is characterized as: the larger α is, the larger is the fraction of the orbits of confined chaotic particles. In case of $\alpha \cong 16$ a majority of the particles are chaotic, although most particles do not exhibit complete ergodicity in that the phase space density is not uniform even for the cases where α is large.

The stochastic nature of the orbits of fusion produced particles is studied in two distinct statistical approaches. The first method is to construct the surface of section of particle orbits at $z = 0$ and to study the structure of the surface of section appearing in this Lorentz plot.[19] For a given confined particle, the orbit intersects with surface $z = 0$ many times during a computational run. The runs have typically 10^2 intersections through $z = 0$. The radial position of the n-th section r_n and its associated velocity $\dot{r}_n$, or alternatively, r_n and r_{n+1} provide phase space of the orbit at the $z = 0$ section. [The Poincare return map[20] or Lorentz plot[19] (r_n, r_{n+1})]. The map (r_n, r_{n+1}) or the phase space plot $(r_n, \dot{r}_n)$ for various n's provides the map. From these maps one can study the regularity or diffusion structure of the orbits. Figure 15.4 shows an example of the Lorentz map and the phase space map for the same orbit and is an example of a relatively regular orbit.

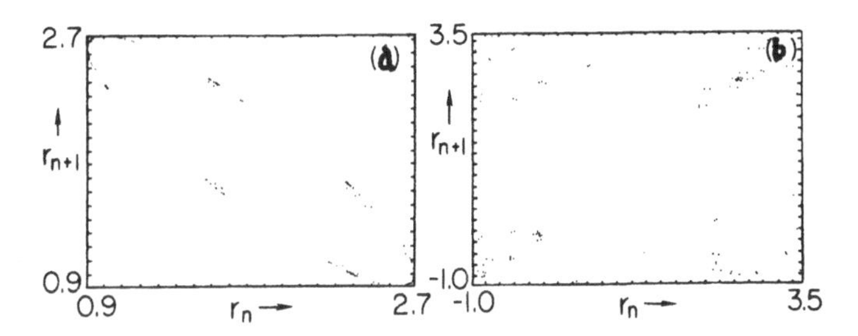

FIGURE 15.4 Lorentz map of particles confined in a field reversed confinement configuration [Regular (a) and stochastic (b) orbits] (Ref. 14)

The dimensionality of the measure of the region covered by the orbit can decide whether orbits are chaotic or regular. As is the case, the Hausdorff dimension[20] of a design D_H is defined as

$$D_H = \lim_{\varepsilon \to 0} \frac{\ell n N(\varepsilon)}{\ell n(1/\varepsilon)}, \tag{15.27}$$

where ε is the width of phase space cube and $N(\varepsilon)$ is the number of cubes covering the region. Alternatively, the index of correlation can also yield the dimension of the region

$$D_I = \ell n \frac{1}{n^2} \sum_{i,j=1}^{n} \theta(r - |\mathbf{X}_i - \mathbf{X}_j|) , \tag{15.28}$$

where the argument of the logarithm is called the correlation integral, θ is the step function, $\mathbf{X}_i$ and $\mathbf{X}_j$ are points in an m-dimensional phase space or vectors in a sample space[21] $\mathbf{X}_i = (x_i, x_{i+p}, \ldots, X_{i+(m-1)p})$, and n is the number of vectors under consideration. Swinney and others utilized such a technique.[21-23] In general, $D_I \leqq D_H$, but if D_H (or D_I) are equal to 1, we call the orbit regular, while if D_H (or D_I) is equal to 2, the chaotic orbit. For the particle simulation Eq. (15.28) may be a more rigorous and practical method for the digitalization needed for determining the domain.

In practice, however, we can qualitatively determine the nature of stochasticity by inspection, which roughly corresponds to D_H. If the attractor is diffuse, the orbit is chaotic; while if the attractor is on or almost on a curve, the orbit is regular. Here an attractor means a phase space region towards which the orbital points asymptotically precipitate in a long range time scale. For the Poincare map, let

$$r_{n+1} = f(r_n) , \tag{15.29}$$

then

$$r_{n+s} = [f(r_n)]^s . \tag{15.30}$$

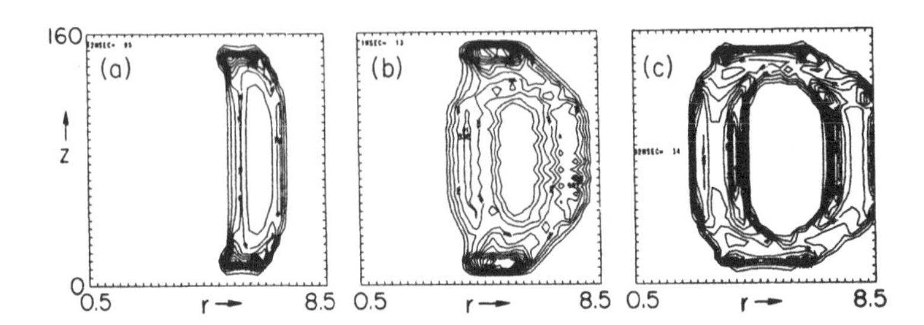

FIGURE 15.5 Ergodicity check. The probability of particle occupying a region of phase space (Ref. 14)

Although rigorously speaking $D_H(f^s) = D_H(f)$, for practical purposes

$$D_H(f^s) \geq D_H(f) \tag{15.31}$$

for weakly irregular orbits, since the larger the s, the more sensitive the test. For the $\alpha = 5$ case all confined particles in the simulation are regular, regardless of the choice of s ($s = 1$ or 4). For $\alpha = 9$, $D_H(f^4)$ and $D_H(f^1)$ are different for a fair number of particles, as it is for the case where $\alpha = 16$. Overall, fewer particles are regular, as α increases.

The second approach is to study the phase space density of a given orbit. If the particle follows an ergodic orbit, the number of particles dN in the phase space volume element $dp_\theta v dv$ is

$$dN(\theta, r, z, v_z) = F(p_\theta, v)\frac{d^3 r d^3 v}{dpvdv} = F(p_\theta, v)\frac{d\theta dr dz dv_z}{\left[v^2 - \frac{1}{r^2}(p-U)^2 - v_z^2\right]^{1/2}}, \tag{15.32}$$

where $F(p_\theta, v)$ is a function of only p_θ and v. After integrating Eq. (15.32) over θ and v_z, we get

$$dN(r, z) = 2\pi^2 F(p_\theta, v) dr dz , \tag{15.33}$$

which is independent of r and z for a given particle orbit. This simulation examines the validity of the completely ergodic assumption by checking if Eq. (15.32) is indeed obtained. Figure 15.5 shows density contours for three representative particle orbits in r, z for the cases where $\alpha = 5$, 9 and 16, respectively. As is clear for all these figures, the density is not uniform, and thus inconsistent with Eq. (15.32). In particular the densities near the turning points of the orbit are very much higher than those in other areas, indicating that orbits are nearly regular, since then the density at the turning point would theoretically be divergent.

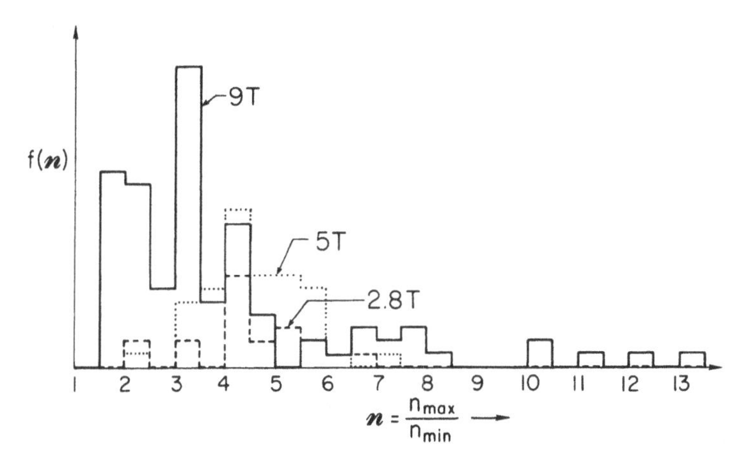

FIGURE 15.6 The ratio of the orbital density maximally occupied to minimally occupied (Ref. 14)

The statistics of the ratio $\mathcal{N}$ of the maximum density in configuration space to the minimum occupied density are shown in Fig. 15.6. Figure 15.6 shows that $f(\mathcal{N})$ has a peak near $\mathcal{N} = 3$ for the case where $\alpha = 16$, and $\mathcal{N} = 4$ for the case where $\alpha = 9$ and 5. As α decreases, the distribution of $f(\mathcal{N})$ is more spread, indicating that the smaller α is, the less ergodicity there is.

15.4 Mapping Methods

A. Magnetic Mirror

In a problem in which an oscillatory (periodic) or nearly oscillatory motion appears one can find an adiabatic invariant by

$$J_i = \oint p_i dq_i, \tag{15.34}$$

where p and q are the i-th generalized momentum and coordinate.[24] The new variable J_i (action) and its conjugate variable θ_i (angle) form a new set of variables for the same Hamiltonian. The transformation from (p_i, q_i) to (J_i, θ_i) is called a canonical transformation. In the nearly periodic problem, J_i can be invariant or nearly invariant, while θ_i is a rapidly varying parameter. An example of this is an electron gyration motion in a magnetic field. The electron (or ion) motion perpendicular to the magnetic field is a well-known gyromotion (see Chapter 7), in which the magnetic moment $\mu = mv_\perp^2/2B$ is the action variable and the gyrophase is the angle variable. Since after this canonical transformation one can obtain a nearly invariant (a very slowly changing) variable with an angle variable changing rapidly, one can analyze the problem very efficiently by time-averaging over the period of the angle variable (gyroperiod), as we have seen in Chapters 7 and 9. One should no longer be concerned with the fast variable θ_i: the appropriate time scale (and therefore time step) for the problem is that of variations of the almost invariant variable J_i. For example, since we know that θ_i repeats itself, all we have to know is how J_i changes over each period of θ_i. That is, to compute $J_i^1, J_i^2, J_i^3, \cdots$ where the superscript refers to the first θ_i period, the second and so on. Because of these advantages, many authors prefer to describe the problem by the canonical transformation.[25]

Let us take the example of the motion of a particle in a magnetic mirror.[26] To the zeroth order, the particle motion in a mirror can be characterized by a constant magnetic moment $\mu = mv_\perp^2/2B(z)$ and a constant kinetic energy $\varepsilon = m(v_\parallel^2 + v_\perp^2)/2$. As the strength of the magnetic field increases when the particle travels toward the mirror throat, the perpendicular kinetic energy has to go up in order to keep μ constant. When $mv_\perp^2/2$ becomes equal to the total kinetic energy ε, the particle has to bounce back because $v_\parallel$ has vanished. To the next order, a particle bouncing in a mirror field undergoes jumps in its magnetic moment μ,[27] which may be regarded as the result of resonances of high harmonics of the bounce motion and the gyromotion.[25] These jumps can be calculated by analytic theory,[27] and also by simulation.[28]

Figure 15.7 shows the result of a simulation using the full particle dynamics[28] (see Chapter 4) for the model magnetotail fields (the reverse fields B_x plus the "vertical" field B_z). The particle moves in normalized electromagnetic fields[29] as

$$\frac{dv_x^*}{dt^*} = v_y^* \tag{15.35}$$

$$\frac{dv_y^*}{dt^*} = v_z^* B_x^* - v_x^* + E_y^* \tag{15.36}$$

$$\frac{dv_z^*}{dt^*} = -v_y^* B_x^* \ , \tag{15.37}$$

where $B_x = B_{x0}\tanh(z^*/L^*)$, $B_{x0}^* = B_{x0}/B_{z0}$, $L^* = L/\rho_0$, $E_y^* = E_y/v_0 B_{z0}$, $t^* = t/\tau_0$, and ρ_0 and τ_0 are the representative gyroradius and gyroperiod. Here the B_z field is normalized to be unity. The effect of E_y^* can be absorbed by the Lorentz transformation in the x-direction with velocity E_y^*/B_z^*. The degree of adiabaticity changes with a choice of the field parameters L^* and B_{x0}^*. For fixed $B_{x0}^* = 10.0$ (which may correspond to a geometry of the far plasma sheet in the earth's magnetosphere, $B_{x0} = 20\gamma$ and $B_{z0} = 2\gamma$), the

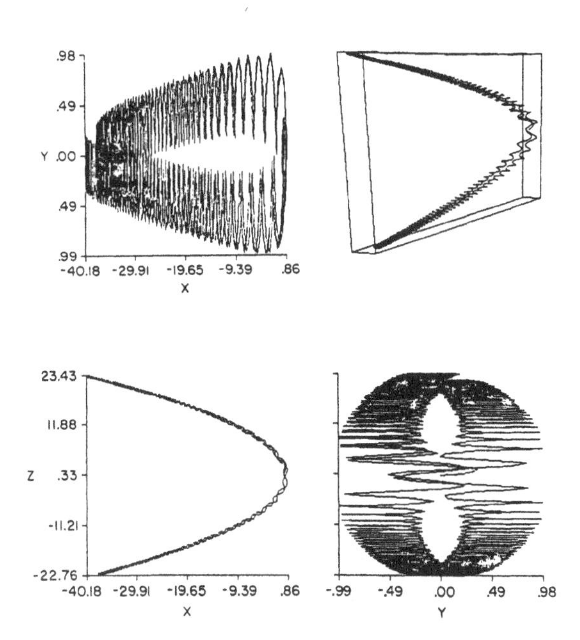

FIGURE 15.7 Trapped particle motion in fields of the geomagnetic tail (Ref. 28)

choice of $L^* < 100$ yields a strongly adiabatic motion of the "figure eight" [Fig. 15.7]. The particle is trapped and always passes through the plasma sheet with the same orientation, and the particle bounces between symmetrical mirror points, as shown in the $z-x$ plane. The displacement of the particle in the y-direction, as it passes through the midplane, is due to the curvature drift which is strongest there. Figure 15.8 shows the magnetic moment with three different L^*'s. The apparent nonadiabatic regions correspond to the passage through the mirror midplane. The three figures demonstrate the transition from strongly adiabatic, trapped trajectories, to weakly adiabatic, partially trapped trajectories. Figure 15.8(a) is for $L^* = 100$ with the curvature radius $R_c = 10$ (at $z = 0$), showing the superadiabaticity;[30] that is, the magnetic moment oscillates as it passes the midplane, but it returns to its nearly original value. Figure 15.8(b) corresponds to $L^* = 65$ and $R_c = 6.5$ (at $z = 0$), exhibiting an apparent random displacement of μ upon the passage at the midplane. Figure 15.8(c) is for the case with $L^* = 20$ ($R_c = 2$ at $z = 0$), showing a large accumulated change in magnetic moment of the order of its own size, eventually entering the loss cone after a few bounce periods.[26] Note that in a single bounce period a particle executes many gyromotions (Fig. 15.7).

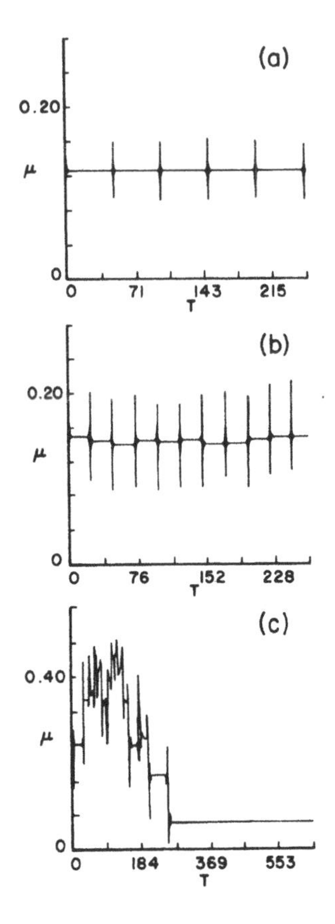

FIGURE 15.8 The temporal variation of the magnetic moment μ for three different curvature radii of the mirror field (Ref. 28)

Figure 15.9(a) shows the change $\Delta\mu$ in magnetic moment μ as a function of the radius of curvature at the midplane is expressed by the formula

$$\Delta\mu = A(p,\mu)\cos\psi\exp(-\beta R_c) \;, \tag{15.38}$$

where $A(p,\mu)$ is a function of the pitch and the phase ψ at the midplane. This dependence Eq. (15.38) agrees well with theory.[27] The change $\Delta\mu$ may be regarded as occurring stochastically, i.e., at each entry to the midplane the phase ψ is uncorrelated with that of the previous entry. Thus the process of changes in μ may be looked upon as the diffusion process

$$\frac{\partial}{\partial t}P(\mu,t) = D_\mu \frac{\partial^2}{\partial \mu^2}P(\mu,t) \;, \tag{15.39}$$

where $P(\mu,t)$ is a one-dimensional probability distribution function[8] and the diffusion coefficient is defined as

$$D_\mu = \frac{1}{2}\frac{\langle \Delta\mu^2 \rangle}{\Delta t} \ . \tag{15.40}$$

The diffusion coefficient, as measured in the simulation, is shown in Fig. 15.9(b) as a function of the radius of curvature.[28]

To study the long-range consequences of such small but finite non-adiabatic effects on the particle transport is of critical importance for investigation of the fusion plasma confinement and the magnetospheric energy flows. Since there exists a clear rule governing how changes in the magnetic moment $\Delta\mu$ and the phase $\Delta\psi$ take place at the midplane as we have seen,[27,28] it is not necessary to advance the particle's dynamics every fine step with a small fraction of gyrophase, as in Ref. 28. One can simply specify these jumps $\Delta\mu$ and $\Delta\psi$ as known functions of μ and ψ (and perhaps other quantities) at the midplane such as Eq. (15.38) and advance $\Delta\mu$ and $\Delta\psi$ only at every entry to the midplane, rather than at every small fraction of the gyroperiod (not the bounce period). In this way, the study of transport is greatly expedited and may become much more accurate, as long as the jump formulas are accurate enough. Otherwise, the great many time steps necessary for the long range transport time scales in the brute force pushing of equations may compromise the subtle physics of the transport via numerical instabilities, error accumulations, a lack of statistics, and so on.

According to Ref. 31, these jump conditions for μ and ψ from the n-th transit at the midplane to the $n+1$-st transit are expressed as

$$\mu_{n+1} \cong \mu_n + \alpha\mu_{n+1}^{1/2} \exp\left[-\kappa(\mu_{n+1})/\varepsilon\right] \cos\psi_n \ , \tag{15.41}$$

$$\psi_{n+1} \cong \psi_n + D(\mu) \ . \tag{15.42}$$

Here the mirror field is given as $B_z(z, r=0) = B_0(1 + z^2/L^2)$, $\varepsilon \equiv v/\Omega_0 L$ (Ω_0: the gyrofrequency at $z=0$), $\alpha \equiv (3\pi/4)\rho_0/L$,

$$D(\mu) = \frac{\pi}{2}\frac{L\Omega_0}{\sqrt{2B_0}}\left(\mu_{n+1}^{-1/2} + v^2/2B_0\mu_{n+1}^{3/2}\right) \ ,$$

μ the normalized magnetic moment $(v_\perp^2/B)/(v^2/B_0)$ and

$$\kappa(\mu) = \frac{1}{4\mu}\left[\left(\lambda + \frac{1}{\lambda}\right)\ell n\frac{1+\lambda}{1-\lambda} - 2\right] \ , \tag{15.43}$$

with $\lambda = \mu^{1/2} = (v_\perp/v)_0$.

Equations (15.41) and (15.42) constitute a set of finite differencing equations to advance the full quantities necessary to describe the long-range evolution (transport time scale evolution) of a particle in the mirror field with the

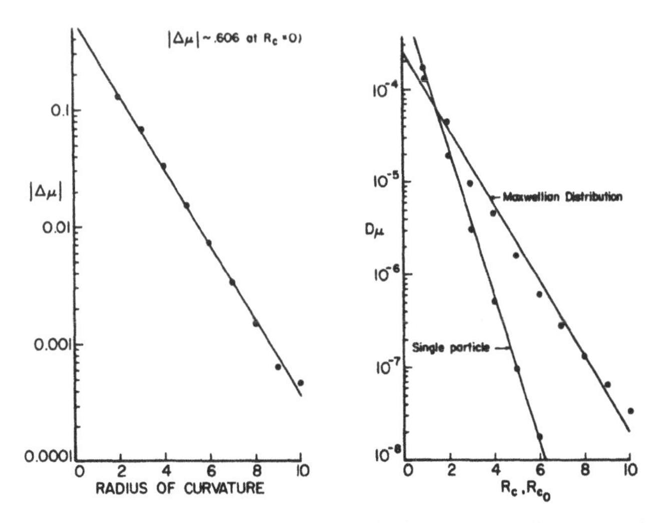

FIGURE 15.9 The jump $\Delta\mu$ in a mirror bounce and the diffusive evolution of the magnetic moment μ in time (Ref. 28)

suffix representing each passage of the midplane; these equations are equivalent to the full particle equations, including the gyromotion as far as the transport is concerned. This method is equivalent to Poincare's surface of section method, as we discussed in Sec. 15.3. In the present equations, the time step is half of the bounce period. Such equations were called maps (Sec. 15.3). [A simple one-dimensional map takes the form of $x_{n+1} = x_n + f(x_n)$, where f is an arbitrary function.] Thus one comes to the realization that the finite difference equations (Chapter 3) in the particle simulation is a map of very high dimensionality (the dimensions typically $6N$ with N being the number of particles we follow). In the transport problem under consideration, after many layers of coarse-graining of faster motions and freezing of the ambient fields, we have in effect carried out the reduction of many degrees of freedom to but a small few and the enhancement of the time step to push by a huge amount. Whenever this reduction of the system of many degrees of freedom to that of a few degrees of freedom is plausible or justifiable, a great deal of simplification of the problem and thus a deeper penetration into the problem become possible. In fact some recent significant progress in understanding nonlinear physics has been in the field of such systems with a few degrees of freedom.

One can linearize this map, Eqs. (15.41)-(15.42), in μ and rescale the variable μ to obtain a standard map of the type considered by Taylor and

Chirikov:[25]

$$I_{n+1} = I_n + K \cos \psi_n \ , \tag{15.44}$$

$$\psi_{n+1} = \psi_n + I_{n+1} \ , \tag{15.45}$$

where the mapping parameter K is a function of μ (or I, which is a function of μ)

$$K = (\pi/4)\varepsilon^{-1}(\alpha/\mu^2)(\mu + 3)\exp(-\kappa/\varepsilon)$$

and $I = (\partial D(\mu)/\partial\mu)(\mu - \mu_0)$, where μ_0 is the value of μ at resonance. In such a standard map, I is considered as the generalized momentum (or action) and ψ the generalized coordinate (or angle,) which are updated upon each half bounce period.

A general theory[25] predicts that the particle motion is very stochastic (or diffusive) when $|K|$ is much larger than unity and it is less so when $|K|$ is less than unity. This may be understood as follows. The finite differencing equations, Eqs. (15.44)-(15.45) can be exactly written in a set of differential forms:

$$\frac{dI}{dt} = \frac{2\pi K}{T} \cos \psi \left[\sum_{n=-\infty}^{\infty} \delta(t + 2\pi n) \right] \ , \tag{15.46}$$

$$\frac{d\psi}{dt} = \frac{I}{T} \ , \tag{15.47}$$

where T is the bounce time, and the index indicates the n-th *alias* (see Chapter 4). This system of differential equations is equivalent to the Hamiltonian system

$$H(J, \psi, t) = \frac{1}{2} J^2 + K \sum_{n=-\infty}^{\infty} \cos(\psi - nt) \ , \tag{15.48}$$

where $J \equiv I/T$. The mapping Eqs. (15.46)-(15.47) is measure (phase space volume) preserving, i.e.,. the system Eq. (15.78) is conservative. We can now see that if $K \gtrsim 1$, the potential terms interfere with each other (with different n's), while if $K \lesssim 1$, the potential terms are separated from each other. The separatrices of these particle orbits show an island structure and we call that islands overlap[32] when $K \gtrsim 1$ [see Fig. 15.10]. See the drift wave turbulence in Chapter 13. An example of Eqs. (15.46)-(15.47), where K is the given function of μ, gives rise to a mapping shown in Fig. 15.10, in which the left half ($\mu \gtrsim 0.5$) corresponds to a non-stochastic (regular) case and the right half corresponds to a stochastic case because of the large K value.

Extensive studies have been conducted in more complex geometries such as tandem mirrors[31,33] and tokamaks.[34,35] In the latter example[34] the transport process is sensitively determined by the so-called banana particles[36] particularly for high energy particles. Another example of transport study using

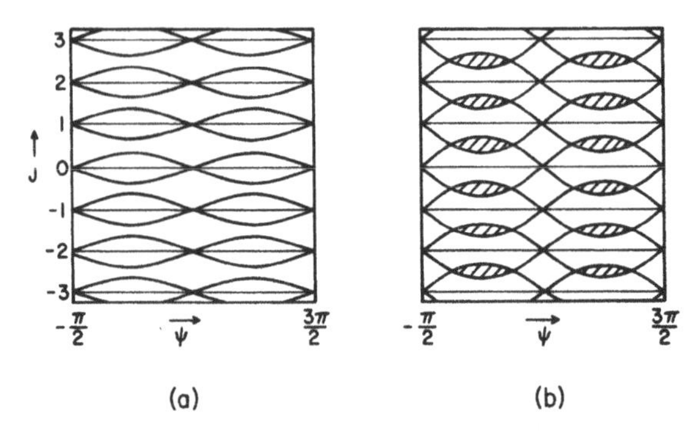

FIGURE 15.10 Island structure in phase space (ψ, J). a) not overlapped b) overlapped

mappings is by Meiss et al.[37] as applied to anomalous transport in toroidal geometry such as a tokamak with fluctuating magnetic fields.

Some of the chaotic behaviors described by the mapping exhibit a structure that has structures within structures within structures *ad infinitum*, the kind of hierarchical display we have discussed repeatedly in Chapters 1, 13, and 14. In this case we have islands within islands within islands *ad infinitum*, which have been discussed by the renormalization group theoretical technique,[38,39] the tree (branching) method[40] etc.

B. Transport in Accelerators

The modern high energy accelerator may be considered as a series of electromagnetic components (such as magnets) that interact with the particle beam. In this regard it possesses elements similar to the electron microscope, in which the electron beam is manipulated by a series of electromagnetic components. These interactions are studied by the discipline of (charged) beam optics,[41]which is closely similar to (light) optics.[42] In a circular colliding accelerator such as the proposed Superconducting Super Collider (SSC) high energy hadron beams typically have to circulate in the machine more than 10^8 turns. In order to study a great many accumulated influences of tiny kicks by each component of the accelerator (i.e., the transport of the beam), it is necessary to devise a very precise algorithm to compute the transport. Dragt has developed an elaborate elegant method based on a Hamiltonian formulation of beam optics.[43] In what follows we discuss his method. The principle is the same as we discussed for mirrors, Eqs. (15.41)-(15.45). After all among many components of accelerators belong the mirror fields including the quadrupole. Potential differences are that in accelerators the interaction between beam particles are often not important, while in plasma confinement machines it is often vitally important. However, the map equations in the latter thus reduced often do not retain those features any more, leading to mathematical equivalence in these two systems. This commonality is the reason that we treat these two seemingly remote problems in the same section.

In optics (see Fig. 15.11) a light ray originates at the initial point P^i ($\mathbf{r}^i$) moving in the initial direction $\hat{s}^i = \mathbf{r}^i/|\mathbf{r}^i|$, passes through an optical device, and arrives at the final point p^f ($\mathbf{r}^f$) in a direction $\hat{s}^f = \mathbf{r}^f/|\mathbf{r}^f|$. The fundamental problem in geometrical optics is to determine the final quantities $\mathbf{r}^f$ from the initial quantities $\mathbf{r}^i$ with the function of the optical properties such as the index of refraction $n(\mathbf{r})$. The optical path length

$$S = \int_{z^i}^{z^f} n(x,y,z)\left[1+(x')^2+(y')^2\right]^{1/2} dz \ , \tag{15.49}$$

where the primes refer to differentiation with respect to the coordinate z shown in Fig. 15.11. Fermat's principle dictates that action S be an extremum for an actual ray. The ray path, therefore, satisfies the Euler-Lagrange equations ($x_i = x$ or y for $i = 1$ or 2)

$$\frac{d}{dz}\left(\frac{\partial \mathcal{L}}{\partial x_i}\right) - \frac{\partial \mathcal{L}}{\partial x_i} = 0 \ , \tag{15.50}$$

where the Lagrangian is defined by

$$\mathcal{L} = n(x,y,z)\left[1+(x')^2+(y')^2\right]^{1/2} \ . \tag{15.51}$$

The corresponding Hamiltonian is

$$\mathcal{H} = -\left[n^2(\mathbf{r}) - r_x^2 - p_y^2\right]^{1/2} \tag{15.52}$$

with

$$p_i = n(\mathbf{r})x_i' / \left[1+(x')^2+(y')^2\right]^{1/2} \ , \tag{15.53}$$

which acts as a conjugate momentum.

Let us now consider a beam optics example of a horizontally defocussing magnetic quadrupole of angle ϕ_0 [see Fig. 15.12].[44] This particular component consists of a part of the entire grid of a given accelerator. The effect of this component can be represented in the so-called transfer map $\mathcal{M}_1$. The entire orbit in the machine can be obtained by a map $\mathcal{M}$ of successive products of such a map $\mathcal{M}_1$. Many turns n in a circular machine, for example, can be given by $\mathcal{M}^n$. In the cylindrical coordinates, (ρ, ϕ, z), a charged particle moving in an external vector potential $\mathbf{A}(\rho, \phi, z)$ is described by the Lagrangian

$$\mathcal{L}(\rho, \dot{\rho}, \phi, \dot{\phi}, z, \dot{z}) = -mc^2\sqrt{1-\beta^2} + q\mathbf{v}\cdot\mathbf{A} \ , \tag{15.54}$$

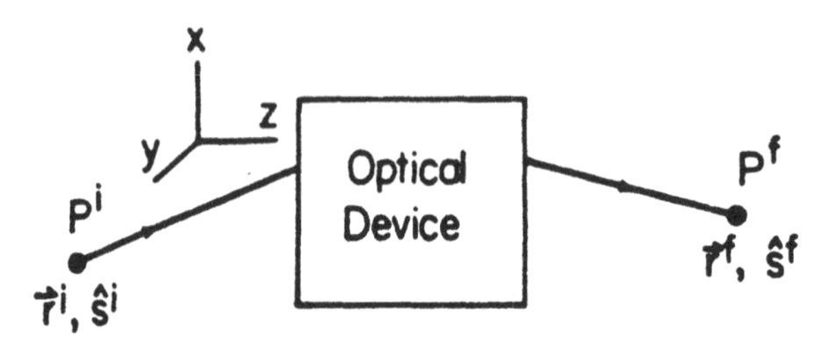

FIGURE 15.11 Ray dynamics in geometrical optical systems (Ref. 44)

where $\beta^2 = v^2/c^2$. The Hamiltonian is (with $\mathbf{A} = A_\phi \hat{\phi}$)

$$\mathcal{H} = \left\{ m^2c^4 + c^2 \left[\rho_p^2 + \left(\frac{p_\phi - q\rho A_\phi}{\rho} \right)^2 + p_z^2 \right] \right\}^{1/2} \tag{15.55}$$

The trajectory of a particle moving through this magnetic field may be conveniently described by the effective Hamiltonian transformed from variables $(\rho, p_\rho, z, p_z, \phi, p_\phi)$ to $(\rho, p_\rho, z, p_z, t, p_t = -\mathcal{H})$.

$$K(\rho, p_\rho, z, p_z, t, p_t) = -p_\phi = -\frac{\rho}{c} \left\{ p_t^2 - m^2c^4 - c^2 \left[p_\rho^2 + p_z^2 \right] \right\}^{1/2} + q\rho A_\phi \,. \tag{15.56}$$

We expand the Hamiltonian around the zeroth order equilibrium orbit ("design orbit") with $r = \rho - p_0$, $T = t - t_0$, $p_r - p_\rho$, $p_T = p_t = p_{t0}$ with the subscripts referring to the design orbit, obtaining

$$K(r, P_r, z, P_z, T, P_z) = -\frac{(r + \rho_0)}{c} \left\{ (p_{t0} + p_T)^2 - m^2c^4 - c^2(p_r^2 - p_z^2) \right\}^{1/2}$$

$$-\frac{q}{2}(r + \rho_0)^2 B - \frac{r m \rho_0}{p_0}(p_T + p_{t0}) \,. \tag{15.57}$$

With dimensionless variables $\phi \to \phi$, $r \to R = r/\ell$, $K \to \kappa = K/(\ell p_0)$, $p_r \to P_R = p_t/p_0$, $T \to \tau = T/(\ell/c)$, $z \to Z = z/\ell$, $P_T \to P_t = P_T/(\rho_0 c)$, $p_z \to P_z = p_z/\ell$, we write the new variables collectively by $\boldsymbol{\xi} = (R, P_R, Z, P_z, \tau, P_\tau)$. The trajectory equations are then described[43] by

$$\frac{d\boldsymbol{\xi}}{d\phi} = -:k:\boldsymbol{\xi}(\phi) \,. \tag{15.58}$$

Here the symbol :: indicates the Lie operation[43] which is an abbreviated expression of the Poisson bracket

$$: H : z_i \equiv - [z_i, \mathcal{H}] = \sum_{j=1}^{n} \left(\frac{\partial Z_i}{\partial q_j} \frac{\partial \mathcal{H}}{\partial p_j} - \frac{\partial Z_i}{\partial p_j} \frac{\partial \mathcal{H}}{\partial q_j} \right) . \tag{15.59}$$

The differential form of the orbit description, Eq. (15.58), can be integrated over ϕ (the time-like variable). This leads to an exponential solution in the operator space called the Lie operator:

$$\boldsymbol{\xi}(\phi_0) = \mathcal{M}_1 \boldsymbol{\xi}(0) , \tag{15.60}$$

where the Lie operator (or transformation) is given by

$$\mathcal{M}_1 = e^{-\phi_0 : \kappa :} , \tag{15.61}$$

using the result of Campbell-Baker-Hausdorff (see Chapter 1). Thus Eq. (15.61) describes the impact of the bend magnet on orbits near the design orbit when the particle exists from the machine end ($\phi = \phi_0$) (see Fig. 15.11). If the particle transits this component n times, the orbit, for example, is given by

$$\boldsymbol{\xi} = \mathcal{M}_1^n \boldsymbol{\xi}(0) = \exp(-n\phi_0 : \kappa :) \boldsymbol{\xi}(0) . \tag{15.62}$$

This operation $\mathcal{M}_1$ or its n successive operations $\mathcal{M}_1^n$ can be looked upon as a map and n successive maps (of six dimensions). This transformation $\mathcal{M}_1$ is called the symplectic (area-preserving) map of the orbit from $\boldsymbol{\xi}(0)$ to $\boldsymbol{\xi}(\phi_0)$. This is simply a manifestation of the energy preserving nature of the accelerator component or, more formally speaking, the Hamiltonian nature of the orbital equations.

Next we discuss a perfect quadrupole with length ℓ. Following the procedures similar to the above, the Lie operator in the quadrupole order is given as[43]

$$\exp(-\ell : H_2) = \exp(-\ell : H_2^t :) \exp(-\ell : H_2^y :) \exp(-\ell : H_2^z :) \tag{15.63}$$

where

$$H_2^t = \frac{p_t^2}{2} \frac{m^2}{p_0^3} , \quad H_2^y = \frac{p_y^2}{2p_0} + \frac{q a_2}{2} y^2 , \quad H_2^z = \frac{p_z^2}{2p_0} - \frac{q a_2}{2} z^2 , \tag{15.64}$$

with p_0 the design orbit for p_t and a_2 the coefficient of the vector potential around the design orbit position. The actual transfer matrix takes the form

$$\exp\left(-\ell : H_2^t :\right) \begin{pmatrix} t \\ p_t \end{pmatrix} = \begin{pmatrix} 1 & \ell m^2 / p_0^3 \\ 0 & 1 \end{pmatrix} \begin{pmatrix} t \\ p_t \end{pmatrix} ,$$

$$\exp\left(-\ell : H_2^y :\right) \begin{pmatrix} y \\ p_y \end{pmatrix} = \begin{pmatrix} \cos k\ell & \frac{1}{k p_0} \sin k\ell \\ -k p_0 \sin k\ell & \cos k\ell \end{pmatrix} \begin{pmatrix} y \\ p_y \end{pmatrix} ,$$

$$\exp\left(-\ell : H_2^z :\right) \begin{pmatrix} z \\ p_z \end{pmatrix} = \begin{pmatrix} \cosh k\ell & \frac{1}{k p_0} \sinh k\ell \\ k p_0 \sinh k\ell & \cosh k\ell \end{pmatrix} \begin{pmatrix} z \\ p_z \end{pmatrix} , \tag{15.65}$$

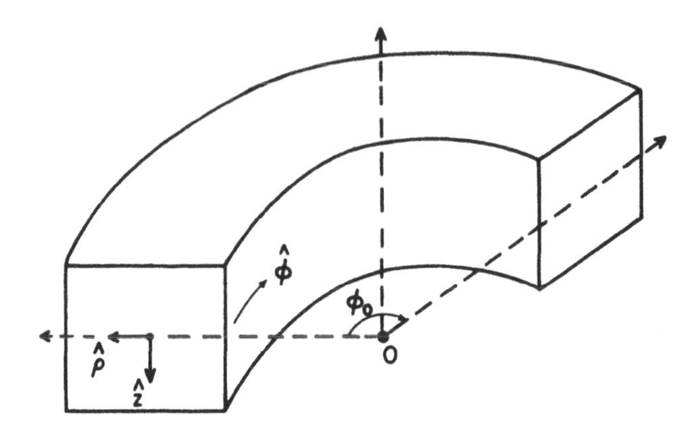

FIGURE 15.12 Circular machine (accelerator) geometry (Ref. 43)

In general one has higher order terms $\exp(-\ell{:}H_3{:})\exp(-\ell{:}H_4{:})$ etc. The operators, Eq. (15.65), may be looked upon as generalization of beam optical matrices.[41]

Finally, let us consider an example of the beam-beam interaction, which we have neglected so far. We look at a circular machine with a collision section in which a beam-beam interaction takes place and with a ring section in which the beam is simply transported. In such a case the particle mapping in one revolution of the machine may be represented by

$$\mathcal{M} = \mathcal{M}_R \mathcal{M}_B \tag{15.66}$$

where $\mathcal{M}_R$ is the Lie operator and $\mathcal{M}_B$ the beam-beam collision operator. The impulse approximation for the momentum-coordinate (p, q)

$$\mathcal{M}_B q = q$$

$$\mathcal{M}_B p = p + u(q) \ , \tag{15.67}$$

where $u(q)$ is a deflection function. Equation (15.67) is symplectic. The deflection function $u(p)$ is proportional to the electrostatic force exerted by the beam. Examples of this mapping are shown in Fig. 15.13. When the effect of the beam-beam interaction is strong, the orbits eventually become stochastic, as exemplified by a spread in the phase space of Fig. 15.13(c).

In this chapter, as a closing chapter, we discussed the transport process, generally the longest time scale of the system. The method of computational description, correspondingly, is most coarse-grained and "distilled" to the limit. We only covered a few computational techniques, the Monte-Carlo

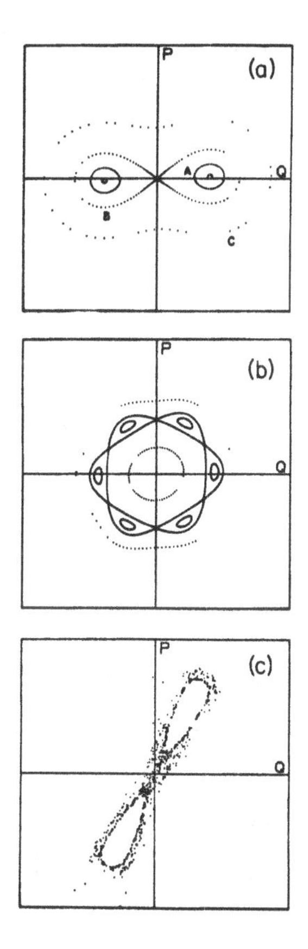

FIGURE 15.13 Phase space plots of an example of map $\mathcal{M}_B$ (Ref. 43)

method, the Fokker-Planck model, the particle simulation method, and the mapping method. The Monte-Carlo method is a very popular and powerful method, but generally handles a stationary system. So we only cite it for reference sake. General transport equations derived by various ways such as a moment method (Problem 1) can be integrated in time by the general methods we discussed in Chapters 3 and 9. We did not dwell on this, as this is not new in terms of numerics beyond what we already presented, although in practice this is perhaps the most often used method. The particle simulation technique is a natural extension to the transport time scale from the earlier discussions. We discussed the mapping method only to the extent that the reader can pick up the technique from where we left off. We also clarified the natural link between the particle simulation and the mapping. In these methods the "particle pushers" are identical, except for the degrees of freedom, dimensionality and the time scales, and the allowance of the possibility of field variations in the particle simulation case. Thus we can naturally peek through the effects of differences and similarities in the different level of plasma simulation hierarchy we mentioned in Chapter 1 by inspecting the two methods, particle simulation and mapping. And, of course, the richness of these techniques and their results is just the reflection of the plasma's exhibition of rich and hierarchical behaviors, a kaleidoscope of nonlinear physics of a many-body system with long range forces.

Problems

1. Describe a strategy for a transport code that is based on moment equations of the kinetic equation, such as the continuity equation, the momentum equation, the temperature equation and so forth, where the density, momentum, temperature etc. obey transport time-scale variations in time and space.

2. The transport equations by nature most often take the form of the diffusion equation. Discuss appropriate time advancement algorithms, explicit or implicit, for the transport equations. See also Chapter 3.

3. Survey applications of the Monte-Carlo method in various fields of physics.

4. Describe the advantages and limitations of the mapping method.

5. Write down the Chirikov overlap condition for drift wave turbulence. Also write down the conditions for the large amplitude Langmuir waves and for the ion Bernstein waves.[45]

6. Prove the properties of the Lie operator: :

$$:f:(ag + bh) =: f:g + b:f:h \ ,$$

$$:f:(gh) = g(:f:)h + h(:f:)g \ ,$$

$$:f:([g,h]=[g,(:f:h)] + [(:f:g),h]$$

$$:f::g:-:g::f:=:[f,g]: , e^{:f:}(gh) = (e^{:f:}g)(e^{:f:}h) \ ,$$

where f, g, and h are functions of phase space variables, a and b are numbers, and where [,] means the Poisson brackets.

7. Prove that Lie transformations $\mathcal{M}$ are symplectic maps,[44] where $\mathcal{M} = e^{:f:}$. Here the symplectic map $\mathcal{M}$ satisfies $\mathcal{M}J\mathcal{M}^t = J$ with J being

$$J = \begin{pmatrix} 0 & I \\ -I & 0 \end{pmatrix}$$

and prove that the Hamiltonian flow generates a symplectic map: $dz_i(t)/dt = -:H:z_i(t)$ has a solution $z_i(t) = \mathcal{M}z_i(0)$ with $\mathcal{M} = e^{-t:H:}$, where H is the Hamiltonian.

8. Compare the charged particle beam optics with geometric optics, then discuss an example of how to treat a plasma lense (such as a self-focussing device[46]) for both optics. Describe these problems in terms of the Hamiltonian dynamical formalism.

References

1. L.D. Landau, Zhur. Eksptl. Teoret. Fiz. **7**, 203 (1937).
2. R. Balescu, Phys. Fluids **3**, 52 (1960); A. Lenard, Ann. Phys. (N.Y.) **10**, 390 (1960).
3. Los Alamos Science, no. 15 (1987).
4. N. Metropolis, A. Rosenbluth, M.N. Rosenbluth, A. Teller, and E. Teller, J. Chem. Phys. **21**, 1087 (1953).
5. K. Binder, ed. *Monte Carlo Methods in Statistical Physics* (Springer, Berlin, 1979); J.E. Hirsch, in *Supercomputers* ed. by F.A. Matsen and T. Tajima (University of Texas Press, Austin, 1986) p. 171.
6. M. Crentz, Phys. Rev. Lett. **50**, 1411 (1983).
7. A. Einstein, Ann. Physik **19**, 371 (1906).
8. S. Chandrasekhar, Rev. Mod. Phys. **15**, 1 (1943).
9. M.N. Rosenbluth, W.M. MacDonald, and D.L. Judd, Phys. Rev. **107**, 1 (1957).
10. S. Ichimaru, *Basic Principles of Plasma Physics* (Benjamin, Reading, 1973) Chapter 10.
11. J. Killeen, G.D. Kerbel, M.G. McCoy, and A.A. Mirin, *Computational Methods for Kinetic Models of Magnetically Confined Plasmas* (Springer, New York, 1986) Chapter 2.
12. R.D. Richtmyer and K.W. Morton, *Difference Methods for Initial Value Problems* (Wiley, New York, 1967).
13. J.M. Finn and R.N. Sudan, Nucl. Fusion **22**, 1443 (1982).

14. H.L. Berk, H. Momota, and T. Tajima, Phys. Fluids. **30**, 3548 (1987).

15. M.J. LeBrun and T. Tajima, to be submitted to J. Comput. Phys.

16. H.L. Berk and H. Weitzner, Phys. Fluids **24**, 1758 (1981).

17. D.W. Kerst, in *Handbook der Physik*, ed. S. Flugge, vol. 44 (Springer Verlag, Berlin, 1959) p. 13.

18. V.I. Arnold, *Mathematical Methods of Classical Mechanics* translated by K. Vogtmann and A. Weinstein (Springer, New York, 1978).

19. E.N. Lorentz, J. Atm. Sci. **20**, 130 (1963).

20. A.J. Lichtenberg and M.A. Lieberman, *Regular and Stochastic Motion* (Springer, New York, 1983).

21. J.D. Farmer, E. Ott and J.A. York, Physica **7D**, 153 (1983).

22. R.H. Simoi, A. Wolf and H.L. Swinney, Phys. Rev. Lett. **49**, 245 (1982).

23. P. Grassberger and I. Procaccia, Phys. Rev. Lett. **31**, 31 (1983).

24. L.D. Landau and I.M. Lifshitz, *Mechanics* (Pergamon, 1960) Chapter 1.

25. B.V. Chirikov, Phys. Reports **52**, 263 (1979); J.B. Taylor, Culham report (1968, unpublished).

26. T.K. Fowler, in *Fusion* ed. by E. Teller (Academic, New York, 1981) vol. 1, p. 291.

27. R.H. Cohen, G. Rowlands and J.H. Foote, Phys. Fluids **21**, 627 (1978).

28. P.C. Gray, and L.C. Lee, J. Geophys. Res. **87**, 7445 (1982).

29. J.S. Wagner, J.R. Kan, and S.-I. Akasofu, J. Geophys. Res. **84**, 89 (1979).

30. M.N. Rosenbluth, Phys. Rev. Lett. **29**, 408 (1972).

31. R.H. Cohen, in *Intrinsic Stochasticity in Plasmas*, eds. G. Laval and D. Gresilion (Editions de Physiques Orsays 1979) p. 181.

32. M.N. Rosenbluth, R.Z. Sagdeev, J.B. Taylor, and G.M. Zaslavsky, Nucl. Fusion **6**, 297 (1966); J.R. Cary and R.J. Littlejohn, Ann. Phys. (N.Y.) **151**, 1 (1983).

33. Y.H. Ichikawa, T. Kamimura and C.F.F. Karney, Physica **6D**, 233 (1983).

34. R.J. Goldston, R.B. White and A.H. Boozer, Phys. Rev. Lett. **47**, 647 (1981).

35. A.H. Boozer and R.B. White, Phys. Rev. Lett. **49**, 786 (1982).

36. A.A. Galeev and R.Z. Sagdeev, Sov. Phys. JETP **26**, 233 (1968).

37. J.D. Meiss, J.R. Cary, D.F. Escande, R.S. MacKay, I.C. Percival, and J.L. Tennyson, *Plasma Physics and Controlled Nuclear Fusion Research* (International Atomic Energy Agency, Vienna, 1985) vol. 3, p. 441.

38. D.F. Escande, Phys. Reports **121**, 165 (1985).

39. R.S. MacKay, *Renormalization in Area Preserving Maps*, Ph.D. dissertation (Princeton Univ. 1982).

40. J.D. Meiss and E. Ott, Physica **20D**, 387 (1986).

41. for example, J.D. Lawson, *The Physics of Charged Particle Beams* (Clarendon, Oxford, 1977).

42. for example, M. Born and Wolf, *Optics* (Pergamon, New York, 1975).

43. A.J. Dragt, in *Physics of High Energy Particle Accelerators*, AIP Conference Proc. No. 87 (AIP, New York, 1982) p. 159; A.J. Dragt, R.D. Ryne, L.M. Healy, F. Neri, D.R. Douglass, E. Forest, *A Program for Charged Particle Beam Transport Based on Lie Algebra Methods*, University of Maryland Physics Department Technical Report (University of Maryland, College Park, 1987). *Lie Algebraic Methods for Particle Accelerator Theory* (Ph.D. Dissertation Univ. of Maryland, 1982).

44. D.R. Douglass, *Lie Algebraic Methods for Particle Accelerator Theory* (Ph.D. Dissertaion, Univ. of Maryland, 1982).

45. S. Riyopoulos and T. Tajima, Phys. Fluids **29**, 4161 (1986).

46. D.C. Barnes, T. Kurki-Suonio, and T. Tajima, IEEE Trans. Plas. Soc. **PS-15**, 154 (1987).

EPILOGUE: NUMERICAL LABORATORY

We have surveyed various algorithms and methods of computational physics of plasmas (and perhaps other many-body systems) in Chapters 1-11. We have further fine-tuned our endeavor by implementing these techniques in actual physics applications in Chapters 12-15. These elements discussed in Chapters 1-15 constitute building blocks of our system of computational physical simulation. Just like an experimental laboratory, our numerical laboratory has to consist of many "equipments" and "instruments." These components in our numerical laboratory are very numerous and rich in variety, reflecting the hierarchical nature of plasmas. Because of this complexity it has become increasingly important to make a total system as a tree-like structure in which elements of "roots," a "trunk," "branches," and "leaves" are interchangeable or reducible depending on necessary tasks of investigation at hand. This is the direction of synthesis, as opposed to the various analysis tools as shown in Chapters 1-15. This means that we have to make each algorithmic program as a module and we stack them up. This also means that we need a manager system which manages the total system. We have begun this task of building a numerical research laboratory. So far it is in an infantile stage.

It is vitally important for students to learn computational physics by practice. For this purpose we have also begun to build a numerical classroom laboratory. It is the author's typical practice to give students a certain set of codes at the beginning of the semester of this graduate course (PHY 391T) so that they can get familiar with the codes and obtain meaningful and sometimes original computational results. I have made available some of these classroom laboratory codes to the reader. They can be electronically accessible to him as described in the following. (I reserve the right to be notified if the reader obtains those codes and tries to publish results based on these, in order to avoid unnecessary misinterpretation. However, unauthorized access for any purpose other than copying these codes is against Federal Laws and punishable. I am also very happy to hear from the reader any criticism or detection of mistakes in the text or codes). Available codes are: electrostatic codes in 1D and 2D, relativistic electromagnetic code in 1D, MHD particle code in 2D, magnetoinductive guiding center code in 2D, along with some utility programs. Examples of classroom laboratory work are shown in a paper.[1] These programs are generic ones but have been directly or indirectly developed by many of my colleagues, my students and myself. These include: J.M. Dawson, J.N. Leboeuf, F. Brunel, A.T. Lin, J. Wagner, J. Geary, E. Zaidman, M.LeBrun, and J. Schutkeker.

Since I am no longer teaching and researching at the University of Texas, please write to me directly at:

Kansai Research Establishment
Japan Atomic Energy Research Institute
8-1, Umemidai, Kizu-cho, Souraku-gun
Kyoto 619-0215 Japan

or by email: tajima@apr.jaeri.go.jp

I will attempt to contact my former students and advise on the availability of the computer codes referenced in this book.

We have also made an effort to make available our codes to undergraduate students for personal computers. W. Miner, J. Wiley, W. Eubank, and W. Saphir have worked on it and diskettes should become available soon from Kinko's. Examples of this application are shown in a paper.[2]

Happy computing!

[1] T. Tajima et al., Am. J. Phys. 53, 365 (1985).
[2] S. Eubank, W. Miner, T. Tajima, and J. Wiley, Am. J. Phys. 57, 457 (1989).

SUBJECT INDEX

AUTHOR INDEX

CREDITS

Figure 2.3 from C. K. Birdsall, et al., *Method in Computational Physics,* vol. 9, p. 251. Copyright ©1970, Academic Press, Inc.

Figure 5.1 from T. Tajima, Y. C. Lee, *Journal of Computational Physics,* vol. 42, p. 406. Copyright ©1981, Academic Press, Inc.

Figure 5.6 from P. M. Morse and J. Feshbach, *Mathematical Method of Physics* (New York: McGraw-Hill, 1953), p. 699. Reprinted by permission of the publisher.

Figures 6.5 and 6.6 from F. Brunel, et al., *Journal of Computational Physics,* vol. 43, p. 268. Copyright ©1981, Academic Press, Inc.

Figure 6.10 from S. A. Orszag, *Journal of Fluid Mechanics,* vol. 49, p. 75. Copyright ©1971 Cambridge University Press. Reprinted with permission of Cambridge University Press.

Figures 7.6, 7.7, and 7.8 from J. L. Geary, et al., *Computer Physics Communications,* vol. 42, p. 321. Copyright ©1986 North-Holland Physics Publishing.

Figure 11.1 reprinted from "The Influence of Computational Fluid Dynamics on Experimental Aerospace Facilities," 1983, with permission of the National Academy Press, Washington, D. C.

Figure 11.5 from E. Oran Brigham, *The Fast Fourier Transform,* ©1974, p. 107. Adapted by permission of Prentice-Hall, Inc., Englewood Cliffs, NJ.

Figure 12.10 from T. Tajima, *Lascr and Particle Beams,* vol. 3, p. 351. Copyright ©1985 Cambridge University Press. Reprinted with permission of Cambridge University Press.

Figures 13.1, 13.2, 13.3, and 13.4 from R. D. Sydora, et al., *Physics of Fluids,* vol. 28, p. 528, Copyright ©1985. Reprinted by permission of the American Institute of Physics.

Figures 13.5, 13.6, and 13.7 from R. D. Sydora, et al., *Physics Review Letters,* vol. 57, p. 3269. Copyright ©1986. Reprinted by permission of the American Physical Society.

Figure 13.10 from T. Tajima and J. N. Lebouf, *Physics of Fluids,* vol. 23, p. 884. Copyright ©1980. Reprinted by permission of the American Institute of Physics.

Figures 13.11 and 13.12 from W. Horton, et al., *Physics of Fluids,* vol. 30, p. 3485. Copyright ©1987. Reprinted by permission of the American Institute of Physics.

Figures 14.3, 14.4, and 14.5 from I. Katamuma and T. Kamimura, *Physics of Fluids,* vol. 23, p. 2500. Copyright ©1980. Reprinted by permission of the American Institute of Physics.

Figures 14.9 and 14.10 from F. Brunel, et al., *Physics Review Letters,* vol. 49, p. 323. Copyright ©1982. Reprinted by permission of the American Physical Society.

Figures 14.11, 14.12, 14.13, 14.14, 14.15, 14.17, 14.18, 14.19, 14.20, 14.21, and 14.22 reprinted courtesy of *The Astrophysical Journal,* published by the University of Chicago Press; ©1987 The American Astronomical Society.

Figures 15.2, 15.3, 15.4, 15.5, and 15.6 from H. L. Berk, et al., *Physics of Fluids,* vol. 30, p. 3548. Copyright ©1987. Reprinted by permission of the American Institute of Physics.

Figures 15.7, 15.8 and 15.9 from P. C. Gray and L. C. Lee, *Journal of Geophysical Research,* vol. 87, p. 7445. Copyright ©1982 by the American Geophysical Union.

Figures 15.12 and 15.13 from A. J. Dragt, *Physics of High Energy Particle Accelerators,* American Institute of Physics Conference, New York (1982), Proceedings #87, p. 159. Copyright ©1982. Reprinted by permission of the American Institute of Physics.

Printed in the United States
by Baker & Taylor Publisher Services